Abiotic Stress Tolerance in Crop Plants

Abiotic Stress Tolerance in Crop Plants

Editor: Pauline Gibbs

STATES
ACADEMIC PRESS
www.statesacademicpress.com

States Academic Press,
109 South 5th Street,
Brooklyn, NY 11249, USA

Visit us on the World Wide Web at:
www.statesacademicpress.com

ISBN: 978-1-63989-012-5 (Hardback)

Cataloging-in-Publication Data

Abiotic stress tolerance in crop plants / edited by Pauline Gibbs.
p. cm.
Includes bibliographical references and index.
ISBN 978-1-63989-012-5
1. Crops--Effect of stress on. 2. Crops--Physiology. I. Gibbs, Pauline.
SB112.5 .A25 2022
571.2--dc23

Table of Contents

Preface VII

Chapter 1 **Elevated ozone deteriorates grain quality of *Japonica* rice cv. koshihikari, even if it does not cause yield reduction** 1
Hiroko Sawada, Keita Tsukahara, Yoshihisa Kohno, Keitaro Suzuki, Nobuhiro Nagasawa and Masanori Tamaoki

Chapter 2 **Proteomic and glycomic characterization of rice chalky grains produced under moderate and high-temperature conditions in field system** 11
Kentaro Kaneko, Maiko Sasaki, Nanako Kuribayashi, Hiromu Suzuki, Yukiko Sasuga, Takeshi Shiraya, Takuya Inomata, Kimiko Itoh, Marouane Baslam and Toshiaki Mitsui

Chapter 3 **Association mapping of yield and yield-related traits under reproductive stage drought stress in rice (*Oryza sativa* L.)** 27
B. P. Mallikarjuna Swamy, Noraziyah Abd Aziz Shamsudin, Site Noorzuraini Abd Rahman, Ramil Mauleon, Wickneswari Ratnam, Ma. Teressa Sta. Cruz and Arvind Kumar

Chapter 4 **High resolution mapping of QTLs for heat tolerance in rice using a 5K SNP array** 40
Shanmugavadivel PS, Amitha Mithra SV, Chandra Prakash, Ramkumar MK, Ratan Tiwari, Trilochan Mohapatra and Nagendra Kumar Singh

Chapter 5 **Association of SNP haplotypes of *HKT* family genes with salt tolerance in Indian wild rice Germplasm** 51
Shefali Mishra, Balwant Singh, Kabita Panda, Bikram Pratap Singh, Nisha Singh, Pragati Misra, Vandna Rai and Nagendra Kumar Singh

Chapter 6 **Comparative leaf and root transcriptomic analysis of two rice *Japonica* cultivars reveals major differences in the root early response to osmotic stress** 64
Elena Baldoni, Paolo Bagnaresi, Franca Locatelli, Monica Mattana and Annamaria Genga

Chapter 7 **Mapping QTLs using a novel source of salinity tolerance from Hasawi and their interaction with environments in rice** 84
M. Akhlasur Rahman, Isaac Kofi Bimpong, J. B. Bizimana, Evangeline D. Pascual, Marydee Arceta, B. P. Mallikarjuna Swamy, Faty Diaw, M. Sazzadur Rahman and R. K. Singh

Chapter 8 **Exploring traditional aus-type rice for metabolites conferring drought tolerance** 101
Alberto Casartelli, David Riewe, Hans Michael Hubberten, Thomas Altmann, Rainer Hoefgen and Sigrid Heuer

Chapter 9 **Inoculation with the endophyte *Piriformospora indica* significantly affects mechanisms involved in osmotic stress in rice** 117
Muhammad Abu Bakar Saddique, Zulfiqar Ali, Abdus Salam Khan, Iqrar Ahmad Rana and Imran Haider Shamsi

Chapter 10 **Suppression of *OsMDHAR4* enhances heat tolerance by mediating H_2O_2-induced stomatal closure in rice plants** 129
Jianping Liu, Xinjiao Sun, Feiyun Xu, Yingjiao Zhang, Qian Zhang, Rui Miao, Jianhua Zhang, Jiansheng Liang and Weifeng Xu

Chapter 11 ***OsRACK1A*, encodes a circadian clock-regulated WD40 protein, negatively affect salt tolerance in rice** 141
Dongping Zhang, Yuzhu Wang, Jinyu Shen, Jianfeng Yin, Dahong Li, Yan Gao, Weifeng Xu and Jiansheng Liang

Chapter 12 **Overexpression of the Tibetan Plateau annual wild barley (*Hordeum spontaneum*) *HsCIPKs* enhances rice tolerance to heavy metal toxicities and other abiotic stresses** 156
Weihuai Pan, Jinqiu Shen, Zhongzhong Zheng, Xu Yan, Jianxin Shou, Wenxiang Wang, Lixi Jiang and Jianwei Pan

Chapter 13 **The transcription factor OsbHLH035 mediates seed germination and enables seedling recovery from salt stress through ABA-dependent and ABA-independent pathways, respectively** 169
Hung-Chi Chen, Wan-Hsing Cheng, Chwan-Yang Hong, Yu-Sen Chang and Men-Chi Chang

Chapter 14 **Constitutive expression of *REL1* confers the rice response to drought stress and abscisic acid** 186
Jiayan Liang, Shaoying Guo, Bo Sun, Qing Liu, Xionghui Chen, Haifeng Peng, Zemin Zhang and Qingjun Xie

Chapter 15 **Progressive drought alters architectural and anatomical traits of rice roots** 197
Mohamed Hazman and Kathleen M. Brown

Chapter 16 **Salinity tolerance in Australian wild *Oryza* species varies widely and matches that observed in *O. sativa*** 213
Yoav Yichie, Chris Brien, Bettina Berger, Thomas H. Roberts and Brian J. Atwell

Permissions

List of Contributors

Index

Preface

The negative effect of non-living factors on various living organisms in a particular environment is known as abiotic stress. It can affect the growth and productivity of crops. Salt stress, drought stress, phosphate starvation in plants and serpentine soils are some of the sources of abiotic stress in plants. Chemical priming is one of the methods used to increase the tolerance of abiotic stress in crop plants. In this method, stress-inducing chemical vaccinations are given to plants to help them prepare defense mechanisms to fight when the actual abiotic stress occurs. This book provides comprehensive insights into the field of abiotic stress tolerance in plants. It discusses the fundamentals as well as modern approaches of this discipline. Through this book, we attempt to further enlighten the readers about the new concepts in this field.

The researches compiled throughout the book are authentic and of high quality, combining several disciplines and from very diverse regions from around the world. Drawing on the contributions of many researchers from diverse countries, the book's objective is to provide the readers with the latest achievements in the area of research. This book will surely be a source of knowledge to all interested and researching the field.

In the end, I would like to express my deep sense of gratitude to all the authors for meeting the set deadlines in completing and submitting their research chapters. I would also like to thank the publisher for the support offered to us throughout the course of the book. Finally, I extend my sincere thanks to my family for being a constant source of inspiration and encouragement.

Editor

1

Elevated Ozone Deteriorates Grain Quality of *Japonica* Rice cv. Koshihikari, Even if it Does Not Cause Yield Reduction

Hiroko Sawada[1,2], Keita Tsukahara[1], Yoshihisa Kohno[3], Keitaro Suzuki[4], Nobuhiro Nagasawa[5] and Masanori Tamaoki[1*]

Abstract

Background: It is becoming clear that ozone affects not only grain yield but also grain quality in rice. However, the biochemical mechanisms responsible for ozone-induced changes in appearance quality or components are poorly understood. We analyzed appearance quality and starch composition in the rice cultivars "Koshihikari" (*japonica*) and "Kasalath" (*indica*) grown under elevated ozone conditions.

Results: Elevated ozone significantly increased the proportion of immature (mainly chalky) kernels in "Koshihikari" but not in "Kasalath". Scanning electron microscopy of transverse sections of kernels showed that endosperm starch granules of "Koshihikari" ripened under elevated ozone were loosely packed with large spaces and contained irregular rounded granules. Amylose content was increased in "Koshihikari" kernels with ozone exposure, but was unchanged in "Kasalath" kernels. Distribution analysis of amylopectin chain length showed that ozone induces a decrease of long-side chains and alterations of short side-chains in "Koshihikari" kernels. Furthermore, *Starch Synthase (SS) IIIa* transcript levels in "Koshihikari" caryopses were decreased by elevated ozone.

Conclusions: The *japonica* cultivar "Koshihikari" showed significant deterioration in appearance quality of kernels caused by abnormal starch accumulation due to exposure to ozone. The alteration patterns of amylose and amylopectin in ozone-exposed rice kernels are similar to those in rice kernels harvested from *SSIIIa*-deficient mutants. These findings suggest that the increase of chalky kernels in ozone-treated "Koshihikari" is partly attributable to the repressed expression of *SSIIIa* involved in amylopectin side-chain elongation with ozone exposure. Elevated ozone reduced appearance quality in "Koshihikari" although it did not impair starch properties contributing to the eating quality of cooked rice.

Keywords: Grain Quality, Ozone Stress, Rice, Starch

Background

Tropospheric ozone is a major gaseous pollutant generated by the photochemical reactions of precursor gases such as nitrogen oxides with volatile organic compounds that are mainly emitted from factories and car exhausts. Recently, ozone concentrations have rapidly increased in the developing Asian countries. For example, during 1982–2003, ozone concentrations in Beijing, China increased from a daily maximum of approximately 80 μg m^{-3} (about 40 nL L^{-1}, ppb) to 250 μg m^{-3} (125 ppb) (Shao et al. 2006). High concentrations of ozone have also been observed frequently at an agricultural site in the Yangtze River Delta, China, with the highest values of daily 1-h maximum and monthly 7-h mean being 144 and 67 ppb, respectively, during 2007–2011 (Tang et al. 2013). In the 2020s under "Current legislation" scenario, which incorporates expected economic development and planned emission controls in individual countries, the annual average ozone levels showed maximum increases of 8–12 ppb in India, Pakistan, Bangladesh, China, and Southeast Asia comparing with the 1990s (Dentener et al. 2005).

Elevated ozone concentrations reduce the growth and yield of crop plants, including rice (*Oryza sativa* L.),

* Correspondence: mtamaoki@nies.go.jp
[1]Center for Environmental Biology and Ecosystem Studies, National Institute for Environmental Studies, Onogawa 16-2, Tsukuba, Ibaraki 305-8506, Japan
Full list of author information is available at the end of the article

which is the most important food crop in Asia (Kobayashi et al. 1995; Yonekura et al. 2005; Sawada and Kohno 2009). It is becoming clear that ozone influences not only grain yield but also grain quality of rice by increasing protein and reducing starch concentrations and absolute amounts of nutrient elements (Frei et al. 2012; Huang et al. 2012; Zheng et al. 2013). Moreover, elevated ozone increases the proportion of chalky grain, which is undesirable to the majority of consumers in the Far East (Wang et al. 2012, 2014). Wang et al. (2012) pointed out that the premature senescence of rice under ozone stress will conduce to incomplete filling of kernels, resulting in more kernels with a chalky appearance. However, the biochemical mechanisms responsible for ozone-induced changes in the components or appearance quality of rice kernels are poorly understood. It is noteworthy that deterioration in the grain quality occurs at a relatively low ozone level, even one not sufficient to reduce grain yield. Thus, impairment of rice grain quality by elevation of tropospheric ozone concentration is of greater concern than a decline in rice yield, especially in East Asia.

Recently, the grain quality of rice in Japan has often been impaired by high temperatures during the grain-filling period. The chalky phenotype of rice kernels is a typical symptom found in high-temperature ripened plants. Detailed observation has shown that the chalky region of a rice kernel contains many small amyloplasts containing small, single starch granules and shows numerous interspaces among the amyloplasts (Zakaria et al. 2002). Yamakawa et al. (2007) reported that transcription of genes for the biosynthesis of starch was suppressed and expression of genes for starch-consuming enzymes was induced by high temperature. Thus, inhibition of starch accumulation in kernels occurs during ripening under high temperature and results in chalkiness of the rice grain. However, the effect of elevated ozone concentration on starch accumulation in cereal kernels has been little reported. To our knowledge, only one study has measured starch synthesis enzyme activities in ozone-exposed wheat (Zhang et al. 2013). Thus, study of the effects of ozone elevation on rice quality is needed.

In this study, we analyzed the appearance quality and starch composition in the major Japanese rice cultivar "Koshihikari" (*japonica*) and the traditional Indian cultivar "Kasalath" (*indica*) under elevated ozone concentrations to clarify the mechanism of ozone-induced chalky kernel formation and irregular starch accumulation in rice kernels.

Results

Effect of Ozone on Grain Yield and Quality in Two Rice Cultivars

Grain yields were lowered to 4.2 and 5.2 % in "Koshihikari" and "Kasalath", respectively, in plants grown under elevated ozone from that in plants grown under ambient air, although the difference was not statistically significant (Table 1). With ozone exposure, the filled kernel percentage was significantly decreased in both cultivars ($P < 0.05$). Thousands-grain weight was significantly reduced under elevated ozone only in "Koshihikari" ($P < 0.001$).

The appearance of ozone-treated "Koshihikari" kernels was severely chalky (Additional file 1). Imaging analysis data for rice kernels showed that the proportion of immature kernels in "Koshihikari" increased with ozone treatment by 24 %, a value equal to the rate of decrease in perfect grain (Fig. 1). Among the immature kernels, "milky-white kernels," which develop a chalky region at the center of the endosperm, were increased substantially in "Koshihikari" with ozone elevation. The proportion of "white-based/backed kernels" was also significantly higher in the kernels harvested from ozone-exposed rice ($P < 0.05$). To evaluate the grain quality of "Kasalath," kernels were classified into the following categories by visual inspection: perfect, immature, or damaged kernels. In "Kasalath," most of kernels were classified into immature or damaged without ozone treatment, and no significant difference in the proportion of kernel types was observed between elevated ozone and ambient air (Fig. 1 and Additional file 1).

Appearance Traits of Kernels and Starch Granules in Ozone-Exposed "Koshihikari"

Grain shape dimensions in "Koshihikari" (average length, width, and thickness for each sample) were measured with a grain image analyzer (Table 2). Kernels from rice grown under elevated ozone were significantly shorter and thinner than those grown under ambient air ($P < 0.01$), although the average kernel width was not affected by ozone treatment. The average kernel density was significantly decreased in ozone-treated rice because the decrease in the average kernel weight (Table 1, 95.3 % of AA) was greater than the decrease in the average kernel volume.

Given that the reduction in grain density suggested that starch accumulation in the endosperm was inhibited by elevated ozone levels, we observed a transverse section of the rice kernels using visual and scanning electron microscopy (SEM) (Fig. 2). As indicated by the appearance data shown in Fig. 1, ozone induced chalky kernels in "Koshihikari". SEM observation showed that translucent kernels of "Koshihikari" ripened under ambient air were filled with densely packed and similar sized granules with sharp edges. In contrast, the endosperm starch granules of "Koshihikari" ripened under elevated ozone conditions were loosely packed with large air spaces and contained irregular rounded granules. In "Kasalath", a small chalky region was observed at the center of kernels from plants grown under ambient air, and the chalky region expanded to the kernel edge with ozone-exposure. Detailed observation using SEM showed that large starch granules were surrounded by a number of small and irregularly shaped starch granules in "Kasalath" kernels ripened under elevated ozone levels. Although

Table 1 Harvest parameters of two rice cultivars grown under AA and O_3

Cultivars		Panicle number (/plant)	Total number of kernel (/plant)	Number of filled kernel (/plant)	Filled kernel percentage (%)	1000-grain weight (g)	Grain yield [a] (g/plant)	Relative yield (%)
Kos	AA	9.0 ± 0.9	642 ± 89.5	587 ± 83.2	91.2 ± 0.9	24.6 ± 0.2	14.4 ± 2.1	100.0
	O_3	8.5 ± 0.5	670 ± 36.7	591 ± 33.2	88.2 ± 1.0	23.4 ± 0.1	13.8 ± 0.8	95.8
		ns	ns	ns	*	***	ns	
Kas	AA	9.1 ± 0.6	1234 ± 78.7	1175 ± 75.1	95.2 ± 0.4	17.9 ± 0.1	21.0 ± 1.4	100.0
	O_3	7.8 ± 0.5	1218 ± 67.1	1134 ± 61.5	93.1 ± 0.3	17.6 ± 0.1	19.9 ± 1.0	94.8
		ns	ns	ns	***	ns	ns	

Values show mean ± SE ($n = 12$). [a]grain yield was determined as rough rice grain weight in this report
Asterisks denote significant differences between ambient air (AA) and elevated ozone (O_3) treatment (Student's *t*-test, *$P < 0.05$, ***$P < 0.001$). *ns* not significant, *Kos* Koshihikari, *Kas* Kasalath

the interspaces among large starch granules were filled with small starch granules, there were many small air spaces among small starch granules. A similar tendency was observed in the endosperm in "Kasalath" grown under ambient air, although fewer voids among large starch granules were detected than those in kernels grown under elevated ozone.

Alterations of the Starch Component and Amylopectin Fine Structure in Ozone-Exposed "Koshihikari" and "Kasalath"

SEM observation showed that elevated ozone affected starch accumulation in the endosperm of "Koshihikari" kernels. We accordingly investigated the effects of elevated ozone on starch and amylose contents of kernels in "Koshihikari" and "Kasalath" (Fig. 3). Content of starch was not affected by ozone in kernels of both cultivars. Amylose content was increased in "Koshihikari" but unchanged in "Kasalath" kernels grown under elevated ozone levels.

To evaluate the effects of elevated ozone on starch components in more detail, the side-chain length distribution of amylopectin was determined by the fluorescence capillary electrophoresis (FCEP) method (Fig. 4). In "Koshihikari" kernels, elevated ozone levels resulted in increased numbers of chains with a degree of polymerization (DP) 8–12, 19–22, and 25–30 and decreased numbers of chains with DP 7, 13–18, and 31–52. The alteration pattern in length distribution of amylopectin side chains in ozone-exposed "Kasalath"

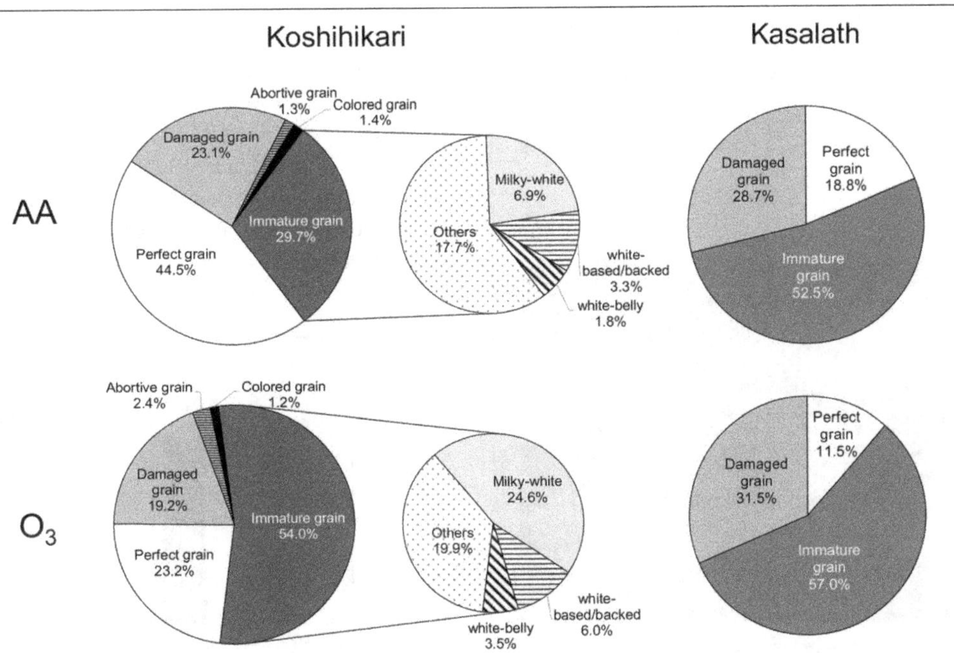

Fig. 1 Appearance quality of dehulled kernels. The proportions of perfect (translucent), immature, damaged, abortive, and colored kernels were determined using a grain image analyzer, ES-1000 (Shizuoka Seiki) for "Koshihikari". Immature kernels were further classified into milky-white, white-belly, and white-based/backed kernels with the ES-1000. The proportions of perfect, immature, and damaged kernels were determined by visual inspection in "Kasalath". *AA* ambient air, O_3 elevated ozone

Table 2 Size of rice kernels in "Koshihikari" grown under AA and O_3

	Length (mm)	Width (mm)	Thickness (mm)	Volume (mm^3)	Density (mg/mm^3)
AA	4.96 ± 0.02	2.89 ± 0.01	1.93 ± 0.01	14.51 ± 0.14	1.69 ± 0.01
O_3	4.89 ± 0.02	2.89 ± 0.01	1.90 ± 0.01	14.06 ± 0.12	1.66 ± 0.01
% of AA	98.7 **	100^{ns}	98.2 **	96.9 *	98.4 *

Values show mean ± SE ($n = 12$). Asterisks denote significant differences between ambient air (AA) and elevated ozone (O_3) treatment (Student's *t*-test, *$P < 0.05$, **$P < 0.01$). *ns* not significant

kernels differed from that in "Koshihikari" kernels. In "Kasalath" kernels, chains with DP 7–10 and 15–18 were increased and chains with DP 12–14 and 24–29 were decreased.

Expression Levels of Genes Involved in Starch Synthase in Ozone-Exposed Rice Cultivars

Given that the content of long chains of amylopectin with DP > 30 was reduced in "Koshihikari" kernels under elevated ozone levels from those in kernels grown under ambient air, we evaluated the expression in rice caryopses of the *STARCH SYNTHASE (SS) IIIa* gene, which plays an important role in generating long chains of amylopectin, and of other genes involved in starch synthesis (*SSI* and *GRANULE BOUND (GB) SSI*) (Fig. 5 and Additional file 2). On 12 day after flowering (DAF), the mRNA level of *SSIIIa* in ozone-exposed "Koshihikari" was reduced to 30 % of that of plants grown under ambient air. The reduction in the gene expression in "Koshihikari" by ozone was increased to 19 % at 23 DAF. The *SSIIIa* transcript level in "Kasalath" was higher than that in "Koshihikari", but was not affected by ozone treatment. The levels of *SSI* and *GBSSI* transcripts were not changed significantly by ozone treatment in either cultivar (Additional file 2).

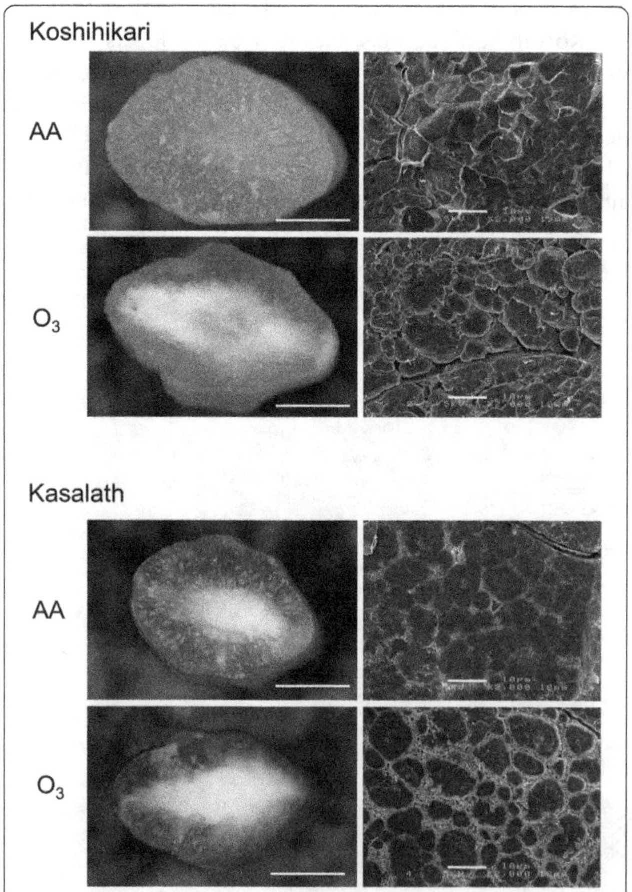

Fig. 2 Transverse sections of kernels grown under AA and O_3 in "Koshihikari" and "Kasalath". Stereoscopic microscope images are shown on the left and scanning electron microscope (SEM) images are shown on the *right*. *Bars* in stereoscopic images, 10 mm; *bars* in SEM images, 10 μm. *AA* ambient air, O_3 elevated ozone

Analysis of the Physicochemical Properties of Rice Kernels in Ozone-Exposed "Koshihikari" and "Kasalath"

To evaluate the effect of the alteration of starch structure on physicochemical properties, we analyzed the starch pasting properties of ground kernels using a rapid viscoanalyzer (RVA; Table 3 and Additional file 3). Maximum viscosity increased slightly in "Koshihikari" grown under elevated ozone ($P < 0.1$). The property of breakdown showed contrasting responses to ozone in the two cultivars: the breakdown value was increased significantly in ozone-exposed "Koshihikari" but decreased in "Kasalath" ($P < 0.05$). The setback value decreased and peak time increased significantly in ozone-treated "Kasalath" ($P < 0.05$). The pasting temperature increased slightly in kernels of ozone-treated "Kasalath" ($P < 0.1$).

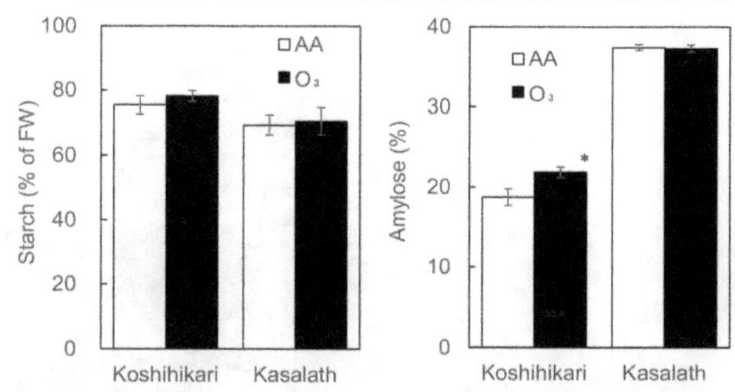

Fig. 3 Effects of elevated ozone on starch and amylose content in "Koshihikari" and "Kasalath". Values are mean ± SE ($n = 4$). *Asterisk* indicates a significant difference between AA and O_3 treatment according to Student's *t*-test ($P < 0.05$). *AA* ambient air, O_3 elevated ozone

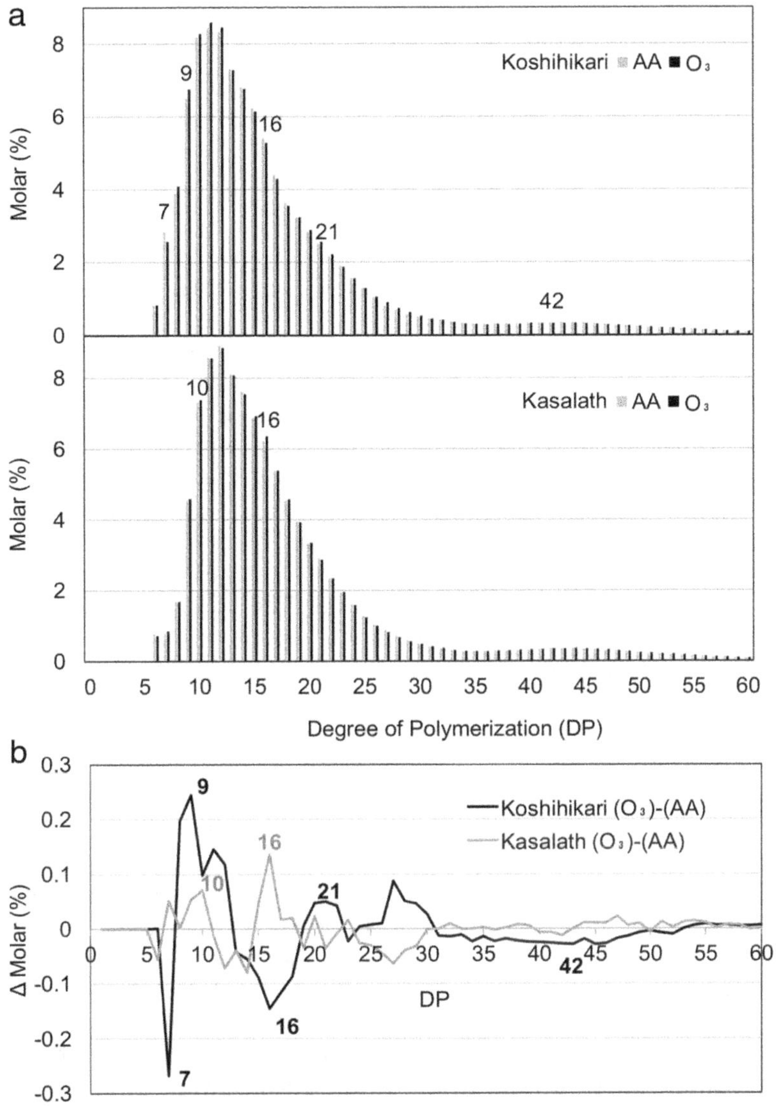

Fig. 4 Chain-length distribution patterns of endosperm amylopectin in kernels grown under AA and O_3. (**a**: *top*) "Koshihikari" and (**a**: *bottom*) "Kasalath". Debranched amylopectin extracted from AA-grown plants (*gray bars*) or O_3 (*black bars*) was analyzed using the FCEP method, and the relative peak area of the chromatogram is shown for individual DP. **b** Comparison of differences in the chain-length distribution pattern of amylopectin (delta molar %) between "Koshihikari" and "Kasalath". The difference in relative peak area in B between AA and O_3 is shown. The *numbers* in the plots are DP values. *AA* ambient air, *O_3* elevated ozone

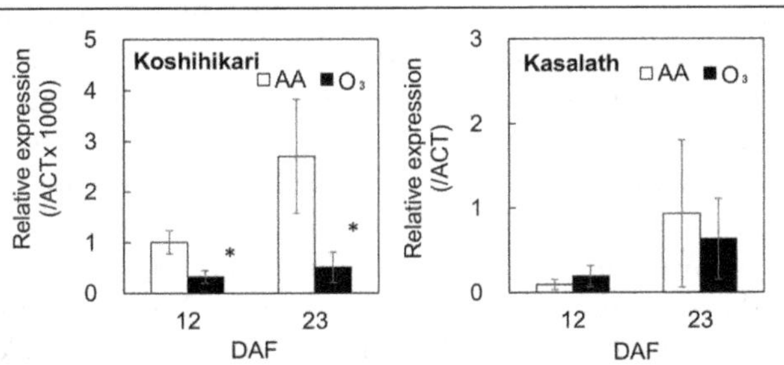

Fig. 5 Relative expression levels of *SSIIIa* transcript in rice caryopses grown under AA and O_3. Values are mean ± SE ($n = 3$). *Asterisk* indicates a significant difference between AA and O_3 treatment by Student's *t*-test ($P < 0.05$). *DAF* day after flowering, *AA* ambient air, *O_3* elevated ozone

Discussion

The appearance quality of rice kernels is important, given that chalky kernels are readily broken during milling, reducing the market price. In this study, we showed that marked deterioration in appearance quality in the *japonica* cultivar "Koshihikari" was induced by an elevated level of ozone that did not lead to grain yield reduction (Table 1, Fig. 1). In contrast, no significant change was detected in the proportion of chalky kernels in "Kasalath" grown under elevated ozone, suggesting that the appearance quality of "Kasalath" is insensitive to elevated ozone.

Table 3 Pasting properties by RVA of the flours in two rice cultivars grown under AA and O_3

		Maximum viscosity (cP)	Minumum viscosity (cP)	Breakdown (cP)	Final viscosity (cP)	Setback (cP)	Peak Time (min)	Pasting Temp (°C)
Kos	AA	3511 ± 39	1636 ± 43	1875 ± 12	2577 ± 37	941 ± 13	6.24 ± 0.04	66.4 ± 0.03
	O_3	3605 ± 15	1633 ± 23	1972 ± 21	2542 ± 22	909 ± 19	6.26 ± 0.01	66.4 ± 0.04
		*	ns	**	ns	ns	ns	ns
Kas	AA	2429 ± 53	1334 ± 42	1095 ± 12	2932 ± 54	1597 ± 14	6.12 ± 0.02	70.9 ± 0.25
	O_3	2307 ± 41	1455 ± 40	852 ± 19	2943 ± 44	1488 ± 11	6.33 ± 0.03	71.8 ± 0.15
		ns	ns	***	ns	***	***	*

Values show mean ± SE ($n = 3$). Asterisks denote significant differences between ambient air (AA) and elevated ozone (O_3) treatment (Student's *t*-test, *$P < 0.1$, **$P < 0.05$ and ***$P < 0.01$). *ns* not significant, *Kos* Koshihikari, *Kas* Kasalath

The deterioration of appearance quality found in ozone-exposed "Koshihikari" kernels occurred as the result of the marked increase in milky-white kernels under elevated ozone (Figs. 1 and 2). The chalky appearance is derived from a change in light refraction resulting from the occurrence of numerous air spaces between loosely packed starch granules (Tashiro and Wardlaw 1991). The determinations of kernel weight and kernel dimensions shown in Tables 1 and 2 suggest a lower density of kernels in ozone-exposed "Koshihikari," indicating an increase in air spaces in the endosperm.

To clarify the internal structure of the rice endosperm, transverse sections of rice kernels were observed with SEM (Fig. 2). Endosperm starch of "Koshihikari" grown under elevated ozone contained small and irregular rounded granules resulted in the creation of large air spaces between themselves. In contrast, the numbers of air spaces in "Kasalath" endosperm were fewer than those in "Koshihikari" under elevated ozone. This is the first report of ozone-induced abnormal accumulation of starch in rice kernels revealed by SEM observation. Although alteration of amyloplast structure observed in ozone-exposed kernels was similar to that in kernels ripened under high temperature, small holes in the surface of starch granules, found in kernels ripened under high temperature, were not observed in kernels exposed to ozone (Tashiro and Wardlaw 1991; Zakaria et al. 2002). This finding indicates that the underlying mechanism of the deterioration of appearance quality of rice grain differs between ozone exposure and high-temperature stress.

Previous study showed that grain chalkiness is occurred as a consequence of alterations in amylopectin structure (Yamakawa et al. 2007). We found that abnormal starch accumulation was induced by elevated ozone in "Koshihikari" and "Kasalath". In the *japonica* cultivar "Koshihikari", amylose content increased significantly with elevated ozone, although starch content was not affected by ozone (Fig. 3). Distribution analysis of amylopectin chain-length showed a decrease in long side chains of DP > 30 as well as alterations of short side chains of DP < 20 in the kernels of ozone-treated "Koshihikari". In contrast, amylose content was not changed by ozone and amylopectin chain length tended to increase in both short and long side-chains in kernels of the *indica* cultivar "Kasalath". In previous reports, elevated ozone decreased or did not change the amylose content in hybrid *indica* rice grain (Wang et al. 2012, 2014; Zheng et al. 2013), but there are no reports of the distribution pattern of amylopectin chain length in kernels ripened under elevated ozone. The effect of ozone elevation on starch composition appears to differ between *japonica* and *indica* rice cultivars. Interestingly, high-temperature treatment reduced the amylose content of *japonica* rice, a response differing from that to the ozone-induced change in the kernel amylose content. Moreover, rising temperature during the grain-ripening period resulted in a decrease in short side chains of DP 10–19 and an increase of longer side chains of DP 21–32 and > 41 (Yamakawa et al. 2007), suggesting that the pattern of alteration in distribution of amylopectin chain length under high temperature differs from that under elevated ozone. Taken together, our results suggest that chalky kernels found in ozone-exposed "Koshihikari" are generated by a novel mechanism that differs from that of the induction of chalky kernels under high temperature.

The chain-length distribution pattern of amylopectin in kernels from ozone-exposed "Koshihikari" was similar to that in kernels from *SSIIIa*-deficient rice mutants (Fujita et al. 2007). We accordingly evaluated *SSIIIa* expression in "Koshihikari" and "Kasalath" caryopses during the grain-filling period (Fig. 5). The transcript level of *SSIIIa* was decreased by ozone in "Koshihikari", whereas no such decrease was detected in "Kasalath". Previous studies showed that the *ssIIIa* mutants result in a chalky interior appearance in kernels and an irregular shape of starch granules caused by insufficient crystallization of starch (Fujita et al. 2007; Ryoo et al. 2007). Furthermore, the authors suggested that SSIIIa plays an important role in the elongation of long-B chains (DP > 30) connecting the amylopectin cluster. We thus propose that the chalky phenotype in ozone-exposed "Koshihikari" kernel is attributed to loosely packed starch granules, which is caused by the reduction of long chains of amylopectin resulting from the decrease in the expression of *SSIIIa*. On the other hand, "Kasalath" showed high percentage of chalky grain (Fig. 1), and the *SSIIIa* transcript levels in the cultivar were higher than those in "Koshihikari" with or without ozone exposure (Fig. 5). Xu et al. (2015) showed that *Indica* rice was inferior in the appearance quality compared with *japonica* rice regardless of their growth conditions, suggesting that the inherent high chalky grain in "Kasalath" is independent from the expression levels of *SSIIIa*. Although further studies, such as expression analysis of other genes involved in starch synthesis or metabolism, are needed to determine the role of *SSIIIa* on the alteration of starch structure under elevated ozone, our results indicate the presence of a novel mechanism of ozone-induced deterioration of the appearance quality in *japonica* rice.

The viscosities of pasting starch of rice kernels were analyzed to evaluate the effect of elevated ozone on eating quality (Table 3 and Additional file 3). Chamura et al. (1979) reported a negative correlation between amylose content and maximum viscosity among rice cultivars. In "Koshihikari", both the amylose content and maximum viscosity were increased by ozone treatment, suggesting that the increase of amylose content alone could not account for the reduction in eating quality by ozone. Igarashi (2010) showed that the molar ratio of short/long unit-chains of amylopectin decreased when the temperature during grain filling was high, resulting in a decrease in breakdown and increases in both minimum viscosity and pasting temperature. In the present study, the molar ratio of shorter unit-chains (DP $\leq$ 30) to long unit-chains (DP > 30) of amylopectin was increased in ozone-exposed "Koshihikari" kernels (11.0 ± 0.2 under ambient air and 11.8 ± 0.2 under elevated ozone), whereas those in "Kasalath" kernels were decreased (12.7 ± 0.2 under ambient air and 11.5 ± 0.2 under elevated ozone). The correlations between the change in the molar ratio of amylopectin and the change of pasting properties occurring under ozone elevation are consistent with those reported by Igarashi (2010). This result suggests that the chain distribution of amylopectin affects the alteration of pasting properties rather than amylose content in rice kernels under elevated ozone concentration.

In general, higher values of maximum viscosity and breakdown and lower values of setback are closely associated with better palatability of rice (Suzuki 1979; Chikubu et al. 1983). Elevated ozone significantly reduced the appearance quality of "Koshihikari" kernels, but may have improved its eating quality by increasing the values of the maximum viscosity and breakdown. In contrast, "Kasalath" kernels, which did not change in appearance quality under elevated ozone levels, showed reduced breakdown values and increased pasting temperature, suggesting detrimental effects on eating quality. However, eating quality should be evaluated by further analysis of other components, such as protein and amino acid content, and of physicochemical properties such as texture and appearance of cooked rice (Suzuki et al. 2006), that could also affect eating quality.

Conclusions

The *japonica* cultivar "Koshihikari" showed deteriorations in appearance quality under ozone stress, which occurred at a level of ozone insufficient to cause grain yield reduction. The deterioration in grain quality was induced by abnormal starch accumulation under elevated ozone levels as well as by high-temperature stress. However, the alteration patterns of amylose and amylopectin in ozone-exposed "Koshihikari" were similar to those in *SSIIIa*-deficient rice mutants rather than those in kernels ripened under high-temperature. Furthermore, the *SSIIIa* transcript in "Koshihikari" caryopses was suppressed by elevated ozone levels. Thus, the increase in chalky kernels in "Koshihikari" may result from the decreased expression, under elevated ozone, of *SSIIIa* involved in amylopectin side-chain elongation.

Methods

Plant Materials and Ozone Treatment

Seeds of rice cultivars "Koshihikari" (a *japonica* cultivar) and "Kasalath" (an *indica* cultivar) were sown in seedling boxes and grown for 3 weeks in a glasshouse under ambient air. The seedlings were transplanted into pots (0.05-m^2 surface area and 0.015-m^3 volume) (four pots of each cultivar) on June 6 and grown until September 30, 2014 in commercial-type vinyl greenhouses located in the experimental field of the National Institute for Environmental Studies (Tsukuba, Ibaraki). Fertilizer (N:P:K 8:8:5) was supplied at 150 g m^{-2} before transplanting. The dimensions of the exposure chamber were 3.15-m width, 1.8-m height, and 16.2-m length and the long axes of the chambers were in north–south orientation. Rice plants were exposed daily to ambient air (AA) or ozone-supplemented air (O_3) through air ducts from June 25 to September 19, 2014. The experiment was arranged in a randomized complete block design within the whole plot with four replicates of AA and O_3 chambers. Ozone was generated using an electrical discharge ozone generator (Model SG-1, Dylec Inc., Ibaraki, Japan) from industrial-grade oxygen of 99.5 % purity as the source gas. Concentrations of ozone were continuously monitored at three points in each greenhouse at 10-min intervals using a UV absorption ozone analyzer (Model 1210-6, Dylec Inc.). Daily mean ozone concentrations for 7 h (10:00–17:00 Japanese Standard Time) were 36 (AA) and 71 (O_3) ppb (Additional file 4). The average air temperature and relative humidity in the each greenhouse from June to September were 26.3 °C and 60.3 % (AA), and 26.6 °C and 59.9 % (O_3), respectively. The average of air temperature, relative humidity, and light intensity (photosynthetically active radiation) in both glasshouses were not significantly different throughout the growing period.

Measurement of Yield and Grain Quality Traits

Rice cultivars were harvested on September 30, 2014. All yield data were determined by measurements of 12 individual plants for each treatment. Kernels were separated from panicles and categorized into two groups: filled grain and unfilled grain, using an automatic seed-sorting machine (FV-459A, Fujiwara Seisakusho KK, Tokyo, Japan). The filled kernels (rough rice) were weighed to determine the grain yield. Filled kernels in "Koshihikari" were unhusked and appearance quality was determined by a rice grain image analyzer (ES-1000; Shizuoka Seiki Co., Ltd., Shizuoka, Japan) that classified kernels into perfect, immature, damaged, abortive, and colored kernels. Immature kernels, is a term according to the official inspection standard of the Japan Food Agency, means "insufficient filling" kernels, were further classified into milky-white, white-belly, and white-based/white-back kernels. Kernel physical attributes (kernel length, width, and thickness) were measured using another rice image analyzer (Satake Engineering Corp., Tokyo, Japan). Kernel volume (V, mm^3) was calculated as follows:

$$V = 4/3 \times L/2 \times D/2 \times T/2 \times \pi$$

where L, D, and T is the length (mm), width (mm), and thickness (mm) of the kernels, respectively. Bulk density was calculated by dividing kernel weight (mg, the date are presented in Table 1) by kernel volume (mm^3) (Fukumori and Mishima 2002).

To determine the grain quality of "Kasalath," 100 unhusked kernels were categorized by visual inspection as perfect, immature, or damaged kernels, because *indica*-type kernels were not adapted to analysis by the grain image analyzer.

For measurement of amylose and starch content, unhusked brown rice kernels were polished to remove the embryo and pericarp with a test mill (Pearlest; Kett Electric Laboratory, Tokyo, Japan), after which these kernels were ground into powder with a food mill (Milser, IFM-800DG, Iwatani Corporation, Tokyo, Japan).

Scanning Electron Microscopy (SEM)

Brown rice kernels were cut transversely with a razor blade and the cracked surfaces were coated with osmium metal. The surfaces were viewed with a JSM-6320F scanning electron microscope (JEOL Ltd., Tokyo, Japan) at an acceleration voltage of 5 kV.

Determination of Amylose Content in Kernels

Apparent amylose content was measured by an iodine colorimetric method (Yamakawa et al. 2007). Twenty milligrams of polished rice powder was gelatinized by treatment with 0.1 mL of 95 % ethanol and 0.9 mL of 1 M NaOH and was kept for 10 min in boiling water. After the addition of 5 mL of distilled water, the solution was homogenized and filled to 10 mL with distilled water. An aliquot (1 mL) of the solution was added to 0.2 mL of 1 M acetic acid, 0.2 mL of 0.2 % (*w/v*) I_2, 2 % (*w/v*) KI, and 8.6 mL of distilled water.

After incubation at 27 °C for 20 min, A620 was measured using a SmartSpec Plus spectrophotometer (Bio-Rad Laboratories Inc., Hercules, CA, USA). The apparent amylose concentration was estimated by the method of Juliano (1971) from the base calibration line, which was obtained from the absorbance values by changing the ratio of potato amylose (type III, Sigma Chemical Co., St. Louis, MO, USA) and waxy rice powder in the iodine solution.

Determination of Starch Content in Kernels

Twenty milligrams of polished rice powder was homogenized with 4 mL of dimethyl sulfoxide and 1 mL of HCl (8 mol L^{-1}) at 60 °C for 30 min. The reaction solution was adjusted to pH 4–5, and the starch concentration was determined using the enzymatic method (F-kit starch, Roche Diagnostics, Mannheim, Germany).

Determination of the Distribution of α-Glucan Chain Length of α-Polysaccharides

Chain length distribution analyses of α-glucan was performed according to the method of Fujita et al. (2012). Dehulled rice kernels (0.3–1.0 g) were ground with a mortar and pestle and the powder was suspended in 3–4 mL of distilled water. The suspension was centrifuged at 600 × *g* at 20 °C for 10 min. Three volumes of methanol were added to the supernatant, and the mixture was kept at 4 °C overnight. The precipitate was collected by centrifugation at 3000 × *g* at 4 °C for 10 min. The precipitate was washed by suspension in 2 mL of ice-cold methanol followed by centrifugation at 10,000 × *g* at 4 °C for 10 min. The sample was dried in a centrifugal vacuum evaporator. The chain length distributions of α-glucans from endosperm were analyzed using the FCEP method of O'Shea and Morell (1996) and Fujita et al. (2001) in a P/ACE MDQ Carbohydrate System (Beckman Coulter, Fullerton, CA, USA).

Quantitative Polymerase Chain Reaction (PCR) Analysis

Total RNA was isolated from developing caryopses taken at 12 and 23 DAF of "Koshihikari" and "Kasalath" grown under AA and O_3 conditions, using an RNeasy Plant Mini Kit (Qiagen, Valencia, CA, USA) following the manufacturer's instructions. A 1-μg aliquot of total RNA was reverse-transcribed using random hexamers and ReverTra

Ace® qPCR Master Mix with gDNA Remover (TOYOBO CO., LTD., Osaka, Japan) in a 20 μL reaction volume, and 2 μL of the reaction mixture was used for quantitative real-time PCR. Quantitative real-time PCR was performed in a LightCycler 480 System (Roche Applied Science, Mannheim, Germany) using KOD SYBR® qPCR Mix (TOYOBO) according to the manufacturer's specifications. A fragment of cDNA was amplified with the PCR primers 5′-GCCTGCCCTGGACTACATTG-3′and 5′-GCAAACATATGTACACGGTTCTGG-3′ for *SSIIIa (GenBank accession no: AY100469)*, 5′-GGGCCTTCATGGATCAACC-3′ and 5′-CCGCTTCAAGCATCCTCATC -3′ for *SSI (GenBank accession no: D16202)*, and 5′-AACGTGGCTGCTCCTTGAA -3′ and 5′-TTGGCAATAAGCCACACACA -3′ for *GBSSI (GenBank accession no: X62134)*. As an internal standard for cDNA amounts, a fragment of actin 1 cDNA *(GenBank accession no: AB047313)* was amplified with PCR primers 5′-CTTCATAGGAATGGAAGCTGCGGGTA-3′ and 5′-CGACCACCTTGATCTTCATGCTGCTA-3′. The relative expression level of each gene was calculated by dividing the value of each gene by the value of the actin 1 signal. Three independent biological samples were used with the gene-specific primers.

Pasting Properties

The pasting properties of rice flours were determined with a rapid viscoanalyzer (RVA) (RVA Rice Master, Newport Scientific, Sydney, Australia) by the method of Toyoshima et al. (1997). The properties were expressed as the starting gel point temperature, peak viscosity (maximum viscosity), breakdown viscosity, (the difference between peak and minimum viscosity), and setback viscosity (the difference between minimum and final viscosity).

Statistics

Software (IBM SPSS Statistics version 22; IBM) was used for statistical analyses. To assess the statistical significance of treatment differences, *t*-tests (with *P* set at 0.05) were used. Statistical analyses were performed for individual plant data for yield, yield components, and quality, and for individual pot data for starch analysis.

Additional files

Additional file 1: Visual images of dehulled kernels. Black and white arrowheads indicate severely chalky and almost translucent grains, respectively. AA, ambient air; O_3, elevated ozone.

Additional file 2: Relative expression levels of *SSI* and *GBSSI* transcript in rice caryopses grown under AA and O_3 at day after flowering 12. Values are mean ± SE ($n = 3$). AA, ambient air; O_3, elevated ozone.

Additional file 3: Pasting properties of rice kernels of "Koshihikari" and "Kasalath". The viscosity value at each temperature point is the average of three replications. The thin line indicates the change in temperature during measurement with a RVA. AA, ambient air; O_3, elevated ozone.

Additional file 4: Daily ozone concentrations for AA and O_3 exposure chambers. Values represent mean ozone concentration each hour of the day averaged from 25 June to 19 September in 2014. AA, ambient air; O_3, elevated ozone.

Abbreviations

AA: ambient air; DAF: day after flowering; DP: degree of polymerization; SEM: scanning electron microscopy; SS: starch synthase.

Competing Interests

The authors declare that they have no competing interests.

Authors' Contributions

HS, YK, and MT conceived and designed the experiments. HS, KT, KS, NN, and MT performed the experiments. HS, KS, and MT analyzed the data. HS and MT wrote the paper. All authors read and approved the final manuscript.

Acknowledgements

We thank Dr. Motohiko Kondo, Dr. Fukuyo Tanaka, and Dr. Yoshinari Ohwaki for critical advice throughout the study. Chain-length distribution analysis was performed at the Biotechnology Center, Akita Prefectural University. We would like to thank Enago (www.enago.jp) for the English language review. This work was supported by JSPS KAKENHI (Grant no. 26-40147).

Author details

[1]Center for Environmental Biology and Ecosystem Studies, National Institute for Environmental Studies, Onogawa 16-2, Tsukuba, Ibaraki 305-8506, Japan. [2]JSPS Research Fellow, Japan Society for the Promotion of Science, Chiyoda-ku, Tokyo 102-0083, Japan. [3]Central Research Institute of Electric Power Industry, Abiko, Chiba 270-1194, Japan. [4]NARO Institute of Crop Science, Tsukuba, Ibaraki 305-8518, Japan. [5]Department of Agribusiness, Akita Prefectural University, Minami akita-gun, Akita 010-0444, Japan.

References

Chamura S, Kaneko H, Saito Y (1979) Effect of temperature at ripening period on the eating quality of rice: effect of temperature maintained in constant levels during the entire ripening period (in Japanese with English abstract). Jpn J Crop Sci 48:475–482

Chikubu S, Watanabe S, Sugimoto T, Sakai F, Taniguchi Y (1983) Relation between palatability evaluations of cooked rice and physicochemical properties of rice (In Japanese with English abstract). J Jpn Soc Starch Sci (Denpun Kagaku) 44: 333–341

Dentener F, Stevenson D, Cofala J, Mechler R, Amann M, Bergamaschi P, Raes F, Derwent R (2005) The impact of air pollutant and methane emission controls on tropospheric ozone and radiative forcing: CTM calculations for the period 1990–2030. Atmos Chem Phys 5:1731–1755

Frei M, Kohno Y, Tietze S, Jekle M, Hussein MA, Becker T, Becker K (2012) The response of rice grain quality to ozone exposure during growth depends on ozone level and genotype. Environ Pollut 163:199–206

Fujita N, Hasegawa H, Taira T (2001) The isolation and characterization of a waxy mutant of diploid wheat (*Triticum monococcum* L.). Plant Sci 160:595–602

Fujita N, Yoshida M, Kondo T, Saito K, Utsumi Y, Tokunaga T, Nishi A, Satoh H, Park JH, Jane JL, Miyao A, Hirochika H, Nakamura Y (2007) Characterization of SSIIIa-deficient mutants of rice: the function of SSIIIa and pleiotropic effects by SSIIIa deficiency in the rice endosperm. Plant Physiol 144:2009–2023

Fujita N, Hanashiro I, Suzuki S, Higuchi T, Toyosawa Y, Utsumi Y, Itoh R, Aihara S, Nakamura Y (2012) Elongated phytoglycogen chain length in transgenic rice endosperm expressing active starch synthase IIa affects the altered solubility and crystallinity of the storage α-glucan. J Exp Bot 63:5859–5872

Fukumori I, Mishima T (2002) Correlation between breed improvement of ancient rice and changing molecular structure of starch (In Japanese with English abstract). The bulletin of the Experimental Farm, Faculty of Bioresources Mie University 13:30–36

Huang YZ, Sui LH, Wang W, Geng CM, Yin BH (2012) Visible injury and nitrogen metabolism of rice leaves under ozone stress, and effect on sugar and protein contents in grain. Atmos Environ 62:433–440

Igarashi T (2010) Effect of temperature during grain filling on molecular structure of rice starch from Hokkaido cultivars, and new eating quality evaluation method of rice (In Japanese with English abstract). Report of Hokkaido Prefectural Agricultural Experiment Stations 127:1–63

Juliano BO (1971) A simplified assay for milled-rice amylose. Cereal Sci Today 16: 334–340,360

Kobayashi K, Okada M, Nouchi I (1995) Effects of ozone on dry matter partitioning and yield of Japanese cultivars of rice (*Oryza sativa* L.). Agric Ecosyst Environ 53: 109–122

O'Shea MG, Morell MK (1996) High resolution slab gel electrophoresis of 8-amino-1,3,6-pyrenetrisulfonic acid (APTS) tagged oligosaccharides using a DNA sequencer. Electrophoresis 17:681–686

Ryoo N, Yu C, Park CS, Baik MY, Park IM, Cho MH, Bhoo SH, An G, Hahn TR, Jeon JS (2007) Knockout of a starch synthase gene *OsSSIIIa/Flo5* causes white-core floury endosperm in rice (*Oryza sativa* L.). Plant Cell Rep 26:1083–1095

Sawada H, Kohno Y (2009) Differential ozone sensitivity of rice cultivars as indicated by visible injury and grain yield. Plant Biol 11(Suppl 1):70–75

Shao M, Tang X, Zhang Y, Li W (2006) City clusters in China: air and surface water pollution. Front Ecol Environ 4:353–361

Suzuki H (1979) Amylography and alkali viscography of rice. In: Proceedings of the workshop on chemical aspects of rice grain quality, International Rice Research Institute, Los Banos., pp 261–282

Suzuki K, Okadome H, Nakamura S, Ohtsubo K (2006) Quality evaluation of various "new characteristic rice" varieties based on physicochemical measurements (In Japanese with English abstract). Nippon Shokuhin Kagaku Kaishi 53:287–295

Tang H, Liu G, Zhu J, Han Y, Kobayashi K (2013) Seasonal variations in surface ozone as influenced by Asian summer monsoon and biomass burning in agricultural fields of the northern Yangtze River Delta. Atmos Res 122:67–76

Tashiro T, Wardlaw I (1991) The effect of high temperature on kernel dimensions and the type and occurrence of kernel damage in rice. Aust J Agric Res 42: 485–496

Toyoshima H, Okadome H, Ohtsubo K, Suto M, Horisue N, Inatsu O, Narizuka A, Aizaki M, Okawa T, Inouchi N, Fuwa H (1997) Cooperative test on the small-scale rapid method for the gelatinization properties test of rice flours with a rapid-visco-analyser (RVA). Nippon Shokuhin Kogyo Gakkaishi 44:579–584

Wang Y, Yang L, Han Y, Zhu J, Kobayashi K, Tang H, Wang Y (2012) The impact of elevated tropospheric ozone on grain quality of hybrid rice: a free-air gas concentration enrichment (FACE) experiment. Field Crop Res 129:81–89

Wang Y, Song Q, Frei M, Shao Z, Yang L (2014) Effects of elevated ozone, carbon dioxide, and the combination of both on the grain quality of Chinese hybrid rice. Environ Pollut 189:9–17

Xu Q, Chen W, Xu Z (2015) Relationship between grain yield and quality in rice germplasms grown across different growing areas. Breed Sci 65:226–232

Yamakawa H, Hirose T, Kuroda M, Yamaguchi T (2007) Comprehensive expression profiling of rice grain filling-related genes under high temperature using DNA microarray. Plant Physiol 144:258–277

Yonekura T, Shimada T, Miwa M, Arzate A, Ogawa K (2005) Impacts of tropospheric ozone on growth and yield of rice. J Agric Meteol 60:1045–1048

Zakaria S, Matsuda T, Tajima S, Nitta Y (2002) Effect of high temperature at ripening stage on the reserve accumulation in seed in some rice cultivars. Plant Prod Sci 5:160–168

Zhang R, Hu H, Zhao Z, Yang D, Zhu X, Guo W, Zhu J, Kobayashi K (2013) Effects of elevated ozone concentration on starch and starch synthesis enzymes of Yangmai 16 under fully open-air field conditions. J Integr Agric 12:2157–2163

Zheng F, Wang X, Zhang W, Hou P, Lu F, Du K, Sun Z (2013) Effects of elevated O_3 exposure on nutrient elements and quality of winter wheat and rice grain in Yangtze River Delta, China. Environ Pollut 179:19–26

2

Proteomic and Glycomic Characterization of Rice Chalky Grains Produced Under Moderate and High-temperature Conditions in Field System

Kentaro Kaneko[1], Maiko Sasaki[1], Nanako Kuribayashi[1], Hiromu Suzuki[1], Yukiko Sasuga[2], Takeshi Shiraya[2,3], Takuya Inomata[1], Kimiko Itoh[1], Marouane Baslam[1] and Toshiaki Mitsui[1,2*]

Abstract

Background: Global climate models predict an increase in global mean temperature and a higher frequency of intense heat spikes during this century. Cereals such as rice (*Oryza sativa* L.) are more susceptible to heat stress, mainly during the gametogenesis and flowering stages. During periods of high temperatures, grain filling often causes serious damage to the grain quality of rice and, therefore, yield losses. While the genes encoding enzymes involved in carbohydrate metabolism of chalky grains have been established, a significant knowledge gap exists in the proteomic and glycomic responses to warm temperatures in situ. Here, we studied the translucent and opaque characters of high temperature stressed chalky grains of 2009 and 2010 (ripening temperatures: 24.4 and 28.0 °C, respectively).

Results: Appearance of chalky grains of both years showed some resemblance, and the high-temperature stress of 2010 remarkably extended the chalking of grain. Scanning electron microscopic observation showed that round-shaped starch granules with numerous small pits were loosely packed in the opaque part of the chalky grains. Proteomic analyzes of rice chalky grains revealed deregulations in the expression of multiple proteins implicated in diverse metabolic and physiological functions, such as protein synthesis, redox homeostasis, lipid metabolism, and starch biosynthesis and degradation. The glycomic profiling has shown slight differences in chain-length distributions of starches in the grains of 2009-to-2010. However, no significant changes were observed in the chain-length distributions between the translucent and opaque parts of perfect and chalky grains in both years. The glucose and soluble starch contents in opaque parts were increased by the high-temperature stress of 2010, though those in perfect grains were not different regardless of the environmental changes of 2009-to-2010.

Conclusion: Together with previous findings on the increased expression of α-amylases in the endosperm, these results suggested that unusual starch degradation rather than starch synthesis is involved in occurring of chalky grains of rice under the high-temperature stress during grain filling period.

Keywords: Amylase, Chaperones, Endosperm, Environment, Heat shock protein, High-temperature stress, Late embryogenesis abundant (LEA) protein, Grain chalkiness, Starch granule, Stress-related protein

* Correspondence: t.mitsui@agr.niigata-u.ac.jp
[1]Graduate School of Science and Technology, Niigata University, Niigata 950-2181, Japan
[2]Department of Applied Biological Chemistry, Niigata University, Niigata 950-218, Japan
Full list of author information is available at the end of the article

Background

Global climate change is one of the most serious environmental threats we face today. Since the early 20th century, the average surface temperature of the Earth has unusually increased by about 0.8 °C coupled with the rapid warming of 0.6 °C over the past three decades. Climate change is also projected to have significant impacts on crop production (IPCC 2013). In 80 % of the rice harvested areas in Japan climate variability was more important and the explanation is on account of temperature variability (Ray et al. 2015). For every 1 °C increase in temperature, there was a 6.6 % decrease in yield from the current values for early rice, 5.2 % for late rice, and 8.2 % for single-cropped rice (Defeng and Shaokai 1995). The abnormal high temperature during rice endosperm development and grain filling periods can change the chemical ingredients of rice caryopses such as starch and storage proteins and the contents of fatty acid, thus causing a decrease in grain yield, quality and, hence, price. The occurrence of chalky grains of rice is increased by high-temperature stress during grain filling (Nagato and Ebata 1965). Morita et al. (2016) showed that chalky grains of *japonica* cultivars have been produced under temperatures more than 26 °C during the grain-filling period. Daily mean air temperatures of 26 °C during grain filling are becoming frequent in Japan (Usui et al. 2014). Many research groups have studied morphological characteristics of chalky grains, showing that abnormal starch granules were loosely packed in the chalky grains (Evers and Juliano 1976; Tashiro and Wardlaw 1991; Kim et al. 2000; Lisle et al. 2000; Singh et al. 2003; Ishimaru et al. 2009). Kernels with chalky have a lower density of starch granules than do vitreous ones and are more prone to breakage during milling. The surface of round-shaped starch granules in chalky grains caused by high-temperature stress had small pits occasionally (Tashiro and Wardlaw 1991). It should be noted that the feature of change of granule surface was similar to that occurring during germination (Fuwa et al. 1977). The environmental temperature at the grain filling stage has been reported to influence the starch composition in rice grains (Asaoka et al. 1984; Inouchi et al. 2000; Lisle et al. 2000; Umemoto and Terashima 2002; Cheng et al. 2005; Yamakawa et al. 2007; Mitsui et al. 2016). High temperature caused a reduction in the amylose contents and changed the fine structure of amylopectin (Asaoka et al. 1984; Inouchi et al. 2000; Cheng et al. 2005), suggesting that the unusual expression of the starch synthesizing enzymes is a possible key factor causing the chalky grains of rice (Nishi et al. 2001; Tanaka et al. 2004). On the other hand, surface of the starch granules showed clear 'erosion' with multiple small pits suggesting an attack by α-amylases (Zakaria et al. 2002; Iwasawa et al. 2009); the suppression of α-amylase genes by RNA interference improved the appearance quality of rice grains ripened under heat stress (Hakata et al. 2012). However, the mechanism of grain chalkiness under high-temperature stress is considerably complicated and still remains poorly understood: a disorder of photosynthesis, translocation efficiency, source-sink relationship, and protein expression in ripening seeds may involve in such chalking mechanism. Many quantitative trait loci (QTLs) controlling grain appearance quality have been identified in populations derived from crosses between *japonica* cultivars (Tabata et al. 2007; Ebitani et al. 2008; Kobayashi et al. 2013; Ishimaru et al. 2016), between *japonica* and *indica* cultivars (He et al. 1999; Wan et al. 2005; Ishimaru et al. 2016), *indica* cultivars (Mei et al. 2013), and between *O. sativa* and *O. glaberrima* (Li et al. 2004). Therefore, understanding of the mechanisms of grain chalking is indispensable to develop a strategy for reducing the high rate of chalky grains under the likely scenario of global warming (Lin et al. 2014).

On the other hand, in addition to starch, proteins account for 6–10 % of the dry mass and are important for the nutrition, cooking, and brewing quality of rice grains (Bressani et al. 1971; Hamaker 1994). Therefore, the importance of rice endosperm proteins should not be underestimated. The lack of perfect correlation of mRNA and protein levels during heat treatment has pushed the researchers to explore insights into the proteomic basis, since this technique provides a more direct assessment of the actual proteins performing the signaling, enzymatic, regulatory and structural functions encoded by the genome and transcriptome (Echevarría-Zomeño et al. 2015). Considerable work on rice grains proteomics has been carried out using two-dimensional polyacrylamide gel electrophoresis (2D-PAGE) and gel-free-based shotgun procedures (Koller et al. 2002; Lin et al. 2005, 2014; Xu et al. 2008; Lee and Koh 2011). Non-redundant proteins of 4,172 with a wide range of molecular weight (5.2–611 kDa) and pI values (pH 2.9–12.6) in developing and mature grains of rice has been identified by using a label-free shotgun technique (Lee and Koh 2011). The ontology categories of 52 including the carbohydrate metabolic process, transport, localization, lipid metabolic process, and secondary metabolic process were enriched. Expression analyzes of functionally categorized protein groups showed dynamic changes of metabolisms during rice grain development. As a noteworthy observation, proteins involved in glycolysis, citric acid cycle, lipid metabolism, and proteolysis accumulated at higher levels in mature grain than those of developing stages (Lee and Koh 2011). This probably indicates that the preparation of materials required in germination occurred until the seeds were fully matured and dried.

Proteomic information of rice grains in the anthesis, ripening, and maturing stages under heat stress has been gradually accumulated. The anthesis and early ripening stages are known to be highly sensitive to heat stress. Gel-based proteomic analyzes of different genotype anthers prepared from rice plants treated with high (38 °C) and control (29 °C) temperature at anthesis were carried out (Jagadish et al. 2008). Both cold (19 kDa) and heat (24 kDa) shock proteins were significantly up-regulated in a heat-tolerant genotype N22, these possibly contributing to the greater heat tolerance of N22. Heat stress (35/30 °C day/night) during an early stage of caryopsis development reduced the expression of starch granule-bound starch synthase (Wx) and prolamin, but enhanced the expression of dnaK-type hsp70 and glutelins in comparison with those in control temperature (30/25 °C) (Lin et al. 2005). In addition, heat stress response of several different cultivars including high-chalky types were analyzed, the results showing that sHSP was positively correlated with the appearance of chalky kernels (Lin et al. 2005). In recent studies, accumulation of all classes of storage proteins was increased at early filling stage under heat stress (35/30 °C), whereas the prolamin accumulation was decreased at maturation and desiccation stages (Lin et al. 2010). On the other hand, Lin et al. (2014) showed that up-regulation of proteins involved in starch accumulation and down-regulation of ER proteins, PDIL 2–3 and BiP, were observed in the chalky tissue of notched-belly mutant appeared regardless of the environmental stress, employing a comparative proteomic analysis by iTRAQ (isobaric tags for relative and absolute quantification). Understanding protein expression patterns and its respective posttranslational modifications in chalky grain is of fundamental importance for targeting rice quality under temperatures variability.

In the present investigation, we characterized chalky and perfect grains of rice harvested in Niigata, central Japan, of 2009 and 2010 employing proteomic and glycomic techniques to provide a better understanding of the mechanisms of chalky formation under hot season in field conditions. 2010 was the hottest year since Japan began keeping records; on the other hand, 2009 was an average crop year. Our results provide a comprehensive view of proteome and glycomic characterization under temperature variability. We consider which proteins, possibly, involved in the grain chalking under the paddy field conditions.

Results and discussion

Rice grains under high temperature showed chalky appearance

Perfect and chalky grains of rice (cv. *Koshihikari*) were harvested at a paddy field of Sanjo city (Niigata, Japan) in 2009 and 2010. The average temperature of heading and ripening period of Koshihikari in 2010 (28.0 °C) was much higher than that in 2009 (24.4 °C). The 1,000-kernel weight of chalky grains in 2010 (15.61 g) was apparently small compared with the 2009 chalky grains (16.97 g), but a variation of the 1,000-kernel weight of the perfect grains in 2009 and 2010 (21.62 and 21.99 g, respectively) was indistinguishable. The volumes (length x width x thickness) of 2009 perfect, 2010 perfect, 2009 chalky, and 2010 chalky grains were 29.15, 28.73, 23.85 and 23.84 mm^3, respectively. Thus, the size of the chalky grain was smaller than that of the perfect grain, whereas there was no difference between the sizes in 2009 and 2010 chalky grains. The decrease in rice grain length and width might be related with the reduction in average endosperm cell area observed under high temperature (Morita et al. 2005). Consequently, it has been suggested that the flow of C and N to the grain (Mohammadi et al. 2010), and insufficient supply of photosynthates from source to sink organ and carbohydrate deficit (Liu et al. 2009; Kanno and Makino 2010; Shi et al. 2013) caused by increased temperature could be causes of chalkiness.

The ratios of the opaque region to the whole area in 2009 and 2010 chalky grains were estimated to be 61.1 ± 10.2 and 78.4 ± 13.6 % ($n = 50$), respectively. These results strongly suggest that the high-temperature stress further extended the chalking of grain. Starch granule morphologies of the perfect and chalky grains were analyzed by scanning electron microscopy (SEM). In the perfect grain (Fig. 1a), the starch granules were tightly packed in the endosperm cells, and the shape of starch granule was polygonal with sharp edges (Fig. 1Ca). In the chalky grain (Fig. 1b), the starch granules in the translucent part of chalky grains had similar tight packing and shape (Fig. 1Cb) compared to the perfect grain. While in the opaque part, the starch granules were loosely packed, and some granules had a round shape with several small pits (Fig. 1c; c-1, 2, 3 and 4). The numerous pits were frequently observed upon the surface of round-shaped starch granules in the chalky endosperm caused by high-temperature stress of grain filling period (Evers and Juliano 1976, Tashiro and Wardlaw 1991). It is noteworthy that the present result of SEMs observation confirmed display signs of pitting on the starch granules in the chalky grains (Fig. 1c; c-1, 2, 3 and 4). The chalky zone sometimes occurs midway between the center and peripheral part of the endosperm in the developing kernels under high-temperature conditions. The ring-shaped chalkiness could be a cell-specific event associated with the disruption of starch accumulation. Similar observations made for cell layers in rice endosperm showed that inadequate starch accumulation occurs from the center towards outward in the endosperms (Wada et al. 2014). The increase of the frequency of pitting in starch granules with high temperature is presumably due to the premature autolysis of starch. This is similar to the premature autolysis reported in the case of wheat and maize kernels subjected to high temperatures (Tashiro and Wardlaw 1990, 1991; Commuri and Jones 1999).

For analyzing changes in the proteome and glycome of grains, the translucent and opaque parts of chalky grains and the central parts of perfect grains were prepared as

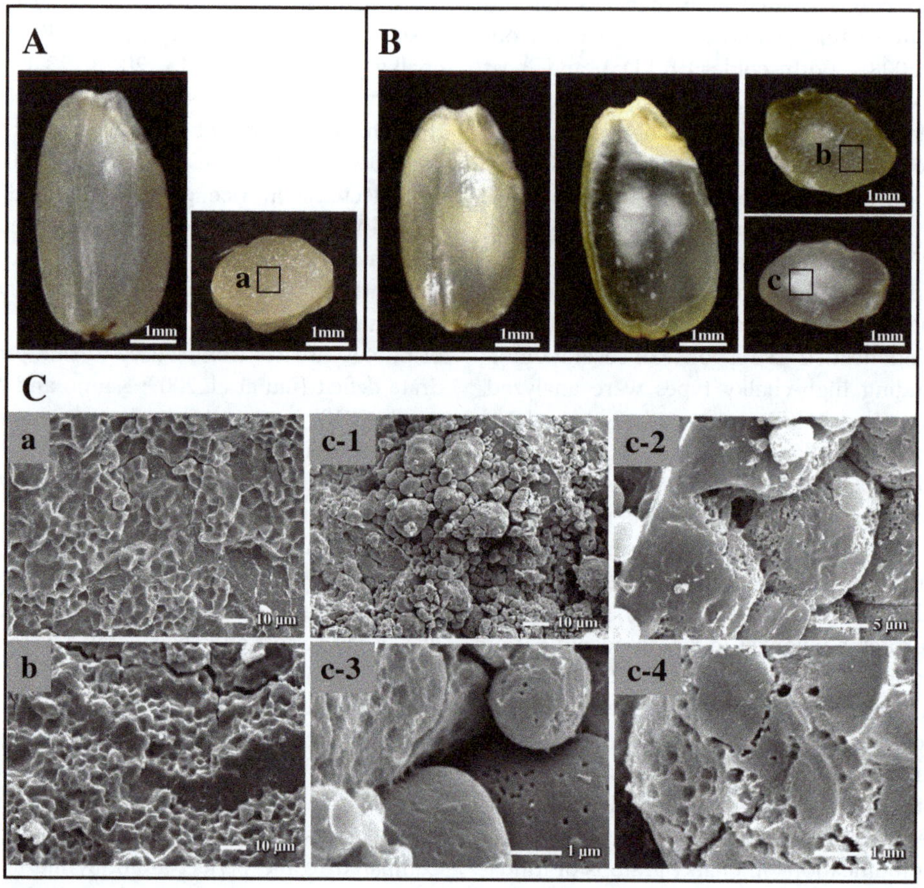

Fig. 1 Morphological characteristics of chalky grain of rice. (A) Perfect grain. Left and right panels show pictures of whole grain and horizontal slice, respectively. (B) Chalky grain. Left, whole grain; middle, vertical slice; right, horizontal slices. (C) SEM pictures. Center part of perfect grain (a), and translucent (b) and opaque (c) parts of chalky grain were subjected to SEM observation. Magnifications were x1,000 (a,b,c-1), x4,000 (c-2) and x10,000 (c-3, c-4), respectively

follows: central part of perfect grain in 2009 (PG 2009), translucent part of chalky grain in 2009 (tCG 2009), opaque part of chalky grain in 2009 (oCG 2009), central part of perfect grain in 2010 (PG 2010), translucent part of chalky grain in 2010 (tCG 2010), and opaque part of chalky grain in 2010 (oCG 2010).

Proteomic profile of chalky grains caused under different temperature conditions

To characterize the proteome involved in the mechanism of grain chalking, we carried out a quantitative shotgun proteomic analysis of starchy endosperms prepared from the opaque part of chalky grains (oCG 2009 and oCG2010) and the corresponding part of perfect grains harvested in 2009 and 2010 (PG 2009 and PG2010). The extracted proteins were trypsin-digested and labeled by iTRAQ (isobaric tag for relative and absolute quantitation), followed by tandem mass spectrometry (MS/MS) analysis. Analysis of protein extracts from the different samples resulted in the detection 938 of roteins (Fig. 2a, Additional file 1: Table S1). This analysis revealed that the expression of 61 proteins, 6.5 % of all identified proteins, were deregulated (more than 2.0-fold difference relative to PG control; P value < 0.05) in the oCG 2010 (Fig. 2b, Additional file 1: Table S1). Among this population, 41 genes were up-regulated and 20 genes were down-regulated.

To determine the biological processes affected by high temperature, an analysis of proteins using the InterPro-scan was carried out. This research revealed that chalkiness clearly leads to the differential expression of proteins and suggest that the energy and metabolic pathways are highly disrupted; other categories of proteins may involve in protecting the cellular damage under high-temperature stress (Fig. 2). The annotation of protein sequences using InterProScan permitted to fall them into 33 categories. In the molecular functional group, the identified proteins; ribosomal, redox, LEA and HSP were ranked at the top of the category occupancy, suggesting that the relevant functions were important in the response to high-temperature stimuli. It is thus conceivable that, as illustrated in Fig. 2b, differential expression of these proteins are partially the consequence of the apparition of chalky tissue. Under

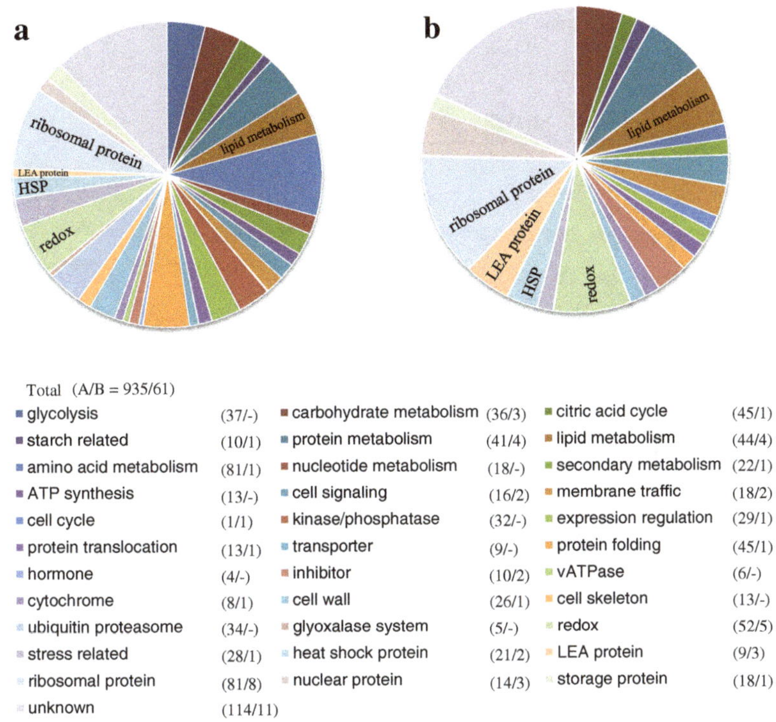

Fig. 2 Proteome of perfect and chalky grains of rice. Center part of perfect grains and opaque part of chalky grains harvested in 2009 and 2010 were subjected to protein extraction, followed by in solution shotgun proteomic analysis with iTRAQ labeling. (A/B): **a**, Total identified proteins (935); **b** ≧ Two-fold up- or down-regulated proteins in the chalky grains of 2010 (61)

high-temperature stress, reactive oxygen species are generated, and cell membrane integrity is often lost leading to cell death. The heat-stressed plants signal for reprogramming cellular metabolism, either by increasing or decreasing the transcriptional and translational events. As shown in Fig. 3, we classified the detected stress-related proteins involved in oCG during 2009 and 2010 into four major groups according to their physiological functions: late embryogenesis abundant (LEA) (Fig. 3a), heat shock protein (HSP) (Fig. 3b), glutathione redox regulation (Fig. 3c) and other stress-related proteins (Fig. 3d).

The results suggested that in both oCG of 2009 and 2010 LEA proteins were markedly accumulated, being more abundant in oCG 2010 hot season (+ ~ 4 °C than 2009). In oCG 2010, Q75LD9, Q8S7U3, like other six identified LEA proteins, increased in abundance, while Q9AWZ5 slightly decreased in abundance The majority of LEA proteins display a preponderance of hydrophilic and charged amino acid residues (Xiao et al. 2007). Also, considerable evidence suggests that LEA proteins are involved in desiccation resistance (Ingram and Bartels 1996), a variety of mechanisms for achieving this end have been proposed including protecting cellular structures from the effects of water loss by retention of water, sequestration of ions, direct protection of other proteins or membranes, or renaturation of unfolded proteins (Cuming 1999; Bray et al. 2000; Olvera-Carrillo et al. 2011).

HSPs can be used to stabilize protein conformation, prevent aggregation, and, therefore, maintain non-native proteins in a competent state for subsequent refolding in plants under heat stress (Morimoto 2002; Wang et al. 2004; Zi et al. 2013). The significant up-regulation of the HSPs is a key part of the heat shock response and is induced primarily by heat shock factors (HSFs) such as temperatures variability, drought, salinity, cold and chemicals (Scarpeci et al. 2008; Al-Whaibi 2011; Liao et al. 2013). Among the HSPs illustrated in this study, the amount of the HSP 81–1, HSP 81–3, and 70 kDa HSP involved strong up-regulated in oCG 2010 -hot season-, while they were down-regulated in oCG 2009 -moderate temperature season-. The HSP70 family, group of molecular chaperones, such as BiP1 and BiP2 maintains polypeptides in an unfolded state (Lin et al. 2014). There is evidence that changes in BiP protein levels induce endoplasmic reticulum stress, and BiP overexpression affects the accumulation of seed storage proteins and, then, results in an opaque phenotype in the whole endosperm of rice (Yasuda et al. 2009). In addition, small heat-shock protein (sHSPs) encoding genes (HSP16.9A, HSP17.9A, HSP18.1) were up-regulated in oCG 2010, although the oCG 2009 moderately accumulated these sHSP (Fig. 3b). Among the various plant Hsps (i.e. HSP100, HSP90, HSP70, and HSP20), sHSPs have been identified to expressed in maximal amounts under high-temperature stress. They are also

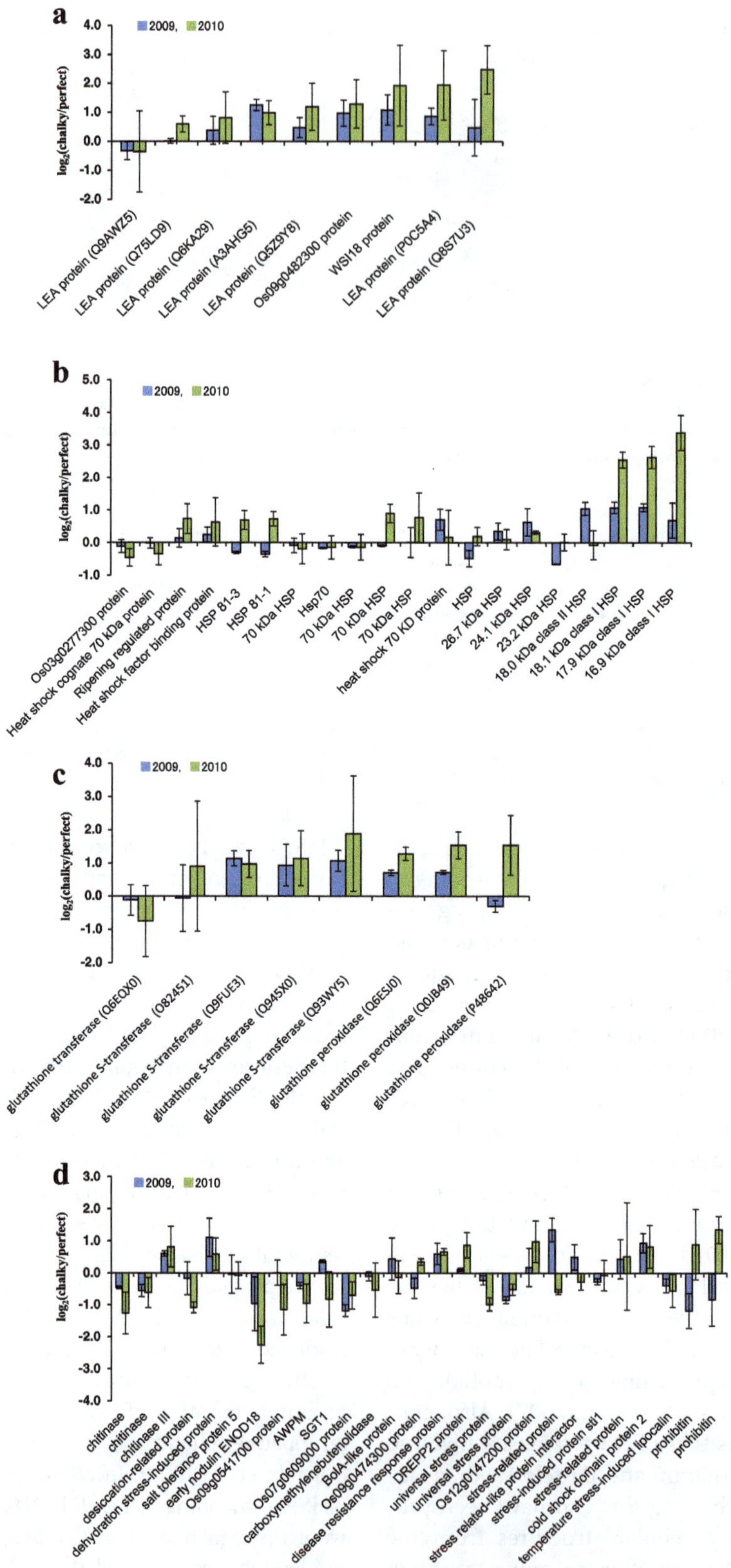

Fig. 3 (See legend on next page.)

(See figure on previous page.)
Fig. 3 Proteome of stress-related proteins in chalky grains. Center part of perfect grains and opaque part of chalky grains harvested in 2009 and 2010 were subjected to protein extraction, followed by in solution shotgun proteomic analysis with iTRAQ labeling. **a** Late embryogenesis abundant proteins; **b** Heat shock proteins; **c** Glutathione redox regulation; **d** other stress-related proteins. Proteins were categorized by NCBI databases. Values are represented as mean ± s.d. ($n = 3$)

up-regulated in rice caryopses during the grain milky stage (Lin et al. 2005; Liao et al. 2013). Our data were supported by a number of previous studies. HSP70 was reported to potentially be involved in a repair function after desiccation rather than biochemical stabilization in the dry state for *Richtersius coronifer* (Zi et al. 2013), whereas sHSPs associate with nuclei, cytoskeleton, and membranes, and as molecular chaperones they bind partially denatured proteins, thereby preventing irreversible protein aggregation during stress. sHSPs were positively correlated with the appearance of chalky kernels (Lin et al. 2005). In corroboration with another report (Das et al. 2015), the present results support the view that the induction of sHSPs suggests their function in re-establishing normal protein confirmation and thus cellular homeostasis. Besides, this study revealed that the high temperature in 2010 (+ ~ 4 °C) increased the abundance of high molecular weight HSPs. The HSP81.3 and 81.1, of the HSP90 gene family, are associated with different polypeptides serving a general mode of cellular activities. Moreover, the expression of the HSP90 genes and mRNA accumulation in plants and calli were strongly induced by high temperature (Milioni and Hatzopoulos 1997). HSP data indicate that the HSP90 (HSP81-1, HSP81-3) and 70 kDa HSP may play an important role at the chalky tissue formation. It is believed that this diversification of HSP reflects an adaptation to tolerate the heat stress in chalky rice grains. Moreover, the physiological role of chaperones like HSP90 in plants remains poorly understood. Needless to say, further research will be necessary to study the possible positive correlation between the encoding proteins of high molecular HSPs and chalky endosperm.

As to the possible mechanisms explaining the high content of chalky grains in 2010, it is worth to note that the levels of some glutathione redox isoforms in oCG 2010 were, all of them, higher than oCG 2009 (Fig. 3c). Glutathione, implicated in the antioxidant defense through the ascorbate/GSH cycle, plays a key role in maintaining the homeostasis of reactive oxygen species (ROS). Glutathione S-transferases (GSTs) are ubiquitous enzymes encoded by a large family of genes, which play an important role in cellular detoxification to a wide variety of endobiotic and xenobiotic substrates by conjugating the tripeptide glutathione. GSTs have been found to be differentially regulated by dehydration (Bianchi et al. 2002). However, antioxidant systems cannot completely prevent the deleterious effects of ROS. In this study, nine of the identified proteins are implicated in redox homeostasis-related functions in chalky grains, including glutathione transferase (GT), GSTs and glutathione peroxidase (GPX). Therefore, this would indicate that the ROS scavenging system may be activated in rice chalky kernels to alleviate such oxidative damage and to enhance high-temperature tolerance, mostly in 2010. Two proteins, GPX and GST, which are involved in the glutathione-ascorbate cycle for removing H_2O_2, showed the same expression patterns (Figs. 3c). GST and GPX can reduce H_2O_2 to its corresponding hydroxyl compounds to remediate oxidative membrane damage, and their expression showed the increases in abundance in chalky tissues. Some of these proteins were consistent with those observed in the rice (Lin et al. 2014) and maize (Luo et al. 2010) in response to stress in developing kernel tissues. Thus, our results, in conjunction with those previously reported (Liu et al. 2010; 2011; Lin et al. 2014), show that the up-regulation of these proteins, more pronounced in high-temperature conditions, suggest a close relation between redox homeostasis and the enhancement of chalky grains frequency. Recent investigation revealed that a heat-tolerant cultivar of rice exhibits a characteristic high expression of superoxide dismutase MSD1. In addition, the grain quality of transgenic overexpressor plants with the maize *Ubiquitin-1* promoter fused to *MSD1* was significantly improved in comparison with the wild type under heat stress after heading (Shiraya et al. 2015). We infer that the timely enhancement of H_2O_2 level by MSD1 under high-temperature stress is probably important, which acts as a signal that rapidly can promote the expression of stress-response proteins (Shiraya et al. 2015; Mitsui et al. 2016).

Among various additional chalky grain stress-related proteins, the expression of prohibitin and DREPP2 proteins were most remarkably increased under a high temperature -oCG 2010- (Fig. 3d). Moreover, a characteristic behavior of expression of prohibitin which may have multiple functions including mitochondrial chaperone activity (Tatsuta et al. 2005; Van Aken et al. 2009) was up-regulated in oCG 2010, while it was down-regulated under oCG 2009. To our knowledge, this study is the first to provide insights into the differential expression of prohibitin proteins in opaque chalky part collected from rice grains following heat stress season. Prohibitin have been shown to play central roles in cell cycle regulation, receptor-mediated signaling at the cell surface, aging, apoptosis, and plant development and senescence (Berger and Yaffe 1998; Coates et al. 1997; McClung et al. 1992, 1995; Nijtmans et al. 2002; Piper et al. 2002; Chen et al. 2005). Recent studies using muscovy ducks suggested changes in the abundance of the mitochondrial protein chaperone, prohibitin, into thermal tolerance related to energy metabolism (Zeng et al. 2013).

The biosynthesis of starch is the major determinant of yield in cereal grains (Emes et al. 2003). Furthermore, the main constituent of the rice grain is starch, thus the grain filling capacity is determined mainly by the starch-synthesizing capability of endosperm. Proteins associated with carbohydrate metabolism, especially starch synthesis, showed altered expression patterns induced by high temperature. Among them, the α-amylase isoform, Amy3E (AmyII-3), was dramatically up-regulated in oCG 2010 (Fig. 4a). As expected, the expression level of granule-bound starch synthase 1 (GBSS1), starch branching enzyme (SBE) BEIIb, soluble starch synthases (SS), and sucrose synthases (SuSy) were down-regulated in the chalky grains (Fig. 4a, b). It has been observed that the expression of starch-synthesis-related genes repressed under high-temperature conditions (Yamakawa et al. 2007). In the growing rice grains, the concerted functioning of multiple forms of granule-bound (GBSS) and SS, SBE and debranching enzymes (DBE), together with the ADPG supplying strength, determine the overall grain-filling capacity (Su 2000). Thus, the impaired balance of starch biosynthesis and degradation at oCG 2010 may increase the rate of grains to appear chalky because of the imperfect filling of endosperm cells with starch granules.

Noteworthy, high-temperature stress can lead to changes in the other proteins involved in lipid (Fig. 2b, Additional file 1: Table S1) and glycolysis metabolic processes (Fig. 2b), suggesting that primary metabolism might have been inhibited in chalky grains. Indeed, we found the lipid metabolism-related proteins, caleosin and sterol carrier protein, were differentially expressed in oCG, which may aid the mechanisms of chalkiness formation. The down-regulation of these proteins might retard the metabolic speed of fatty acid synthesis, under high-temperature conditions, in chalky grains. Because caleosin is associated with the endoplasmic reticulum and/or oil bodies (OBs) in seed embryo development (Næsted et al. 2000), promotes specific interaction of OBs with vacuoles and facilitates access to triacylglycerides, to serve as energy source, by lipases (Poxleitner et al. 2006). Likewise, sterol carrier protein, peroxisomal lipid protein, might affect the transfer of lipids between membranes, β-

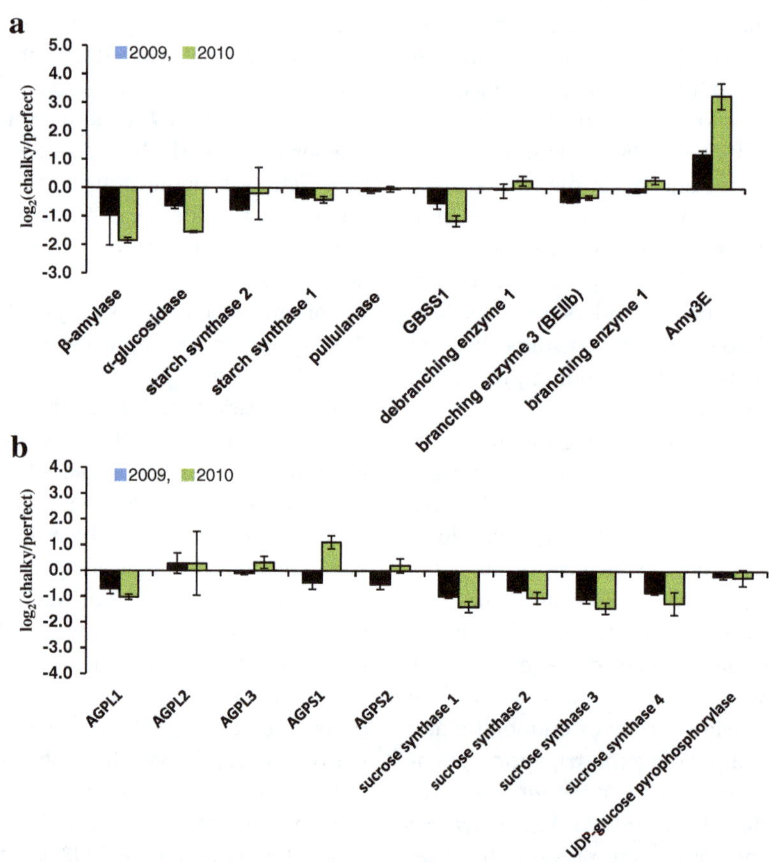

Fig. 4 Proteome of starch metabolism in chalky grains. Center part of perfect grains and opaque part of chalky grains harvested in 2009 and 2010 were subjected to protein extraction, followed by in solution shotgun proteomic analysis with iTRAQ labeling. **a** starch synthesis and degradation; **b** ADP-glucose supply. Values are represented as mean ± s.d. ($n = 3$)

oxidation and glyoxylate pathways, and hence the normal development and morphology of grains (Zheng et al. 2008). Thus, down-regulation of lipid metabolism related proteins probably contribute to the chalkiness formation in grains of 2009-to-2010. Further detailed investigations, including lipid profiling may provide us with additional clues to the biological function of those proteins in chalkiness formation.

Glycomic characteristics of chalky grains caused under different temperature conditions

Starch quality is an important parameter in determining the quality of rice grains. Numerous investigations have revealed that the environmental temperature at the grain filling stage apparently alters the starch composition in rice grains (Mitsui et al. 2016). Furthermore, the content and fine structure of amylose and amylopectin in starch affect the physicochemical characteristics (such as viscosity) and texture properties of the rice grains (Muench et al. 1997). We have previously determined amylose contents and chain-length distributions of starches prepared from the translucent and opaque parts of perfect and chalky grains of Koshihikari harvested in 2009 and 2010 using a fluorescence labeling followed by HPLC size-exclusion chromatography. These data have suggested that the contents of amylose chain-length distributions between the translucent and opaque parts of rice grains in both seasons were similar. Although, slight differences in the contents of amylose and long B chains of amylopectin were observed in 2009 and 2010 (Tsutsui et al. 2013). To confirm the results, we conducted a further analysis of chain-length distribution of amylopectin of PG 2009, tCG 2009, oCG 2009, PG 2010, tCG 2010, and oCG 2010 using the fluorescence capillary electrophoresis (FCEP) method. As shown in Fig. 5, these analyzes revealed clearly that the differences in the chain-length distribution of starch among the perfect grains and the translucent and opaque parts of chalky grains were below the detection limit. In oCG and tCG 2010, the relative amounts of amylopectin middle-size chains (DP20-30) slightly increased while short-chain DP5-15 decreased, although no differences between two parts in the same chalky grains were observed (Fig. 5). Therefore, the reason for the appearance of loosely-packed round-shaped starch granules of the chalky grains must be attributed to factors other than chain-length distribution of amylopectin.

High-temperature stress decreased the amylose contents and the weight ratio of A+ short B chains to long B chains of amylopectin (Asaoka et al. 1984; Inouchi et al. 2000). In line with these conclusions, previous results of our group obtained by analyzing the starch composition of Koshihikari grains harvested in moderate and high-temperature conditions were not contradictory to the conclusions mentioned above (Tsutsui et al. 2013). Furthermore, transcriptomic analysis of developing rice seeds has demonstrated that the expression of several starch synthesis-related genes including *GBSSI, BEIIb*, ADP-glucose pyrophosphorylase (*AGPS2b, AGPS1 and AGPL2*) and ADP-glucose translocator (*BT1-2*) were decreased under high-temperature condition (Yamakawa et al. 2007). It is well known that the *amylose-extender* (*ae*) mutant of rice, that is deficient in *BEIIb* gene, exhibited a severe chalky phenotype of grain. The *ae* mutant revealed that the mutation in the gene for BEIIb specifically altered the structure of amylopectin in the endosperm by reducing short chains with degree of polymerization of 17 or less, with the greatest decrease in chains with degree of polymerization of 8 to 12 and enriched in long chains with DP more than 19 (Nishi et al. 2001). The chalkiness of *ae* mutant was alleviated by manipulation of BEIIb activity (Tanaka et al. 2004; Abe et al. 2014), suggesting that the unusual expression of *BEIIb* could be one of the key factors causing the grain chalkiness. However, Yamakawa et al. (2007) showed that the reduction of amylase content and amylopectin side chain by high temperature was not correlated to the grain chalkiness in rice cultivars ranging from high-temperature tolerance to high temperature sensitive. Thus, the relevance of the starch fine structure to the chalkiness of grain under high-temperature stress is obscure.

The function of starch synthesis-degradation in stem seems to serve as an important regulatory role of rice grain filling. The starch synthesis in grains starts from sucrose translocated from leaf cells. The contents of soluble starch in the opaque parts (oCG 2009 and oCG 2010) were remarkably high in comparison with the corresponding translucent parts of perfect grains (PG 2009 and PG 2010) (Fig. 6a), indicating that amylolytic enzyme exists and works in the opaque parts of chalky grains. Soluble sugar contents including sucrose were significantly increased in oCG 2010 (Fig. 6b). The sucrose contents of oCG under high temperature were higher than PG and those in 2009. It is thus likely that the downregulation of SuSy proteins under heat stress is one reason for the accumulation of sucrose in chalky grains. Another striking alteration involving downregulation of the sucrose transporter (*OsSUT1*) under heat stress favored the accumulation of sucrose in these organs (Phan et al. 2013). Another possible reason explaining the accumulation of sucrose is to alleviate water stress as a consequence of osmotic adjustment. Recent findings during rice grain filling, under simultaneous occurrence of high temperature and water deficit, showed that sugars to be synthesized to starch kept accumulating in vacuoles and cytosol in the cells (Wada et al. 2014), so that the cells could be expanded by maintaining water volume in size and cell Ψp (Morgan 1977; Meyer and Boyer 1981), followed by starch accumulation.

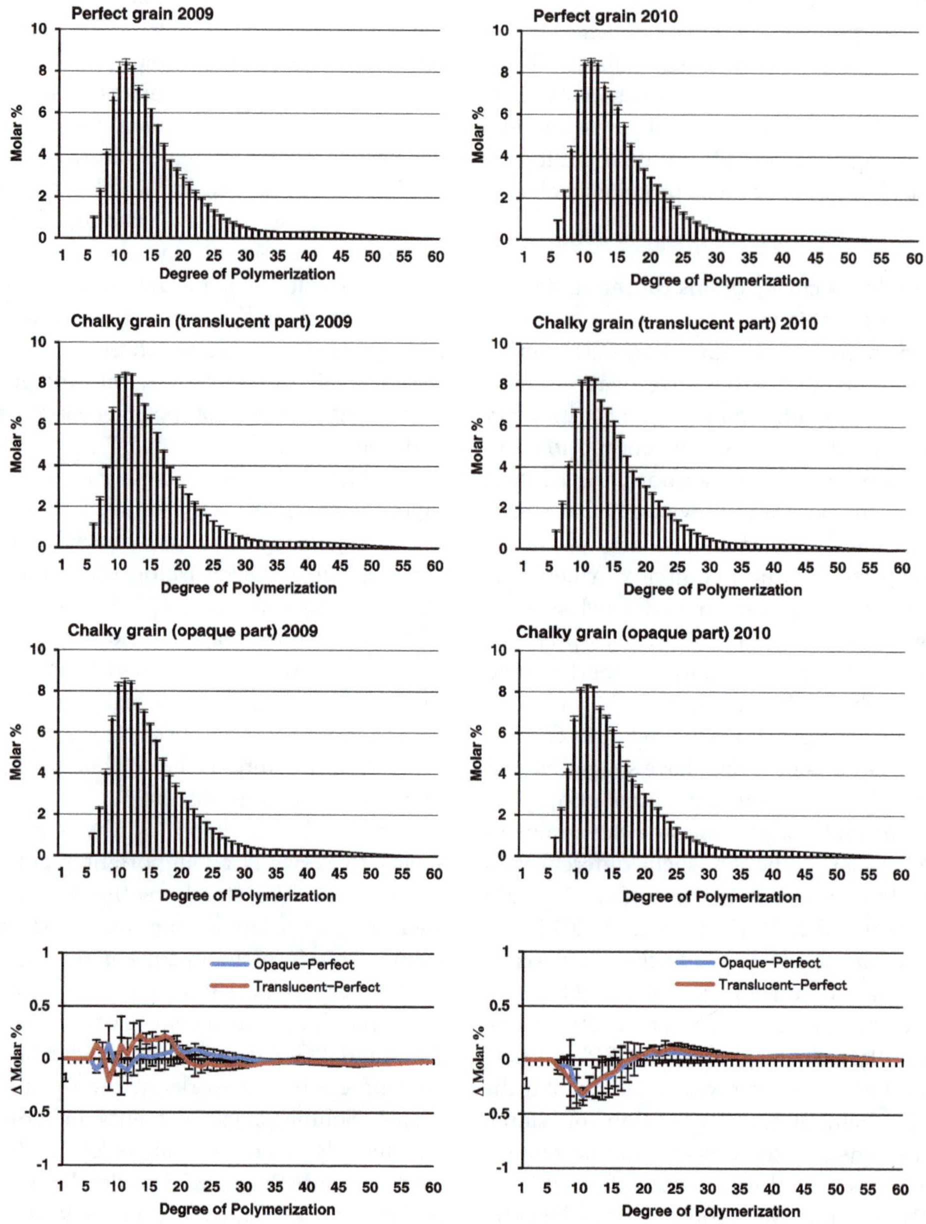

Fig. 5 Chain-length distributions of perfect and chalky grain amylopectins of rice. Rice grains were harvested in 2009 and 2010 that the average temperatures of grain filling periods were 24.4 and 28.0 °C, respectively. Center part of perfect grains and translucent and opaque parts of chalky grains were subjected to starch extraction, followed by APTS labeling and capillary electrophoretic analysis. Bottom panels show differences in the chain distributions of amylopectins in perfect and chalky grains. Blue lines, differences between the opaque part of chalky grain and perfect grain; red lines, differences between the translucent part of chalky grain and perfect grain. Values are represented as mean ± s.d. ($n = 3$)

We further determined the amounts of α-amylase isoforms in PG 2009, oCG 2009, PG 2010, and oCG 2010 by Western blotting (Fig. 7a, b). The results showed that AmyII-3 (Amy3E) and AmyII-4 (Amy3D) were highly expressed in oCG 2010 compared with PG 2010, and AmyI-1 (Amy1A) and AmyII-4 were significantly expressed in oCG 2009 compared with PG 2009 (Fig. 7b). It is noteworthy that in developing seeds of rice, the expression of α-amylase genes *Amy1A*, *Amy1C*, *Amy3D* and *Amy3E* were induced under high-temperature stress and its suppression, through RNA interference (RNAi) strategy, ameliorated rice grain damage such as chalkiness (Hakata et al. 2012). Asatsuma et al. (2005) reported that ectopic overexpression of α-amylases such as Amy1A and Amy3D produced chalky grains even under ambient temperature. As the extent of the decrease in chalky grains was highly correlated to decreases in the expression of Amy1A, Amy1C, Amy3A and Amy3B. Furthermore, studies have

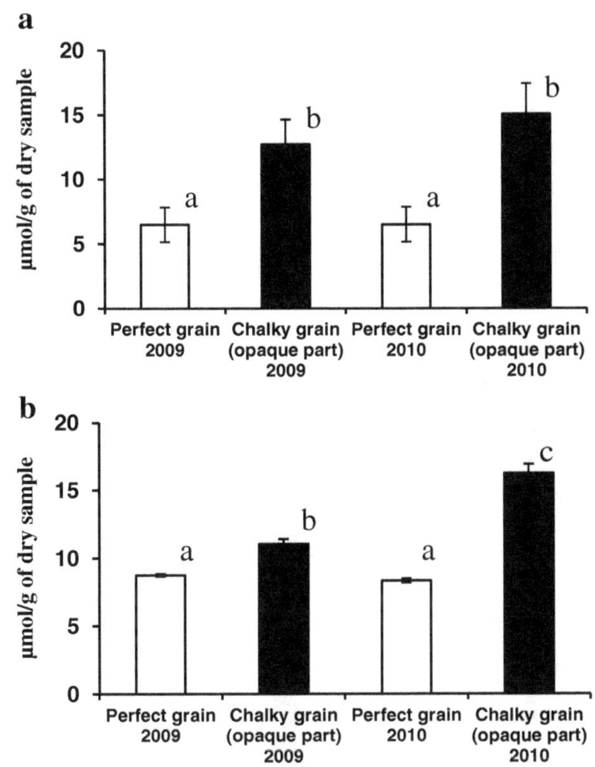

Fig. 6 Soluble starch with glucose (**a**) and sucrose (**b**) contents in the perfect and chalky (opaque part) grains harvested in 2009 and 2010. Values are represented as mean ± s.d. (n = 6–12). Histograms with the same letter do not differ significantly (p ≤0.05)

demonstrated that AmyI-1 and AmyII-4 proteins existed in the outer layers (100 to 80 % fractions) of rice grain (cv. Koshihikari), while α-glucosidase and AmyII-3 were mainly detected in the inner layers (80 to 0 % fractions) by immunoblotting with the specific antibodies. Likewise, starch-component profiles are impacted by changes in air temperature. The overall experimental results revealed that the degradation of starch accumulating in the developing grains by amylase under high temperature is an another layer of regulation causing the chalkiness.

Conclusions

In this study, the higher coverage of rice grains proteome by the iTRAQ/Shotgun strategy offered a good opportunity to discover more stress-related proteins involved in chalkiness tissue. Under field conditions, the differentially expressed proteins in opaque chalky grains under moderate (2009) and heat stress (2010) suggest three highly enriched functional terms, i.e. LEA proteins, HSPs, and glutathione peroxidase and S-transferase. These analyzes give us a better understanding about the chalky dynamics with global-mean warming of roughly 4 °C in rice grains. Furthermore, the formation of chalky grain under high-temperature season is also triggered by alterations in proteins associated with carbohydrate metabolism and starch structure. Thus, down-regulation of genes involved in starch biosynthesis and the involvement of α-amylase isoforms in the central opaque area of chalky grains were elucidated under elevated temperature. Our results provide new insights into proteome and glycome characterization in perfect and chalky grain in high-temperature scenarios at the field. The proteins identified here provide a basis to elucidate further the molecular mechanisms underlying the chalkiness under elevated temperature and may reveal useful targets in climate change scenarios studies and strategies.

Methods

Plant materials

Rice grains (*Oryza sativa* L. cv. Koshihikari) grown at Sanjo city (Niigata, Japan) were harvested in two consecutive years, 2009 and 2010. The average temperatures during the heading and ripening period of Koshihikari in the area of 2009 and 2010 were 24.4 and 28.0 °C, respectively. Grain quality (chalky or perfect) was determined with a rice grain grader (RGQI20A, Satake, Hiroshima, Japan). "Perfect" grains (PG) that exhibited a transparency and "chalky" grains (tCG) that contain opaque part(s) (oCG) within the endosperm were selected from these harvested grains on a viewer (Fujicolor light box New-5,000 Inverter, Fuji film Co., Tokyo, Japan). All the samples were stored in a temperature range of 4 to 10 °C until the experiment was carried out.

Scanning electron microscopy (SEM)

Brown rice grain was cracked with a razor blade, and the cracked surface was coated with gold for 90 s using a vaporizer (IB-3, EIKO, Tokyo, Japan) and subjected to a Scanning Electron Microscope (JSM6510LA, JEOL, Tokyo, Japan). Observation conditions were as follows: acceleration voltage, 10 kV; magnification, 1,000-10,000. SEM analysis was run using three biological replicates with at least four technical repetitions of developing grains per replication of the mounted specimens.

Proteomic analysis

Two hundred mg of starchy endosperm prepared from the opaque part of chalky grains or the corresponding part of perfect grains were ground in liquid nitrogen to fine powder and then suspended in 7 M urea, 2 M thiourea, 3 % (w/v) CHAPS, 1 % (v/v) Triton X-100, and 10 mM DTT. The suspensions were centrifuged at 10,000× g at 4 °C for 5 min. The supernatants were mixed with 1/10 volume of 100 % (w/v) trichloroacetic acid, incubated on ice for 15 min, and centrifuged at 10,000× g at 4 °C for 15 min. The resulting protein precipitates were washed 3 times in ice-cold acetone and resuspended in 8 M urea. Protein concentration was determined by the Pierce 660 nm Protein

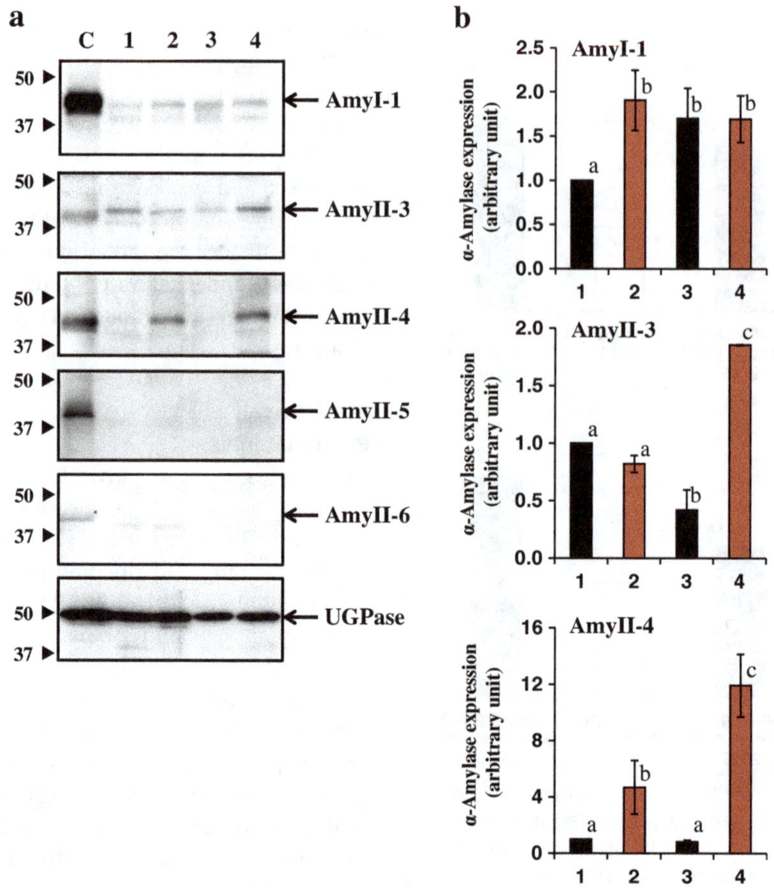

Fig. 7 Changes in the expression of α-amylase isoforms in central opaque area of chalky grains harvested in 2009 and 2010. **a** Immunoblotting images. Center part of perfect grain and opaque part of chalky grain were subjected to protein, followed by SDS-PAGE and immunoblotting with specific antibodies. C, germinating seeds after 5 days of imbibition; lane 1, perfect grain in 2009; lane 2, chalky grain in 2009; lane 3, perfect grain in 2010; lane 4, chalky grain in 2010. **b** Quantitation of α-amylase isoforms expression. Amount of each α-amylase isoform in perfect grains was normalized to 1 unit. The perfect proteins of AmyII-5 and AmyII-6 were not detected. Values are represented as mean ± s.d. ($n = 3$). Histograms with the same letter do not differ significantly ($p \leq 0.05$)

Assay Kit (Thermo Fisher Scientific) using bovine serum albumin (BSA) as a standard.

The procedure of quantitative shotgun proteomic analysis was essentially identical as described previously (Shiraya et al. 2015). Each protein preparation (50 μg) was digested in 20 μl of endoproteinase Lys-C (1 μg μl^{-1}) at 37 °C for 3 h, then diluted to 10 times volume by ultrapure water (18.2 MΩ cm). The diluted samples were further digested in 200 μl of trypsin (1 μg μl^{-1}) at 37 °C for 12 h. iTRAQ labeling of peptides were carried out with 4-plex iTRAQ tags the manufacturer's protocol (Sciex), and the resultant 4 iTRAQ-labeled peptide samples were mixed. iTRAQ analysis was performed by using a DiNa-A-LTQ-Orbitrap-XL system. The iTRAQ labeled peptides were loaded on a trap column (HiQ sil C-18 W-3; 0.5 mm i.d. × 1 mm, 3 μm particle size) with buffer A consisting of 0.1 % (v/v) formic acid and 2 % (v/v) acetonitrile in water using a DiNa-A system (KYA Tech., Tokyo, Japan). A linear gradient from 0 to 33 % buffer B (0.1 % formic acid and 80 % acetonitrile in water) for 600 min, 33 to 100 % B for 10 min and back to 0 % B in 15 min was applied, and peptides eluted from the HiQ sil C-18 W-3 column were directly loaded on a separation column (MonoCap C18 High Resolution 2000; 0.1 mm i.d. x 2000 mm, 2 μm pore size). The separated peptides were introduced into an LTQ-Orbitrap XL mass spectrometer (Thermo Fisher Scientific) with a flow rate of 300 nL min^{-1} and an ionization voltage 1.7–2.5 kV.

Liquid chromatography-MS/MS (LC-MS/MS) spectrometer was operated using Xcalibur 2.0 software (Thermo Fisher Scientific). The mass range selected for MS scan was set to 350–1,600 m/z and the top three peaks were subjected to MS/MS analysis. Full MS scan was detected in the Orbitrap, and the MS/MS scans were detected in the linear ion trap and Orbitrap. The normalized collision energy for MS/MS was set to 35 eV for collision-induced dissociation (CID) and 45 eV for higher-energy C-trap dissociation (HCD). High resolution of Fourier transform mass spectrometer (FTMS) was maintained at 60,000 resolution. Divalent or trivalent ions were subjected to MS/MS analysis in dynamic exclusion mode, and proteins were identified with Proteome Discoverer v. 1.4 software and the SEQUEST HT (Thermo Fisher Scientific) and MsAmanda (Dorfer et al. 2014) search tool using the UniProt (http://www.uniprot.org/) *O. sativa* subsp. *japonica* database (63,535 proteins) with the following parameters: enzyme, trypsin; maximum missed cleavages site, 2; peptide charge, 2+ or 3+; MS tolerance, 5 ppm; MS/MS tolerance, ±0.5 Da; dynamic modification, carboxymethylation (C), oxidation (H, M, W), iTRAQ 4-plex (K, Y, N-terminus). False discovery rates were <1 %.

Western blotting

The starchy endosperm samples (500 mg) were powdered in liquid nitrogen and suspended in 1 mL of 10 mM Tris–HCl (pH 7.5), 1.0 mM $CaCl_2$ and 0.1 % (w/v) Triton X-100 for 15 h at 4 °C. The suspension was centrifuged at 15 000× *g* for 10 min at 4 °C. An aliquot of the supernatant was subjected to sodium dodecyl sulfate - polyacrylamide gel electrophoresis (SDS-PAGE) in 12 % separation gels. After electrophoresis, the separated proteins were transferred to polyvinylidene fluoride membranes (Hybond-P; GE Healthcare, USA). The membranes were incubated in 15 mM phosphate-buffered saline (pH 6.8) containing 0.1 % Tween-20 and 5 % skim milk for blocking, and then reacted with specific antibodies: anti-rice α-amylase isoform AmyI-1 (rabbit serum; 1:10,000), AmyII-3 (mouse serum; 1:500), AmyII-4 (rabbit serum; 1:5,000) (Itayagoshi et al. 2015), AmyII-5, AmyII-6 (rabbit serum; 1:5000) (Nanjo et al. 2004) and UDPglucose pyrophosphorylase (Nanjo et al. 2004). Horseradish peroxidase-conjugated anti-rabbit IgG (Nacalai Tesque, Japan) and anti-mouse IgG (MP Biomedicals, USA) were used as secondary antibodies. The α-amylase immunobloting assay is isoforms specific, as there is no cross-reactivity of the test among isoforms AmyI-1, Amy II-4 (Mitsui et al. 1996) and Amy II-3 (Tsuyukubo et al. 2012), due to antibody specificity used in the test. The immunoreactive bands were visualized using chemiluminescence reagent (Amersham, UK), and quantified by LAS-3000 molecular imager (Fujifilm, Japan) (Asatsuma et al. 2005).

Measurement of chain-length distribution of starch

The method determining chain-length distribution of starch was essentially identical to the procedure described by Fujita et al. (2012). Starchy endosperm (5 mg) was ground with a mortar and pestle and the powder was boiled in 1.5 ml of methanol for 10 min. The suspension was centrifuged at 3,000 x *g* and room temperature for 5 min. The pellet was re-suspended in 1.5 ml of 90 % (v/v) methanol, and centrifuged at 3,000 x *g* for 5 min at room temperature. After removing methanol, the pellet was suspended with 143 μl of ultrapure water, and mixed with 7.5 μl of 5 N NaOH and boiled for 5 min. The α-glucan sample was desalted and subjected to hydrolysis with isoamylase (0.03 U/mg of amylopectin in 40 mM acetate buffer, pH 4.4) at 37 °C for 12 h. The chain length distributions of α-glucans from endosperm were analyzed using the fluorescence capillary electrophoresis (FCEP) method of O´Shea and Morell (1996) in a P/ACE MDQ Carbohydrate System (Beckman Coulters, CA, USA).

Soluble starch and sucrose contents

Briefly, 50 mg of starchy endosperm was treated with 2 ml of 80 % ethanol in boiling water bath for 1 min, and the mixture was centrifuged at 10,000x *g* for 10 min. The ethanol extraction was carried out three times. The ethanol fraction containing soluble glycan and free sugars was dried-up and then subjected to sugar assays. The soluble glycan was hydrolyzed by 5 unit of amyloglucosidase and 1 unit of α-amylase, and the released glucose from soluble glycan was measured by a coupled enzyme reaction using hexokinase (HK) and Glc-6-P dehydrogenase (G6PDH) (Guglielminetti et al. 1995). The assay mixture, composed of 100 mM Tris–HCl (pH 7.6), 3 mM $MgCl_2$, 2 mM ATP, 0.6 mM NAD^+, 1 unit of HK and 1 unit of G6PDH, was incubated at 37 °C for 30 min. After cooling, absorbance was measured at 340 nm using a spectrophotometer (Hitachi U-2900). Sucrose was first broken down using 85 units invertase (in 15 mM sodium acetate, pH 4.6) and the resulting glucose and fructose were assayed as described above.

Abbreviation

ADPG, adenosine diphosphate D-glucose; AGP, ADPG pyrophosphorylase; AGPP, ADPG pyrophosphatase; Amy, amylase; BE, branching enzyme; DBE, debranching enzyme; G6PDH, Glc-6-P dehydrogenase; GBSS, granule bound starch synthase; GPX, glutathione peroxidase; GST, glutathione S-transferases; GT, glutathione transferase; HK, hexokinase; HSPs, heat shock proteins; LEA, late embryogenesis abundant; oCG, opaque chalky grain; PG, perfect grain; PPase, pyrophosphorylase; ROS, reactive oxygen species; SDS, sodium dodecyl sulfate; sHSPs, small HSPs; SPS, sucrose phosphate synthase; SS, starch synthase; SuSy, sucrose synthase; tCG, transparent chalky grain; UDPG, UDPglucose; UGPase, UDP-glucose pyrophosphorylase.

Competing interests

The authors declare that they have no competing interests. The authors declare that the research was conducted in the absence of any commercial or financial relationships that could be construed as a potential conflict of interest.

Authors' contributions

TM planned and supervised the project. KK and MS designed the experimental plot. KK, MS, NK, HS, YS, TI, and TS conducted the experiments.

KK, MS, KI, and MB analyzed the data and interpreted the results. TM, MB, and KK wrote the paper. All authors read and approved the final manuscript.

Acknowledgements

This research was supported by KAKENHI Grants-in-Aid for Scientific Research (A) (15H02486) from Japan Society for the Promotion of Sciences and Grant for Promotion of KAAB Projects (Niigata University) by the Ministry of Education,

Culture, Sports, Science, and Technology; Japan. We wish to thank Dr. T. Sawada (Akita Prefectural University) for analyzing chain-length distribution of starch, and Dr. K. Tsutsui (Aichi University of Education) for helpful discussion.

Author details

[1]Graduate School of Science and Technology, Niigata University, Niigata 950-2181, Japan. [2]Department of Applied Biological Chemistry, Niigata University, Niigata 950-218, Japan. [3]Present address: Niigata Crop Research Center, Niigata Agricultural Research Institute, Nagaoka 940-0826, Japan.

References

Abe N, Asai H, Yago H, Oitome NF, Itoh R, Crofts N, Nakamura Y, Fujita N (2014) Relationships between starch synthase I and branching enzyme isozymes determined using double mutant rice lines. BMC Plant Biol 14:80. doi:10.1186/1471-2229-14-80

Al-Whaibi MH (2011) Plant heat-shock proteins: A mini review. J King Saud Univ - Sci 23:139–150. doi:10.1016/j.jksus.2010.06.022

Asaoka M, Okuno K, Sugimoto Y, Kawakami J, Fuwa H (1984) Effect of environmental temperature during development of rice plants on some properties of endosperm starch. Starch/Stärke 36:189–193. doi:10.1002/star.19840360602

Asatsuma S, Sawada C, Itoh K, Okito M, Kitajima A, Mitsui T (2005) Involvement of α-amylase I-1 in starch degradation in rice chloroplasts. Plant Cell Physiol 46:858–869. doi:10.1093/pcp/pci091

Berger K, Yaffe M (1998) Prohibitin family members interact genetically with mitochondrial inheritance components in *Saccharomyces cerevisiae*. Mol Cell Biol 18:4043–4052. doi:10.1128/MCB.18.7.4043

Bianchi MW, Roux C, Vartanian N (2002) Drought regulation of GST8, encoding the Arabidopsis homologue of ParC/Nt107 glutathione transferase/ peroxidase. Physiol Plant 116:96–105. doi:10.1034/j.1399-3054.2002.1160112.x

Bray EA, Bailey-Serres J, Weretilnyk E (2000) Responses to abiotic stresses. In: Buchanan BB, Gruissem W, Jones RL (eds) Biochemistry and Molecular Biology of Plants. American Society of Plant Physiologists, Rockville, pp 1158–1203

Bressani R, Elias LG, Juliano BO (1971) Evaluation of the protein quality and milled rices differing in protein content. J Agric Food Chem 19:1028–1034

Chen JC, Jiang CZ, Reid MS (2005) Silencing a prohibitin alters plant development and senescence. Plant J 44:16–24. doi:10.1111/j.1365-313X.2005.02505.x

Cheng F, Zhong L, Zhao N, Liu Y, Zhang G (2005) Temperature induced changes in the starch components and biosynthetic enzymes of two rice varieties. Plant Growth Regul 46:87–95

Coates PJ, Jamieson DJ, Smart K, Prescott AR, Hall PA (1997) The prohibitin family of mitochondrial proteins regulate replicative lifespan. Curr Biol 7:607–610. doi:10.1016/S0960-9822(06)00261-2

Commuri PD, Jones RJ (1999) Ultrastructural characterization of maize (*Zea mays* L.) kernels exposed to high temperature during endosperm cell division. Plant Cell Environ 22:375–385. doi:10.1046/j.1365-3040.1999.00424.x

Cuming AC (1999) Lea proteins. In: Shewry PR, Casey R (eds). Seed Proteins. Kluwer Academic Publishers, Dordrecht, The Netherlands, pp 753–780

Das S, Krishnan P, Mishra V, Kumar R, Ramakrishnan B, Singh HK (2015) Proteomic changes in rice leaves grown under open field high temperature stress conditions. Mol Biol Rep 42:1545–1558. doi:10.1007/s11033-015-3923-5

Defeng Z, Shaokai M (1995) Rice production in China under current nd future climates. In: Malthews RB, Kropff MJ, Bachelet D, van Laar HH (eds). Modelling the impact of climate change on rice production in Asia. IRRI/CAB International, Wallingford, pp 217–235

Dorfer V, Pichler P, Stranzl T, Stadlmann J, Taus T, Winkler S, Mechtler K (2014) MS amanda, a universal identification algorithm optimised for high accuracy tandem mass spectra. J Proteome Res 13(8):3679–3684. doi:10.1021/pr500202e

Ebitani T, Yamamoto Y, Yano M, Funane M (2008) Identification of quantitative trait loci for grain appearance using chromosome segment substitution lines in rice. Breed Res 10:91–99. doi:10.1111/j.1744-7909.2009.00822.x

Echevarría-Zomeño S, Fernández-Calvino L, Castro-Sanz AB, López JA, Vázquez J, Castellano MM (2015) Dissecting the proteome dynamics of the early heat stress response leading to plant survival or death in Arabidopsis. Plant Cell Environ. doi:10.1111/pce.12664

Emes MJ, Bowsher CG, Hedley C, Burrell MM, E. Scrase-Field ESF, Tetlow IJ (2003) Starch synthesis and carbon partitioning in developing endosperm. J Exp Bot 54:569–75. doi:10.1093/jxb/erg089

Evers D, Juliano BO (1976) Varietal differences in surface ultrastructure of endosperm cells and starch granules of rice. Starch/Stärke 28:160–166

Fujita N, Hanashiro I, Suzuki S, Higuchi T, Toyosawa Y, Utsumi Y, Itoh R, Aihara S, Nakamura Y (2012) Elongated phytoglycogen chain length in transgenic rice endosperm expressing active starch synthase IIa affects the altered solubility and crystallinity of the storage α-glucan. J Exp Bot 63:5859–5872. doi:10.1093/jxb/ers235

Fuwa H, Glover DV, Sugimoto Y (1977) Scanning electron microscopic observations of degradation of starch granules in germinating kernels of several maize (*Zea mays* L.) endosperm mutants in four inbred and one hybrid background and their normal counterparts. J Jpn Soc Starch Sci 24:99–111

Guglielminetti L, Yamaguchi J, Perata P, Alpi A (1995) Amylolytic activities in cereal seeds under aerobic and anaerobic conditions. Plant Physiol 109: 1069–1076

Hakata M, Kuroda M, Miyashita T, Yamaguchi T, Kojima M, Sakakibara H, Mitsui T, Yamakawa H (2012) Suppression of α-amylase genes improves quality of rice grain ripened under high temperature. Plant Biotechnol J 10:1110–7. doi:10.1111/j.1467-7652.2012.00741.x

Hamaker BR (1994) The influence of rice protein on rice quality. In: Marshall WE, Wadsworth JI (eds) Rice science and technology. Dekker, New York, pp 177–194

He P, Li SG, Qian Q, Ma YQ, Li JZ, Wang WM, Chen Y, Zhu LH (1999) Genetic analysis of rice grain quality. Theor Appl Genet 98:502–508

Ingram J, Bartels D (1996) The molecular basis of dehydration tolerance in plants. Annu Rev Plant Physiol Plant Mol Biol 47:377–403

Inouchi N, Ando H, Asaoka M, Okuno K, Fuwa H (2000) The effect of environmental temperature on distribution of unit chains of rice amylopectin. Starch/Stärke 52:8–12. doi:10.1002/(SICI)1521-379X(200001) 52:1<8::AID-STAR8>3.0.CO;2-Q

IPCC (2013) Climate Change 2013: The Physical Science Basis. Summary for Policymakers. Contribution of Working Group I to the Fifth Assessment Report of the Intergovernmental Panel on Climate Change. In: Stocker TF, Qin D, Plattner G-K et al. (eds). Cambridge, Cambridge University Press, pp 1-27

Ishiamaru T, Hirabayashi H, Sasaki K, Ye C, Kobayashi A (2016) Breeding efforts to mitigate damage by heat stress to spikelet sterility and grain quality. Plant Prod Sci 19:12–21

Ishimaru T, Horigane AK, Ida M, Iwasawa N, San-oh YA, Nakazono M, Nishizawa NK, Masumura T, Kondo M, Yoshida M (2009) Formation of grain chalkiness and changes in water distribution in developing rice caryopses grown under hightemperature stress. J Cereal Sci 50:166–174

Itayagoshi S, Mizusawa S, Kawakami O, Shibukawa H, Takamatsu T, Sasaki M, Kaneko K, Mitsui T (2015) Suppressive effects of low seed-soaking temperatures on germination of long-term-stored rice seeds. Plant Prod Sci 18:455–463

Iwasawa N, Umemoto T, Hiratsuka M, Nitta Y, Matsuda T, Kondo M (2009) Structural characters of milky-white rice grains caused by high temperature and shading during grain-filling. Jpn J Crop Sci 78:322–323. doi:10.14829/jcsproc.227.0.330.0

Jagadish SVK, Craufurd PQ, Wheeler TR (2008) Phenotyping parents of mapping populations of rice (*Oryza sativa* L.) for heat tolerance during anthesis. Crop Sci 48:1140–1146

Kanno K, Makino A (2010) Increased grain yield and biomass allocation in rice under cool night temperature. Soil Sci Plant Nutr 56:412–417

Kim SS, Lee SE, Kim OW, Kim DC (2000) Physicochemical characteristics of chalky kernels and their effects on sensory quality of cooked rice. Cereal Chem 77:376–379

Kobayashi A, Sonoda J, Sugimoto K, Kondo M, Iwasawa N, Hayashi T, Tomita K, Yano M, Shimizu T (2013) Detection and verification of QTLs associated with heat-induced quality decline of rice (*Oryza sativa* L.) using recombinant inbred lines and near-isogenic lines. Breed Sci 63:339–346. doi:10.1270/jsbbs.63.339

Koller A, Washburn MP, Lange BM, Andon NL, Deciu C, Haynes PA, Hays L, Schieltz D, Ulaszek R, Wei J, Wolters D, Yates JR (2002) Proteomic survey of metabolic pathways in rice. Proc Natl Acad Sci U S A 99:11564–11566. doi:10.1073/pnas.172183199

Lee J, Koh HJ (2011) A label-free quantitative shotgun proteomics analysis of rice grain development. Proteome Sci 9:61. doi:10.1186/1477-5956-9-61

Li J, Xiao J, Grandillo S, Jiang L, Wan Y, Deng Q, Yuan L, McCouch SR (2004) QTL detection for rice grain quality traits using an interspecific backcross population derived from cultivated Asian (*O. sativa* L.) and African

Liao JL, Zhou HW, Zhang HY, Zhong PA, Huang YJ (2013) Comparative proteomic analysis of differentially expressed proteins in the early milky stage of rice grains during high temperature stress. J Exp Bot 65:655–671. doi:10.1093/jxb/ert435

Lin SK, Chang MC, Tsai YG, Lur HS (2005) Proteomic analysis of the expression of proteins related to rice quality during caryopsis development and the effect of high temperature on expression. Proteomics 5:2140–2156. doi:10.1002/pmic.200401105

Lin CJ, Li CY, Lin SK, Yang FH, Huang JJ, Liu YH, Lur HS (2010) Influence of high temperature during grain filling on the accumulation of storage proteins and grain quality in rice (*Oryza sativa* L.). J Agric Food Chem 58:10545–10552

Lin Z, Zhang X, Yang X, Li G, Tang S, Wang S, Ding Y, Liu Z (2014) Proteomic analysis of proteins related to rice grain chalkiness using iTRAQ and a novel comparison system based on a notched-belly mutant with white-belly. BMC Plant Biol 14:163. doi:10.1186/1471-2229-14-163

Lisle AJ, Martin M, Fitzgerald MA (2000) Chalky and translucent rice grains differ in starch composition and structure and cooking properties. Cereal Chem 77:627–632

Liu L, Hu X, Song J, Zong X, Li D (2009) Over-expression of a *Zea mays* L. protein phosphatase 2C gene (ZmPP2C) in *Arabidopsis thaliana* decreases tolerance to salt and drought. J Plant Phys 166:531–542

Liu X, Guo T, Wan X, Wang H, Zhu M, Li A, Su N, Shen Y, Mao B, Zhai H, Mao L, Wan J (2010) Transcriptome analysis of grain-filling caryopses reveals involvement of multiple regulatory pathways in chalky grain formation in rice. BMC Genomics 11:730. doi:10.1186/1471-2164-11-730

Liu X, Wan X, Ma X, Wan J (2011) Dissecting the genetic basis for the effect of rice chalkiness, amylose content, protein content, and rapid viscosity analyzer profile characteristics on the eating quality of cooked rice using the chromosome segment substitution line population across eight environments. Genome 54:64–80

Luo M, Liu J, Lee RD, Scully BT, Guo B (2010) Monitoring the expression of maize genes in developing kernels under drought stress using oligo-microarray. J Integr Plant Biol 52:1059–1074. doi:10.1111/j.1744-7909.2010.01000.x

McClung JK, King RL, Walker LS, Danner DB, Nuell MJ, Stewart CA, Dell'Orco RT (1992) Expression of prohibitin, an antiproliferative protein. Exp Gerontol 27:413–417

McClung JK, Jupe ER, Liu XT, Dell'Orco RT (1995) Prohibitin: potential role in senescence, development, and tumor suppression. Exp Gerontol 30:99–124

Mei DY, Zhu YJ, Yu YH, Fan YY, Huang DR, Zhuang JY (2013) Quantitative trait loci for grain chalkiness and endosperm transparency detected in three recombinant inbred line populations of indica rice. J Integr Agr 12:1–11

Meyer RF, Boyer JS (1981) Osmoregulation, solute distribution and growth in soybean seedlings having low water potentials. Planta 151:482–489

Milioni D, Hatzopoulos P (1997) Genomic organization of hsp90 gene family in Arabidopsis. Plant Mol Biol 35:955–961. doi:10.1023/A:1005874521528

Mitsui T, Yamaguchi J, Akazawa T (1996) Physicochemical and serological characterization of rice α-amylase isoforms and identification of their corresponding genes. Plant Physiol 110:1395–1404

Mitsui T, Yamakawa H, Kobata T (2016) Molecular physiological aspects of chalking mechanism in rice grains under high-temperature stress. Plant Prod Sci 19:22–29

Mohammadi R, Mozaffar RM, Yousef A, Mostafa A, Amri A (2010) Relationships of phenotypic stability measures for genotypes of three cereal crops. Can J Plant Sci 90:819–830

Morgan JM (1977) Differences in osmoregulation between wheat genotypes. Nature 270:234–235

Morimoto RI (2002) Dynamic remodeling of transcription complexes by molecular chaperones. Cell 110:281–284

Morita S, Yonemaru J, Takanashi J (2005) Grain growth and endosperm cell size under high night temperatures in rice (*Oryza sativa* L.). Ann Bot 95: 695–701

Morita S, Wada H, Matsue Y (2016) Countermeasures for heat damage in rice grain quality under climate change. Plant Prod Sci 19:1–11

Muench DG, Wu Y, Zhang Y, Li X, Boston RS, Okita TW (1997) Molecular cloning, expression and subcellular localization of a BiP homolog from rice endosperm tissue. Plant Cell Physiol 38:404–412. doi:10.1093/oxfordjournals.pcp.a029183

Næsted H, Frandsen GI, Jauh GY, Hernandez-Pinzon I, Nielsen HB, Murphy D, Rogers JC, Mundy J (2000) Caleosins: Ca^{2+} binding proteins associated with lipid bodies. Plant Mol Biol 44:463–476

Nagato K, Ebata M (1965) Effects of high temperature during ripening period on the development and the quality of rice kernels. Jpn J Crop Sci 34:59–66

Nanjo Y, Asatsuma S, Itoh K, Hori H, Mitsui T (2004) Proteomic identification of α-amylase isoforms encoded by RAmy3B/3C from germinating rice seeds. Biosci Biotechnol Biochem 68:112–118

Nijtmans L, Artal-Sanz M, Grivell L, Coates P (2002) The mitochondrial PHB complex: roles in mitochondrial respiratory complex assembly, ageing and degenerative disease. Cell Mol Life Sci 59:143–155

Nishi A, Nakamura Y, Tanaka N, Satoh H (2001) Biochemical and genetic analysis of the effects of amylose-extender mutation in rice endosperm. Plant Physiol 127:459–472. doi:10.1104/pp.010127

O'Shea MG, Morell MK (1996) High resolution slab gel electrophoresis of 8-amino-1,3, 6-pyrenetrisulfonic acid (APTS) tagged oligosaccharides using a DNA sequencer. Electrophoresis 17:681–688s

Olvera-Carrillo Y, Reyes JL, Covarrubias AA (2011) Late embryogenesis abundant proteins: Versatile players in the plant adaptation to water limiting environments. Plant Signal Behav 6:586–589. doi:10.4161/psb.6.4.15042

Phan TTT, Ishibashi Y, Miyazaki M, Tran MHT, Okamura K, Tanaka S, Nakamura J, Yuasa T, Iwaya-Inoue M (2013) High temperature-induced repression of the rice sucrose transporter (*OsSUT1*) and starch synthesis-related genes in sink and source organs at milky ripening stage causes chalky grains. J Agron Crop Sci 199:178–188. doi:10.1111/jac.12006

Piper P, Jones G, Bringloe D, Harris N, MacLean M, Mollapour M (2002) The shortened replicative life span of prohibitin mutants of yeast appears to be due to defective mitochondrial segregation in old mother cells. Aging Cell 1: 149–157

Poxleitner M, Rogers SW, Samuels AL, Browse J, Rogers J (2006) A role for caleosin in degradation of oil-body storage lipid during seed germination. Plant J 47:917–933. doi:10.1111/j.1365-313X.2006.02845.x

Ray DK, Gerber JS, MacDonald GK, West PC (2015) Climate variation explains a third of global crop yield variability. Nat Commun 6:5989. doi:10.1038/ncomms6989

Scarpeci TE, Zanor MI, Valle EM (2008) Investigating the role of plant heat shock proteins during oxidative stress. Plant Signal Behav 3:856–7

Shi W, Muthurajan R, Rahman H, Selvam J, Peng S, Zou Y, Jagadish KSV (2013) Source-sink dynamics and proteomic reprogramming under elevated night temperature and their impact on rice yield and grain quality. New Phytol 197:825–837. doi:10.1111/nph.12088

Shiraya T, Mori T, Mruyama T, Sasaki M, Takamatsu T, Oikawa K, Kaneko K, Itoh K, Ichikawa H, Mitsui T (2015) Golgi/plastid-type manganese superoxide dismutase involved in heat-stress tolerance during grain filling of rice. Plant Biotechnol J 13:1251–1263. doi:10.1111/pbi.12314

Singh N, Sodhi NS, Kaur M, Saxena SK (2003) Physicochemical, morphological, thermal, cooking and textural properties of chalky and translucent rice kernels. Food Chem 82:433–439

Su J-C (2000) Starch synthesis and grain filling in rice. In: Gupt AK and Kaur N (eds). Carbohydrate Reserves in Plants - Synthesis and Regulation. Amsterdam, The Netherlands, Elsevier, pp 107-124

Tabata M, Hirabayashi H, Takeuchi Y, Ando I, Iida Y, Ohsawa R (2007) Mapping of quantitative trait loci for the occurrence of white-back kernels associated with high temperatures during the ripening period of rice (*Oryza sativa* L.). Breed Sci 57:47–52

Tanaka N, Fujita N, Nishi A, Satoh H, Hosaka Y, Ugaki M, Kawasaki S, Nakamura Y (2004) The structure of starch can be manipulated by changing the expression levels of starch branching enzyme IIb in rice endosperm. Plant Biotechnol J 2:507–16. doi:10.1111/j.1467-7652.2004.00097.x

Tashiro T, Wardlaw IF (1990) The response to high temperature shock and humidity changes prior to and during the early stages of grain development in wheat. Aust J Plant Physiol 17:551–561

Tashiro T, Wardlaw IF (1991) The effect of high temperature on the accumulation of dry matter, carbon and nitrogen in the kernel of rice. Aust J Plant Physiol 18:259–265

Tatsuta T, Model K, Langer T (2005) Formation of membrane-bound ring complexes by prohibitins in mitochondria. Mol Biol Cell 16:248–259. doi:10.1091/mbc.E04-09-0807

Tsutsui K, Kaneko K, Hanashiro I, Nishinari K, Mitsui T (2013) Characteristics of opaque and translucent parts of high temperature stressed grains of rice. J Appl Glycosci. doi:10.5458/jag.jag.JAG-2012014

Tsuyukubo M, Ookura T, Tsukui S, Mitsui T, Kasai M (2012) Elution behavior analysis of starch degrading enzymes during rice cooking with specific antibodies. Food Sci Technol Res 18:659–666

Umemoto T, Terashima K (2002) Activity of granule-bound synthase is an

important determinant of amylose content in rice endosperm. Funct Plant Biol 29:1121–1124
Usui Y, Sakai H, Tokida T, Nakamura H, Nakagawa H, Hasegawa T (2014) Heat-tolerant rice cultivars retain grain appearance quality under free-air CO_2 enrichment. Rice 7:6. doi:10.1186/s12284-014-0006-5
van Aken O, Zhang BT, Carrie C, Uggalla V, Paynter E, Giraud E, Whelan J (2009) Defining the mitochondrial stress response in *Arabidopsis thaliana*. Mol Plant 2:1310–1324. doi:10.1093/mp/ssp053
Wada H, Masumoto-Kubo C, Gholipour Y, Nonami H, Tanaka F, Erra-Balsells R, Tsutsumi K, Hiraoka K, Morita S (2014) Rice chalky ring formation caused by temporal reduction in starch biosynthesis during osmotic adjustment under foehn-induced dry wind. PLoS One 9:e110374. doi:10.1371/journal.pone.0110374
Wan XY, Wan JM, Weng JF, Jiang L, Bi JC, Wang CM, Zhai HQ (2005) Stability of QTLs for rice grain dimension and endosperm chalkiness characteristics across eight environments. Theor Appl Genet 110:1334–1346
Wang W, Vinocur B, Shoseyov O, Altman A (2004) Role of plant heat-shock proteins and molecular chaperones in the abiotic stress response. Trends Plant Sci 9:244–252
Xiao B, Huang Y, Tang N, Xiong L (2007) Over-expression of a LEA gene in rice improves drought resistance under the field conditions. Theor Appl Genet 115:35–46. doi:10.1007/s00122-007-0538-9
Xu SB, Li T, Deng ZY, Chong K, Xue Y, Wang T (2008) Dynamic proteomic analysis reveals a switch between central carbon metabolism and alcoholic fermentation in rice filling grains. Plant Physiol 148:908–925
Yamakawa H, Hirose T, Kuroda M, Yamaguchi T (2007) Comprehensive expression profiling of rice grain filling-related genes under high temperature using DNA microarray. Plant Physiol 144:258–277. doi:10.1104/pp.107.098665
Yasuda H, Hirose S, Kawakatsu T, Wakasa Y, Takaiwa F (2009) Overexpression of BiP has inhibitory effects on the accumulation of seed storage proteins in endosperm cells of rice. Plant Cell Physiol 50:1532–43. doi:10.1093/pcp/pcp098
Zakaria S, Matsuda T, Tajima S, Nitta Y (2002) Effect of high temperature at ripening stage on the reserve accumulation in seed in some rice cultivars. Plant Prod Sci 5:160–168. doi:10.1626/pps.5.160
Zeng T, Jiang X, Li J, Wang D, Li G, Lu L, Wang G (2013) Comparative proteomic analysis of the hepatic response to heat stress in Muscovy and Pekin ducks: insight into thermal tolerance related to energy metabolism. PLoS One 8:e76917. doi:10.1371/journal.pone.0076917
Zheng BS, Rönnberg E, Viitanen L, Salminen TA, Lundgren K, Morita T, Edqvist J (2008) Arabidopsis sterol carrier protein-2 is required for normal development of seeds and seedlings. J Exp Bot 59:3485–3499. doi:10.1093/jxb/ern201
Zi J, Zhang J, Wang Q, Zhou B, Zhong J, Zhang C, Qiu X, Wen B, Zhang S, Fu X, Lin L, Liu S (2013) Stress responsive proteins are actively regulated during rice (*Oryza sativa*) embryogenesis as indicated by quantitative proteomics analysis. PLoS One 8:e74229. doi:10.1371/journal.pone.0074229

3

Association Mapping of Yield and Yield-related Traits Under Reproductive Stage Drought Stress in Rice (*Oryza sativa* L.)

B. P. Mallikarjuna Swamy[1], Noraziyah Abd Aziz Shamsudin[1,2], Site Noorzuraini Abd Rahman[1,3], Ramil Mauleon[1], Wickneswari Ratnam[2], Ma. Teressa Sta. Cruz[1] and Arvind Kumar[1*]

Abstract

Background: The identification and introgression of major-effect QTLs for grain yield under drought are some of the best and well-proven approaches for improving the drought tolerance of rice varieties. In the present study, we characterized Malaysian rice germplasm for yield and yield-related traits and identified significant trait marker associations by structured association mapping.

Results: The drought screening was successful in screening germplasm with a yield reduction of up to 60% and heritability for grain yield under drought was up to 78%. There was a wider phenotypic and molecular diversity within the panel, indicating the suitability of the population for quantitative trait loci (QTL) mapping. Structure analyses clearly grouped the accessions into three subgroups with admixtures. Linkage disequilibrium (LD) analysis revealed that LD decreased with an increase in distance between marker pairs and the LD decay varied from 5–20 cM. The Mixed Linear model-based structured association mapping identified 80 marker trait associations (MTA) for grain yield (GY), plant height (PH) and days to flowering (DTF). Seven MTA were identified for GY under drought stress, four of these MTA were consistently identified in at least two of the three analyses. Most of these MTA identified were on chromosomes 2, 5, 10, 11 and 12, and their phenotypic variance (PV) varied from 5% to 19%. The *in silico* analysis of drought QTL regions revealed the association of several drought-responsive genes conferring drought tolerance. The major-effect QTLs are useful in marker-assisted QTL pyramiding to improve drought tolerance.

Conclusion: The results have clearly shown that structured association mapping is one of the feasible options to identify major-effect QTLs for drought tolerance-related traits in rice.

Keywords: Drought, Genetic diversity, Association mapping, Genes

Background

Rice is the primary food source for more than half of the world's population and contributes 30–50% of the daily caloric intake (Fairhust and Dobermann, 2002). Among different rice ecosystems, rainfed upland and rainfed lowland rice occupy 30% of total rice area but contribute only 21% of total rice production. Drought is one of the most severe climate-related risks for rice production in rainfed areas of Asia and Africa (Pandey, 2007). With limited options for expanding rice area and the existing plateau in the yield potential of irrigated rice, a further increase in rice production has to come from highly vulnerable, less productive drought-prone rainfed lowland and upland rice areas (Khush, 1997). These areas received much less attention during the Green Revolution and even now most of the varieties grown in these areas are ones that were developed for high-input irrigated conditions. These varieties are highly susceptible to the various abiotic and biotic stresses prevalent in low-input rainfed environments. Thus, there is an urgent need to develop climate-smart rice varieties with multiple abiotic and biotic stress tolerance, and with improved grain

* Correspondence: a.kumar@irri.org
[1]Plant Breeding Division, International Rice Research Institute (IRRI), DAPO Box 7777 Metro Manila, Philippines
Full list of author information is available at the end of the article

quality and high yield potential that are suitable for rainfed areas (Kamoshita et al. 2008; Lafitte et al. 2004).

Rapid and precise exploitation of the abundant genetic diversity available within rice germplasm is highly critical to ensuring sustainable rice production and global food security in the ever-changing climatic conditions (McCouch et al. 2012; Voss-Fels and Snowdon 2016). The recent advances in rice biotechnological tools have been very helpful in unraveling the genetic basis of complex traits within rice germplasm to identify major genes/QTLs for use in rice breeding (Thomson et al. 2010; Swamy et al. 2014; Swamy and Kumar, 2013). For improving drought tolerance, several major-effect grain yield QTLs under drought have been identified and successfully used in marker-assisted breeding (MAB) (Swamy et al. 2011). But, most of the QTL studies were on biparental or multiparent populations, which are limited by the allelic diversity within the selected parents. In addition, population development is time-consuming and mapping resolution is low (Kumar et al. 2014; Pascual et al. 2016). Also, the drought QTLs identified from different studies and meta-analysis of *qDTYs* have clearly shown that only a few major-effect *qDTYs* have been consistently and repeatedly identified, indicating limited exploration of genetic resources to identify novel major-effect *qDTYs* (Swamy et al. 2011; Kumar et al. 2014).

Marker-assisted pyramiding of *qDTYs* into elite rice varieties has shown that they have synergistic relationships in particular combinations and their effect varies with the genetic background. Introgression of four *qDTYs* into popular mega-variety IR64 and three *qDTYs* in the background of MR219 showed that *qDTYs* have a better effect in different combinations in different genetic backgrounds (Swamy et al. 2013; Noraziyah et al. 2016a). This further emphasizes the urgent need to identify novel *qDTYs* to improve the drought tolerance of a wide range of drought-susceptible rice varieties.

The availability of genome-wide molecular markers, cheaper genotyping services and advances in statistical analysis have made it possible to explore natural populations to identify significant marker and trait associations, also popularly called genome-wide association studies (GWAS) (Korte and Farlow, 2013). Considering the huge genetic diversity available for multiple traits within rice germplasm, GWAS can be a feasible approach to simultaneously map loci for many traits and the improved mapping resolution helps in precisely identifying the genes/SNPs associated with the traits. There are several successful examples of accurate QTL/gene detection using GWAS in rice and other crop species (Huang et al. 2010; Han and Huang, 2013; Wu et al. 2015; Yang et al. 2014; Kumar et al. 2014), but there are only a few association studies for drought-related traits in rice (Vasant 2012; Courtois et al. 2013; Vannirajan et al. 2012; Muthukumar et al. 2015).

The present study was undertaken with the objectives of screening Malaysian rice germplasm for drought tolerance, determining the population structure, doing association mapping of yield and yield-related traits under drought stress and non-stress (NS) conditions, and carrying out reference genome-based analysis of *qDTY* physical regions.

Results

Drought Screening

Drought was imposed in the dry season (DS) by draining out all the water one month after transplanting. Data on rainfall and water table depth were recorded in both trials. The rainfall was less than 86.2 mm during the 2011DS and was 264.1 mm during the 2012DS (Additional file 1: Figure S1 and Additional file 2: Figure S2). The water table depth reached less than 100 cm before flowering during both seasons. The results showed that there was moderate to severe drought stress during the reproductive stage of the crop in both seasons. The NS counterpart was completely flooded from transplanting until the crop reached physiological maturity.

Phenotypic Analysis

Phenotypic traits measured under both NS and drought stress conditions and in both years showed wide variation (Table 1). The overall mean reduction in GY under drought stress compared with NS was 60% in the 2011DS, 51% during the 2012DS and 53% in combined analysis. In general, there was a reduction in PH and delay in flowering under drought stress compared with NS, indicating the effect of drought stress on plant growth. The heritability (H^2) for DTF and PH was very high: it varied from 84.0% to 86.0% for DTF and from 86.0% to 87.0% for PH under drought stress in different seasons; whereas, under non-stress, it varied from 83.6% to 87.0% for DTF and from 81.7% to 89.0% for PH. For GY, H^2 varied from 72.0% to 78.0% under drought stress and from 63.0% to 69.0% under non-stress. Table 2 presents the correlation among the traits under drought stress and NS. Nine out of 18 correlations for three traits under both drought stress and NS were found significant. There was a significant negative correlation between DTF and GY under drought stress in both seasons and also in the combined analysis. Also, PH and GY were negatively correlated under drought stress; whereas, under non-stress, the correlation between PH and GY was significant only in 2012. The coefficient of variation (CV) was moderate to high for all three traits under both drought stress and NS, indicating wide variation.

Table 1 Various statistical parameters for yield and yield-related traits under non-stress and drought stress

		Mean		Range		SD		CV (%)		H^2 (%)	
Trait	Year/season	NS	S	NS	S	NS	S	NS	S	NS	S
DTF (days)	2011DS	93.0	95.0	55.0–124.0	60.0–127.0	12.8	12.1	6.0	6.5	87.0	86.0
	2012DS	88.9	96.1	77.0–120.0	65.0–123.0	9.4	10.6	8.4	6.5	83.6	84.0
	Combined	90.9	95.5	60.0–124.0	62.0–124.0	10.9	11.3	7.3	6.5	85.3	83.5
PH (cm)	2011DS	148.0	94.0	68.7–148.7	55.3–121.3	19.5	12.1	7.1	9.5	89.0	87.0
	2012DS	106.8	87.1	64.3–172.0	64.7–129.7	19.9	13.9	15.2	7.8	81.7	86.0
	Combined	101.0	80.9	68.7–154.3	58.8–119.8	19.3	13.3	11.9	8.5	85.3	86.3
GY (Kg/ha)	2011DS	1736.0	687.7	633.8–3689.0	57.0–1577.0	684.6	327.1	34.2	25.6	63.0	72.0
	2012DS	5766.1	2831.5	930.6–10850	100.0–5615.0	2086.5	1161.3	26.4	24.3	69.0	78.0
	Combined	3832.9	1794.9	672.4–10737	98.0–5295.3	2478.2	1339.9	23.5	25.4	65.4	75.4

Genotypic Analysis

Out of the 125 SSR markers genotyped in the 75 Malaysian genotypes, 119 (95.2%) were found to be polymorphic. The number of SSR markers varied on different chromosomes: with highest number of 21 markers on chromosome 2, followed by 17 markers on chromosome 1, 11 markers each on chromosomes 9 and 12, on other chromosomes number of markers varied from 4 to 9 and lowest of four markers on chromosome 8, with an average density of one marker for every 3 MB. In all, 119 SSR markers amplified 1524 alleles, out of which 482 (31.6%) were in POP1, 483 (31.7%) were in POP2 and 559 (36.7%) were in POP3. Among the different SSR markers, seven were bi-allelic, 17 were tri-allelic and 12 were tetra-allelic, while 77 SSR markers amplified 5–10 alleles, three markers (RM13, RM6374 and RM180) amplified 11 alleles, two markers (RM335 and RM491) amplified 12 alleles and one marker (RM440) amplified a maximum of 13 alleles in the 75 rice genotypes.

Table 2 Correlation among yield and yield-related traits

Trait/Season	Environment	PH	GY
DTF			
2011DS	S	0.256**	−0.025
2011DS	NS	0.051	−0.174
2012DS	S	−0.246**	−0.494**
2012DS	NS	0.449**	0.063
Combined	S	−0.175	−0.456**
Combined	NS	0.404**	0.023
PH		–	
2011DS	S	–	−0.152
2011DS	NS	–	−0.009
2012DS	S	–	−0.144
2012DS	NS	–	0.195*
Combined	S	–	0.327**
Combined	NS	–	0.310**

** significant at 1%, * significant at 5%

Additional file 3: Table S1 presents the overall and chromosome-wise allelic frequency, number of alleles, allelic diversity, PIC and number of subpopulation-specific alleles. Allele frequencies, genetic diversity and PIC were highest on chromosome 5 and lowest on chromosome 7. The PIC values varied from 0.026 to 0.840; seven markers (RM21, RM12460, RM12182, RM17524, RM108, RM553 and RM219) had the highest PIC of more than 0.8016. Gene diversity varied from 0.263 to 0.856 and the major allelic frequency varied from 0.227 to 0.987.

Population Structure and Kinship

We performed structural analysis using 119 marker genotypic data from 75 genotypes and the analysis was carried out in ten replications with 500,000 burns. The number of subpopulations (K) was determined based on the posterior probability values (LnP(D) and ΔK). For K = 2 to 10, the values of LnP(D) and ΔK continuously increased in all ten replicates and the highest ΔK of 24.2 was reached when K = 3 with an LnP(D) of−19790.04. Thus, the 75 genotypes were grouped into three subpopulations and the results of the three subpopulations were used for further analysis. The molecular diversity analysis also grouped the genotypes into three clusters, thus supporting the results of the structure analysis. Among the three subpopulations, POP1 consisted of 23 genotypes, POP2 consisted of 18 genotypes and POP3 consisted of 34 genotypes. The fixation index (F_{st}) was 0.277, 0.270 and 0.194 in POP1, POP2 and POP3, respectively, and the expected heterozygosity was 0.544, 0.572 and 0.559 in POP1, POP2 and POP3, respectively (Additional file 3: Table S2). Twenty-nine alleles differentiated POP1 and most of them were on chromosomes 1, 2, 4, 5 and 9; in POP2, there were 57 population-specific alleles, with most of them on chromosomes 1, 2, 5, 6, 7, 8 and 11; whereas 61 alleles differentiated POP3, with most of them on chromosomes 1, 2, 7, 9, 11 and 12 (Additional file 3: Table S2). The bar diagram shows the distribution of genotypes within and between the populations (Fig. 1). The allelic frequency divergence between POP1 and POP2 was 0.194, between POP1 and POP3 it was 0.120 and between POP2 and POP3 it was 0.197.

Linkage Disequilibrium (LD)

LD analysis in the whole population showed 7072 LD pairs, out of which 2264 (32%) pairs were significant ($P > 0.05$) (Additional file 3: Table S3). There were overall 2071 inter-chromosomal LD pairs and 193 significant intra-chromosomal LD pairs (Additional file 3: Table S3). Chromosomes 1, 2, 3, 6, 9 and 10 had more than 10 intra-chromosomal LD pairs, whereas chromosomes 2, 6, 9, 10 and 12 had more than 200 significant inter-chromosomal LD pairs. Chromosome 2 had the highest number of significant intra- and inter-chromosomal LD pairs, while chromosome 7 had the lowest number. The LD scatter plot showed a reduction in the number of significant LD pairs as the interval distances between marker pairs increased. The LD (R^2) plotted against cM is presented in Fig. 2. There was a sharp decline in LD decay for the linked markers at 5–20 cM. Overall, the r^2 varied from 0.005 to 0.31, with an average r^2 of the locus pairs of 0.03. Significant pairs of linked loci ($r^2 > 0.2$) showed an average distance of 5–20 cM, indicating the presence of a higher number of significant LD blocks in this set of germplasm accessions.

Association Analysis

The results of association analysis for GY under drought stress and non-stress are presented in Tables 3 and 4, and for DTF and PH in Additional file 3: Tables S4 and S5. In all, there were 198 and 80 significant marker trait associations (MTA) for all three traits identified by GLM and MLM analyses. MTA identified by MLM are further discussed. Among the QTLs indentified by MLM, 7 and 13 of them were for GY under drought stress and NS, respectively. There were five MTA for GY in the 2012DS, four in combined analysis, while there were only three QTLs in the 2011DS. Four MTA for GY under drought stress were consistently identified in at least two of the three analyses. These MTA identified were on chromosomes 2, 5, 6, 10, 11 and 12. The PV of the QTLs varied from 5 to 19; five MTA had more than 10% PV. It is interesting to note that the novel QTL on chromosome 5 was consistently identified in all the three analyses and had highest PV of 19%. Under NS, seven MTA were identified in the 2011DS, five in combined analysis, while only two MTA were significant in the 2012DS. Almost all these QTLs are highly season-specific; only one locus (RM582 on chromosome 1) was consistently identified in both the 2011DS and combined analysis. These yield QTLs were on chromosomes 1, 2, 3, 5, 6, 7, 8, 10 and 12. The PV of the yield QTLs varied from 12% to 27%. Five QTLs had more than 20% PV. For DTF, 16 and 14 QTLs were identified under drought stress and NS, respectively; whereas, for PH, 13 and 17 QTLs were identified under drought stress and NS, respectively, and the results are provided in Additional file 3: Tables S4 and S5.

Genes Within *qDTYs* Regions

All the *qDTYs* identified by both GLM and MLM during both the 2011DS and 2012DS were mapped into the physical genome and binned as meta-QTLs (described in the data of the methods section), resulting in 20 QTL bins or meta-QTLs representing 18.1 MB of the reference genome (Additional file 4: Table S6). The physical reference genome of variety Nipponbare was used to survey the genes in the meta-QTLs (IRGSP 1.0; Kawahara et al. 2013). We restricted the gene survey to a 1 MB region, covering 500 Kb regions toward the left and right side of the peak SSR marker to survey the genes more precisely. Using the Rice Genome Annotation Project (RGAP release 7, http://rice.plantbiology.msu.edu/), information on genes within the meta-QTL regions was gathered. In all, 3222 genes were found within the entire 20 meta-QTL regions, out of which 22% were transposable elements, 27% were uncharacterized expressed proteins and hypothetical proteins, and the remaining 53% were unique genes/gene families with assigned definitive functions (Additional file 5: Table S7). In all, there were

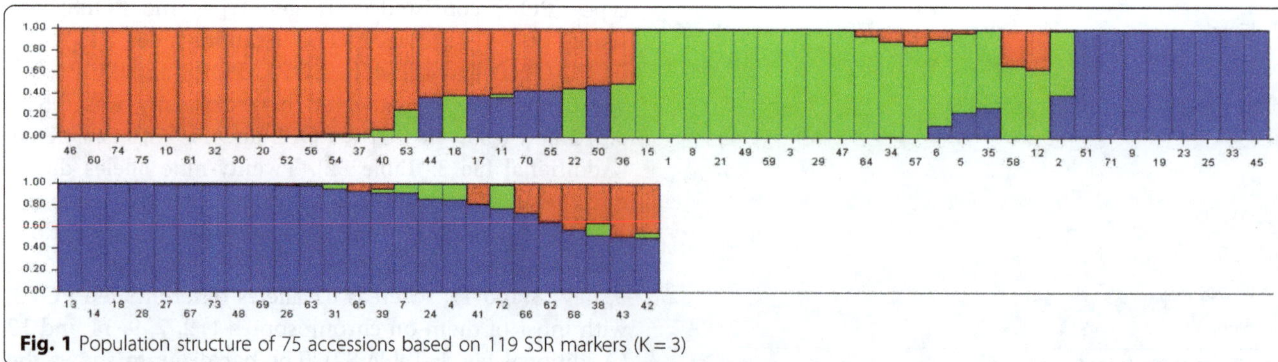

Fig. 1 Population structure of 75 accessions based on 119 SSR markers (K = 3)

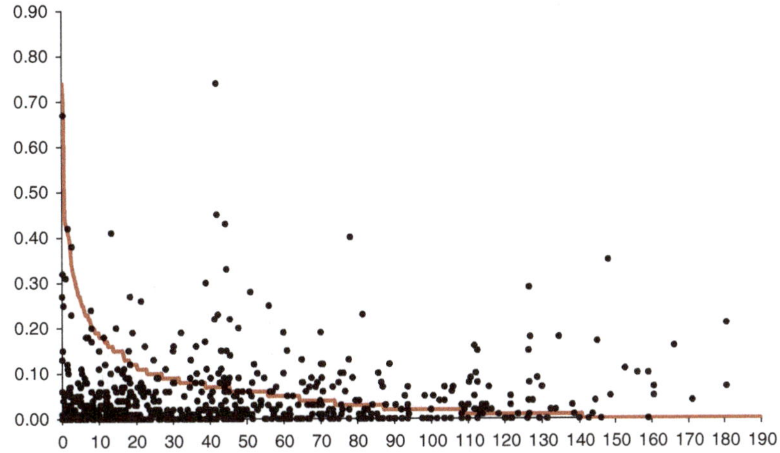

Fig. 2 Distribution LD (R^2) values observed between linked SSR marker pairs as a genetic distance in centiMorgans (cM)

541 unique genes/gene families within the meta-QTLs and predominantly their putative functions were related to stress tolerance, stress signaling, hormonal regulation, growth and development, nutrient uptake, metal carriers and transcription factors. Some of the important genes/gene families with a high frequency of presence in the meta-QTLs were F-box domain proteins, MADS-box proteins, tetra- and pentatricopeptide proteins, no apical meristem proteins (NAM), heat shock proteins (HSPs), cytochrome P450 family, ethylene response factors (AP2/ERFs), auxin response factors (ARFs), brassinosteroids, CaMK, GDSL-like lipases, membrane proteins, hydrolases, kinases, peroxidases, peptidase and dismutase family proteins, zinc finger proteins (ZFPs), leucine zipper motifs (ZIP) and myeloblastosis (MYB) transcription factors. The role of these genes/gene families in abiotic stress tolerance is well known. Enrichment analysis of the list of genes within meta-QTLs against annotation of the RGAP genes for gene ontology (GO) identified a wide range of complex processes, functions and compartments/major sites of biochemical functions involved in abiotic stress tolerance. In all, 63 GO processes, 70 GO functions and 23 compartments related to stress tolerance, stress signaling, growth and development, and key metabolic processes were identified. MapMan pathway analysis for the genes located within the meta-QTLs identified 90 key biochemical pathways that might be conferring drought tolerance. QTARO of the meta-QTLs resulted in 160 curated QTLs significantly enriched in the gene lists (http://qtaro.abr.affrc.go.jp/qtab/table.). More than 60 (38%) curated QTLs were related to abiotic stress tolerance; others were related to growth and development, grain quality traits and disease

Table 3 Markers significantly associated with grain yield under drought stress

Marker	Chr	F-Marker	*p*-Marker	R^2	Year/Season
RM262	2	2.4772	0.0258	7.6	combined
		4.4261	0.0031	7.4	combined
RM26	5	5.3105	8.93E-04	19.1	11DS
		2.9933	0.0246	7.8	12DS
RM249	6	2.4746	0.0259	11.1	12DS
RM25694	10	2.8118	0.0231	13.9	11DS
RM224	11	2.4361	0.0436	5.6	combined
		3.2531	0.011	10.3	12DS
RM552	11	3.6227	0.0173	4.7	combined
		3.9686	0.0115	7.7	12DS
RM28048	12	3.8536	0.0071	14.8	11DS
		2.6494	0.0407	7.0	12DS

Table 4 Markers significantly associated with grain yield under non-stress

Marker	Chr	F-marker	*p*-value	R^2	Year/Season
RM11943	1	3.494	0.0202	12.6	11DS
RM582	1	2.805	0.0101	24.8	11DS
	1	2.255	0.0346	22.1	combined
RM12979	2	2.492	0.0312	17.7	11DS
RM521	2	3.373	0.0233	12.2	11DS
RM555	2	2.563	0.0352	16.2	combined
RM15983	3	2.817	0.0229	16.6	11DS
RM334	5	2.139	0.0391	23.6	combined
RM19637	6	2.876	0.0293	13.8	11DS
RM248	7	2.237	0.0269	26.7	combined
RM408	8	2.821	0.0317	13.6	11DS
RM24932	10	3.442	0.0052	24.0	combined
RM304	10	2.887	0.0148	11.6	12DS
RM491	12	1.927	0.0491	16.1	12DS

resistance. There are differences in specific biological themes that are enriched per QTL bin but, in general, most of them had relevance to abiotic stress tolerance (Additional file 5: Table S7).

Discussion

The adverse effects of climate change such as drought and heat are becoming a major threat to sustainable rice production and productivity (IPCC et al., 2007). In recent years, El Niño-induced drought has become a common phenomenon across many countries of Asia and there is already a clear prediction that El Niño-induced drought will have a significant effect on overall rice stocks in 2016 (Mohanty, 2015). In order to mitigate drought, there have been increased efforts to breed drought-tolerant rice varieties by both conventional and molecular breeding approaches (Kumar et al. 2014). Several major-effect QTLs for GY under drought have been identified for both upland and lowland conditions (Kumar et al. 2014). These QTLs have been successfully used in marker-assisted backcross (MABC) breeding programs to improve the drought tolerance of widely adopted popular but drought-susceptible rice varieties such as IR64, Swarna, MR219 and MRQ74 (Swamy et al. 2013; Swamy et al. 2014; Noraziyah et al. 2016a; Noraziyah et al. 2016b). However, *qDTYs* have different synergistic effects in different genetic backgrounds and meta-analysis of the major *qDTYs* has shown that few hotspot QTL regions are repeatedly being identified. To gain an economic yield advantage of at least 500 kg ha^{-1}, a minimum of two to three *qDTYs* have to be pyramided in elite genetic backgrounds (Swamy and Kumar, 2012); thus, exploration of genetic resources and the identification of several new *qDTYs* by association mapping is one of the most promising approaches.

Association analysis using germplasm collections has been successfully used to identify several major loci for different traits (Han and Huang 2013; Wu et al. 2015; Yang et al. 2014), but only a few association studies exist for drought-related traits in rice (Vasant, 2012; Courtois et al. 2013; Vannirajan et al. 2012; Muthukumar et al. 2015). There is thus large scope to explore germplasm through association mapping to identify loci linked to drought tolerance.

Phenotypic Analysis

Drought stress was successfully created in the trials with controlled irrigation and the screening was supported by less rainfall during the cropping seasons, which resulted in a drastic reduction in the water table. The lesser amount of rainfall and longer dry spells are the key components for successful drought screening at IRRI during the dry season (Vikram et al. 2011). Some 68.8% of the accessions reached 50% flowering by the end of March. The remaining germplasm has early and late flowering time, by 5.1% and 21.6%, respectively. This indicates that the crops received severe stress for about two weeks during panicle development to pre-flowering stage. Jennings et al. (1979) reported that exposure to at least two weeks of drought stress due to rainless days during the vegetative stage and at least one week during the reproductive stage can differentiate susceptible and drought-tolerant genotypes.

The delayed flowering observed under stress in the present study is in agreement with earlier reports (Vikram et al. 2011; Jongdee et al. 2006; Zhao et al. 2010; Atlin et al. 2006; Lafitte and Courtois, 2002; Pantuwan et al. 2002b). According to Pantuwan et al. (2002b), delayed flowering under drought stress could be a good measurement of plant responses and adaptability to drought tolerance, and also an efficient selection criterion for distinguishing drought-susceptible and drought-tolerant genotypes. The negative effect of drought on performance in many traits, including yield, has been reported. Even mild stresses during flowering cause a severe reduction in PH, biomass, spikelet fertility and GY. The reduction is mainly because of water shortage in the plant causing more respiration and reduced photosynthesis, leading to less biomass accumulation and less GY (Fukai et al. 1999; O'Toole, 1982; Boonjung and Fukai, 1996).

The yield reduction in the trials varied from 51% to 60%, indicating that the association panel was subjected to moderate to severe drought stress. A yield reduction of more than 50% was also reported in earlier studies, indicating successful drought screening (Venuprasad et al. 2009; Vikram et al. 2011; Ghimire et al. 2012; Noraziyah et al. 2016a; Noraziyah et al. 2016b). Poor panicle exsertion, reduced number of spikelets and poor grain filling are some of the causes of reduced GY under drought stress (Pantuwan et al. 2002a; O'Toole and Namuco, 1983; Cruz and O'Toole, 1984; Ekanayake et al. 1989). Our results once again reaffirm the use of GY under drought stress as an effective criterion for improving drought tolerance in rice. The CV and ranges for various traits clearly indicated the wide variations for yield and yield-related traits under both drought stress and NS, so there are some good drought-tolerant donor lines in the panel that can be used in drought breeding programs by combining both drought tolerance and high yield potential.

The H^2 for GY under drought stress was more than 70%, indicating that selection for GY under drought stress may be effective for improving drought tolerance. Moderate to high H^2 of GY under drought stress conditions was also observed in many studies (Babu et al. 2003; Atlin et al. 2004; Lafitte et al. 2004; Kumar et al. 2008; Kamoshita et al. 2008). High H^2 in PH and DTF under drought stress was also observed in several studies (Lafitte et al. 2004; Lanceras et al. 2004; Bernier et al. 2007). Gomez et al. (2006) concluded that the high H^2 in these traits was due to the preponderance of additive gene action and was suitable for direct selection in improving drought tolerance in rice. In the present study, the moderate to high H^2 values for yield and yield-related traits indicate that they are genetically controlled by additive gene action and can be used as a selection parameter under both drought stress and non-stress environments. This indicates that the genotypes that have early flowering time may have a smaller yield reduction. This negative correlation was also observed in earlier studies (Vikram et al. 2011, 2016; Zhao et al. 2010). The genotypes with early flowering time have higher GY than the genotypes that flowered later (Pantuwan et al. 2002b).

Genotypic Analysis

There is a wider molecular diversity within the association panel (almost 95% of the SSR markers tested were polymorphic) and in total 119 SSR markers amplified 1524 alleles in the 75 genotypes. The number of alleles per loci varied from 2 to 13, with an average of 8.36 per loci. The amount of diversity available within this association panel is comparable with or even higher than that of earlier reports (Agrama et al. 2007; Yan et al. 2009; Borba et al. 2010; Muthukumar et al. 2015; Raju et al. 2016). In general, higher genetic diversity in any germplasm collection indicates lesser gene flow among the accessions (Rahman et al. 2007; Dinesh Raj et al. 2010). This was also true in our panel as the genetic diversity was very high, perhaps because the germplasm consisted of landraces, breeding lines and introductions with various origins and collected from different geographic regions of Malaysia.

One of the main concerns of association mapping is the spurious marker trait associations due to population structure (Krill et al. 2010). In association mapping studies, it has been a common practice to perform structure analysis in any structured populations to overcome spurious associations (Pritchard et al. 2000). All the 75 genotypes used in our study were from the *indica* subgroup but the structure analysis grouped them into three subgroups ($K = 3$) with admixtures. Nearly 40 (53%) of the genotypes were clearly grouped into one of the subgroups, while 35 (47%) genotypes showed some amount of allelic reshuffling or sharing among different subgroups. Such allelic sharing among different genotypes has been attributed to the accumulation of several spontaneous mutations among genotypes from different geographic areas (Mather et al. 2004; Agrama et al. 2007).

The molecular diversity-based clustering also grouped the genotypes into three clusters, confirming the results of the structure analysis indicating the clear divergence among three subgroups (Fig. 1). This kind of clear differentiation among genotypes of any subspecies may be because of their different growth environments, adaptive traits and evolutionary patterns. The allelic fixation (F_{st}) values were different among the three subgroups and varied from 0.194 to 0.277, suggesting higher genome compatibility among the genotypes, indicating that crosses made between the genotypes from these distinct subgroups would generate superior lines with better adaptability to moderate to severe drought conditions (N'Goran et al. 2000; Raju et al. 2016). The expected heterozygosity was higher in all the three subgroups and it was more than 0.544 among the three subgroups.

LD analysis was carried out for the all the marker pairs and 32% of the pairs were significant ($P > 0.05$). The number of significant intra- and inter-chromosomal LD pairs varied on different chromosomes. Chromosome 2 had the highest number of significant intra- and inter-chromosomal LD pairs, whereas chromosome 7 had the lowest number. There was a reduction in number of significant LD pairs as the interval distances between marker pairs increased. The LD plotted against r^2 and LD plotted against cM is presented in Fig. 2. There was a sharp decline in LD decay for the linked markers at 5-20 cM. Overall, the r^2 varied from 0.005 to 0.31, with an average of 0.03. Significant pairs of linked loci ($r^2 > 0.2$) showed an average distance of 25 cM, indicating the presence of a higher number of significant LD blocks in this set of genotypes. The larger LD blocks within a germplasm collection are highly useful in genome-wide association mapping. There are numerous reports of LD patterns in rice. Olsen et al. (2006) and Mather et al. (2007) reported LD decay occurring at about 1 cM distance, whereas others reported LD decay at 20–30 cM distances using SSR markers (Agrama et al. 2007; Agrama and Eizenga, 2008; Jin et al. 2010; Vannirajan et al. 2012). Several factors such as nature of pollination, geographic isolation, evolutionary pattern, mutation, selection pressure and genetic drift influence the size and number of LD blocks (Gupta et al. 2005). In a predominantly self-pollinated crop species such as rice, larger LD blocks extending over several cM are usually expected (Abdurakhmonov and Abdukarimov, 2008); also, LD varied among different subspecies of rice. The extent of LD in the *indica* subspecies was lesser than in *temperate japonica* or *tropical japonica* (Khush, 1997; Garris et al. 2005; Mather et al. 2007).

Association Analysis

The structured association mapping of yield and yield-related traits revealed that there were 198 and 80 significant marker trait associations (MTA) for all three traits by GLM and MLM analysis and most of these loci were similar to the major loci reported for these traits

(Bernier et al. 2007; Venuprasad et al. 2009; Swamy et al. 2011; Vikram et al. 2011, 2015; Ghimire et al. 2012; Palanog et al. 2014; Courtois et al. 2013; Kumar et al. 2014). In the genotypic analysis, seven out of the 119 SSR markers used were linked to known major-effect drought GY QTLs identified at IRRI (Kumar et al. 2014). Thus confirming the consistency, accuracy and effectiveness of these QTLs, and also most of them were in the drought QTL hotspot regions (Swamy et al. 2013). The existence of such large-effect and consistent QTLs is one of the main reasons for successful MAB for improving the drought tolerance of several popular but drought-susceptible rice varieties (Noraziyah et al. 2016a; Noraziyah et al. 2016b). It is also interesting to note that major-effect QTLs $qDTY_{12.1}$, was present in 10.6% of the genotypes, and a similar result was also reported by Swamy et al. (2011). In addition, we have identified a few new major loci for GY under drought stress on chromosomes 5, 6 and 11 with a PV of 5–19%. It is interesting to that the novel QTL on chromosome 5 was consistently identified in all the three analyses and had highest PV of 19%. These QTLs are useful for MAB in combination with other major-effect *qDTYs*. Eventhough some major effect QTLs were identified for three traits, the results could have been improved by increasing the marker density and the population size. Several consistent major-effect QTLs were also identified for GY under non-stress, but most of them were season-specific except for one locus (RM582) on chromosome 7. The overall results have clearly shown that association mapping is one of the feasible options for identifying major-effect QTLs for GY under drought.

The low marker density of one in every 3 MB and a smaller population size of 75 lines used in this study might have impacted the identification of new QTLs. Recently genome wide association analysis (GWAS) for complex traits using high density single nucleotide polymorphism markers (SNP) are becoming common, it provides high mapping resolution and high accuracy of the QTLs due to better structuring of the populations (Han and Huang, 2013; Wu et al. 2015; Yang et al. 2014). Therefore, we suggest the QTLs identified in the present study may be further reconfirmed by increasing marker density and size of the association panel.

Enrichment Analysis of Genes Within *qDTYs*

The binning of all the *qDTYs* resulted in 20 meta-QTLs spanning 18 MB, which is 4.6% of the reference genome; this clearly shows that binning has been successful in reducing the QTL regions without losing significant information on gene content. The binned multiple QTL regions with less than 10% of the genome size are normally considered useful in identifying the genes/gene families underlying target traits. The materials used in this study are from diverse sources and there is no physical map that exactly matches the genome of these materials. However, there are more than 98% similarities among the genomes of different rice varieties and the available reference genome information can be used in gene assessment in a wide range of germplasm (Schatz et al. 2014). We used the Nipponbare physical map as a reference genome for gene survey, identification of processes, pathways and curation of the QTLs. Enrichment analysis of the genes within QTLs against the genome-wide annotation for functions/processes/cell compartments (GO), pathways (MapMan) and association to previously published QTLs (QTARO) shows significantly enriched ($P < 0.005$) interesting biological processes that may be working toward achieving improved yield under drought (e.g., GO:0005975: carbohydrate metabolic process; GO:0006950: response to stress; MapMan 20.1.5: stress. biotic.regulation of transcription) as well as significant association of the discovered QTLs to previously mapped QTLs for drought tolerance (e.g., 21 drought tolerance genes affecting spikelet fertility, root volume/biomass/length, leaf rolling, stomatal resistance and various yield-correlated traits, in supplemental file < DEF_enrichment_QTL-genes_p005.xlsx>) in the QTARO database. This additional information about these contiguous genes offers further inferences on how a QTL functions to deliver the trait expected, and helps to guide in the selection of which QTLs could be more effective in a breeding program.

Here, we provide glimpses of the important genes/gene families associated with the drought QTL regions and evidence of their role in drought tolerance is discussed. The prominent genes/gene families enriched within the drought QTLs were NAC transcription factors, F-box domain proteins, MADS-box proteins, tetra- and pentatricopeptide proteins, NAC, NAM, HSPs, cytochrome P450 family, AP2/ERFs, ARFs, brassinosteroids, CaAMK, GDSL-like lipases, membrane proteins, hydrolases, kinases, peroxidases, peptidase and dismutase family proteins, ANC, ZFPs, ZIP, and MYB transcription factors (TFs). The role of these genes/gene families in drought stress tolerance has been reported in rice and other crops.

Root traits play an important role in enhancing drought tolerance in rice. Several TFs such as ARF, MYB, ZIP, NAC, NAM and MADS-box genes were involved in changes in root architecture and they help in tolerating drought stress (Okushima et al. 2007; Xie et al. 2000; Zhang and Forde, 1998). *OsNAC1* has been reported to increase the number of lateral roots under drought stress, while *OsNAC10* resulted in thicker roots (Khong et al. 2008; Jeong, 2010). Since grain yield under drought is the most economic component under drought, floral meristic tissue activities, floral organ development, control of flowering time and pollen function play an important role in spikelet development and filling of spikelets to produce grains under drought stress (Hepworth et al. 2006; Imaizumi et al. 2003; Sonneveld et al. 2005). Genes such as *OsGILP*, ZIP, cytochrome P450 family and F-box family genes are involved in floral development and grain production under stress conditions, including drought. Heat shock family proteins improve heat tolerance by improving plant physiological phenomena such as photosynthesis; assimilate partitioning, water and nutrient use efficiency, and membrane stability (Wahid et al. 2007). ERF/AP2 domain proteins such as *OsERF1*, *OsERF2* and *OsDREBs* enhance the osmotic and drought tolerance of rice by modulating the increase in stress-responsive gene expression through different pathways (Dubouzet et al. 2003; Gao et al. 2008; Chen et al. 2008; Mishra et al. 2009; Quan et al. 2010; Fukao and Serres, 2008). The over expression of *OsbZIPs* has increased drought stress tolerance in rice, whereas its down-regulation/knockout leads to higher sensitivity (Xiang et al. 2008; Lu et al. 2009). Zhang et al. (2016) reported that the ZFP and MYB protein families also play an important role in response to water stress, such as the regulation of stomatal movement, synthesis of suberin and cuticular wax and the regulation of flower development (Xiong et al. 2014; Ma et al. 2009; Baldoni et al. 2015). All these evidences clearly indicate that drought QTLs were enriched with several genes/gene families involved in drought stress responses and their coordinated action is essential for conferring drought tolerance and producing yield under drought stress. However, we suggest further in-depth analysis of the entire drought yield QTLs reported to gain more insight into the genes underlying drought QTLs and their molecular mechanisms involved in conferring drought tolerance.

Conclusions

Drought is a severe abiotic stress affecting rice production. The identification and introgression of major-effect QTLs are one of the best and proven approaches to improving the drought tolerance of rice varieties. The accuracy and consistency of QTLs have clearly shown that structured association mapping with genome-wide molecular markers is an attractive option to identify major-effect QTLs for GY under drought stress. Some of the new *qDTYs* identified in this study are useful for MAB in combination with other major *qDTYs*. The *in silico* analysis of QTL regions revealed that several drought-responsive genes were associated with the grain yield under drought.

Methods

Plants Materials

We used 75 rice genotypes in this study. The seeds were collected from the Rice Gene Bank of MARDI, Seberang Perai, Malaysia. The material consisted of Malaysian landraces, breeding lines, varieties, cultivars and introductions (Additional file 3: Table S8). Six drought-tolerant checks (Vandana, Apo, PSBRC-82, UPLRi7, Mokwoo and IR77298-14-1-2-10) and two drought-susceptible checks (IR64 and MTU1010) were included in the experiment.

Genotyping

Molecular work was conducted in the Genotyping Services Laboratory (GSL) of the Plant Breeding Division at IRRI.

DNA Extraction and PCR Amplification

Fresh leaf samples were collected from each accession three weeks after transplanting and were freeze-dried using a lyophilizer. The DNA was extracted by using a modified CTAB protocol. This version of DNA extraction method was developed by Murray and Thompson (1980); it uses cetyl trimethyl ammonium bromide (CTAB). Agarose gel electrophoresis was used to check the quality and quantity of DNA extracted. The concentration of the isolated DNA was estimated based on the band brightness and thickness compared with those of the reference λ DNA. The DNA samples were diluted with 1x TE into an equal concentration of 25 ng/ul.

Amplification of SSR markers using polymerase chain reaction (PCR) was done with a 15 μl reaction mixture that contained 3 μl of DNA template, 1.5 μl of 10x PCR buffer, 2.0 μl of $MgCl_2$, 0.5 μl each of forward and reverse primers, 0.5 μl of 1 mM dNTP and 0.5 μl of Taq DNA polymerase (1:20 homemade). A drop of mineral oil was added on each well to prevent the mixture from evaporating and the plate was covered with PCR sealing film. Finally, 10x loading buffer was added to the PCR product prior to loading (1.3 μl of 10x loading buffer for every 10 μl of PCR product). Amplification reaction was carried out in a 96-well PCR plate in a thermocyler. The following PCR profile was used for SSR amplification: initial denaturation at 94 °C for 5 min and then 35 cycles of denaturation at 94 °C for 30 s, annealing at 55 °C for 30 s and extension at 72 °C for 30 s; and final extension at 72 °C for 5 min and storage at 10 °C forever. The PCR products were resolved using high-resolution 8% polyacrylamide gel electrophoresis (PAGE) as described by Sambrook and Russell (2001). The gel was run in 1x TBE at 95 volts for 1 to 3 hours depending on the product size of the SSR marker. Gels were stained with SYBR Safe^{TM} DNA gel stain and were viewed after 20 minutes.

Marker Analysis

We used 119 highly polymorphic SSR markers for genotyping all 75 genotypes. Out of 119 markers, 45 markers were linked to major-effect grain yield QTLs under drought such as $qDTY_{1.1}$, $qDTY_{2.1}$, $qDTY_{2.2}$, $qDTY_{3.1}$, $qDTY_{4.1}$, $qDTY_{9.1}$, $qDTY_{10.1}$, $qDTY_{12.1}$, $qDTY_{1.2}$ and $qDTY_{2.3}$ and all the remaining markers were randomly distributed within the genome (Additional file 3: Table S9). The details of marker information such as chromosome number, position (cM), expected product size and annealing temperature were obtained from the Gramene database (http://www.gramene.org).

Phenotyping Under Reproductive Stage (RS) Drought Stress and Non-stress (NS) The phenotyping was conducted at the International Rice Research Institute (IRRI), Los Baños, Laguna, Philippines, during the 2011DS and 2012DS. IRRI is located at 14°13'N latitude, 121°15'E longitude, at an altitude of 21 m above mean sea level. The soil type of the experimental field is a Mahaas clay loam, isohyperthermic mixed typic tropudalf. RS and NS trials were established in an alpha lattice design with two replications in two-row plots with 2 m in length and plant spacing of 20 cm between rows and 20 cm between plants with standard agronomic practices. Both RS and NS trials were planted in the 2011DS and repeated in the 2012 DS. The NS trial was managed like irrigated lowland where no water stress was employed. The purpose of the NS trial was to serve as a source of seed for the succeeding trial. Conversely, the RS trial refers to a trial in which drought stress is artificially imposed during the reproductive stage. In the RS trial, water was drained in the field at four weeks after transplanting to start the imposition of drought stress. Perforated PVC pipes were strategically installed in six locations in the field to monitor for parching water table daily. The parching water table was measured in all pipes until the crops reached 50% maturity. When the water table reached 100 cm below the soil surface for about three weeks and 70% of the plants showed severe leaf rolling and seemed beyond recovery, irrigation was immediately applied through flooding for 6 hours and the water was drained again after the field was saturated.

After transplanting, 2–5 cm of standing water was maintained in the field. Drought was imposed at 30 days after transplanting (DAT) by draining the water in the field and withholding it until soil moisture tension reached -70 kPa at 0.2-m depth. The fields were re-irrigated by flooding for 24 hours. This was repeated during harvesting. The soil moisture level below the soil surface was regularly monitored and measured through the observation wells strategically installed in the field. Fertilizer was applied three times at 10, 25 and 45 DAT at the rate of 90-30-30 kg NPK ha^{-1}.

Data recording and analysis Observations were recorded at different stages of crop growth until maturity from both RS and NS trials. Phenotypic data recorded were days to 50% flowering (DTF), plant height (PH) and grain yield (GY). The phenotypic observations were analyzed using PB Tools to estimate the trial mean, range, SD, CV, broad-sense heritability (H^2) and genetic advance (GA). Correlations among the traits under drought stress and NS were estimated using STAR (v17.0).

Association Analysis

Allelic Diversity and Cluster Analysis

Allele frequency, genetic diversity, polymorphic information content (PIC) and molecular diversity were assessed using PowerMarker (v3.25). The linkage disequilibrium between marker pairs was tested at 1% significance level by PowerMarker (v3.25).

Population Structure

Based on the genotyping data from 119 SSR markers that were evenly distributed across all 12 chromosomes, population structure was estimated with the model-based (Bayesian) cluster software STRUCTURE 2.2 (Pritchard et al. 2000). The software was set to have length of burn-in period of 500,000 followed by 500,000 Markov chain Monte Carlo (MCMC) reps after burn-in. The optimum number of populations was inferred by running an admixture ancestry model with correlated frequencies starting from two populations, $K = 2$ to $K = 10$, with three replications at each K. The optimum value of $K = 3$ was determined, thus indicating that all the germplasm could be divided into three subgroups: SG1, SG2 and SG3.

Association Mapping

Association between markers and three traits (DTF, PH and GY) under RS drought stress and NS was calculated using TASSEL2.1. We used generalized linear model (GLM) and Mixed Linear Model (MLM) functions for the analysis. The marker and trait association was declared significant when $P < 0.05$.

In Silico Analysis of qDTY Regions

All the SSRs reported to be associated with a *qDTY* as a union set (all markers in two-season trials as one set) were mapped to the IRGSP 1.0 genome by aligning the SSR primer sequences using BLAT (Kent, 2002) with parameter tile size 11 for maximum sensitivity. Since this is a single-marker regression QTL mapping, we estimated the physical interval size of a QTL based on most of the original drought QTLs to be around a 1-MB region. Regions defined by a single SSR marker were extended to include 500 kb left and 500 kb right of each individual marker. We call these QTL genome intervals associated bins (they are not true QTLs), and they represent +/−0.5 mb to the left and right of the region flanked by the SSR forward and reverse primer, except for three bins that were merged from two adjacent SSR regions, which may be slightly bigger than 1 mb or as small as 20 kb only (Additional file 4: Table S6). Twenty associated bins that represent 18 megabases of the reference genome were determined and 3222 RGAP genes were located within these associated bins, which were used for the enrichment analyses against the annotation of all the RGAP genes with gene ontology (GO) terms (RGAP release 7), MapMan pathway annotation (MapMan URL) and QTARO QTL association (http://qtaro.abr.affrc.go.jp/).

Additional files

Additional file 1: Figure S1. Daily rainfall during the dry season experiment period from February to May in 2011.

Additional file 2: Figure S2. Daily rainfall during the dry season experiment period from February to May in 2012.

Additional file 3: Table S1. Various population parameters measured in the germplasm. **Table S2.** Subpopulation-specific statistical parameters. **Table S3.** Number of intra- and inter-chromosomal linkage disequilibrium (LD) pairs. **Table S4.** Markers significantly associated with DTF and PH under drought stress. **Tables S5.** Markers significantly associated with DTF and PH under non stress. **Table S8.** List of *Oryza* accession used in this study assemble in their origin. **Table S9.** List of random and specific microsatellite markers group according to the chromosome number.

Additional file 4: Table S6. List of genes in association bins for grain yield under drought.

Additional file 5: Table S7. Enrichment analysis of drought resposive genes in the drough QTLs.

Abbreviations

CV: Coefficient of variation; DS: Dry season; DTF: Days to 50% flowering; GSL: Genotyping Services Laboratory; GY: Grain yield; IRRI: International Rice Research Institute; LD: Linkage disequilibrium; NS: Non-stress; PH: Plant height; POP: Population; *qDTY*: Drought yield QTL

Acknowledgments

The authors would like to thank the Global Partnership Initiative for Plant Breeding Capacity (GIPB) for its research funding support and the Faculty of Science and Technology (FST) and International Rice Research Institute (IRRI) for the laboratory, field and glasshouse facilities support.

Authors' Contributions

BPMS, NAAS, SNB and RM carried out the genotyping and phenotyping studies and contributed to statistical analysis and manuscript draft preparation, and AK conceptualized the study, supervised the genotyping and phenotyping, and contributed to manuscript preparation. All authors read and approved the final manuscript.

Competing Interests

The authors declare that they have no competing interests.

Author details

[1]Plant Breeding Division, International Rice Research Institute (IRRI), DAPO Box 7777 Metro Manila, Philippines. [2]Faculty of Science and Technology, Universiti Kebangsaan Malaysia, 43600 Bangi, Selangor, Malaysia. [3]MARDI, Seberang Perai, P.O. Box No. 20313200 Kepala Batas, Pulau Pinang, Malaysia.

References

Abdurakhmonov I, Abdukarimov A (2008) Application of association mapping to understanding the genetic diversity of plant germplasm resources. Int J Plant Genomics 2008:1–18

Agrama HA, Eizenga GC (2008) Molecular diversity and genome-wide linkage disequilibrium pattern in worldwide rice and its wild relatives. Euphytica 160: 339–355. doi:10.1007/s10681-007-9535-y

Agrama HA, Eizenga GC, Yan W (2007) Association mapping of yield and its components in rice cultivars. Mol Breed 19:341–356

Atlin GN, Lafitte HR, Venuprasad R, Kumar R, Jongdee B (2004) Heritability of rice yield under reproductive-stage drought stress, correlations across stress levels and effects of selection: implications for drought tolerance breeding. In: Poland D, Sawkins JM, Ribaut M, Hoisington D (eds) Proceedings of a Workshop Held at Cuernavaca on Resilient Crops for Water Limited Environments, Mexico, May 24–28. International Maize and Wheat Improvement Center, Mexico, pp 85–87

Atlin GN, Lafitte HR, Tao D, Laza M, Amante M, Courtois B (2006) Developing rice cultivars for high-fertility upland systems in the Asian tropics. Field Crops Res 97:43–52. doi:10.1016/j.fcr.2005.08.014

Babu RC, Nguyen BD, Chamarerk V, Shanmugasundaram P, Chezhian P, Jeyaprakash P, Ganesh SK, Palchamy A, Sadavisam S, Sarkarung S, Wade LJ, Nguyen HT (2003) Genetic analysis of drought resistance in rice by molecular markers: association between secondary traits and field performance. Crop Sci 43:1457–1469

Baldoni E, Genga A, Cominelli E (2015) Plant MYB transcription factors: their role in drought response mechanisms. Int J Mol Sci 16:15811–15851. doi:10.3390/ijms160715811

Bernier J, Kumar A, Ramaiah V, Spaner D, Atlin G (2007) A large-effect QTL for grain yield under reproductive-stage drought stress in upland rice. Crop Sci 47:507–518. doi:10.2135/cropsci2006.07.0495

Boonjung H, Fukai F (1996) Effects of soil water deficit at different growth stage on rice growth and yield under upland conditions. 2. Phenology, biomass production and yield. Field Crops Res 48:47–55

Borba TC, Brondani RV, Breseghello F, Coelho ASG, Mendonça JA, Rangel PHN, Brondani C (2010) Association mapping for yield and grain quality traits in rice (*Oryza sativa* L.). Genet Mol Biol 33(3):515–524

Chen JQ, Meng XP, Zhang Y, Xia M, Wang XP (2008) Over-expression of *OsDREB* genes leads to enhanced drought tolerance in rice. Biotechnol Lett 30:2191–2198

Courtois B, Audebert A, Dardou A, Roques S, Ghneim-Herrera T (2013) Genome-wide association mapping of root traits in a japonica rice panel. PLoS One 8(11):78037. doi:10.1371/journal.pone.0078037

Cruz RT, O'Toole JC (1984) Dryland rice response to an irrigation gradient at flowering stage. Agron J 76:178–183

Dinesh Raj R, Kiran AG, Thomas G (2010) Population genetic structure and conservation priorities of *Oryza rufipogon* Griff. populations in Kerala. India. Curr Sci 98:65–68

Dubouzet JG, Sakuma Y, Ito Y, Kasuga M, Dubouzet EG, Miura S, Shinozaki K, Shinozaki KY (2003) *OsDREB* genes in rice, *Oryza sativa* L., encode transcription activators that function in drought, high salt and cold responsive gene expression. Plant J 33:751–763

Ekanayake IJ, De Datta SK, Steponkus PL (1989) Spikelet fertility and flowering response of rice to water stress at anthesis. Ann Bot 63:257–264

Fairhust TH, Dobermann A (2002) Rice in the global food supply. Better Crops Int 16:3–6

Fukai S, Pantuwan G, Jongdee B, Cooper M (1999) Screening for drought resistance in rainfed lowland rice. Field Crops Res 64:61–74

Fukao T, Serres JB (2008) Submergence tolerance conferred by *Sub1A* is mediated by SLR1 and SLRL1 restriction of gibberellin responses in rice. Proc Natl Assoc Sci USA 105(43):16814–16819

Gao S, Zhang H, Tian Y, Li F, Zhang Z, Lu X, Chen X, Huang R (2008) Expression of TERF1 in rice regulates expression of stress-responsive genes and enhances tolerance to drought and high-salinity. Plant Cell Rep 27:1787–1795

Garris AJ, Tai TH, Coburn J, Kresovich S, McCouch S (2005) Genetic structure and diversity in *Oryza sativa* L. Genetics 169:1631–1638

Ghimire KH, Quiatchon LA, Vikram P, Swamy BPM, Dixit S, Ahmed H, Hernandez JE, Borromeo TH, Kumar A (2012) Identification and mapping of a QTL (qDTY1.1) with a consistent effect on grain yield under drought. Field Crops Res 131:88–96. doi:10.1016/j.fcr.2012.02.028

Gomez SM, Kumar SS, Jeyaprakash P, Suresh R, Biji KR, Boopathi NM, Price AH, Babu RC (2006) Mapping QTLs linked to physio-morphological and plant production traits under drought stress in rice (*Oryza sativa* L.) in the target environment. Am J Biochem Biotechnol 2(4):161–169

Gupta P, Rustgi S, Kulwal P (2005) Linkage disequilibrium and association studies in higher plants: present status and future prospects. Plant Mol Biol 57(4): 461–485

Han B, Huang X (2013) Sequencing-based genome-wide association study in rice. Curr Opin Plant Biol 16:133–138

Hepworth SR, Klenz JE, Haughn GW (2006) UFO in the *Arabidopsis* inflorescence apex is required for floral-meristem identity and bract suppression. Planta 223:769–778

Huang X, Wei X, Sang T, Zhao Q, Feng Q, Zhao Y (2010) Genome-wide association studies of 14 agronomic traits in rice landraces. Nat Genet 42:961–967

Imaizumi T, Tran HG, Swartz TE, Briggs WR, Kay SA (2003) FKF1 is essential for photoperiodic-specific light signalling in *Arabidopsis*. Nature 426:302–306

IPCC et al (2007) Climate Change 2007: Impacts, Adaptation and Vulnerability. In: Parry ML, Canziani OF, Palutikof JP, Hanson CE (eds) Contribution of Working Group II to the Fourth Assessment Report of the Intergovernmental Panel on Climate Change. Cambridge University Press, Cambridge, p 976

Jennings PR, Coffman WR, Kauffman HE (1979) Upland adaptability. In: Rice improvement. International Rice Research Institute (IRRI), Los Baños

Jeong JS (2010) Root-specific expression of OsNAC10 improves drought tolerance and grain yield in rice under field drought conditions. Plant Physiol 153:185–197

Jin L, Lu Y, Xiao P, Sun M, Corke H, Bao J (2010) Genetic diversity and population structure of a diverse set of rice germplasm for association mapping. Theor Appl Genet 121(3):475–487. doi:10.1007/s00122-010-1324-7

Jongdee B, Pantuwan G, Fukai S, Fischer K (2006) Improving drought tolerance in rainfed lowland rice: an example from Thailand. Agric Water Manage 80:225–240

Kamoshita A, Babu RC, Boopathi N, Fukai S (2008) Phenotypic and genotypic analysis of drought-resistance traits for development of rice cultivars adapted to rainfed environments. Field Crops Res 109:1–23

Kawahara Y, de la Bastide M, Hamilton JP, Kanamori H, McCombie WR, Ouyang S, Schwartz DC, Tanaka T, Wu J, Zhou S, Childs KL, Davidson RM, Lin H, Quesada-Ocampo L, Vaillancourt B, Sakai H, Lee SS, Kim J, Numa H, Itoh T, Buell CR, Matsumoto T (2013) Improvement of the *Oryza sativa* Nipponbare reference genome using next generation sequence and optical map data. Rice 6(1):4. doi:10.1186/1939-8433-6-4

Kent WJ (2002) BLAT: the BLAST-like alignment tool. Genome Res 4:656–664

Khong GN, Richaud F, Coudert Y, Pati PK, Santi C, Périn C, Breitler JC, Meynard D, Vinh DN, Guiderdoni E, Gantet P (2008) Modulating rice stress tolerance by transcription factors. Biotechnol Genet Eng Rev 25:381–404

Khush GS (1997) Origin, dispersal, cultivation and variation of rice. Plant Mol Biol 35:25–34

Korte A, Farlow A (2013) The advantages and limitations of trait analysis with GWAS: a review. Plant Methods 9(1):29

Krill AM, Kirst M, Kochian LV, Buckler ES, Hoekenga OA (2010) Association linkage analysis of aluminum tolerance genes in maize. PLoS One 5(4):1992–1998

Kumar A, Bernier J, Verulkar S, Lafitte HR, Atlin GN (2008) Breeding for drought tolerance: direct selection for yield, response to selection and use of drought-tolerant donors in upland and lowland adapted populations. Field Crops Res 107:221–231

Kumar A, Dixit S, Ram T, Yadaw RB, Mishra KK, Mandal NP (2014) Breeding high-yielding drought-tolerant rice: genetic variations and conventional and molecular approaches. J Exp Bot doi. doi:10.1093/jxb/eru363

Lafitte HR, Courtois B (2002) Interpreting cultivar × environment interactions for yield in upland rice: assigning value to drought-adaptive traits. Crop Sci 42: 1409–1420

Lafitte HR, Price AH, Courtois B (2004) Yield response to water deficit in an upland rice mapping population: association among traits and genetic markers. Theor Appl Genet 109:1237–1246

Lanceras JC, Pantuwan G, Boonrat J, Toojinda T (2004) Quantitative trait loci associated with drought tolerance at reproductive stage in rice. Plant Physiol 135:384–399

Lu G, Gao CC, Zheng XN, Han B (2009) Identification of OsbZIP72 as a positive regulator of ABA response and drought tolerance in rice. Planta 229:88–91

Ma Q, Dai X, Xu Y, Guo J, Liu Y, Chen N, Xiao J, Zhang D, Xu Z, Zhang X (2009) Enhanced tolerance to chilling stress in OsMYB3R-2 transgenic rice is mediated by alteration in cell cycle and ectopic expression of stress genes. Plant Physiol 150:244–256. doi:10.1104/pp.108.133454

Mather DE, Hyes PM, Chalmers KJ, Eglinton JK, Matus I, Richardson K (2004) Use of SSR marker data to study linkage disequilibrium and population structure in *Hordeum vulgare*: prospects for association mapping in barley. In: Spunar J, Janikova J (eds) Proceedings from the 9th International Barley Genetics Symposium. Czech Journal of Genetics and Plant Breeding, Brno, pp 302–307

Mather KA, Caicedo AL, Polato NR, Olsen KM, McCouch S, Purugganan MD (2007) The extent of linkage disequilibrium in rice (*Oryza sativa* L.). Genetics 177(4): 2223–2232

McCouch SR, McNally KL, Wang W, Sackville Hamilton R (2012) Genomics of gene banks: a case study in rice. Am J Bot 99(2):407–423

Mishra M, Das R, Pandey GK (2009) Role of ethylene responsive factors (Erfs) in abiotic stress mediated signaling in plants. E J Biol Sci 1(1):2076–9954, ISSN: 2076–9946

Mohanty S (2015) How long can the rice market defy El Niño? Rice Today 14:38–39

Murray MG, Thompson WF (1980) Rapid isolation of high molecular weight plant DNA. Nucleic Acids Res 8:4321–4326

Muthukumar C, Subathra T, Aiswarya J, Gayathri V, Chandra Babu R (2015) Comparative genome-wide association studies for plant production traits under drought in diverse rice (*Oryza sativa* L.) lines using SNP and SSR markers. Curr Sci 109:139–147

N'Goran JAK, Laurent V, Risterucci AM, Lanaud C (2000) The genetic structure of cacao populations (*Theobroma cacao* L.) revealed by RFLP analysis. Euphytica 115:83–90

Noraziyah AAS, Swamy BPM, Wickneswari R, Teressa SC, Anitha R, Kumar A (2016a) Marker-assisted pyramiding of drought yield QTLs into a popular Malaysian rice cultivar, MR219. BMC Genet 17:30

Noraziyah AAS, Swamy BPM, Wickneswari R, Sta. Cruz MA, Nitika S, Anitha R, Kumar A (2016b) Pyramiding of drought yield QTLs into a high quality Malaysian rice cultivar MRQ74 improves yield under reproductive stage drought. Rice 2016:9–21.

O'Toole JC (1982) Adaptation of rice to drought-prone environments. In: Drought resistance in crops, with emphasis on rice. International Rice Research Institute, Manila, pp 195–213

O'Toole JC, Namuco OS (1983) Role of panicle exsertion in water stress induced sterility. Crop Sci 23:1093–1097

Okushima Y, Fukaki H, Onoda M, Theologis A, Tasaka M (2007) ARF7 and ARF19 regulate lateral root formation via direct activation of LBD/ASL genes in *Arabidopsis*. Plant Cell 19:118–130

Olsen KM, Caicedo AL, Polato N, McClung A, McCouch S, Purugganan MD (2006) Selection under domestication: evidence for a sweep in the rice Waxy genomic region. Genetics 173:975–983

Palanog AD, Swamy BPM, Shamsudin NAA, Dixit S, Hernandez JE, Boromeo TH, Sta Cruz PC, Kumar A (2014) Grain yield QTLs with consistent-effect under reproductive-stage drought stress in rice. Field Crop Res 161:46–54

Pandey S (2007) Economic costs of drought and rice farmers' coping mechanisms. Int Rice Res Notes 1:5–11

Pantuwan G, Fukai S, Cooper M, Rajatasereekul S, O'Toole JC (2002a) Yield response of rice (*Oryza sativa* L.) to drought under rainfed lowlands: 1. Grain yield and yield components. Field Crops Res 73:153–168

Pantuwan G, Fukai S, Cooper M, Rajatasereekul S, O'Toole JC (2002b) Yield response of rice (*Oryza sativa* L.) to drought under rainfed lowlands: 2. Selection of drought resistant genotypes. Field Crops Res 73:169–180

Pascual L, Albert E, Sauvage C, Duangjit J, Bitton BF, Desplat N, Brunel D, Le Paslier M, Ranc N, Bruguier L, Chauchard B, Verschave P, Causse M (2016) Dissecting quantitative trait variation in the resequencing era: complementarity of bi-parental, multi-parental and association panels. Plant Sci 242:120–130

Pritchard J, Stephens M, Donnelly P (2000) Inference of population structure using multilocus genotype data. Genetics 155:945–959

Quan R, Hu S, Zhang Z, Zhang H, Zhang Z, Huang R (2010) Overexpression of an ERF transcription factor TSRF1 improves rice drought tolerance. Plant Biotechnol J 8:476–488

Rahman SN, Islam MS, Alam MS, Nasiruddin KM (2007) Genetic polymorphism in rice (*Oryza sativa* L.) through RAPD analysis. Indian J Biotechnol 6:224–229

Raju BR, Rajanna MP, Sheshshayee MS, Mohankumar MV, Udayakumar M, Sumanth KK, Prasad TG (2016) Discovery of QTLs for water mining and water use efficiency traits in rice under water-limited condition through association mapping. Mol Breed 36:35

Sambrook J, Russell DW (2001) Molecular cloning: a laboratory manual (3-volume set), Cold Spring Harbor. Cold Spring Harbor Laboratory Press, New York

Schatz MC, Maron LG, Stein JC, Hernandez Wences A, Gurtowski J, Biggers E, Lee H, Kramer M, Antoniou E, Ghiban E, Wright MH, J-m C, Ware D, McCouch SR, McCombie WR (2014) Whole genome de novo assemblies of three divergent strains of rice, *Oryza sativa*, document novel gene space of aus and indica. Genome Biol 15:506

Sonneveld T, Tobutt KR, Vaughan SP, Robbins TP (2005) Loss of pollen-S function in two self-compatible selections of *Prunus avium* is associated with deletion/mutation of an S haplotype-specific F-box gene. Plant Cell 17:37–51

Swamy BPM, Kumar A (2012) Sustainable rice yield in water short drought prone environments: conventional and molecular approaches. In: Lee Teang S (ed) Irrigation systems and practices in challenging environments. Intech Press, Croatia, pp 149–168

Swamy BPM, Kumar A (2013) Genomics-based precision breeding approaches to improve drought tolerance in rice. Biotechnol Adv 31(8):1308–1318

Swamy BPM, Vikram P, Dixit S, Ahmed HU, Kumar A (2011) Meta-analysis of grain yield QTL identified during agricultural drought in grasses showed consensus. BMC Genomics 12:319

Swamy BPM, Ahmed HU, Henry A, Mauleon R, Dixit S, Vikram P (2013) Physiological and gene expression analyses reveal multiple QTLs enhance the yield of rice mega-variety IR64 under drought. PLoS One 8:e62795

Swamy BPM, Kaladhar K, Ashok Reddy G, Viraktamath BC, Sarla N (2014) Mapping and introgression QTLs for yield and related traits in two backcross populations derived from *O. sativa* cv Swarna and two accessions of *O. nivara*. J Genet 93:643–654

Thomson MJ, Ismail AM, McCouch SR, Mackill MJ, Pareek A, Sopory SK, Bohnert HJ, Govindjee (2010) Marker assisted breeding. In: Abiotic Stress Adaptation in Plants: Physiological, molecular and genomic foundation. Springer, Dordrecht, pp 451–469

Vannirajan C, Vinod KK, Pereira A (2012) Molecular evaluation of genetic diversity and association studies in rice (*Oryza sativa* L.). J Genet 91:9–19

Vasant DV (2012) Genome wide association mapping of drought resistance traits in rice (*Oryza sativa* L.). Master Thesis submitted to Tamil Nadu Agricultural University, Coimbatore, India.

Venuprasad R, Bool ME, Dalid CO, Zhao D, Espiritu M, Cruz MTS, Amante M, Kumar A, Atlin GN (2009) Genetic loci responding to two cycles of divergent selection for grain yield under drought stress in a rice breeding population. Euphytica 167:261–269. doi:10.1007/s10681-009-9898-3

Vikram P, Swamy B, Dixit S, Ahmed H, Cruz MTS, Singh A, Kumar A (2011) qDTY1. 1, a major QTL for rice grain yield under reproductive-stage drought stress with a consistent effect in multiple elite genetic backgrounds. BMC Genet 12:89. doi:10.1186/1471-2156-12-89

Vikram P, Swamy BPM, Dixit S, Singh R, Singh BP, Miro B, Kohli A, Henry A, Singh NK, Kumar A (2015) Drought susceptibility of modern rice varieties: an effect of linkage of drought tolerance with undesirable traits. Sci Reports 5:14799

Vikram P, Swamy BPM, Dixit S, Trinidad J, Sta Cruz MT, Maturan PC (2016) Linkages and interaction analysis of major effect drought grain yield QTLs in rice. PLoS One 11(3):e0151532

Voss-Fels K, Snowdon RJ (2016) Understanding and utilizing crop genome diversity via high-resolution genotyping. Plant Biotechnol J 14:1086–1094

Wahid A, Gelani S, Ashraf M, Foolad MR (2007) Heat tolerance in plants: an overview. Environ Exp Bot 61:199–223

Wu JH, Feng FJ, Lian XM, Teng XY, Wei HB, Yu HH, Xie WB, Yan M, Fan PQ, Li Y, Ma XS, Liu HY, Yu SB, Wang GW, Zhou FS, Luo LJ (2015) Genome-wide association study (GWAS) of mesocotyl elongation based on re-sequencing approach in rice. BMC Plant Biol 15:218

Xiang Y, Tang N, Du H, Ye HY, Xiong LZ (2008) Characterization of OsbZIP23 as a key player of the basic leucine zipper transcription factor family for conferring abscisic acid sensitivity and salinity and drought tolerance in rice. Plant Physiol 148:1938–1952

Xie Q, Frugis G, Colgan D, Chua NH (2000) *Arabidopsis* NAC1 transduces auxin signal downstream of TIR1 to promote lateral root development. Genes Dev 14:3024–3036

Xiong H, Li J, Liu DJ, Zhao Y, Guo X, Li Y, Zhang H, Ali J, Li Z (2014) Overexpression of OsMYB48-1, a novel MYB-related transcription factor, enhances drought and salinity tolerance in rice. PLoS One 9, e92913. doi:10.1371/journal.pone.0092913

Yan WG, Li Y, Agrama HA, Luo D, Gao FY, Lu XJ, Ren GY (2009) Association mapping of stigma and spikelet characteristics in rice (*Oryza sativa* L.). Mol Breed 24(3):277–292

Yang W, Guo Z, Huang CL, Duan LF, Chen GX, Jiang N, Fang W, Feng H, Xie WB, Lian XM, Wang GW, Luo QM, Zhang QF, Liu Q, Xiong LZ (2014) Combining high-throughput phenotyping and genome-wide association studies to reveal natural genetic variation in rice. Nature Comm 5:5087

Zhang HM, Forde BG (1998) An *Arabidopsis* MADS box gene that controls nutrient-induced changes in root architecture. Science 279:407–409

Zhang X, Zhang B, Li MJ, Yin XM, Huang LF, Cui YC, Wang ML, Xia X (2016) OsMSR15 encoding a rice C2H2-type zinc finger protein confers enhanced drought tolerance in transgenic *Arabidopsis*. J Plant Biol 59(3):271–281

Zhao X, Qin Y, Sohn JK (2010) Identification of main effects, epistatic effects and their environmental interactions of QTLs for yield traits in rice. Genes Genomics 32:37–45

4

High Resolution Mapping of QTLs for Heat Tolerance in Rice Using a 5K SNP Array

Shanmugavadivel PS[1,2], Amitha Mithra SV[1], Chandra Prakash[1], Ramkumar MK[1], Ratan Tiwari[3], Trilochan Mohapatra[4] and Nagendra Kumar Singh[1*]

Abstract

Background: Heat stress is one of the major abiotic threats to rice production, next to drought and salinity stress. Incidence of heat stress at reproductive phase of the crop results in abnormal pollination leading to floret sterility, low seed set and poor grain quality. Identification of QTLs and causal genes for heat stress tolerance at flowering will facilitate breeding for improved heat tolerance in rice. In the present study, we used 272 F_8 recombinant inbred lines derived from a cross between Nagina22, a well-known heat tolerant *Aus* cultivar and IR64, a heat sensitive popular *Indica* rice variety to map the QTLs for heat tolerance.

Results: To enable precise phenotyping for heat stress tolerance, we used a controlled phenotyping facility available at ICAR-Indian Institute of Wheat and Barley Research, Karnal, India. Based on 'days to 50% flowering' data of the RILs, we followed staggered sowing to synchronize flowering to impose heat stress at uniform stage. Using the Illumina infinium 5K SNP array for genotyping the parents and the RILs, and stress susceptibility and stress tolerance indices (SSI and STI) of percent spikelet sterility and yield per plant (g), we identified five QTLs on chromosomes 3, 5, 9 and 12. The identified QTLs explained phenotypic variation in the range of 6.27 to 21. 29%. Of these five QTLs, two high effect QTLs, one novel (*qSTIPSS9.1*) and one known (*qSTIY5.1/qSSIY5.2*), were mapped in less than 400 Kbp genomic regions, comprising of 65 and 54 genes, respectively.

Conclusions: The present study identified two major QTLs for heat tolerance in rice in narrow physical intervals, which can be employed for crop improvement by marker assisted selection (MAS) after development of suitable scorable markers for breeding of high yielding heat tolerant rice varieties. This is the first report of a major QTL for heat tolerance on chromosome 9 of rice. Further, a known QTL for heat tolerance on chromosome 5 was narrowed down from 23 Mb to 331 Kbp in this study.

Keywords: Rice, Heat tolerance, Nagina22, QTL mapping, SNP

Background

Rice is a major staple food crop for nearly half of the world population. The global population is projected to grow from seven to nine billion by 2050 and to reach ten billion before 2100 (United Nations 2011). To ensure food security to the added population, rice production has to increase by 0.6 to 0.9% annually until 2050 (Carriger and Vallee 2007). However, rise in global average temperature to the tune of 0.5 °C in the twentieth century and future projections in the range of 1.4–5.8 °C by the end of this century (IPCC 2007), will be detrimental to crop yield (Lobell et al. 2011). Declining farmland resources coupled with global warming have forced rice cultivation to marginal environments and beyond the normal rice season. This in turn has exposed the rice crop to higher day temperature (>33 °C) adversely impacting seed set (Nakagawa et al. 2002; Prasad et al. 2006; Jagadish et al. 2010a).

Heat stress alters the initiation and duration of developmental phases, especially the duration from floral/ panicle initiation to anthesis/panicle exertion in plants (Sato et al. 1973). Heat stress during flowering and anthesis can lead to failure in fertilization due to pollen or ovule sterility (Matsui et al. 1997). Early reproductive processes viz., micro- and megasporogenesis, pollen and

* Correspondence: nksingh4@gmail.com
[1]ICAR-National Research Centre on Plant Biotechnology, New Delhi, India
Full list of author information is available at the end of the article

stigma viability, anthesis, pollination, pollen tube growth, fertilization, and early embryo development are all highly susceptible to heat stress (Giorno et al. 2013). Flowering stage is crucial for crop productivity as heat stress during this phase causes reduced pollen fertility and low seed set in rice (Jagadish et al. 2010a). Anthesis processes, including anther dehiscence, pollination, and pollen germination are most sensitive to high temperature stress in rice. The main cause of spikelet sterility induced by high temperature is anther indehiscence (Matsui et al. 1999). High temperature inhibits swelling of pollen grains, which is a driving force for anther dehiscence in rice. Successful anther dehiscence depends on several parameters, including rupturing of septa, expansion of locule walls, pollen swelling, and rupturing of stomium (Liu et al. 2006).

Enhanced heat tolerance in rice is required at flowering stage to avoid spikelet sterility. Since 1985, germplasm screening for high temperature stress tolerance has been carried out by different research groups worldwide (Sarwar and Avesi 1985; Matsui and Omasa 2002; Jagadish et al. 2007, 2008). Heat tolerance at flowering stage in rice is attributed to multiple genes with cumulative effects on trait expression, otherwise called as quantitative trait loci (QTL; Cao et al. 2003; Xiao et al. 2011a; Ye et al. 2012). The discovery of genes/QTLs for enhanced tolerance to high temperature stress has practical implications in agriculture. Mapping of QTLs for heat tolerance in rice was first reported by Cao et al. (2003) based on percent spikelet fertility using doubled haploid population derived from IR64/Azucena cross. Thereafter many research groups have mapped QTLs for heat stress tolerance using F_2, back cross inbred lines (BIL) and recombinant inbred lines (RIL) populations, evaluated at the time of heading in controlled environment conditions (Chang-Lan et al. 2005; Chen et al. 2008; Zhang et al. 2008, 2009; Jagadish et al. 2010b; Xiao et al. 2011a; Ye et al. 2012, 2015; Cheng et al. 2012; and Poli et al. 2013). Some of these studies created high temperature condition for phenotyping by late planting in open field (Xiao et al. 2011b; Tazib et al. 2015; and Zhao et al. 2016). Almost all of these studies employed RFLP or SSR markers, except Ye et al. (2012, 2015), who used 300 SNP markers for the QTL mapping. IR64 has been used as one of the parental lines in generating mapping populations for mapping heat stress tolerance QTLs in some studies (Cao et al. 2003; Ye et al. 2012, 2015), while Nagina22 and its derived mutant lines have been used as parents in generating mapping population by other researchers (Buu et al. 2014; Poli et al. 2013). There is a report on using IR64/Nagina22 derived F_2 population for mapping heat tolerance QTLs (Ye et al. 2012).

Mapping of QTLs for heat stress tolerance using stress tolerance indices, which compare the performance of genotypes under control and stress condition, have not been reported earlier, but it has been utilized for mapping salt stress tolerance (Fernandez 1992; Pandit et al. 2010; Tiwari et al. 2016). The relative performance of genotypes under stress and control conditions can be used as an indicator to identify and map QTLs, which can be further used in breeding crop varieties for stress tolerance, rather than mapping QTLs based on phenotypic performance in stress environment alone (Raman et al. 2012). This has practical relevance since genotypes with low yield potential under control condition quite often show higher tolerance to stress than high yielding genotypes. Genomic regions governing salinity stress tolerance was successfully mapped in rice using stress indices (Pandit et al. 2010, Kumar et al. 2015; Tiwari et al. 2016). The present study focused on identification of QTLs for heat stress tolerance at flowering stage in a RIL mapping population derived from Nagina22/IR64 cross using controlled phenotyping facility for imposing heat stress, using stress tolerance indices for normalization of intrinsic differences in yield potential and high density SNP mapping. High density linkage map is expected to result in finding QTLs flanked by closely linked markers that can be readily used in breeding programmes for marker assisted selection.

Methods

Plant Materials

We used 272 $F_{7:8}$ RILs developed through single seed descent method from a cross between Nagina22 (N22), a heat tolerant cultivar (Mohapatra et al. 2014; Prakash et al. 2016) and IR64, a heat susceptible cultivar (Jagadish et al. 2010). To achieve synchronized flowering, the RILs were first phenotyped for days to 50% flowering and grouped into three categories as early, medium and late flowering types and then were sown in a staggered manner for synchronization of their flowering time. This exercise enabled us to impose heat treatment at a uniform stage in the population that in turn minimized the interference of phenological differences in analysis.

Heat Stress Treatment

The RILs along with the two parental lines were direct sown in the controlled temperature phenotyping facility at ICAR-Indian Institute of Wheat and Barley Research, Karnal, India in an augmented design for exposing them to high temperature at flowering stage. A plant-to-plant distance of 15 cm and row-to-row distance of 20 cm was maintained. The structure was kept open from sowing to till the flowering stage, where the experimental RILs were grown in a condition similar to that prevailing outside the green house (Fig. 1). Heat stress was imposed on plants during flowering time by closing the shutter. The temperature inside the structure was programmed to be 5 °C higher than the air temperature outside the structure (Additional file 1: Figure. S1). Relative humidity of 70% was maintained inside the facility. Heat stress was imposed continuously for 10 days including night time. After the treatment, the structure was kept open until harvest. The same RIL population was also raised outside the green house to phenotype their performance under control or ambient conditions.

Phenotyping of the F_8 RILs for Heat Stress Tolerance

Five individual plants from each RIL were sampled and harvested separately. Main panicle from each plant was used for analysing spikelet fertility by counting the number of filled and empty spikelets. The remaining panicles were collected separately from each plant and utilised for calculating yield potential of RILs under heat stress. In the same way, five plants from each RIL were harvested from the control experiment for evaluating the performance of genotypes under ambient conditions, used for computing their stress response index. The response of genotypes to heat stress was expressed as stress susceptibility index (SSI) given by Fisher and Mauer (1978) and stress tolerant index (STI) given by Fernandez (1992). SSI assesses the reduction in yield caused by stress as compared to favourable environment. The heat tolerant genotypes would have lower SSI value, which indicates lower difference in their yield across control and stress while it would be vice-versa for the susceptible genotypes. Thus, SSI helps to identify more tolerant lines. STI, on the other hand, helps to identify genotypes that produce higher yield in control as well as stress conditions, which is more desirable for practical reasons. The tolerant lines have higher STI value.

Fig. 1 Phenotyping of F_8 RIL mapping population under non-stress and heat stress conditions at ICAR-IIWBR. (**a**) At vegetative stage under non-stress condition (**b**). At reproductive stage under non-stress (**c**). At vegetative stage under non-stress condition inside controlled environment structure with roof top open (**d**). With reproductive stage heat stress for 10 days in controlled environment structure with roof top covered

Genotyping of RILs

Genomic DNA of all RILs, and the two parents was isolated from pooled young green leaves from plants grown in a row in the field by CTAB method (Doyle and Doyle 1990). DNA was quantified using NanoDrop 8000 spectrophotometer (Thermo Scientific, USA) and concentration of DNA was adjusted to a minimum of 50 ng/µl and approximately 200 ng of DNA from each genotypes were used for hybridization in Illumina Infinium® II genotyping assay. A customized array with 5246 SNPs in abiotic stress responsive genes of rice employing Illumina Infinium® II design probes and dual color channel assays (Infinium HD Assay Ultra, Illumina), was used for genotyping, following the manufacturer's protocol (Kumar et al. 2015).

SNP Genotype Calling

SNP genotyping data obtained from the array were analysed using Genome Studio V2010.1 (Illumina Inc.). SNPs were called using genotyping module integrated in the software where individual SNP is viewed as GenoPlots. Data quality was confirmed with internal controls and QC functions such as GenTrain and GenCall scores. After calling the data automatically, the SNPs were rescored and checked for their presence in a canonical cluster to get a GenTrain score > 0.7. The samples with call rates of <0.89 and the SNPs with norm R values <0.2 were removed from further analysis.The genotype calls from parental lines Nagina22 and IR64 were converted into AA and BB, respectively and similarly SNP calls of segregating genotypes were transformed in concordance with either parental type and used for the construction of framework linkage map.

QTL Mapping and Epistatic Interaction Network

QTL mapping was carried out using QTL IciMapping software v4.0 (Meng et al. 2015). Segregation pattern of each SNP in the RIL population was analysed using chi-square test with statistical significance at P value of 0.01. The redundant markers with identical scores were removed since they cannot provide any additional information. Markers with correlation coefficient of 1 were deleted by choosing missing proportion option. The genetic distance (cM) between SNP markers was converted to physical distance (kb) with 1 cM equal to 260 kbp (Chen et al. 2002; Tiwari et al. 2016). QTLs for heat stress tolerance were mapped using BIP functionality available in the QTL IciMapping software. Inclusive composite interval mapping of additive and dominant QTL (ICIM-ADD) mapping method was chosen along with the following parameters viz., window size of 1 cM, 500 permutations and type I error at 0.05 to call for QTL. The LOD threshold was set at 3.5 to accept the call as a significant QTL. The epistatic interaction network was determined for all the four traits studied using MQM algorithm in R/QTL (Browman et al. 2003).

In-silico Identification of Non-synonymous SNPs in Genes Present in the Mapped QTL Intervals

Non-synonymous SNPs between N22 and IR64 were identified using Rice SNP-Seek Database (Mansueto et al. 2017) with Nipponbare as reference genome and N22 and IR64 as query genomes. The gene locus id was given as input to retrieve the non-synonymous SNPs between N22 and IR64 for further analysis.

Results

Phenotypic Variation and Correlation Among Traits in the RIL Population

The parents and RILs were analysed for their phenotypic performance under ambient and heat stress environments. Heat sensitive parent IR64 showed very high spikelet sterility (95.94%) under heat stress as compared to the tolerant parent N22 (67.45%; Table 1). The yield reduction under heat stress was to the tune of 66% in N22 while it was 86% in IR64. Though the RIL population mean for percent spikelet sterility and yield was skewed towards IR64, it showed transgressive segregation and had high coefficient of variation (CV), more than 20% under both control and stress conditions (Table 1). Since SSI and STI are better indicators of plant performance under stress, we used these indices for mapping QTLs for heat tolerance. The RIL population exhibited transgressive segregation for all the four parameters analysed, namely SSI and STI of both percent spikelet sterility and yield per plant. The STI for percent spikelet sterility ranged from 0.38 to 14 while SSI for percent spikelet sterility ranged from −0.11 to 5.92. Similarly, the STI for yield ranged from 0.005 to 1.34 while SSI for yield ranged from 0.0078 to 1.37 (Table 2). High CV was observed for all the traits in a range of 0.32 (SSI for yield per plant) to 0.85 (STI of yield per plant), suggesting that all the four traits were suitable for QTL mapping. As expected, significant negative correlation was observed between STI and SSI for percent spikelet sterility and, STI for percent spikelet sterility and STI for yield (Fig. 2). The only positive correlation was between SSI for percent spikelet sterility and SSI for yield per plant, which is expected.

Table 1 Performance of the parents and their recombinant inbred lines under control and heat stress

	Percent spikelet sterility		Yield per plant (g)	
	Control	Heat stress	Control	Heat stress
Nagina22	6.38	67.45	6.41	2.18
IR64	14.53	95.94	9.48	1.33
RILs	2.17–78.01	15.40–100.00	1.89–24.12	0.064–12.65
Mean	16.56	81.8	9.47	2.65
Range	75.84	84.60	22.23	12.59
SD[a]	12.85	16.4	3.19	2
CV[b]	77.60	20.05	33.7	75.5

[a]Standard Deviation
[b]Coefficient of variation

SNP Marker Segregation and Framework Linkage Map

Out of the 5246 SNPs in the stress responsive genes genotyped using Illumina Infinium chip, 1512 were polymorphic between N22 and IR64 and segregated in the RIL population (28.82% polymorphism; Fig. 3). The highest number of polymorphic markers was present on chromosome 1 (203 SNPs) while the lowest was on chromosome 9 (66). In terms of proportion of polymorphic markers, chromosome 5 had the highest (0.35) proportion while chromosome 11 had the least (0.23). Thirty eight percent of polymorphic markers did not segregate as per the expected Mendelian segregation of 1:1 at cut-off probability of 0.01 and hence they were removed from further analysis. Another 117 markers that were redundant and played no role in improving the resolution of genetic map were also removed from further analysis. The highest number of redundant markers was found on chromosome 5 (24.76%) while the lowest number was on chromosome 8 (3.92%; Table 3). After removing the redundant markers, Chromosome 1 had the maximum number of markers (127) while chromosomes 4 and 7 had the lowest number of 31 markers. Finally, 824 markers were included in the framework linkage map used for QTL mapping.

QTLs for Heat Tolerance Traits

Using inclusive composite interval mapping (ICIM) approach, a total of five QTLs for stress tolerance and stress susceptibility index for yield and percent spikelet sterility were identified (Fig. 4). One QTL each for SSI and STI of percent spikelet sterility and two and one QTLs for SSI and STI for yield, respectively were mapped on four different chromosomes, namely chromosomes 3, 5, 9, and 12 with phenotypic variation explained ranging from 6.37 to 21.29% (Table 4). *qSSIY5.2* was the major QTL identified in this study for yield explaining 21.29% of the phenotypic variation while *qSTIPSS9.1* was the major QTL for percent spikelet sterility which explained 16.05% phenotypic variation. Except for *qSTIY5.1*, the heat susceptible parent IR64, contributed the heat sensitivity allele for both the

Table 2 Phenotypic variation for heat stress tolerance indices in Nagina22/IR 64 RIL mapping population

S. No.	Traits	RILs						N22	IR64
		Minimum	Maximum	Range	Mean	SD[a]	CV[b]		
1	SSI for % spikelet sterility	−0.10919	5.9246	6.03379	1.6576	1.2916	0.77919	2.422	1.418
2	STI for % spikeletsterility	0.38193	13.9979	13.6159	4.5501	3.1156	0.68473	1.576	5.106
3	STI for yield per plant	0.00469	1.33855	1.33386	0.2805	0.2395	0.85383	0.3314	0.066
4	SSI for yield per plant	0.00782	1.37249	1.36467	0.9764	0.3084	0.31585	0.9142	1.1918

[a]Standard Deviation
[b]Coefficient of variation

spikelet sterility and yield QTLs. *qSTIY5.1* and *qSSIY5.1* are one and the same as they are present in the same genomic region. As expected, the tolerance allele of *qSSIY5.1* locus was contributed by N22 while the sensitivity allele was from IR64. Further, the additive effect of the trait enhancing allele from N22 was twice as that of IR64 (Table 4). Analysis of physical positions of the identified QTLs revealed that *qSTIY5.1/qSSIY5.1* was in a small interval of 331 kbp on chromosome 5. The other major QTL identified, *qSTIPSS9.1,* was also located in a small interval of 394 kbp on chromosome 9. The largest interval of 1067.5 kbp was for the QTL *qSSIY3*.1, which explained 6.45% of phenotypic variations of SSI for yield (Table 4). *qSSIPSS12*.1, a minor QTL for percent spikelet sterility was mapped between markers SNP14876 and SNP14892, explaining 6.37% of the phenotypic variation. As expected, the sterility enhancing allele for this QTL was from susceptible parent IR64. Thus, for all the four QTLs identified in this study, the heat tolerance allele was from the heat tolerant variety N22.

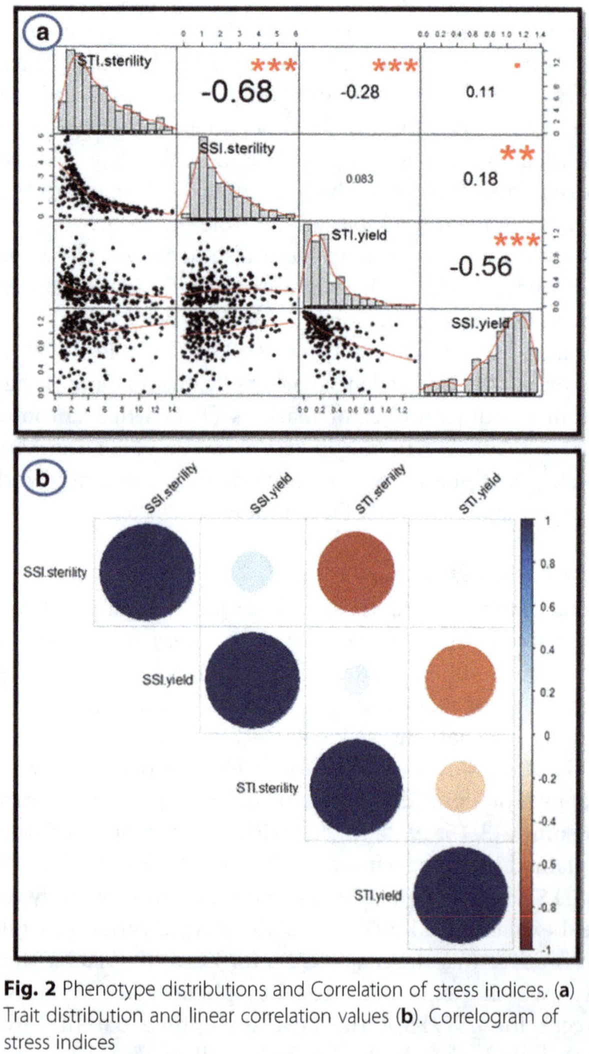

Fig. 2 Phenotype distributions and Correlation of stress indices. (**a**) Trait distribution and linear correlation values (**b**). Correlogram of stress indices

Candidate Genes for Heat Tolerance Located in the QTL Intervals

The genes located in the four genomic regions for the identified QTLs were extracted (Additional file 2: Table S1). Probable candidate genes for heat tolerance located in the major QTL interval *qSTIPSS9.1* for spikelet fertility were *PTC1*, glycosyltransferase, microtubule associated protein, annexin and *HSFs*. Similarly, candidate genes for heat tolerance index in the narrowed down QTL interval *qSTIY5.1/qSSIY5.2* included, trehalose synthase, trehalose-6-phosphate synthase, auxin response factor and calcineurin B-like protein-interacting protein kinase (CIPKs). In the QTL interval *qSSIPSS12.1,* 134 genes were identified including lipases, laccase, isoflavonereductase, cyclopropane-fatty-acyl-phospholipid synthase, *OGR1,* wall associated receptor kinase and pentatricopeptide protein coding genes. The *qSSIY3.1* QTL harboured 142 genes including, TFs, signalling genes and floral organ developmental genes, tesmin/TSO1, Crinkly4 receptor-like kinase, LIM domain containing protein gene and pectin methylesterase inhibitor coding gene. The majority of these genes are reported to be involved in pollen grain development, pollen tube growth and fertilization in rice.

Digenic Interactions of QTLs for Heat Tolerance

Four significant digenic interactions involving seven SNPs were identified for three traits (Fig. 5) namely, STI for sterility and SSI and STI for yield on chromosomes 2, 3, 4, 5, 9 and 12 (Table 5). Interestingly, STI for

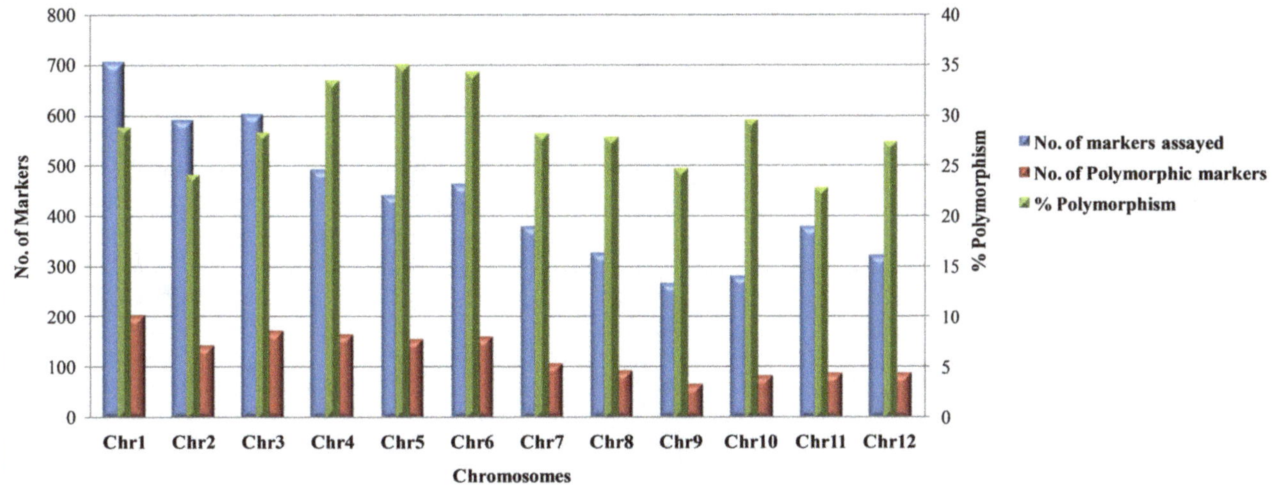

Fig. 3 Chromosome-wise polymorphism survey using 5K SNP array and polymorphic SNPs in Nagina22 x IR64 mapping population

yield had two digenic interactions involving a common SNP (7118) from a locus encoding dehydrogenase (LOC_Os04g52280). One of these interactions involved an SNP (8401) from the major and common QTL identified in the study, *qSTIY5.1/qSSIY5.2*. SNP8401 is present in a gene encoding WD40 domain, G-beta repeat domain containing protein (LOC_Os05g44320). All other SNPs showing epistasis were in uncharacterized expressed protein coding genes (Table 5).

Non-synonymous SNPs in the mapped QTL intervals

The maximum number of SNPs (71 SNPs in 26 genes) were observed in *qSSIPSS12.*1 genomic interval while the lowest number of SNPs (21 SNPs in 14 genes) were in *qSTIY5*.1 /*qSSIY5*.1 region (Table 6). Ty3-gypsy subclass retrotransposon protein encoding gene (LOC_Os12g17290) had the highest number of SNPs (8). Approximately, 77% of the observed substitutions were base transitions. SNPs were present in expressed genes, transcription factor coding genes, transposon related genes and protein and enzyme coding genes. *qSTIPSS9*.1 genomic region had 28 SNPs including 23 transitions and 5 transversions in 16 different genes. *qSSIPSS12*.1 interval had 71 SNPs in 26 genes including laccase precursor protein coding gene, expressed genes, Cyclopropane-fatty-acyl-phospholipid synthase coding genes and transposon related genes (Additional file 2: Table S2).

Discussion

QTLs for heat tolerance have been mapped on different chromosomes of rice by different research groups during the last decade (Cao et al. 2003; Chen et al. 2008; Zhang et al. 2008, 2009; Jagadish et al. 2010; Xiao et al. 2011; Ye et al. 2012, 2015) (Additional file 2: Table S3). In the current study, using a reasonably large RIL population, high density SNP map and phenotyping under controlled facility, we identified four heat tolerant QTLs in rice, of which three were novel namely *qSTIPSS9.1*, *qSSIPSS12.1* and *qSSIY3.1*. Among these, *qSTIPSS9.1* was the major QTL for percent spikelet sterility. Further, we also identified a known major effect QTL, *qSSIY5*.1/*qSTIY5*.1 for both the indices of yield. Zhang et al. (2008) have reported this QTL in a RIL mapping population derived from a cross Zhongyouzao8 x Toyonishiki between SSR markers, RM405 and RM274 flanking a 23 Mb interval. In their study, this QTL explained 10.7% phenotypic variation for spikelet fertility under heat stress while it was for SSI/STI for yield in our study. Further, the QTL interval was narrowed down to a 331

Table 3 Selection of polymorphic markers for QTL mapping

	Chr1	Chr2	Chr3	Chr4	Chr5	Chr6	Chr7	Chr8	Chr9	Chr10	Chr11	Chr12	Total
No. of Markers after Chi square test at 0.01 significant level	152	120	94	33	105	108	34	51	51	63	62	68	941
% Segregation distortion	25.12	15.49	44.7	79.87	31.81	31.64	68.22	43.95	22.72	24.09	27.9	22.72	37.76
No. of markers per chromosome after removing redundant markers	127	107	83	31	79	98	31	49	43	58	58	60	824
% Redundant markers	16.44	10.83	11.7	6.06	24.76	9.26	8.82	3.92	15.68	7.93	6.45	11.76	12.43
No. of markers used for QTL analysis	127	107	83	31	79	98	31	49	43	58	58	60	824

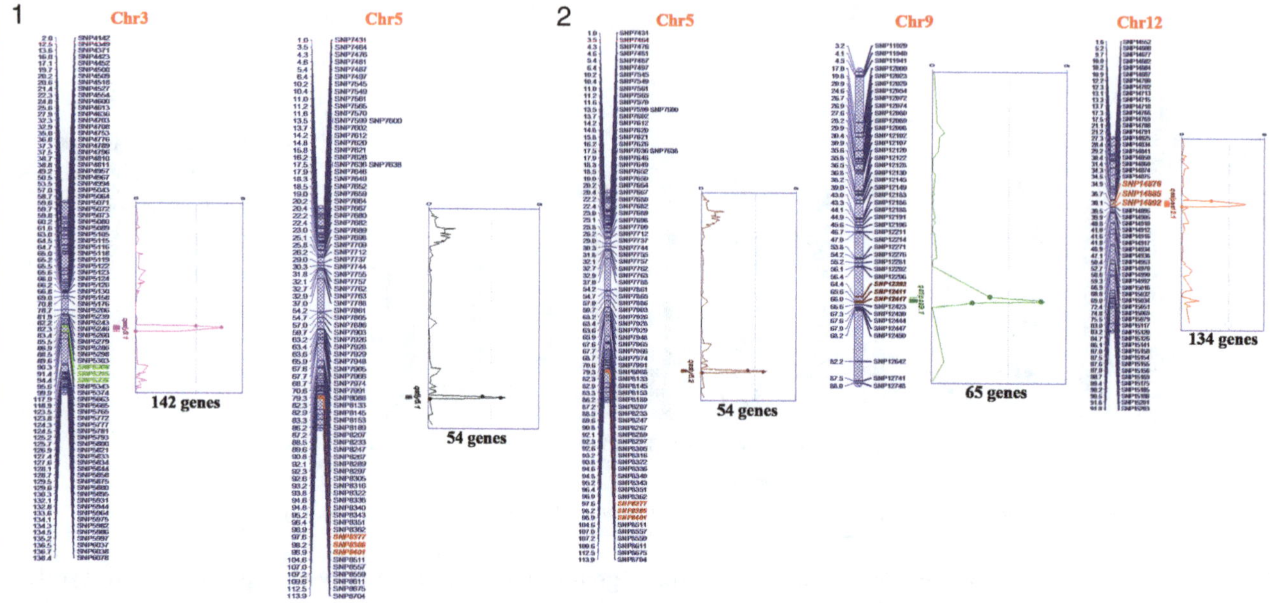

Fig.4 QTLs identified for heat tolerance in rice in the mapping population derived from Nagina22 and IR64

kbp region comprising of 54 genes in our study. This was because earlier studies have used either SSR markers (maximum 264) or less than 300 SNP markers for mapping QTLs for heat tolerance whereas we have used more than 800 SNPs and 272 RILs to achieve a much higher resolution (Buu et al. 2014; Ye et al. 2012, 2015). Our ability to identify QTLs in such narrow intervals could be attributed to the use of 5K SNP array comprising of SNPs from abiotic stress responsive genes (Kumar et al. 2015). Some important candidate genes located in the high effect and minor QTLs identified in the present study are discussed below for their probable role in enhanced spikelet fertility and yield under heat stress.

There were 65 genes in the QTL region, *qSTIPSS9.1*, including transporters, transcription factors such as *HSF* (*OsHsfB4c*), *PHD*-finger domain containing TF (*PTC1*), *bHLH*, and *C2H2* zinc finger, transcriptional regulators, glycosyltransferase microtubule associated protein, and annexin (Additional file 2: Table S1). Tapetum, the innermost cell layer of the anther wall, plays a crucial role in anther development, microspore/pollen formation, and pollen wall formation. During late pollen development, tapetal degeneration triggered by an apoptosis-like process is essential for viable pollen formation (Li et al. 2006). *PERSISTENT TAPETAL CELL1* (*PTC1*) present in the *qSTIPSS9*.1 QTL region encodes a PHD-finger protein that controls programmed tapetal development and degradation to ensure functional pollen formation in rice (Li et al. 2011a, b). *PTC1* is expressed specifically in tapetal cells and microspores during anther development in stages 8 and 9 and initiates a typical apoptosis-like cell death, thereby ensuring proper pollen grain development (Li et al. 2011). Loss of function of *PTC1* displayed uncontrolled tapetal cell proliferation and swelling, delayed DNA fragmentation, and pollen wall development, causing complete male sterility (Li et al. 2011). Timely initiation of tapetal programmed cell death is essential for the regulated release of wall materials from the tapetum to the developing microspore including carbohydrate, lipid molecules, and other nutrients. This gene might be responsible for maintaining higher fertility in N22 under heat stress by timely initiation of PCD in N22 tapetal cell to ensure more fertile pollen grains than in the susceptible parent IR64. This gene otherwise named as a *tms9–1/OsMS1* is

Table 4 QTLs for heat stress indices identified in RIL mapping population derived from N22 and IR64

Trait Name	QTL name	Chr.	Left Marker	Right Marker	Physical position (Mb)	Interval (Kb)	LOD	PVE (%)	Add
STI for % spikelet sterility	*qSTIPSS9.1*	9	SNP12393	SNP12417	16.75–17.14	393.828	4.21	16.05	-1.25
SSI_ for % spikelet sterility	*qSSIPSS12.1*	12	SNP14876	SNP14892	9.06–9.90	840.288	3.88	6.37	−0.33
STI for yield per plant	*qSTIY5.1*	5	SNP8377	SNP8401	25.45–25.79	331.586	4.55	9.01	0.07
SSI for yield per plant	*qSSIY3.1*	3	SNP5308	SNP5336	23.52–24.59	1067.507	4.04	6.45	−0.08
SSI for yield per plant	*qSSIY5.1*	5	SNP8377	SNP8401	25.45–25.79	331.586	3.51	21.29	−0.14

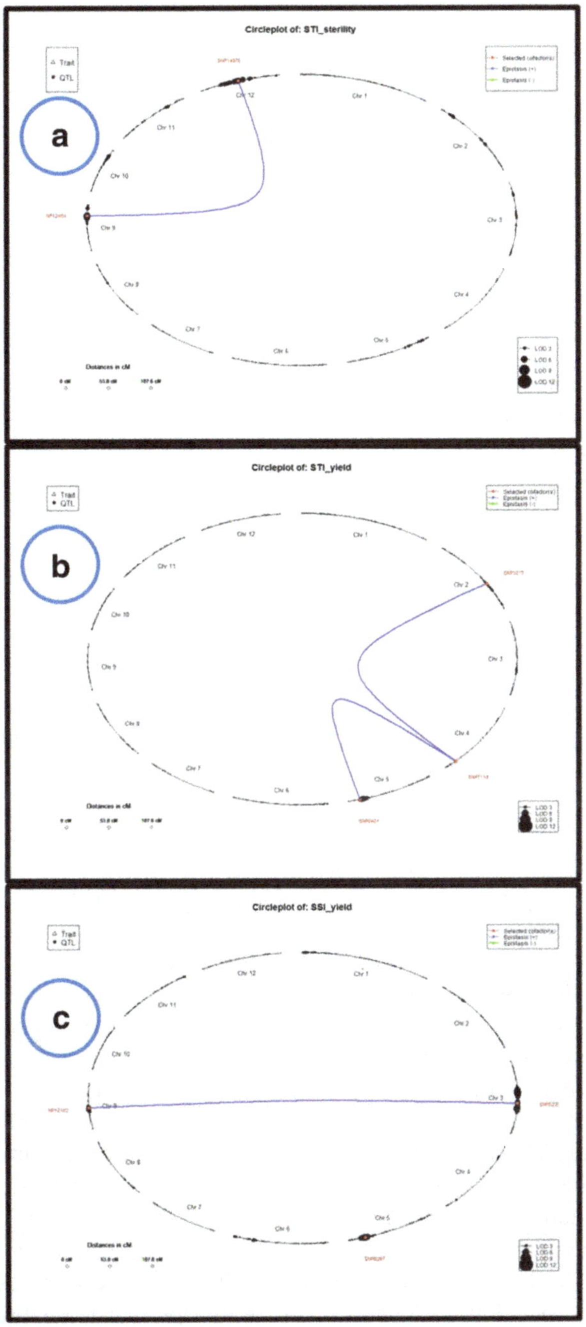

Fig. 5 Epistatic interaction network between SNP marker loci. **a** STI for sterility (**b**). STI for yield (**c**). SSI for yield

responsible for thermo-sensitive genic male sterility in *HengnongS*-1, one of the oldest and often-used TGMS line in *indica* two-line hybrid rice breeding programs in China (Qi et al. 2014). Also, *Arabidopsis thaliana MALE STERILITY1* (*MS1*) gene encodes for a protein homologous to the PHD-finger class of transcription factor and has been demonstrated to be involved in tapetal development and pollen wall biosynthesis (Yang et al. 2007).

Glycosyltransferase attaches a single or multiple sugars to different bio-molecules and highly expresses in mature pollen grains and is involved in mature pollen grain formation in rice (Moon et al. 2013). *GLYCOSYLTRANSFERASE1* (*OsGT1*) of rice present in *qSTIPSS9.1* is involved in pollen wall formation, especially, the exine and intine construction and pollen maturation. The *osgt1* mutant failed to produce mature pollen grains since its pollen had disrupted intine structure owing to low levels of starch and protein (Moon et al. 2013). Similarly, *uneven pattern of exine* 1 (*upex1*) gene of *Arabidopsis* encodes GT31 family glycosyltransferase in Arabidopsis and might be involved in galactosylation of arabinogalactan proteins (AGPs). The mutant of *UPEX1* exhibit defective and irregular exine pattern and suggests that primexine localized AGPs could play a role in sporopollenin adhesion and patterning in early microspore wall development (Li et al. 2016). HSFs are main players in imparting heat stress response by activating transcription of downstream genes including HSPs (Guo et al. 2008). *AtHsfB4* has a role in root development in Arabidopsis and involved in early stage of heat shock (Lohmann et al. 2004; Begum et al. 2013). A similar *OsHsf* has been identified in this major QTL on chromosome 9, which is yet to be characterized in rice.

Microtubule-associated proteins play a crucial role in the regulation of microtubule dynamics, and important for plant cell and organ development (Liu et al. 2013). The 65-kD microtubule-associated protein (MAP65) family member in Arabidopsis (*AtMAP65*–1) is ubiquitously expressed during the cell cycle and in all plant organs and tissues with the exception of anthers and petals (Smertenko et al. 2004). However, Microtubule-associated protein *MAP65*–1a (LOC_Os09g27700) of rice is expressed in anther and pistil. This might be indicative of its role in reproductive organ development in rice and hence is a good candidate for further studies in rice. Annexin functions to counteract oxidative stress, maintain cell redox homeostasis, and enhance drought tolerance (Szalonek 2015). Down-regulation of *Arabidopsis* annexin5 (*Ann5*) in transgenic *Ann5*-RNAi lines caused sterile pollen grains. Ann5 is involved in pollen grain development, germination and pollen tube growth through the promotion of endo-membrane trafficking modulated by calcium (Zhu et al. 2014). *TaAnn10* in wheat is highly expressed in floral bracts, pistil, anthers and immature endosperm and it correlates with anther development. But it fails to be induced by low temperature in thermosensitive genic male sterile lines, suggesting that specific down-regulation of *TaAnn10* is associated with cold induced male sterility in wheat (Xu et al. 2016). The relative expression levels of *TaAnn10* in the stamen strongly correlated with male fertility in recovery lines (Xu et al. 2016). One such

Table 5 Epistatic interaction network

S. No.	Traits	Interaction	SNPs	Chr.	Locus	Physical position (bp)	Annotation
1	STI sterility	SNP12453-SNP14970	SNP12453	9	LOC_Os09g29160.1	17,733,252–17,731,555	expressed protein
			SNP14970	12	LOC_Os12g24090.1	13,706,157–13,705,072	expressed protein
2	STI yield	SNP8401-SNP7118	SNP8401	5	LOC_Os05g44320.1	25,793,638–25,788,753	WD domain, G-beta repeat domain containing protein, expressed
			SNP7118	4	LOC_Os04g52280.1	31,069,545–31,075,601	dehydrogenase, putative, expressed
		SNP7118-SNP3275	SNP3275	2	LOC_Os02g37380.1	22,580,613–22,579,160	expressed protein
3	SSI yield	SNP5239-SNP12183	SNP5239	3	LOC_Os03g38450.1	21,344,825–21,346,870	expressed protein
			SNP12183	9	LOC_Os09g18230.1	11,184,420–11,171,639	expressed protein

annexin 10 (*OsANN10*/ LOC_Os09g27990) is present in the QTL interval *qSTIPSS9.1*.

A common response of organisms to drought, salinity, and temperature stresses is the accumulation of sugars and compatible solutes including trehalose. The increased trehalose accumulation correlates with elevated capacity for photosynthesis under both stress and non-stress conditions in rice (Garg et al. 2002). Trehalose-6-phosphate synthase (TPS) plays an important role in trehalose metabolism and signalling. Overexpression of the trehalose-6-phosphate synthase gene *OsTPS1* enhances the tolerance of rice seedling to cold, high salinity and drought stress without other significant phenotypic changes (Li et al. 2011). Similarly, the over-expression of trehalose-6-phosphate phosphatase in maize ears increases both kernel set and harvest index in drought stress condition. Increase in yield to the tune of 9% to 49% under non-drought or mild-drought conditions, and 31% -123% under more severe drought conditions, relative to yields from non-transgenic controls was observed (Nuccio et al. 2015). Similarly, trehalose concentration increased upon 4 h of heat stress at 40 °C and 4 days after cold stress at 4 °C in *Arabidopsis thaliana* (Kaplan et al. 2004). Over-expression of *ScTPS1* and *ScTPS2* under stress associated *rd29A* promoter provided protection against drought, salt, freezing, and heat stress (Miranda et al. 2007). Hence, trehalose synthase and trehalose phosphate synthase are a probable candidate genes underlying QTL *qSSIY5.1*/*qSTIY5.1* for yield under heat stress.

Our analysis for non-synonymous SNPs between the parents in the candidate genes like *PTC1*, *tms9–1*/*OsMS1*, *OsGT1*,*MAP65*–1a (LOC_Os09g27700), *OsANN10*, trehalose synthase, trehalose phosphate synthase, *OsCR4*, pectin methyl esterase and tesmin could not find allelic variants. This could be due to low coverage of the sequence data in either or both of the parents or lack of variation present in coding regions in the above genes. Further, the InDel polymorphism for these genes is not known. Alternatively, there could be variations in promoter, intron-exon junctions and UTR regions, which are not yet known. Hence, future effort is required to deep sequence these regions in the parents to identify polymorphisms, if any, in candidate genes. Nevertheless, the listed SNPs between N22 and IR64 can be utilised for fine mapping and functional validation of the QTLs.

The interaction network analysis showed evidence for the involvement of the major QTL, *qSSIY5.1*/*qSTIY5*.1, in digenic interactions, strengthening the role of this region in imparting heat tolerance. Though most of the SNPs involved in epistasis were in genes encoding uncharacterized expressed proteins, two SNPs were from known proteins coding genes namely, WD domain, G-beta repeat domain containing gene (LOC_Os05g44320) and dehydrogenase gene (LOC_Os04g52280). The WD40 protein is reported to play a role in diverse protein-protein interactions or protein-DNA interactions by acting as scaffolding molecule and promoting protein activity and thus functioning as a positive regulator of plant responses to various abiotic stresses such as salinity, osmotic and dehydration stress in plants (Mishra et al. 2012; Kong et al. 2015). Further, it is involved in various biological process, viz., signal transduction, gene transcriptional regulation, protein modifications, cytoskeleton assembly, vesicular trafficking, DNA damage and repair, cell death and cell cycle progression (Zhang and Zhang 2015).

Table 6 Non-synonymous SNPs between N22 and IR64 in the mapped QTL region for heat stress tolerance

S. No.	QTLs	No. of genes with SNPs	No. of SNPs	Non-synonymous SNPs between N22 and IR64	
				Transition	Transversion
1	*qstipss9.1*	16	28	23	5
2	*qssipss12.1*	26	71	58	13
3	*qstiy5.1 & qssiy5.2*	14	21	16	5
4	*qssiy3.1*	26	46	30	16

Conclusions

Present study using Nagina22/IR64 RIL mapping population and a 5K SNP genotyping chip, identified a major novel QTL *qSTIPSS9*.1 for reproductive stage heat tolerance in a 394 kbp region of rice chromosome 9. The study also confirmed the presence of a known major QTL for heat tolerance on chromosome 5 (*qSTIY5.1/qSSIY5.1*), which was narrowed down from 23 Mb in the original study to a much smaller interval of 331 kbp. This QTL was also involved in digenic interaction. Though the polymorphism survey in the candidate genes using the available data did not produce any trait linked variation, the SNPs identified could be useful in fine mapping. Further sequencing and functional validation is required for the identification of actual genes in these QTL regions responsible for the heat tolerance. Nonetheless, the two major QTLs identified here can be employed directly for crop improvement by marker assisted selection (MAS) after development of suitable scorable markers for breeding of high yielding heat tolerant rice varieties.

Acknowledgements
The first author is grateful to the Department of Science and Technology, New Delhi, India for providing financial assistance in the form of INSPIRE-fellowship. The research work was funded by National Agricultural Innovation Program (NAIP) under the project "Bioprospecting of genes and allele mining for abiotic stress tolerance".

Authors' contributions
The research material developed by TM was advanced and maintained by PSS, CP and SVAM. Phenotyping was carried out by PSS and MKR. Genotyping was carried out by SVAM. Marker analysis was carried out by PSS and CP. QTL mapping was carried out by PSS and SVAM. *In-silico* search for putative genes in QTL intervals was carried out by PSS. Epistatic interaction was found by CP. PSS compiled and interpreted the results. RT provided and maintained the phenotypic facility. PSS and SVAM drafted the manuscript. NKS and TM supervised the work. NKS edited and approved the manuscript. All the authors read and approved the manuscript.

Competing interests
The authors declare that they have no competing interest.

Author details
[1]ICAR-National Research Centre on Plant Biotechnology, New Delhi, India . [2]Present address, Division of Plant Biotechnology, ICAR-Indian Institute of Pulses Research, Kanpur 208 024, India. [3]ICAR-Indian Institute of Wheat and Barley Research, Karnal 132 001, India. [4]Indian Council of Agricultural Research, Krishi Bhavan, New Delhi 110 001, India.

References

Begum T, Reuter R, Schöffl F (2013) Overexpression of *AtHsfB4* induces specific effects on root development of Arabidopsis. Mech Dev 130(1):54–60. doi:10.1016/j.mod.2012.05.008
Broman KW, Wu H, Sen Ś, Churchill GA (2003) R/qtl: QTL mapping in experimental crosses. Bioinformatics 19(7):889–890
Buu BC, Ha PTT, Tam BP, Nhien TT, Hieu NV, Phuoc NT, Minh LT, Giang LH, Lang NT (2014) Quantitative trait loci associated with heat tolerance in rice (*Oryza sativa L.*) Plant Breed Biotech 2:14–24
Cao L, Zhao J, Zhan X, Li D, He L, Cheng S (2003) Mapping QTLs for heat tolerance and correlation between heat tolerance and photosynthetic rate in rice. Chin J Rice Sci 17:223–227
Carriger S, Vallee D (2007) More crop per drop. Rice Today 6:10–13
Chang-Lan Z, Ying-Hui X, Chun-Ming W, Ling J, Hu-Qu Z, Jian-Min W (2005) Mapping QTL for heat-tolerance at grain filling stage in rice. Rice Sci 12(1):33–38
Chen M, Presting G, Barbazuk WB, Goicoechea JL, Blackmon B, Fang G, Kim H, Frisch D, Yu Y, Sun S, Higingbottom S, Phimphilai J, Phimphilai D, Thurmond S, Gaudette B, Li P, Liu J, Hatfield J, Main D, Farrar K, Henderson C, Barnett L, Costa R, Williams B, Walser S, Atkins M, Hall C, Budiman MA, Tomkins JP, Luo M, Bancroft I, Salse J, Regad F, Mohapatra T, Singh NK, Tyagi AK, Soderlund C, Dean RA, Wing RA (2002) An integrated physical and genetic map of the rice genome. Plant Cell 14(3):537–545
Chen Q, Yu S, Li C, Mou T (2008) Identification of QTLs for heat tolerance at flowering stage in rice. Sci Agric Sin 41:315–321
Cheng L, Wang J, Uzokwe V, Meng L, Wang Y, Sun Y, Zhu L, Xu J, Li Z (2012) Genetic analysis of cold tolerance at seedling stage and heat tolerance at anthesis in rice (*Oryza sativa* L.) J Integ Agriculture 11:359–367
Doyle JJ, Doyle JL (1990) Isolation of plant DNA from fresh tissue. Focus 12:13–15
Fernandez GCJ (1992) Effective selection criteria for assessing plant stress tolerance. In: Kus EG (ed) Adaptation of Food Crop Temperature and Water Stress. Proceeding of 4th International Symposium, Asian Vegetable and Research and Development Center, Shantana, Taiwan, pp 257–270
Fischer RA, Maurer R (1978) Drought resistance in spring wheat cultivars.I. Grain yield responses. Aust J Agric Res 29:892–912
Garg AK, Kim JK, Owens TG, Ranwala AP, Choi YD, Kochian LV, Wu RJ (2002) Trehalose accumulation in rice plants confers high tolerance levels to different abiotic stresses. PNAS 99(10):15898–15903. doi:10.1073/pnas.252637799
Giorno F, Wolters-Arts M, Mariani C, Ivo R (2013) Ensuring reproduction at high temperatures: the heat stress response during anther and pollen development. Plants 2:489–506. doi:10.3390/plants2030489
Guo J, Wu J, Ji Q, Wang C, Luo L, Yuan Y, Wang Y, Wang J (2008) Genome-wide analysis of heat shock transcription factor families in rice and Arabidopsis. J Genet Genomics 35(2):105–118. doi:10.1016/S1673-8527(08)60016-8
IPCC (2007) Summary for Policymakers. In: Solomon S, Qin D, Manning M, Chen Z, Marquis M, Averyt KB, Tignor M, Miller HL (eds) Climate change 2007: The Physical Science Basis. Contribution of working group I to the fourth assessment report of the intergovernmental panel on climate change. Cambridge University Press, Cambridge, pp 1–18
Jagadish SVK, Cairns J, Lafitte R, Wheeler TR, Price AH, Craufurd PQ (2010b) Genetic analysis of heat tolerance at anthesis in rice. Crop Sci 50:1633–1641
Jagadish SVK, Craufurd PQ, Wheeler TR (2007) High temperature stress and spikelet fertility in rice (*Oryza sativa L.*) J Exp Bot 58(7):1627–1635
Jagadish SVK, Muthurajan R, Oane R, Wheeler TR, Heuer S, Bennett J, Craufurd PQ (2010a) Physiological and proteomic approaches to address heat tolerance during anthesis in rice (*Oryza sativa* L.) J Exp Bot 61:143–156. doi:10.1093/jxb/erp289
Jagadish SVK, Craufurd PQ, Wheeler TR (2008) Phenotyping parents of mapping populations of rice (*Oryza sativa* L.) for heat tolerance during anthesis. Crop Sci 48:1140–1146
Kaplan F, Kopka J, Haskell DW, Zhao W, Schiller KC, Gatzke N, Sung DY, Guy CL (2004) Exploring the temperature-stress metabolome of Arabidopsis. Plant Physiol 136(4):4159–4168
Kong D, Li M, Dong Z, Ji H, Li X (2015) Identification of TaWD40D, a wheat WD40 repeat-containing protein that is associated with plant tolerance to abiotic stresses. Plant Cell Rep 3:395–410
Kumar V, Singh A, Mithra SVA, Krishnamurthy SL, Parida SK, Jain S, Tiwari KK, Kumar P, Rao AR, Sharma SK, Khurana JP, Singh NK, Mohapatra T (2015) Genome-wide association mapping of salinity tolerance in rice (*Oryza sativa* L). DNA Res 22(2):133–145. doi:10.1093/dnares/dsu046
Li H, Yuan Z, Vizcay-Barrena G, Yang C, Liang W, Zong J, Wilson ZA, Zhang D (2011a) PERSISTENT TAPETAL CELL1 encodes a PHD-finger protein that is required for tapetal cell death and pollen development in rice. Plant Physiol 156(2):615–630. doi:10.1104/pp.111.175760
Li HW, Zang BS, Deng XW, Wang XP (2011b) Overexpression of the trehalose-6-phosphate synthase gene *OsTPS1* enhances abiotic stress tolerance in rice. Planta 234(5):1007–1018. doi:10.1007/s00425-011-1458-0
Li WL, Liu Y, Douglas CJ (2016) Role of glycosyltransferases in pollen wall primexine formation and exine patterning. Plant Physiol. 173;(1):167–82.
Li N, Zhang DS, Liu HS, Yin CS, Li XX, Liang WQ, Yuan Z, Xu B, Chu HW, Wang J, Wen TQ, Huang H, Luo D, Ma H, Zhang DB (2006) The rice tapetum

degeneration retardation gene is required for tapetum degradation and anther development. Plant Cell 18:2999–3014

Liu C, Qi X, Zhao Q, Yu J (2013) Characterization and functional analysis of the potato pollen-specific Microtubule-Associated Protein SBgLR in Tobacco. PLoS One 8(3):e60543. doi:10.1371/journal.pone.0060543

Liu JX, Liao DQ, Oane R, Estenor L, Yang XE, Li ZC, Bennett J (2006) Genetic variation in the sensitivity of anther dehiscence to drought stress in rice. Field Crop Res 97:87–100

Lobell DB, Schlenker W, Costa-Roberts J (2011) Climate trends and global crop production since 1980. Science 333:616–620

Lohmann C, Eggers-Schumacher G, Wunderlich M, Schöffl F (2004) Two different heat shock transcription factors regulate immediate early expression of stress genes in Arabidopsis. Mol Gen Genomics 271(1):11–21

Mansueto L, Fuentes RR, Borja FN, Detras J, Abriol-santos M, ChebotarovD, … Alexandrov N (2017) Rice SNP-seek database update: new SNPs, InDels, and queries. Nucleic Acids Res 45:D1075-D1081.

Matsui T, Omasa K, Horie T (1997) High temperature induced spikelet sterility of japonica rice at flowering in relation to air humidity and wind velocity conditions. Japan Journal of Crop Science 66:449–455

Matsui T, Omasa K, Horie T (1999) Rapid swelling of pollen grains in response to floret opening unfolds locule in rice. Plant Production Science 2(3):196–199

Matsui T, Omasa K (2002) Rice (*Oryza sativa* L.) cultivars tolerant to high temperature at flowering: anther characteristics. Ann Bot 89(6):683–637

Meng L, Li H, Zhang L, Wang J (2015) QTL IciMapping: integrated software for genetic linkage map construction and quantitative trait locus mapping in bi-parental populations. Crop J 3:265–279. doi:10.1016/j.cj.2015.01.001

Miranda JA, Avonce N, Suárez R, Thevelein JM, Van Dijck P, Iturriaga G (2007) A bifunctional TPS-TPP enzyme from yeast confers tolerance to multiple and extreme abiotic-stress conditions in transgenic Arabidopsis. Planta 226(6):1411–1421

Mishra AK, Puranik S, Prasad M (2012) Structure and regulatory networks of WD40 protein in plants. J Plant Biochem Biotechnol 21:32–39

Mohapatra T, Robin S, Sarla N, Sheshashayee M, Singh AK, Singh K, Singh NK, Amitha Mithra SV, Sharma RP (2014) EMS induced mutants of upland rice variety Nagina22: generation and characterization. Proc Indian National Science Academy 80:163–172

Moon S, Kim SR, Zhao G, Yi J, Yoo Y, Jin P, Lee SW, Jung KH, Zhang D, An G (2013) Rice *GLYCOSYLTRANSFERASE1* encodes a glycosyltransferase essential for pollen wall formation. Plant Physiol 161(2):663–675. doi:10.1104/pp.112.210948

Nakagawa H, Horie T, Matsui T (2002) Effects of climate change on rice production and adaptive technologies. In: Mew TW, Brar DS, Peng S, Dawe D, Hardy B (eds) Rice science: innovations and impact for livelihood China: International Rice Research Institute, pp 635–657

Nuccio ML, Wu J, Mowers R, Zhou HP, Meghji M, Primavesi LF, Paul MJ, Chen X, Gao Y, Haque E, Basu SS, Lagrimini LM (2015) Expression of trehalose-6-phosphate phosphatase in maize ears improves yield in well-watered and drought conditions. Nat Biotech 33:862–869. doi:10.1038/nbt.3277

Pandit A, Rai V, Bal S, Sinha S, Kumar V, Chauhan M, Gautam RK, Singh R, Sharma PC, Singh AK, Gaikwad K, Sharma TR, Mohapatra T, Singh NK (2010) Combining QTL mapping and transcriptome profiling of bulked RILs for identification of functional polymorphism for salt tolerance genes in rice (*Oryza sativa* L.) Mol Gen Genomics 284(2):121–136. doi:10.1007/s00438-010-0551-6

Poli Y, Basava RK, Panigrahy M, Vinukonda VP, Dokula NR, Voleti SR, Desiraju S, Neelamraju S (2013) Characterization of a Nagina22 rice mutant for heat tolerance and mapping for yield traits. Rice 6:36

Prakash C, Mithra SVA, Singh PK, Mohapatra T, Singh NK (2016) Unravelling the molecular basis of oxidative stress management in a drought tolerant rice genotype Nagina22. BMC Genomics 17:774. doi:10.1186/s12864-016-3131-2

Prasad PVV, Boote KJ, Jr LHA, Sheehy JE, Thomas JMG (2006) Species, ecotype and cultivar differences in spikelet fertility and harvest index of rice in response to high temperature stress. Field Crop Res 95:398–411

Qi Y, Liu Q, Zhang L, Mao B, Yan D, Jin Q, He Z (2014) Fine mapping and candidate gene analysis of the novel thermo-sensitive genic male sterility tms9-1 gene in rice. Theor Appl Genet 127(5):1173–1182. doi:10.1007/s00122-014-2289-8

Raman A, Verulkar SB, Mandal N, Variar M, Shukla VD, Dwivedi JL, Singh BN, Singh ON, Swain P, Mall AK, Robin S, Chandrababu R, Jain A, Ram T, Hittalmani S, Haefele S, Piepho HP, Kumar A (2012) Drought yield index to select high yielding rice lines under different drought stress severities. Rice 5: 31. doi:10.1186/1939-8433-5-31

Sarwar M, Avesi GM (1985) Evaluation of rice germplasm for high temperature tolerance. Pakistan J Agric Res 6(3):162–164

Sato K, Inaba K, Tosawa M (1973) High temperature injury of ripening in rice plant. The effects of high temperature treatment at different stages of panicle development on the ripening. Proceedings of Crop Science Society of Japan 42:207–213

Smertenko AP, Chang HY, Wagner V, Kaloriti D, Fenyk S, Sonobe S, Lloyd C, Hauser MT, Hussey PJ (2004) The *Arabidopsis* Microtubule-Associated Protein *AtMAP65*-1: molecular analysis of its microtubule bundling activity. Plant Cell 16(8):2035–2047

Szalonek M, Sierpien B, Rymaszewski W, Gieczewska K, Garstka M, Lichocka M, Sass L, Paul K, Vass I, Vankova R, Dobrev P, SzczesnyP MW, Krusiewicz D, Strzelczyk-Zyta D, Hennig J, Konopka-Postupolska D (2015) Potato Annexin *STANN1* promotes drought tolerance and mitigates light stress in transgenic *Solanum tuberosum* L. Plants. PLoS One 10(7):e0132683. doi:10.1371/journal.pone.0132683

Tazib T, Kobayashi Y, Koyama H, Matsui T (2015) QTL analyses for anther length and dehiscence at flowering as traits for the tolerance of extreme temperatures in rice (*Oryza sativa* L.) Euphytica 203:629–642

Tiwari S, Krishnamurthy SL, Kumar V, Singh B, Rao A, Mithra SVA, Rai V, Singh AK, Singh NK (2016) Mapping QTLs for salt tolerance in rice (*Oryza sativa* L.) by Bulked Segregant Analysis of Recombinant Inbred Lines using 50K SNP Chip. PLoS One 11(4):e0153610. doi:10.1371/journal.pone.0153610

United Nations, Department of Economic and Social Affairs, Population Division (2011) World Population Prospects: The 2010 Revision, Volume I: Comprehensive Tables. ST/ESA/SER.A/313

Xiao Y, Pan Y, Luo L, Zhang G, Deng H, Dai L, Liu X, Tang W, Chen L, Wang GL (2011a) Quantitative trait loci associated with seed set under high temperature stress at the flowering stage in rice (*Oryza sativa* L.) Euphytica 178:331–338. doi:10.1007/s10681-010-0300-2

Xiao YH, Pan Y, Luo LH, Deng HB, Zhang GL, Tang WB, Chen LY (2011b) Quantitative trait loci associated with pollen fertility under high temperature stress at flowering stage in rice (*Oryza sativa L*). Rice Sci 18(2):1–7

Xu L, Tang Y, Gao S, Su S, Hong L, Wang W, Fang Z, Li X, Ma J, Quan W, Sun H, Li X, Wang Y, Liao X, Gao J, Zhang F, Li L, Zhao C (2016) Comprehensive analyses of the annexin gene family in wheat. BMC Genomics 17:415. doi:10.1186/s12864-016-2750-y

Yang C, Vizcay-Barrena C, Conner K, Wilson ZA (2007) MALE STERILITY1 is required for tapetal development and pollen wall biosynthesis. Plant Cell 19(11):3530–3548

Ye C, Argayoso MA, Redoña ED, Sierra SL, Laza MA, Dilla CJ, Mo Y, Thomson MJ, Chin J, Delaviña CB, Diaz GQ, Hernandez JE (2012) Mapping QTL for heat tolerance at flowering stage in rice using SNP markers. Plant Breed 131(1):33–41

Ye C, Tenorio FA, Argayoso MA, Laza MA, Koh HJ, Redoña ED, Jagadish KS, Gregorio GB (2015) Identifying and confirming quantitative trait loci associated with heat tolerance at flowering stage in different rice populations. BMC Genet 16:41

Zhang C, Zhang F (2015) The Multifunctions of WD40 Proteins in genome integrity and cell cycle progression. J Genomics 3:40–50

Zhang G, Chen L, Xiao G, Xiao Y, Chen X, Zhang S (2009) Bulked segregant analysis to detect QTL related to heat tolerance in rice using SSR markers. Agric Sci China 8:482–487

Zhang T, Yang L, Jiang K, Huang M, Sun Q, Chen W, Zheng J (2008) QTL mapping for heat tolerance of the tassel period of rice. Mol Plant Breed 6:867–873

Zhao L, Lei J, Huang Y, Zhu S, Chen H, Huang R, Peng Z, Tu Q, Shen X, Yan S (2016) Mapping quantitative trait loci for heat tolerance at anthesis in rice using chromosomal segment substitution lines. Breeding Science Preview. doi:10.1270/jsbbs.15084.

Zhu J, Wu X, Yuan S, Qian D, Nan Q, An L, Xiang Y (2014) Annexin5 Plays a Vital Role in *Arabidopsis* pollen development via Ca^{2+}-dependent membrane trafficking. PLoS One 9(7):e102407

5

Association of SNP Haplotypes of *HKT* Family Genes with Salt Tolerance in Indian Wild Rice Germplasm

Shefali Mishra[1,2], Balwant Singh[1], Kabita Panda[1], Bikram Pratap Singh[1], Nisha Singh[1], Pragati Misra[2], Vandna Rai[1] and Nagendra Kumar Singh[1*]

Abstract

Background: Rice is one of the most important crops for global food security but its productivity is adversely affected by salt stress prevalent in about 30 % of the cultivated land. For developing salt-tolerant rice varieties through conventional breeding or biotechnological interventions, there is an urgent need to identify natural allelic variants that may confer salt tolerance. Here, 299 wild rice accessions collected from different agro-climatic regions of India were evaluated during growth under salt stress. Of these 95 representative accessions were sequenced for members of *HKT* ion transporter family genes by employing Ion Torrent PGM sequencing platform.

Results: Haplotype analysis revealed haplotypes H5 and H1 of *HKT1;5* and *HKT2;3*, respectively associated with high salinity tolerance. This is the first study of allele mining of eight members of *HKT* gene family from Indian wild rice reporting a salt tolerant allele of HKT2;3. HKT1;5 also showed a salt tolerant allele from wild rice. Phylogenetic analysis based on the nucleotide sequences showed different grouping of the *HKT* family genes as compared to the prevailing protein sequence based classification.

Conclusions: The salt tolerant alleles of the HKT genes from wild rice may be introgressed into modern high yielding cultivars to widen the existing gene pool and enhance rice production in the salt affected areas.

Keywords: Allele mining, HKT, Association, Salt stress, NaCl, Wild rice

Background

Rice (*Oryza sativa* L.) is cultivated around the world and consumed by more than 50 % of the global human population (Mohanty 2013). The current status of rice production is 495.63 million tonnes (MT) which increased by only 9 MT in the preceding 4 year block (2011–2014) as compared to 80 MT increase in two such 4 year blocks during 2004–2011 (http://faostat3.fao.org/browse/Q/*/E). In contrast, the global population is projected to increase by 25 % to 9.2 billion by 2050 (Schroeder et al. 2013). Rice production must increase correspondingly by 70 % to fulfil the growing demand (FAOSTAT 2009). Fifteen percent of the land currently used for agricultural practices is at saturation point to maintain environmentally sustainable production, therefore innovations are needed to enhance crop productivity to meet the projected demand (Rockström et al. 2009). However, multiple abiotic stresses specifically salinity changes the soil texture and creates unfavourable conditions for rice production by severe inhibition of plant growth and development (Miller and Donahue 1995). Globally 45 Mha of irrigated and 32 Mha of rain fed agriculture are affected by salinity (http://www.fao.org/soils-portal/en/). Furthermore, irrigation with brackish water, tidal waves and tsunami continue to increase the soil salinity (Schroeder et al. 2013). According to one estimate up to 50 % of the present arable land may be affected by salinity by 2050 (Wang et al. 2003). Therefore, we need to explore natural genetic resources to find novel genes and alleles that can help withstand high salt concentration and maintain crop productivity.

* Correspondence: nksingh4@gmail.com
[1]National Research Centre on Plant Biotechnology, Pusa Campus, New Delhi 110012, India
Full list of author information is available at the end of the article

Salinity is characterised by presence of exchangeable sodium ions (Na^+) which imposes both ion toxicity and osmotic stress on the plant and alters the physiological status and ionic homeostasis of the cell. Crop plants are vulnerable to salinity and further their level of sensitivity and mechanism of tolerance depends on the growth stage of the plant (Hossain et al. 2015). Rice is most sensitive at the seedling stage and beyond 3 dSm^{-1} of electrical conductivity, seedlings are considered under stress and yield loss due to salinity is measured at twelve percent reduction per dSm^{-1} above threshold level (Lutts et al. 1995; Maas and Hoffman 1977). However, in most of the modern high yielding rice varieties, 50 % yield reduction occurs at 6 dSm^{-1} and they become totally unproductive beyond 12 dSm^{-1} (Linh et al. 2012). A limited number of large scale screening of germplasm for salinity stress has been conducted (Platten et al. 2013). Most of the genetic studies have focused on *Saltol* QTL derived from Indian rice landrace Pokkali which provides seedling stage salt tolerance (Thomson et al. 2010). There is very little work on the genetics of reproductive stage salt tolerance in rice (Mohammadi et al. 2013). The present understanding of salt tolerance mechanisms has facilitated exploitation of spatially located membrane transporters (Zhu 2001; Munns and Tester 2008). The known mechanisms include sequestration of ions in vacuole and exclusion of ions from root and leaves (Munns and Tester 2008). Thus, allelic variations in the sequence of ion transporter genes are likely to play a significant role in providing effective tolerance to salt stress.

In the present study we selected eight different transporters of the *HKT* gene family. These genes are further classified into two subfamilies based on their amino acid sequence similarity and differences in their Na^+ and K^+ selectivity (Horie et al. 2001), though product of each of these is a transmembrane protein (Horie et al. 2009; Mäser et al. 2001; Platten et al. 2006). Functionally, products of subfamily 1 [HKT1;1, HKT1;2, HKT1;3, HKT1;4 and HKT1;5] are Na + specific transporters and have S-G-G-G signature while products of subfamily 2 [HKT2;1, HKT2;2, HKT2;3 and HKT2;4] are Na^+-K^+ co-transporters or Na^+-K^+ uniporters with G-G-G-G signature (Jabnoune et al. 2009; Mäser et al. 2002). Oomen et al. 2012 has reported a hybrid of HKT2;1 and HKT2;2 gene as HKT2;2/1 from Nona Bokra which has strong permeability to Na^+ and K^+ even at high external Na^+ concentrations. Spatial localization and differential expression of these genes further enhances their importance. Horie et al. (2001) reported two isoforms of HKT transporters, a Na + transporter OsHKT1 and a Na^+- K^+-coupled transporter OsHKT2, which may act harmoniously in the salt tolerant indica rice. Further studies reported that OsHKT1;4 is expressed around xylem in the leaf sheath while OsHKT1;5 is expressed around the root xylem (Cotsaftis et al. 2011). Similarly, OsHKT2;1 and OsHKT2;4 are expressed in the outer part of the root and in the root hairs and may provide an entry point for Na^+ into plant roots from the soil (Lan et al. 2010; Schachtman and Schroeder 1994). HKT2;1 expression is significantly upregulated in the root cortex under K^+ starvation and high Na^+ concentration (Almeida et al. 2013; Horie et al. 2001). Further, it was proven that the Na^+ influx into K^+-starved rice roots was primarily OsHKT2;1-dependent while at concentration >30 mM NaCl, it was permeable to Na^+ only (Jabnoune et al. 2009). Another class 2 transporter gene *OsHKT2;4* shows 93 % similarity with *OsHKT2;3* and mediates K^+ transport independent of Na^+ concentration (Horie et al. 2011).

The level of salt tolerance provided by known transporter genes in the modern rice cultivars is insufficient and hence new genes or novel allelic variants of these transporters are needed for enhanced salinity tolerance. Crop domestication has selected only a few agronomically desirable alleles and left behind vast pool of genetic diversity in the wild progenitor species due to domestication bottleneck (Tanksley and McCouch 1997). Potential implication of germplasm and crop wild relatives under extreme environment conditions has been reviewed (Mickelbart et al. 2015). Allelic variant of members of HKT transporters such TmHKT1;5-A (Munns et al. 2012), TaKHT1;5-D (Yang et al. 2014) has been introgressed from wild relatives that led to increase in yield of the plant. Allele mining identifies superior alleles from related genotypes that may have been the effect of mutations in the process of evolution. The superior alleles can be used to develop allele specific markers and use them in marker assisted selection and also in tracing the evolution of alleles. Sequencing based allele mining and association analysis is an effective strategy to unravel the hidden potential of wild rice germplasm. Allele mining has been used across the crop species and novel alleles have been identified for abiotic stress tolerance genes in rice (Latha et al. 2004; Negrão et al. 2013; Platten et al. 2013; Singh et al. 2015a), maize (Yu et al. 2010) and barley (Cseri et al. 2011). However, a collection of untapped germplasm is required to mine novel desirable alleles and identify nucleotide sequence variations associated with these alleles (Kumar et al. 2010). India has a wealth of untapped wild rice germplasm that requires hasty expeditions to collect and exploit this fast depleting genetic resource (Singh et al. 2013). The genes already exploited from Indian wild rice include grassy stunt virus resistance from *Oryza nivara* (Khush and Ling 1974), *Bph 19(t)* from *Oryza rufipogon* (Li et al. 2010), *Xa38* from *Oryza nivara* (Bhasin et al. 2012), salinity tolerance genes, inositol methyl transferase (Sengupta et al. 2008) and L-myo-inositol 1-phosphate synthase from wild rice *Porteresia coarctata* (Das-

Chatterjee et al. 2006). The male sterility (MS) gene from *O. rufipogon* was introduced into the cultivated rice, leading to development of high yielding hybrid rice (Yuan et al. 1993). The beneficial alleles derived from wild relatives of rice have been transferred into elite genetic backgrounds leading to enhanced trait performance in rice (McCouch et al. 2007; Xiao et al. 1996; Xiao et al. 1998).

In the present study, 299 accessions of rice comprising *O. rufipogon* and *O. nivara* were screened for their level of salt tolerance. A candidate gene based allele mining was used to find natural allelic variants of agronomically important HKT family genes. Genes were re-sequenced with their promoters from 103 accessions of wild and cultivated rice varieties and their nucleotide and haplotype diversity was analysed. SNP based association analysis was done to link traits for salt tolerance e.g., SPAD for chlorophyll, Na^+ and K^+ concentration in the shoot and root of rice. Further, merit of *HKT1;5* and *HKT2;3* were discussed for their potentiality in salt tolerance of rice.

Fig. 1 Response of wild rice accessions after 15 days of 150 mM of salt stress in hydroponics. **a** Border; **b** NKSWR149; **c** NKSWR112; **d** NKSWR104; **e** NKSWR092; **f** NKSWR119; **g** NKSWR085; **h** NKSWR115; **i** NKSWR132; **j** NKSWR097; **k** Oryza nivara330646; **l** NKSWR143; **m** NKSWR101; **n** NKSWR079; **o** FL478; **p** VSR156; **q** Border (NKSWR143 is tolerant other accessions are sensitive, FL478- tolerant check, VSR156- sensitive check)

Results

Phenotypic Screening of the Wild Rice Germplasm for Salt Tolerance

Based on standard evaluation system (SES) of IRRI (Gregorio et al. 1997), 299 wild rice accessions were screened and classified according to their level of salt tolerance (Fig. 1, Additional file 1: Table S1). Two accessions NKSWR173 and NKSWR202 were found highly tolerant with SES score 1 as compared to the tolerant check FL478 with SES score 3 after 15 days of salt stress, 30 accessions were found tolerant at 10 days of salt stress, of which 14 accessions maintained the tolerance comparable to FL478 till fifteenth day. Total 84 accessions were found moderately tolerant, 115 sensitive and 67 highly sensitive at 10 days of salt treatment. However, at fifteenth day only 28 accessions were moderately tolerant, while 69 accessions became sensitive and 186 highly sensitive. One accession, *O. nivara* 336676 displayed interesting result showing high tolerance up to tenth day and thereafter it became highly sensitive and died within the next 2 days in all three years of the experiment (Additional file 1: Table S1). After this primary screening, 45 selected accessions, including all tolerant lines and representative lines from the other sensitive classes, were further evaluated for salt tolerance parameters such as biomass, chlorophyll content (using SPAD meter) and Na + and K+ concentrations in replicated tests. The Na^+/K^+ ratio in shoot under stress ranged from 0.8 to 2.1 with the maximum value in *O. rufipogon* 336679 and the minimum in CSR27. The lowest Na^+/K^+ ratio (1.3) in the root was observed in a tolerant line NKSWR232 and the highest ratio (3.3) was in NKSWR144. FL478 had highest chlorophyll content while VSR156 and NKS074 had no chlorophyll as the plants died (Additional file 2: Table S2).

Nucleotide Diversity, Haplotype Diversity and Haplotype Networks of the *HKT* Genes

Eight HKT transporter family genes along with their promoters were re-sequenced from 103 accessions of wild rice and cultivated rice varieties and their nucleotide and haplotype diversity was analysed (Table 1). Nucleotide diversity (π) was the highest (0.00169) for *HKT2;3* and the lowest (0.00012) for *HKT1;1* gene. Maximum haplotype diversity (HD) of 0.891 was found in *HKT2;3* gene followed by *HKT1;4* (0.692) and the lowest in *HKT1;1* (0.084) gene (Table 1). Haplotype analysis showed the maximum 23 haplotypes each for the *HKT1;5* and *HKT2;3* genes with 45 and 28 SNP sites, respectively. The *HKT2;4* gene showed the least number of 5 haplotypes. The Tajima's D test, Fu and Li's D* test and Fu and Li's F* statistics were performed to distinguish between random versus non-random mutations. All the test values were analysed for entire gene, as well as separately for the coding and non-coding regions. Except for the coding region of all accessions, which had positive values of Tajima's D test, all other regions of the genes among each

Table 1 Nucleotide and haplotype diversity and tests of neutrality for eight high affinity potassium transporter (HKT) family genes from 95 Indian wild rice accessions and 8 cultivated varieties

Candidate gene (Locus ID)	Region	S	π	θω	H	HD	D	D*	F*
HKT1;1	Coding	8	0.00013	0.00084	10	0.217	−2.10133	−3.97252	−3.94393
(LOC_Os04g51820)	Non-coding	3	0.0001	0.00059	4	0.076	−1.48732	−2.01292	−2.16727
	All	11	0.00012	0.00076	13	0.084	−2.22874	−4.21517	−4.17222
HKT1;2	Coding	0	0	0	1	0	0	0	0
	Non-coding	20	0.00027	0.00135	12	0.387	−2.30631	−4.55086	−4.42942
Chr4:30548314..30545885	All	20	0.00026	0.00131	12	0.387	−2.30631	−4.55086	−4.42942
HKT1;3	Coding	8	0.00019	0.00108	9	0.25	−1.99519	−2.80081	−2.99198
(LOC_Os02g07830)	Non-coding	6	0.00067	0.00137	4	0.164	−1.14077	0.15513	−0.32717
	All	14	0.00037	0.00118	10	0.29	−1.88141	−1.8801	−2.24114
HKT1;4	Coding	2	0.00031	0.00091	3	0.128	−1.01249	−1.0804	−1.23948
(LOC_Os04g51830)	Non-coding	42	0.00059	0.00191	19	0.69	−2.1536	−3.07965	−3.2462
	All	44	0.00057	0.00182	20	0.692	−2.15775	−3.12193	−3.27713
HKT1;5	Coding	8	0.00025	0.00183	8	0.166	−2.08122	−3.61227	−3.65909
(LOC_Os01g20160)	Non-coding	37	0.0012	0.00322	18	0.571	−1.94371	−2.70539	−2.87895
	All	45	0.00094	0.00284	23	0.631	−2.10258	−3.40294	−3.44926
HKT2;1	Coding	5	0.00015	0.00073	5	0.18	−1.66589	−3.23816	−3.20982
(LOC_Os06g48810)	Non-coding	46	0.00042	0.00241	17	0.333	−2.60346	−4.60994	−4.55342
	All	51	0.00035	0.00196	18	0.421	−2.61003	−4.91693	−4.76765
HKT2;3	Coding	24	0.00238	0.00353	23	0.891	−0.96284	−1.72038	−1.71118
(LOC_Os01g34850)	Non-coding	4	0.00013	0.00132	3	0.038	−1.77201	−3.90552	−3.78567
	All	28	0.00169	0.00284	23	0.891	−1.22676	−2.67757	−2.5308
HKT2;4	Coding	0	0	0	1	0	0	0	0
(LOC_Os06g48800)	Non-coding	4	0.0012	0.00448	5	0.198	−1.44156	−1.48409	−1.72999
	All	4	0.0012	0.00448	5	0.198	−1.44156	−1.48409	−1.72999

(*S* SNP/Indel sites, π nucleotide diversity, θω nucleotide diversity with Watterson's estimator, *H* number of haplotypes, *HD* haplotype diversity, *D* Tajima's D statistic, *D** Fu and Li's D* statistic, *F**: Fu and Li's F* statistic.)

class individually and in group deviated significantly from neutrality and Tajima's D test values were negative.

Haplotype analysis of *HKT1;1* gene showed 13 haplotypes with haplotype number H12 having a single highly tolerant accession and a distinguishing guanine residue instead of adenine in the first intron. *HKT1;2* gene has diversified to 13 haplotypes, *HKT1;3* into 10 haplotypes and *HKT1;4* to 20 haplotypes. Network analysis of *HKT1;5* gene showed that out of the total 23 haplotypes of this gene two haplotypes, H5 and H22 included all the tolerant accessions. Haplotype H5 had 13 accessions, including tolerant cultivated varieties FL478 and CSR27 along with sensitive variety VSR156 and five sensitive accessions from Gujrat. A unique haplotype H22 was represented by salt tolerant cultivated rice variety CSR11, which also matched with the published sequence of tolerant Indian rice variety Nona Bokra (Ren et al. 2005). The members of HKT sub-family 2 genes, *HKT2;1* and *HKT2;4* showed 18 and 5 haplotypes, respectively. Further, haplotype analysis of *HKT2;3* gene showed that its haplotype H1 included four tolerant wild rice accessions along with the salt tolerant check varieties FL478 and CSR27. However, another salt tolerant variety CSR11 possessing haplotype H10 grouped separately along with tolerant *O. nivara* accession 330641. A phenotypically unique accession, showing tolerance only up to 10 days of stress followed by extreme sensitivity, had a unique haplotype H5 of the *HKT2;3* gene (Fig. 2).

Evolution of the HKT Family Genes

A phylogenetic tree was constructed using nucleotide sequences of the eight HKT family genes using NJ method. Sorghum *HKT* gene taken from sorghum gene database was included as outgroup. As expected the tree showed total eight clades representing the individual HKT genes that clustered into two major groups. The phylogenetic tree revealed a somewhat different relationship among the HKT transporter genes as compared with their functional classification. The *HKT1;3* gene is at the root of

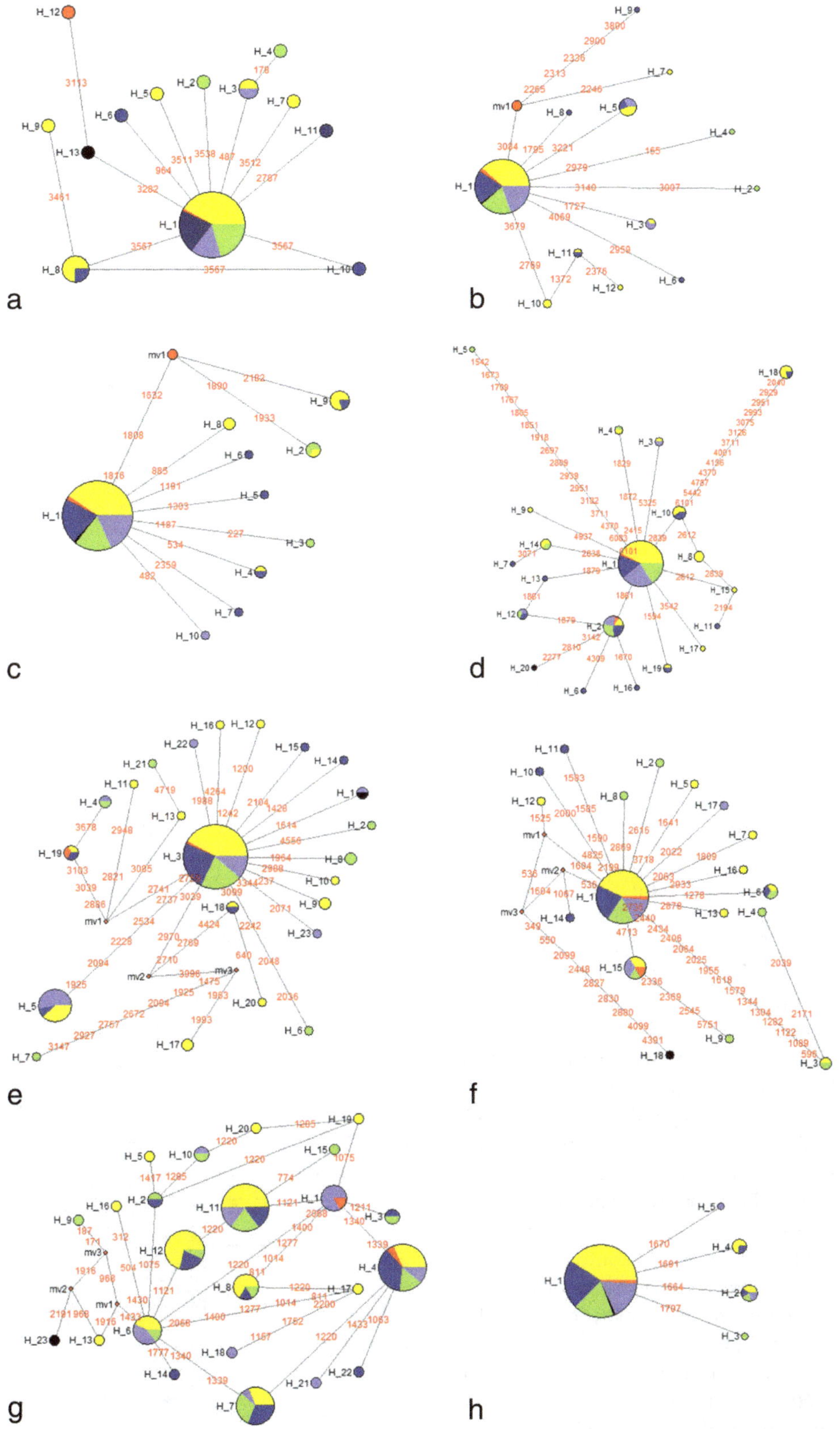

Fig. 2 Haplotype networks of eight HKT transporter genes including (**a**) *HKT1;1* (**b**) *HKT1;2* (**c**) *HKT1;3* (**d**) *HKT1;4* (**e**) *HKT1;5* (**f**) *HKT2;1* (**g**) *HKT2;3* (**h**) *HKT2;4*. Each circle represents a haplotype and size of each circle is proportional to haplotype (allele) frequency. Colour coding represents phenotype class based on standard evaluation system (SES) score. (Red-highly tolerant, Purple-tolerant, Green moderately tolerant, Blue-Sensitive, Yellow-highly sensitive)

divergence as it was most closely related to the sorghum *HKT* gene. A close relationship was observed between the *HKT1;3, HKT1;2* and *HKT2;3.* The remaining five genes were clustered in group I and showed that *HKT1;4* was most primitive followed by *HKT1;5* and *HKT1;2.* The recently evolved genes *HKT2;4* and *HKT2;1* are located in tandem on chromosome 6 and very closely related to each other diverging from a common branching point (Fig. 3).

Association of HKT Gene SNPs with Salt Tolerance

The association study was conducted on 45 accessions including 3 check rice varieties, of which, 18 belonged to tolerant class, 9 moderately tolerant, 7 sensitive and 11 highly sensitive. For association study a Bayesian based analysis of population structure was conducted which used an ad hoc statistics (ΔK), the rate of change in log probability of data between successive K values. It showed that the highest log likelihood was at $K = 3$ suggesting three major sub-population in the analysed Indian wild rice accessions. The individual assignments and Q-matrix in the three different sub-population revealed that sub-population I included 51.4 %, sub-population II 33.8 % and sub-population III 14.8 % of the analysed accessions. The mean Fst values were 0.8981, 0.6154 and 0.7291 for the sub-populations I, II and III, respectively (Fig. 4, Additional file 3: Table S3).

SNP based association analysis showed significant associations with the salt tolerance traits using MLM_Q + K statistical method. Thus, the results were filtered with association values greater than 5 % and p value less than 0.01. A total of 50 SNPs from all the eight *HKT* genes showed significant associations with the salt tolerance traits (Table 2, Additional file 4: Figures S1, S2).

Among the HKT subfamily I members, analysis of associated SNPs of *OsHKT1;1* gene revealed two non-synonymous substitutions, (i) S2094 (Ser_{258} to Asp_{258}) and (ii) S2331 (Ser_{179} to Phe_{179}) that were significantly associated with SPAD chlorophyll content. These two SNPs are spatially located, one (S2094) on the extracellular side and the other on the cytoplasmic side (S2331). In addition, one SNP (S0231) in the intronic region showed highly significant association with sodium concentration in the root. Further, two synonymous SNPs were associated with potassium concentration in the shoot (S1955) and SPAD chlorophyll content (S2546). In case of *HKT1;2* gene three SNPs were associated with potassium concentration in the shoot, two of these (S1430 and S2282) were present in the intronic region while third one (S221) was in the upstream promoter region. In case of *OsHKT1;3* a non-synonymous SNP S337 (Cys_{16} to Arg_{16}) present in the extracellular domain of the protein explained 34 % of phenotypic variance (R^2 value) and was associated with sodium concentration in the root while, three non-synonymous SNPs in the

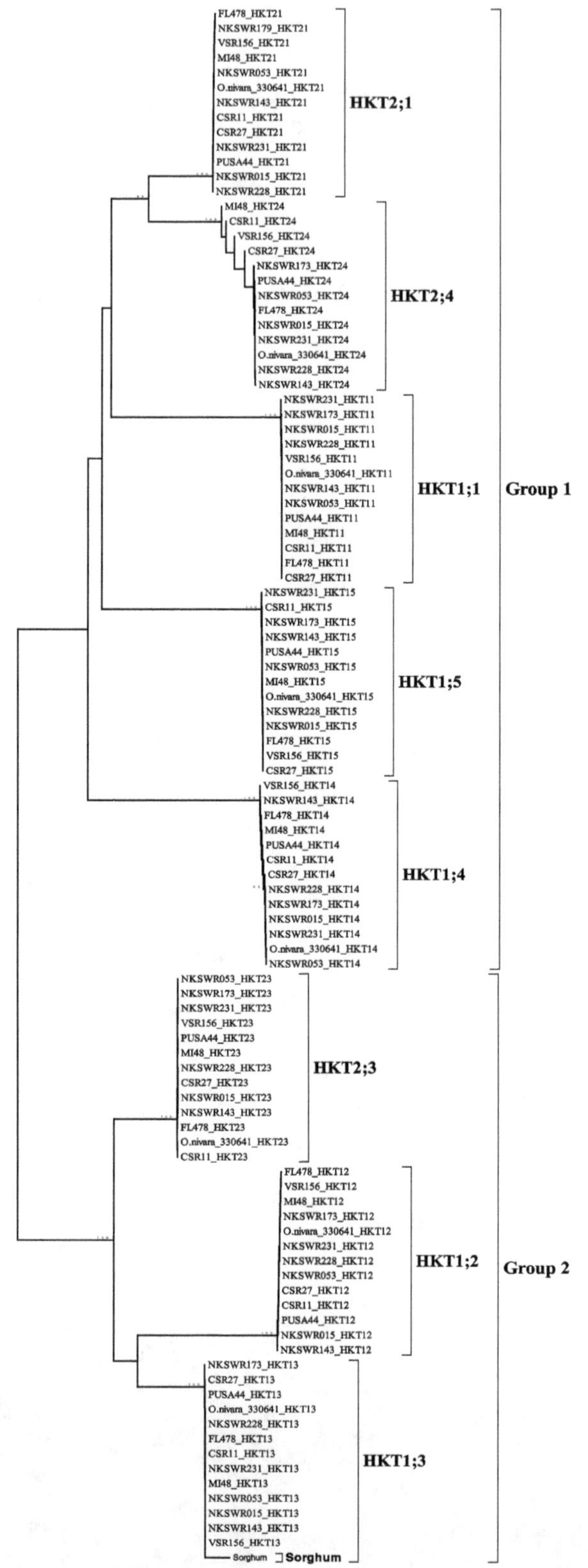

Fig. 3 Minimum evolution phylogenetic tree of eight *HKT* genes re-sequenced from representative wild rice accessions and cultivar checks constructed using MEGA 5.1 software

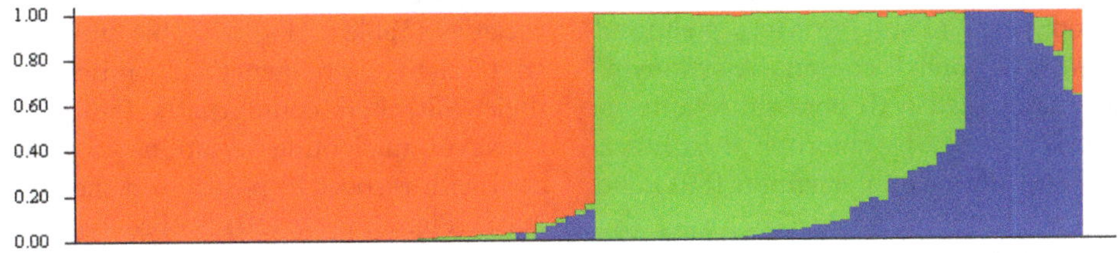

Fig. 4 Distribution of 103 wild rice accessions and cultivated rice into three different sub-populations based on 48 genome wide SNP markers. X-axis shows rice accessions and y-axis represents Fst values (*Blue*: Population I; *Green*: Population II; *Red*: Population III)

Table 2 Association of SNPs with different salt related traits at 10 days of treatment with HKT genes (with R2 > 5 % and $p < 0.01$)

Gene	Trait	SNP	Site	*p*	R2 (%)	SNP type	MAF	Gene	Trait	SNP	Site	*p*	R2 (%)	SNP type	MAF
HKT 1;1	N_R	S0231	30724332	0.003	34.43	Int	37	HKT 1;5	N_S	S0077	11459042	0.002	31.57	Int	36
	K_S	S1955	30726056	0.003	30.75	Syn	5		N_S	S3893	11462858	0.005	33.00	P/A_{140}	20
	SPAD	S2094	30726195	0.003	31.90	S/N_{258}	32		K_S	S0077	11459042	0.003	30.92	Int	36
	SPAD	S2331	30726432	0.003	31.90	S/F_{179}	30		N_R	S0887	11459852	0.003	34.12	Int	13
	SPAD	S2546	30726647	0.003	31.90	Syn	30		N_R	S0919	11459884	0.009	34.42	Int	35
HKT 1;2	K_S	S0221	30545988	0.003	31.83	5′ NC	36		N_R	S0958	11459923	0.009	34.65	Int	35
	K_S	S1430	30547197	0.009	25.16	Int	21		N_R	S0959	11459924	0.003	34.12	Int	13
	K_S	S2282	30548049	0.003	31.54	Int	31		N_R	S0974	11459939	0.008	35.09	Int	6
HKT 1;3	N_R	S0337	4103487	0.005	33.93	C/R_{16}	30		N_R	S1036	11460001	0.002	37.37	Int	5
	SPAD	S0091	4103241	0.009	27.83	5′ NC	12		K_R	S1036	11460001	0.008	26.80	Int	5
	N_R	S0744	4103894	0.008	38.19	Syn	22		N_R	S1379	11460344	0.003	34.44	K/Q_{429}	49
	SPAD	S0744	4103894	0.008	35.21	Syn	22		N_R	S1913	11460878	0.003	34.62	Int	43
	SPAD	S1682	4104832	0.009	27.83	Int	10		N_R	S2043	11461008	0.004	33.62	Int	29
	SPAD	S1762	4104912	0.005	31.75	Int	45		K_S	S2146	11461111	0.002	34.93	Int	45
	SPAD	S2347	4105497	0.003	34.78	A/V_{488}	23		K_S	S2347	11461312	0.006	27.43	Int	28
	SPAD	S2374	4105524	0.004	32.49	R/Q_{497}	46		K_S	S2865	11461830	0.001	36.63	Int	14
	SPAD	S2392	4105542	0.009	27.68	T/M_{503}	13		K_S	S3893	11462858	0.009	31.17	Int	20
HKT 1;4	SPAD	S0886	30734275	0.003	34.72	3′NC	32	HKT 2;1	K_S	S3062	29541671	0.008	32.35	5′ NC	7
	SPAD	S1267	30734656	0.003	34.72	3′NC	32		N_R	S3062	29541671	0.004	39.70	5′ NC	7
	SPAD	S2284	30735673	0.002	37.42	Int	32		SPAD	S4153	29542762	0.009	25.80	5′ NC	38
	SPAD	S2804	30736193	0.003	34.72	Int	32		SPAD	S4377	29542986	0.002	33.96	5′ NC	36
	SPAD	S2974	30736363	0.002	37.03	Int	33		K_S	S5815	29544424	0.002	40.52	5′ NC	21
	SPAD	S3185	30736574	0.002	37.03	Int	32		K_S	S5817	29544426	0.007	32.66	5′ NC	7
	SPAD	S3228	30736617	0.007	36.36	Int	33	HKT 2;3	K_S	S0356	19241884	0.002	34.84	Int	10
	SPAD	S3307	30736696	0.003	34.87	Int	28		SPAD	S0761	19242289	0.003	31.88	I/T_{77}	34
	SPAD	S3544	30736933	0.003	34.72	Int	32		K_S	S1001	19242529	0.009	25.08	I/T_{157}	5
	SPAD	S4678	30738067	0.003	35.09	Int	35		SPAD	S2053	19243581	0.003	31.79	Syn	34
								HKT 2;4	K_S	S0334	29534433	0.004	37.20	Int	37

(*N_S* sodium concentration in shoot, *K_S* potassium concentration in shoot, *N_R* sodium concentration in root, *K_S* potassium concentration in root, *Int* intron, *Syn* synonymous, *NC* non-coding, *MAF* minor allele frequency in %)

cytoplasmic domain, S2374 (Ala_{488} to Val_{488}), S2374 (Arg_{479} to Glu_{479}) and S2392 (Thr_{503} to Met_{503}) showed association with SPAD chlorophyll content. A synonymous SNP (S744) was associated with potassium concentration in the shoot and SPAD chlorophyll content. Interestingly, the SNP in the promoter region (S91) and those in the intronic region (S1682 and S1762) were also associated with the SPAD chlorophyll value. All SNPs identified in the *OsHKT1;4* gene were associated with SPAD chlorophyll content, among them, two were in the downstream promoter region while eight were in the intronic region. Of the two SNPs in the downstream promoter region (S886 and S1267), one was in the consensus GT-1 binding site and the other one in the ACGT sequence. In the *OsHKT1;5* gene total 14 SNPs were associated with the salt tolerance traits. Of which 12 were in the intronic region while remaining 2, S1379 (Lys_{429} Glu_{429}) and S3893 (Pro_{140} Ala_{140}), were non-synonymous. Both the non-synonymous SNPs were associated with sodium concentration: S1379 was located on extracellular side and showed significant association with the concentration of sodium in the root while, the other S3893 was on the cytoplasmic side and showed association with concentration of sodium in shoot, explaining 32 % (R^2 value) of the phenotypic variance (Table 2).

From the HKT subfamily II, *OsHKT2;1* gene had five SNPs in the promoter region showing association with the salt tolerance traits. Three of these (S3062, S5815, and S5817) were associated with shoot potassium concentration, of which S3062 was also associated with root sodium concentration, while the other two, S4153, and S4377, were associated with SPAD chlorophyll content. In *OsHKT2;3* a total of 4 SNPs were associated with the salt tolerance traits. Of the two non-synonymous SNPs, S1001 was associated with potassium concentration in shoot (extracellular side) while S761 was associated with SPAD chlorophyll content (cytoplasmic side), both the SNPs resulted in Isoleucine to Threonine substitution at amino acid positions 157 and 77, respectively. In addition, one synonymous SNP S2053 was associated with SPAD chlorophyll content and another SNP S356 in the intronic region was associated with potassium concentration in shoot. Only one SNP S334 present in the intronic region of *OsHKT2;4* gene was associated with potassium concentration in the shoot (Table 2).

Discussion

Wild rice is expected to play important role in rice improvement in the coming years. In this study we explored untapped diversity of Indian wild rice to identify natural alleles of the HKT transporter family genes. Analysis of nucleotide sequence variations for eight HKT family genes in wild rice showed higher nucleotide and haplotypic diversity as compared to the cultivated rice varieties (Table 1, Fig. 2). This supports the notion that wild relatives are genetically much more diverse than their cultivated counterparts (Hoisington et al. 1999). However, nucleotide variations of a gene is also associated with respect to position in the gene (McNally et al. 2006). Platten et al. (2013) observed comparatively lower nucleotide diversity and haplotypic diversity in cultivated rice and identified relationship between different haplotypes and salt tolerance for *HKT1;5* gene. Our results indicate that nucleotide diversity was quite different from haplotype diversity as only effective SNPs participated in haplotypic group determination (Goodall-Copestake et al. 2012). Differences between nucleotide and haplotype diversity has been measured across different genes such as Sucrose Synthase 3 in rice (Lestari et al. 2011) and *OsDREB1F* in wild and cultivated rice (Singh et al. 2015a).

To understand the natural selection process in the evolutions of HKT genes, Tajima's D test, Fu and Li's D* test and Fu and Li's F* statistics were calculated. Negative values obtained for wild rice population shows an excess of rare polymorphisms which still undergoes some population expansion and positive selection pressure and were further validated by Ka/Ks ratio and Jukes and Counter correction values (>1). However, positive value for coding region in cultivated rice revealed that alleles had evolved either by balancing selection or population bottlenecks (Table 1, Akey et al. 2004).

Among the eight genes only two *HKT1;5* and *HKT2;3* showed tolerant haplotypes. Functionally, HKT1;5 is a Na^+ specific transporter that maintains Na^+/K^+ homeostasis in the leaves under salt stress (Ren et al. 2005; Mickelbart et al. 2015), whereas HKT2;3 regulates K^+ transport independent of Na^+ concentration (Horie et al. 2011). The genotypes having tolerant allele of *HKT1;5* showed lower Na^+/K^+ concentration in the shoot as compared to root (Thomson et al. 2010). Six wild rice accessions that showed lower shoot Na^+ grouped into four separate haplotypic group. In addition, four other accessions that showed low Na^+ concentration in shoot comparable to the tolerant checks fell into separate haplotypic groups indicating allelic variation for the *HKT1;5* gene among the accessions (Fig. 2, Additional file 2: Table S2). An exceptionally higher Na^+ and K^+ ratio was observed in the shoots of a tolerant accession *O. rufipogon* 336679 indicating presence of an alternate mechanism to sustain such a high level of salt concentration in shoot. Mian et al. (2011) identified role of *HKT 2;1* gene in barley and reported increased shoot Na + concentrations and improved biomass production under salt stress. Haplotype analysis of *HKT2;3* gene showed almost even distribution of accessions into different haplotypic groups, but it also formed a separate haplotype for the reference sequence. Of the total 23 observed haplotypes, one was associated with tolerant and two were associated with moderately tolerant genotypes. A unique accession of *O. nivara 336676* showing tolerance only up to tenth day of applying the stress separated in a different haplotype indicating that it is possibly following some different tolerance mechanism which breaks down beyond 10 days of stress. Salt susceptibility of this accessions after 10 days might be due to higher accumulation of Na^+ in root (Additional file 2: Table S2). Another plausible explanation for the tolerance in this accession is presence of root specific tonoplast transporters with some accessory factors associated with the genes for providing energy to transport ions into vacuole but with limited accumulation capacity of ions in vacuole. Functionally, HKT2;3 mediates K^+ transport independent of Na^+ concentration hence, even at high salt concentration the physiological functions that are vital to crop plants relating to K^+ e.g., stomatal movement, CO_2 uptake and cofactor to make enzymes etc. are maintained inside the cell. It also maintains ionic homeostasis (Horie et al. 2007; Roy et al. 2014). However, haplotype analysis revealed uneven distribution of tolerant genotypes among different haplotypic group for different genes. It shows that an unknown type of mechanism might be present among the Indian wild rice that could help to enhance our understanding towards salt tolerance mechanism and be used for improvement of cultivated rice.

Phylogenetic analysis describes the evolutionary relationship among genes and different clustering patterns indicate the level of their functional divergence (Mäser et al. 2001). Here, two genes of phylogenetic group I (*HKT2;1* and *HKT 2;4*) showed high nucleotide similarity with each other. They most likely have evolved from recent duplication of a single gene as also suggested by their tandem location in the rice chromosome 6. Interestingly, in the phylogenetic tree functionally diverse, *HKT1;3* and *HKT1;2* genes grouped together along with *HKT2;3*. This suggests that members of functionally different groups might have evolved from a common ancestor through duplication and subsequent functional divergence.

Population structure analysis revealed three distinct subpopulations in the Indian wild rice accessions with extensive genetic variation between and also within the populations as was also shown recently using a high density 50 K SNP chip (Singh et al. 2015b). Higher average values of Fst indicated that a high level of unshared allelic variation was present in the population (Holsinger and Weir 2009). Here sub-population I comprising more than 50 % of the wild rice accessions was the largest of the three sub-populations.. Linkage disequilibrium (LD) based association analysis is supposed to be the best method for association analysis (Yu and Buckler 2006). Here, LD based association analysis showed strong associations of multiple SNPs in different candidate *HKT* genes with the salt tolerance traits.

Haplotype based association analysis for identification of allelic variants associated with specific traits is considered better over single-allele studies (Johnson et al. 2001). Overall, only three tolerant haplotypes were identified, one for the *HKT1;5* gene and two for the *HKT 2;3* gene, but, multiple SNPs were found to be associated with different salt tolerance traits. It may be because of less number of accessions grouping in many haplotypes resulting in low minor allele frequencies (<2 %), and synonymous or noncoding nature of the SNPs. Negrão et al. (2013) identified allelic variants for *HKT1;5* gene by setting filter for associated SNPs and individually assessing the results. Here total 14 filtered SNPs were found to be associated with traits for *HKT1;5*, and among them 12 SNPs were in the intronic regions, one was synonymous and two were non- synonymous, S1379 (Lys_{429} to Glu_{429}) and S3893 (Pro_{140} to Ala_{140}) leading to amino acid substitutions. Replacement of proline causes changes in the protein backbone that is responsible to introduce tight turns or kinks into alpha helices (Betts and Russell 2003). Conversion of proline to alanine was also found in Pokkali and Nona Bokra alleles of the *HKT1;5* gene which is supposed to increase the substrate specificity (Cotsaftis et al. 2012; Negrão et al. 2013). Cotsaftis et al. (2012) reported change in conformation of protein with change in overall charge.

Two SNPs in the promoter region of *OsHKT1;4* gene, S886 (GT-1 binding site region) and S1267 (downstream to the promoter ACGT sequence), were significantly associated with SPAD chlorophyll content value. It has been reported that in many light-regulated genes like *PHYA* in oat and rice are regulated by GT-1 binding site (Villain et al. 1996). Further, Park et al. (2004) has reported this site to be salt induced. Downstream to the promoter region is the ACGT sequence that is required for etiolation induced expression of *erd1* (early responsive to dehydration) gene in *Arabidopsis* (Simpson et al. 2003), hence it is speculated that *OsHKT1;4* gene may regulate dehydration stress imposed by salinity.

Two haplotypes (H1 and H10) of the *OsHKT2;3* were associated with salt tolerance SES score. In addition, total 4 SNPs were associated with the analysed salt tolerance traits, one synonymous, one intronic and two non-synonymous SNPs, S761 and S1001, both leading to Iso to Thr substitution at amino acid positions 77 and 157, respectively (Table 2). The isoleucine to threonine substitution is known to impact post translation modifications such as phosphorylation. An increase in the phosphorylation level has been observed with increase in Thr in a protein sequence (Vlad et al. 2008). The non-synonymous mutations outside functional domain of genes may alter structure of the protein and consequently its function (Negrão et al. 2013). Both synonymous and non-synonymous SNPs showed significant association with salt tolerance traits, perhaps affecting the RNA splicing, mRNA stability, and post-translational modification of protein function (Negrão et al. 2013). Some of this could be due to close linkage of a non-functional SNP with the functional SNP. A large number of rare alleles and haplotypes were observed for different HKT genes, whose association with trait could not be studied due to low minor allele frequency. Rare alleles contributing to the gain or enhancement of the trait value may be useful for future adaptability of the rice crop, these may involve novel mechanisms of salt tolerance. Introgression of rare alleles through marker-assisted backcross breeding (MABB) techniques may help develop new genetic resources for breeding of rice for tolerance to extreme salt stress. To find out effective rare alleles, bi-parental mapping populations involving these lines will be needed to validate the function of rare alleles and also to understand their genetic control mechanisms (Semagn et al. 2010).

Conclusions

The wild rice accessions screened in this study have been collected from different ecological habitats including salt-affected areas. An accessions (NKSWR173) from upland and another accession (NKSWR202) from upper gangetic plain region showed high level of tolerance phenotypic patterns and may have different salt tolerance mechanisms. Haplotype analysis indicated a substantial level of natural diversity for the HKT family of genes, especially for the HKT1;5 and HKT2;3 genes among the Indian wild rice accessions. The novel haplotypes showing association with salt tolerance may have great impact on rice salinity breeding. A programme for introgression of the identified haplotypes into high yielding but salt sensitive rice varieties has been initiated. It will help breed rice genotypes with higher level of salt tolerance beyond the existing salt tolerant varieties.

Methods

Plant Material

A total of 299 wild rice accessions including 244 new accessions collected from different geographical regions of India along with their passport data and 58 accessions from NBPGR gene bank, New Delhi, India were analysed. In addition, 6 cultivated *O. sativa* (salt tolerant FL478, CSR27, CSR11 with different tolerance mechanisms, MI48 as a moderate check while VSR156 and Pusa 44 as sensitive checks) were used. The collection sites and other detailed information on each wild rice accession is available in a database at http://nksingh.nationalprof.in/ (Additional file 1: Table S1).

Phenotyping for Salt Tolerance

Phenotyping for all the accessions was done at National Phytotron Facility, IARI, New Delhi, India for three times. Experiment was planned according to IRRI protocol with minor modifications (Gregorio et al. 1997). Each set of experiment had three replicates with 10 plants of each accession per replicate. The seeds were germinated in petri plates and then transferred into thermocol trays after two days of germination. Each tray had a positive check and a negative check. The seedlings were allowed to grow for 14 days in Hoagland's solution and on 15th day, they were transferred to Hoagland's solution with 150 mM NaCl concentrations. Rice seedlings were allowed to grow and scoring was done after 10 days and 15 days of stress according to the 1–9 scale of Standard Evaluation System (SES) developed by Gregorio et al. (1997). Most of the SES were same for all the replicates but in case of variation mode was taken. Based on the above data and geographical location 45 representative accessions from each class were selected to evaluate physiological parameters to reduce the cost and labour. Sodium and Potassium ion concentrations from roots and shoots were estimated using a flame photometer LABTRONICS model LT66 after digestion of 0.1 gm of dried plant sample with 1:3 ratio of perchloric acid and nitric acid. Chlorophyll content of leaves was determined using SPAD meter.

PCR Amplification and re-Sequencing

After phenotypic screening of 299 accessions, 103 were selected based on their response to salt stress. All the salt tolerant lines and representative accessions from moderately tolerant and sensitive classes were taken for re-sequencing of *HKT* genes. Genomic DNA was extracted from leaf tissue using the CTAB method described by Murray and Thompson (1980). Eight *HKT* genes were amplified using primer walking method (Additional file 5: Table S4). Nucleotide sequence of the genes were retrieved from NCBI database (http://www.ncbi.nlm.nih.gov/) and primers were designed using Primer3 software (http://bioinfo.ut.ee/primer3-0.4.0/). Specific amplification and validation of the primers was done by NCBI Primer BLAST (http://www.ncbi.nlm.nih.gov/tools/primer-blast/) against *Oryza* taxon and high stringency conditions during PCR amplification. The PCR amplifications were carried out in 25 μl reaction consisting of 1 Unit SpeedStar™HS DNA Polymerase from TAKARA BIO INC, 1x Fast Buffer, 2 μl of dNTP mixture, 0.5 Pico moles of each primers and 80 ng of template DNA. The PCR reaction was carried out in BIORAD Thermal cycler under following conditions: Initial denaturation at 98 °C for 3 min followed by 38 cycles of denaturation for 10 s at 98 °C, annealing for1 min at 64 °C and extension for 1 min at 68 °C, and for final extension for 10 min at 68 °C. The amplified products were further checked by electrophoresis in 1 % agarose gel in 1X TBE buffer.

PCR products were directly sequenced by Ion Torrent PGM sequencer (Life Technologies) after fragmentation, library preparation, purification and cycle sequencing according to manufacturer's instructions. Briefly, the amplicons of different genes of the same genotype were pooled in equimolar concentration, sheared to size of ~200 bp and then barcoded to identify individual accessions. Since each gene has unique sequence no barcoding was needed for individual gene. The sheared and barcoded products were now size selected, pooled together in equimolar ratio and then PCR sequenced. At each step purification was done using Agencourt AMPure XP reagent using Kingfisher Flex. Ion OneTouch2™ was used for emulsion PCR to clone the library on the beads and thereafter for enrichment. The enriched library was loaded on Ion PGM 316 chip and sequencing was performed on Ion PGM™ sequencer.

Sequence Data Analysis

The depth of sequencing obtained was approximately 88x. Coverage analysis and variant caller plugins were run and the sequences were viewed in IGV (Robinson et al. 2011; Thorvaldsdóttir et al. 2012). SamTools mpileup command was used to generate consensus sequences against the reference for each gene and alignment of *OsHKT* genes was done by ClustalW (Thompson et al. 1994) in BioEdit (Hall 2011). Sequences were deposited to NCBI GenBank database (KT795544-KT796361). Nucleotide polymorphisms were analysed using the DnaSP software version 5.10 (Rozas et al. 2003). Level of silent-site nucleotide diversities per site (π) (Nei 1987) and population mutation parameter (θ) (Watterson 1975) was estimated. Sliding window analysis was performed to examine nucleotide polymorphism across the genes in all accessions using DnaSP software. Statistical tests of neutrality such as Tajima's D (Tajima 1989), Fu and Li's D* and F were calculated to examine the selection pressure at this locus. A haplotype network was constructed for comparison of genealogical relationships among the haplotypes using Network software (Bandelt et al. 1999) (http://www.fluxus-engineering.com). The nucleotide sequences were translated into amino acid sequences, and the protein variants were identified as compared with the reference protein. The bootstrap consensus tree inferred from 500 replicates was constructed to represent the evolutionary history of the HKT genes by using the Minimum Evolution Method in MEGA5 (Tamura et al. 2011).

Population Structure and SNP-Trait Association Analysis

Association analysis was performed with the MLM model, considering both kinship (K) and population structure (Q), implemented in TASSEL software. The kinship (K) and population structure (Q) were generated from a genome wide 48-plex Illumina GoldenGate SNP genotyping assay (B. Singh, unpublished) and analysing data with Structure software (Pritchard et al. 2000). To overcome the problems of interpreting the real value of K, a range of ad hoc K values were tested and analysed using Evanno plot (Evanno et al. 2005). The results of estimated likelihood values for a given K (from 2 to 10) in five independent runs were harvested with structure harvester an online program (Earl 2012). For association mapping filtered sites within *OsHKT* genes were used to determine linkage disequilibrium (LD) by correlation between alleles at two loci in TASSEL 5.0 (Bradbury et al. 2007) software and significance of LD among SNPs was determined by Fisher's exact test. The mixed model showed least deviation of observed *P*-values from expected *P*-values in Q-Q plot when compared with that of Q (population structure) or K (kinship) model only. A probability value of 0.01 was used as the threshold for significance of SNP–trait associations. Functions of associated sites in the promoter region was identified by PLACE (Higo et al. 1999).

Additional files

Additional file 1: Table S1. List of wild rice accessions, their place of collection and Standard Evaluation System (SES) score obtained during screening.

Additional file 2: Table S2. Phenotype (SES) score and evaluated salt related traits among Indian wild rice at 10th day of 150mM salt treatment.

Additional file 3: Table S3. Set of overlapping primers used for amplification of HKT genes.

Additional file 4: Figure S1. Linkage Disequilibrium plots for HKT genes **Figure S2.** Q-Q plots of HKT genes obtained after MLM based association of SNP with traits.

Additional file 5: Table S4. List of varieties and their classification in different sub-populations using Structure software.

Competing Interests

The authors declare that they have no competing interests.

Authors' Contributions

SM: Design and conduct of experiments, Sequencing, Data Analysis and Manuscript preparation; BS: Collection and characterization of wild rice, Conduct of experiments and manuscript drafting; KP: Database maintenance; BPS: Collection of Wild Rice; NS: Design of 48-plex SNP genotyping assay; PM: Manuscript editing; VR: Planning of experiment, Manuscript editing; NKS: Wild rice collection, planning and supervision of the experiments, manuscript editing and finalization. All authors read and approved the final manuscript.

Acknowledgements

The financial assistance from Indian Council of Agricultural Research through National Professor B.P. Pal Chair project is gratefully acknowledged. We are thankful to National Bureau of Plant Genetic Resources, New Delhi, India for providing seeds of 58 wild rice accessions.

Author details

[1]National Research Centre on Plant Biotechnology, Pusa Campus, New Delhi 110012, India. [2]Jacob School of Biotechnology and Bioengineering, Sam

Higginbottom Institute of Agriculture, Technology and Sciences, Allahabad 211007, India.

References

Akey JM, Eberle MA, Rieder MJ, Carlson CS, Shriver MD, Nickerson DA, Kruglyak L (2004) Population history and natural selection shape patterns of genetic variation in 132 genes. PLoS Biol 2:1591–1599

Almeida P, Katschnig D, de Boer AH (2013) HKT transporters—state of the art. Int J Mol Sci 14:20359–20385

Bandelt H-J, Forster P, Röhl A (1999) Median-joining networks for inferring intraspecific phylogenies. Mol Biol Evol 16:37–48

Betts MJ, Russell RB (2003) Amino acid properties and consequences of substitutions. Bioinformatics Genet 317:289

Bhasin H, Bhatia D, Raghuvanshi S, Lore JS, Sahi GK, Kaur B, Vikal Y, Singh K (2012) New PCR-based sequence-tagged site marker for bacterial blight resistance gene Xa38 of rice. Mol Breed 30:607–611

Bradbury PJ, Zhang Z, Kroon DE, Casstevens TM, Ramdoss Y, Buckler ES (2007) TASSEL: software for association mapping of complex traits in diverse samples. Bioinformatics 23:2633–2635

Cotsaftis O, Plett D, Johnson AAT, Walia H, Wilson C, Ismail AM, Close TJ, Tester M, Baumann U (2011) Root-specific transcript profiling of contrasting rice genotypes in response to salinity stress. Mol Plant 4:25–41

Cotsaftis O, Plett D, Shirley N, Tester M, Hrmova M (2012) A Two-staged model of Na (+) exclusion in rice explained by 3D modeling of HKT transporters and alternative splicing. PLoS One 7, e39865

Cseri A, Cserháti M, Von KM, Nagy B, Horváth GV, Palágyi A, Pauk J, Dudits D, Törjék O (2011) Allele mining and haplotype discovery in barley candidate genes for drought tolerance. Euphytica 181:341–356

Das-Chatterjee A, Goswami L, Maitra S, Dastidar KG, Ray S, Majumder AL (2006) Introgression of a novel salt-tolerant L-myo-inositol 1-phosphate synthase from *Porteresia coarctata* (Roxb.) Tateoka (PcINO1) confers salt tolerance to evolutionary diverse organisms. FEBS Lett 580:3980–3988

Earl DA (2012) Structure harvester: a website and program for visualizing structure output and implementing the Evanno method. Conserv Genet Resour 4:359–361

Evanno G, Regnaut S, Goudet J (2005) Detecting the number of clusters of individuals using the software STRUCTURE: a simulation study. Mol Ecol 14:2611–2620

FAOSTAT, D. (2009). Food and Agriculture Organization of the United Nations Year book.

Goodall-Copestake W, Tarling G, Murphy E (2012) On the comparison of population-level estimates of haplotype and nucleotide diversity: a case study using the gene cox1 in animals. Heredity 109:50–56

Gregorio GB, Senadhira D, Mendoza RD (1997) Screening rice for salinity tolerance. International Rice Research Institute discussion paper series, Metro Manila

Hall T (2011) BioEdit: an important software for molecular biology. GERF Bull Biosci 2:60–61

Higo K, Ugawa Y, Iwamoto M, Korenaga T (1999) Plant cis-acting regulatory DNA elements (PLACE) database. Nucleic Acids Res 27:297–300

Hoisington D, Khairallah M, Reeves T, Ribaut J-M, Skovmand B, Taba S, Warburton M (1999) Plant genetic resources: what can they contribute toward increased crop productivity? Proc Natl Acad Sci 96:5937–5943

Holsinger KE, Weir BS (2009) Genetics in geographically structured populations: defining, estimating and interpreting FST. Nat Rev Genet 10:639–650

Horie T, Yoshida K, Nakayama H, Yamada K, Oiki S, Shinmyo A (2001) Two types of HKT transporters with different properties of Na + and K+ transport in *Oryza sativa*. Plant J 27:129–138

Horie T, Costa A, Kim TH, Han MJ, Horie R, Leung HY, Miyao A, Hirochika H, An G, Schroeder JI (2007) Rice OsHKT2; 1 transporter mediates large Na + influx component into K + –starved roots for growth. EMBO J 26:3003–3014

Horie T, Hauser F, Schroeder JI (2009) HKT transporter-mediated salinity resistance mechanisms in Arabidopsis and monocot crop plants. Trends Plant Sci 14:660–668

Horie T, Brodsky DE, Costa A, Kaneko T, Schiavo FL, Katsuhara M, Schroeder JI (2011) K+ transport by the OsHKT2; 4 transporter from rice with atypical Na + transport properties and competition in permeation of K+ over Mg2+ and Ca2+ ions. Plant Physiol 156:1493–1507

Hossain H, Rahman M, Alam M, Singh R (2015) Mapping of quantitative trait loci associated with reproductive-stage salt tolerance in rice. J Agron Crop Sci 201:17–31

Jabnoune M, Espeout S, Mieulet D, Fizames C, Verdeil J-L, Conéjéro G, Rodríguez-Navarro A, Sentenac H, Guiderdoni E, Abdelly C (2009) Diversity in expression patterns and functional properties in the rice HKT transporter family. Plant Physiol 150:1955–1971

Johnson GC, Esposito L, Barratt BJ, Smith AN, Heward J, Di Genova G, Ueda H, Cordell HJ, Eaves IA, Dudbridge F (2001) Haplotype tagging for the identification of common disease genes. Nat Genet 29:233–237

Khush GS, Ling K (1974) Inheritance of resistance to grassy stunt virus and its vector in rice. J Hered 65:135–136

Kumar GR, Sakthivel K, Sundaram RM, Neeraja CN, Balachandran SM, Rani NS, Viraktamath BC, Madhav MS (2010) Allele mining in crops: prospects and potentials. Biotechnol Adv 28:451–461

Lan W-Z, Wang W, Wang S-M, Li L-G, Buchanan BB, Lin H-X, Gao J-P, Luan S (2010) A rice high-affinity potassium transporter (HKT) conceals a calcium-permeable cation channel. Proc Natl Acad Sci 107:7089–7094

Latha R, Rubia L, Bennett J, Swaminathan M (2004) Allele mining for stress tolerance genes in *Oryza* species and related germplasm. Mol Biol 27:101–108

Lestari P, Lee G, Ham TH, Woo MO, Piao R, Jiang W, Chu SH, Lee J, Koh HJ (2011) Single nucleotide polymorphisms and haplotype diversity in rice sucrose synthase 3. J Hered 102:735–746

Li R, Li L, Wei S, Wei Y, Chen Y, Bai D, Yang L, Huang F, Lu W, Zhang X (2010) The evaluation and utilization of new genes for brown plant hopper resistance in common wild rice (*Oryza rufipogon* Griff.). Mol Entomol. doi:10.5376/me.2010.01.0001

Linh LH, Linh TH, Xuan TD, Ham LH, Ismail AM, Khanh TD (2012) Molecular breeding to improve salt tolerance of rice (*Oryza sativa* L.) in the Red River Delta of Vietnam. Intern J Plant genom 2012, 949038. doi:10.1155/2012/949038

Lutts S, Kinet J, Bouharmont J (1995) Changes in plant response to NaCl during development of rice (*Oryza sativa* L.) varieties differing in salinity resistance. J Exp Bot 46:1843–1852

Maas E, Hoffman G (1977) Crop salt tolerance: evaluation of existing data. J Irrigat Drainage 15:187

Mäser P, Thomine S, Schroeder JI, Ward JM, Hirschi K, Sze H, Talke IN, Amtmann A, Maathuis FJM, Sanders D (2001) Phylogenetic relationships within cation transporter families of Arabidopsis. Plant Physiol 126:1646–1667

Mäser P, Eckelman B, Vaidyanathan R, Horie T, Fairbairn DJ, Kubo M, Yamagami M, Yamaguchi K, Nishimura M, Uozumi N (2002) Altered shoot/root Na + distribution and bifurcating salt sensitivity in Arabidopsis by genetic disruption of the Na + transporter AtHKT1. FEBS Lett 531:157–161

McCouch SR, Sweeney M, Li J, Jiang H, Thomson M, Septiningsih E, Edwards J, Moncada P, Xiao J, Garris A (2007) Through the genetic bottleneck: *O. rufipogon* as a source of trait-enhancing alleles for O. *sativa*. Euphytica 154:317–339

McNally KL, Bruskiewich R, Mackill D, Buell CR, Leach JE, Leung H (2006) Sequencing multiple and diverse rice varieties. Connecting whole-genome variation with phenotypes. Plant Physiol 141:26–31

Mian A, Oomen RJ, Isayenkov S, Sentenac H, Maathuis FJ, Véry AA (2011) Over-expression of an Na + –and K + –permeable HKT transporter in barley improves salt tolerance. Plant J 68:468–479

Mickelbart MV, Hasegawa PM, Bailey-Serres J (2015) Genetic mechanisms of abiotic stress tolerance that translate to crop yield stability. Nat Rev Genet 16(4):237–251. doi:10.1038/nrg3901

Miller RW, Donahue RL (1995) Soils in our environment. Prentice hall, Upper Saddle River

Mohammadi R, Mendioro MS, Diaz GQ, Gregorio GB, Singh RK (2013) Mapping quantitative trait loci associated with yield and yield components under reproductive stage salinity stress in rice (*Oryza sativa* L.). J Genet 92:433–443

Mohanty S (2013) Trends in global rice consumption. Rice Today 12(1):44–45

Munns R, Tester M (2008) Mechanisms of salinity tolerance. Annu Rev Plant Biol 59:651–681

Munns R, James RA, Xu B, Athman A, Conn SJ, Jordans C, Byrt CS, Hare RA, Tyerman SD, Tester M, Plett D (2012) Wheat grain yield on saline soils is improved by an ancestral Na + transporter gene. Nat Biotechnol 30:360–364

Murray M, Thompson WF (1980) Rapid isolation of high molecular weight plant DNA. Nucleic Acids Res 8:4321–4326

Negrão S, Cecília AM, Pires IS, Abreu IA, Maroco J, Courtois B, Gregorio GB, McNally KL, Margarida OM (2013) New allelic variants found in key rice salt-tolerance genes: an association study. Plant Biotechnol J 11:87–100

Nei M (1987) Molecular evolutionary genetics. Columbia University Press, New York

Oomen RJ, Benito B, Sentenac H, Rodríguez-Navarro A, Talón M, Véry AA, Domingo C (2012) HKT2; 2/1, a K + –permeable transporter identified in a salt-tolerant rice cultivar through surveys of natural genetic polymorphism. Plant J 71:750–762

Park HC, Kim ML, Kang YH, Jeon JM, Yoo JH, Kim MC, Park CY, Jeong JC, Moon BC, Lee JH (2004) Pathogen-and NaCl-induced expression of the SCaM-4 promoter is mediated in part by a GT-1 box that interacts with a GT-1-like transcription factor. Plant Physiol 135:2150–2161

Platten JD, Cotsaftis O, Berthomieu P, Bohnert H, Davenport RJ, Fairbairn DJ, Horie T, Leigh RA, Lin H-X, Luan S (2006) Nomenclature for HKT transporters, key determinants of plant salinity tolerance. Trends Plant Sci 11:372–374

Platten JD, Egdane JA, Ismail AM (2013) Salinity tolerance, Na + exclusion and allele mining of HKT1; 5 in *Oryza sativa* and O. glaberrima: many sources, many genes, one mechanism? BMC Plant Biol 13:32

Pritchard JK, Stephens M, Donnelly P (2000) Inference of population structure using multilocus genotype data. Genetics 155:945–959

Ren Z-H, Gao J-P, Li L-G, Cai X-L, Huang W, Chao D-Y, Zhu M-Z, Wang Z-Y, Luan S, Lin H-X (2005) A rice quantitative trait locus for salt tolerance encodes a sodium transporter. Nature Genet 37:1141–1146

Robinson JT, Thorvaldsdóttir H, Winckler W, Guttman M, Lander ES, Getz G, Mesirov JP (2011) Integrative genomics viewer. Nat Biotechnol 29:24–26

Rockström J, Falkenmark M, Karlberg L, Hoff H, Rost S, Gerten D (2009) Future water availability for global food production: the potential of green water for increasing resilience to global change. Water Resour Res 45, W00A12. doi:10.1029/2007WR006767

Roy SJ, Negrão S, Tester M (2014) Salt resistant crop plants. Curr Opin Biotechnol 26:115–124

Rozas J, Sánchez-DelBarrio JC, Messeguer X, Rozas R (2003) DnaSP, DNA polymorphism analyses by the coalescent and other methods. Bioinformatics 19:2496–2497

Schachtman DP, Schroeder JI (1994) Structure and transport mechanism of a high-affinity potassium uptake transporter from higher plants. Nature 370(6491):655–658

Schroeder JI, Delhaize E, Frommer WB, Guerinot ML, Harrison MJ, Herrera-Estrella L, Horie T, Kochian LV, Munns R, Nishizawa NK (2013) Using membrane transporters to improve crops for sustainable food production. Nature 497:60–66

Semagn K, Bjørnstad Å, Xu Y (2010) The genetic dissection of quantitative traits in crops. Electron J Biotechnol 13:16–17

Sengupta S, Patra B, Ray S, Majumder AL (2008) Inositol methyl tranferase from a halophytic wild rice, Porteresia coarctata Roxb. (Tateoka): regulation of pinitol synthesis under abiotic stress. Plant Cell Environ 31:1442–1459

Simpson SD, Nakashima K, Narusaka Y, Seki M, Shinozaki K, Yamaguchi-Shinozaki K (2003) Two different novel cis-acting elements of erd1, a clpA homologous Arabidopsis gene function in induction by dehydration stress and dark-induced senescence. Plant J 33:259–270

Singh A, Singh B, Panda K, Rai VP, Singh AK, Singh SP, Chouhan SK, Rai V, Singh PK, Singh NK (2013) Wild rices of Eastern Indo-Gangetic plains of India constitute two sub-populations harbouring rich genetic diversity. POJ 6(2):121–127

Singh BP, Jayaswal PK, Singh B, Singh PK, Kumar V, Mishra S, Singh N, Panda K, Singh NK (2015a) Natural allelic diversity in OsDREB1F gene in the Indian wild rice germplasm led to ascertain its association with drought tolerance. Plant Cell Rep 34:993–1004

Singh N, Jayaswal PK, Panda K, Mandal P, Kumar V, Singh B, Mishra S, Singh Y, Singh R, Rai V, Gupta A, Sharma TR, Singh NK (2015b) Single-copy gene based 50 K SNP chip for genetic studies and molecular breeding in rice. Sci Rep 5, 11600. doi:10.1038/srep11600

Tajima F (1989) Statistical method for testing the neutral mutation hypothesis by DNA polymorphism. Genetics 123:585–595

Tamura K, Peterson D, Peterson N, Stecher G, Nei M, Kumar S (2011) MEGA5: molecular evolutionary genetics analysis using maximum likelihood, evolutionary distance, and maximum parsimony methods. Mol Biol Evol 28:2731–2739

Tanksley SD, McCouch SR (1997) Seed banks and molecular maps: unlocking genetic potential from the wild. Science 277:1063–1066

Thompson JD, Higgins DG, Gibson TJ (1994) CLUSTAL W: improving the sensitivity of progressive multiple sequence alignment through sequence weighting, position-specific gap penalties and weight matrix choice. Nucleic Acids Res 22:4673–4680

Thomson MJ, de Ocampo M, Egdane J, Rahman MA, Sajise AG, Adorada DL, Tumimbang-Raiz E, Blumwald E, Seraj ZI, Singh RK (2010) Characterizing the Saltol quantitative trait locus for salinity tolerance in rice. Rice 3:148–160

Thorvaldsdóttir H, Robinson JT, Mesirov JP (2012) Integrative Genomics Viewer (IGV): high-performance genomics data visualization and exploration. Brief Bioinform 14(2):178–192. doi:10.1093/bib/bbs017

Villain P, Mache R, Zhou D-X (1996) The mechanism of GT element-mediated cell type-specific transcriptional control. J Biol Chem 271:32593–32598

Vlad F, Turk BE, Peynot P, Leung J, Merlot S (2008) A versatile strategy to define the phosphorylation preferences of plant protein kinases and screen for putative substrates. Plant J 55:104–117

Wang W, Vinocur B, Altman A (2003) Plant responses to drought, salinity and extreme temperatures: towards genetic engineering for stress tolerance. Planta 218:1–14

Watterson G (1975) On the number of segregating sites in genetical models without recombination. Theor Popul Biol 7:256–276

Xiao J, Li J, Yuan L, Tanksley S (1996) Identification of QTLs affecting traits of agronomic importance in a recombinant inbred population derived from a subspecific rice cross. Theor Appl Genet 92:230–244

Xiao J, Li J, Grandillo S, Ahn SN, Yuan L, Tanksley SD, McCouch SR (1998) Identification of trait-improving quantitative trait loci alleles from a wild rice relative, *Oryza rufipogon*. Genetics 150:899–909

Yang C, Zhao L, Zhang H, Yang Z, Wang H, Wen S, Zhang C, Rustgi S, von Wettstein D, Liu B (2014) Evolution of physiological responses to salt stress in hexaploid wheat. Proc Natl Acad Sci 111:11882–11887

Yu J, Buckler ES (2006) Genetic association mapping and genome organization of maize. Curr Opin Biotechnol 17:155–160

Yu Y-T, Wang R-H, Shi Y-S, Song Y-C, Wang T-Y, Yu L (2010) Haplotypic structure and allelic variation of rab17, an ABA-responsive gene, in a mini core set of Chinese diversified maize inbred lines. Agric Sci China 9:1726–1738

Yuan S-C, Zhang Z-G, He H-H, Zen H-L, Lu K-Y, Lian J-H, Wang B-X (1993) Two photoperiodic-reactions in photoperiod-sensitive genic male-sterile rice. Crop Sci 33:651–660

Zhu J-K (2001) Plant salt tolerance. Trends Plant Sci 6:66–71

6

Comparative Leaf and Root Transcriptomic Analysis of two Rice *Japonica* Cultivars Reveals Major Differences in the Root Early Response to Osmotic Stress

Elena Baldoni[1,2*], Paolo Bagnaresi[3], Franca Locatelli[1], Monica Mattana[1] and Annamaria Genga[1*]

Abstract

Background: Rice (*Oryza sativa* L.) is one of the most important crops cultivated in both tropical and temperate regions and is characterized by a low water-use efficiency and a high sensitivity to a water deficit, with yield reductions occurring at lower stress levels compared to most other crops. To identify genes and pathways involved in the tolerant response to dehydration, a powerful approach consists in the genome-wide analysis of stress-induced expression changes by comparing drought-tolerant and drought-sensitive genotypes.

Results: The physiological response to osmotic stress of 17 *japonica* rice genotypes was evaluated. A clear differentiation of the most tolerant and the most sensitive phenotypes was evident, especially after 24 and 48 h of treatment. Two genotypes, which were characterized by a contrasting response (tolerance/sensitivity) to the imposed stress, were selected. A parallel transcriptomic analysis was performed on roots and leaves of these two genotypes at 3 and 24 h of stress treatment. RNA-Sequencing data showed that the tolerant genotype Eurosis and the sensitive genotype Loto mainly differed in the early response to osmotic stress in roots. In particular, the tolerant genotype was characterized by a prompt regulation of genes related to chromatin, cytoskeleton and transmembrane transporters. Moreover, a differential expression of transcription factor-encoding genes, genes involved in hormone-mediate signalling and genes involved in the biosynthesis of lignin was observed between the two genotypes.

Conclusions: Our results provide a transcriptomic characterization of the osmotic stress response in rice and identify several genes that may be important players in the tolerant response.

Keywords: Hormones, Lignin, *Oryza sativa*, Osmotic stress, RNA-Seq analysis, Transcription factors

Background

Drought is one of the most important environmental constraints affecting plant growth and development and ultimately leads to yield loss. Water deficiency is a global concern because even the most productive agricultural regions can occasionally experience short periods or years of severe drought. Furthermore, irrigation might be restricted in the future because of the competition from other non-agricultural sectors, such as industry and urban areas (Bouman et al. 2007). Moreover, drought will continue to become worse in the next decades because of the potential impact of climate change on rainfall patterns and the need to extend the exploitation of marginal lands (Bates et al. 2008). Therefore, the implementation of water management practices and the development of both drought-tolerant varieties and water-use-efficient crops are key strategies to maintain yields under climate change conditions, extend cultivation to sub-optimal cropping areas and save water for sustainable agriculture. However, the development of drought-tolerant varieties still represents a challenging task, being hampered by the occurrence of genotype × environment interactions, the difficulty of effective drought tolerance screening and a still partial understanding of the

* Correspondence: elena.baldoni@unimi.it; genga@ibba.cnr.it
[1]Institute of Agricultural Biology and Biotechnology - National Research Council, via Bassini 15, 20133 Milan, Italy

molecular mechanisms of plant drought tolerance (Richards 1996; Kumar et al. 2008).

When plants perceive stress signals from their surroundings, a number of physiological, biochemical and molecular modifications occur (Krasensky and Jonak 2012). Some of these responses merely represent a consequence of cell damage, while others correspond to adaptive processes plants have evolved to cope with environmental cues. At the molecular level, the expression of a large number of genes is modulated under water stress conditions (Shinozaki and Yamaguchi-Shinozaki 2007; Qin et al. 2011; Yoshida et al. 2014). These genes encode either proteins with a direct role in protecting cell structures (e.g., metabolic enzymes, late embryogenesis-abundant proteins, detoxification enzymes and chaperones), or proteins with a regulatory function (e.g., transcription factors (TFs), protein kinases and other proteins involved in signal transduction) (Valliyodan and Nguyen 2006; Shinozaki and Yamaguchi-Shinozaki 2007; Hadiarto and Tran 2011). In particular, the identification of genes and pathways involved in the tolerant response to dehydration is clearly a crucial step in the development of drought-tolerant varieties. A powerful approach, which is increasingly being used to discriminate between drought tolerance-related genes and drought-responsive genes, is to perform genome-wide analyses of stress-induced expression changes by comparing drought-tolerant and drought-sensitive genotypes, rather than performing gene expression experiments on single genotypes (Moumeni et al. 2011; Utsumi et al. 2012; Guimaraes et al. 2012; Degenkolbe et al. 2013). This approach has allowed for the identification of genes with a positive function in enhancing drought tolerance and is potentially useful for the development of molecular markers to accelerate breeding programs.

Rice (*Oryza sativa* L.) is one of the most important crops cultivated in both tropical and temperate regions, representing the staple food for a large fraction of the world population. Rice is a high water demanding species, using approximately 40 % of the water diverted for irrigation (Lampayan et al. 2015), and rice cultivation is characterized by a low water-use efficiency and a high sensitivity to water deficit, with yield reductions occurring at lower stress levels compared to most other crops. Rice cultivation relies on cropping systems based on different water regimes, from irrigated systems to rainfed lowland and upland rice fields to deep water fields. The increasingly frequent occurrence of drought and the possible future restrictions of water availability for agricultural purposes are among the major challenges to be met to achieve sustainable rice production. Actually, it is estimated that by 2025, 15 to 20 million hectares of irrigated rice fields will suffer from some degree of water scarcity (Lampayan et al. 2015). For these reasons, the development of new rice cultivars with a better water-use efficiency or an enhanced drought tolerance is a primary goal in rice breeding programs. Currently, an increasing number of studies focuses on the identification of drought responsive genes that are differentially regulated in rice genotypes characterized by a contrasting phenotype in response to stress (Degenkolbe et al. 2009; Lenka et al. 2011; Cal et al. 2013; Degenkolbe et al. 2013; Moumeni et al. 2015).

In the present work, a parallel transcriptomic analysis was conducted on two Italian rice genotypes characterized by a contrasting phenotype in response to osmotic stress. RNA-Sequencing was performed separately on leaves and roots to characterize the specific response of these organs in the considered genotypes. The results of this study may contribute to elucidating the mechanisms involved in the rice response to osmotic stress and to identify genes that are putatively responsible for the stress-tolerant phenotype.

Results and Discussion

Physiological Response to Osmotic Stress

To evaluate the physiological response to osmotic stress of 17 *japonica* rice cultivars, which are currently listed in the Italian National Register, the leaf relative water content (RWC; Table 1) and the leaf electrolyte leakage (EL; Table 2) of plants subjected to 0, 3, 24 and 48 h of 20 % polyethylene glycol (PEG) treatment were measured. The rice cultivars showed different responses to the imposed stress. After 24 and 48 h of treatment, a clear differentiation of the most tolerant and the most sensitive phenotypes was evident. In particular, after 48 h of treatment, Carnaroli, Gigante Vercelli, Loto, Maratelli and Vialone Nano resulted to be the most sensitive cultivars, showing both the lowest RWC (<15 %) and the highest EL (>94 %) values, whereas Augusto and Eurosis resulted to be the most tolerant genotypes, showing both the highest RWC values (>80 %) and the lowest EL values (<40 %). The other cultivars (Thaibonnet, Baldo, Gladio, Koral, Salvo, SISR215, Volano, Arborio, Venere, Asia) exhibited an intermediate phenotype. In a previous study, we investigated the response to osmotic stress of 8 Italian rice cultivars; among them, 6 cultivars analyzed in the present work (Arborio, Augusto, Baldo, Eurosis, Loto and Vialone Nano) were included (Baldoni et al. 2013). In that analysis, Augusto and Eurosis resulted among the most tolerant cultivars, with RWC values > 60 % after 48 h of treatment, whereas Loto and Vialone Nano resulted the most sensitive cultivars, with RWC values < 15 % after 48 h of treatment. The present work, in which a higher number of cultivars was analyzed, confirmed the highly tolerant phenotype for Augusto and Eurosis, and the highly

Table 1 Relative water content (RWC) measurements of rice cultivars

	start		3 h		24 h		48 h	
CV	mean	*p*	mean	*p*	mean	*p*	mean	*p*
Carnaroli	95.3	bcd	68.3	abcd	16.8	abc	14.3	a
Gigante Vercelli	94.4	abcd	66.9	abc	17.2	abc	12.0	a
Loto	91.8	a	64.6	ab	33.9	abcde	11.7	a
Maratelli	92.9	ab	63.4	a	13.7	ab	13.8	a
Vialone Nano	93.3	abc	62.8	a	11.9	a	9.2	a
Thaibonnet	96.7	d	84.0	efg	39.7	cdef	18.5	ab
Baldo	94.6	abcd	78.1	defg	35.7	bcdef	23.6	abc
Gladio	97.0	d	85.6	fg	28.1	abcd	31.5	abc
Koral	92.9	ab	77.9	defg	50.6	defg	27.0	abc
Salvo	96.2	cd	76.8	cdef	30.3	abcde	28.1	abc
SISR215	95.3	bcd	78.4	defg	37.1	cdef	30.4	abc
Volano	93.9	abcd	74.7	bcde	34.4	abcdef	26.7	abc
Arborio	95.7	bcd	79.8	efg	52.8	efg	37.6	bc
Venere	96.4	cd	87.9	g	62.9	gh	44.6	cd
Asia	96.1	cd	85.3	efg	57.0	fgh	63.5	de
Augusto	94.3	abcd	76.9	cdef	76.6	hi	82.4	ef
Eurosis	94.2	abcd	87.0	fg	89.7	i	86.4	f

Data were taken following 0, 3, 24 and 48 h of PEG treatment. Each percentage value is the mean of 5 plants. For each sampling time, data were subjected to a one-way analysis of variance to compare the different varieties. Different letters in the same column show significant differences based on a Tukey's test ($p \leq 0.001$)

Table 2 Electrolyte leakage (EL) measurements of rice cultivars

	Start		3 h		24 h		48 h	
CV	Mean	*p*	Mean	*p*	Mean	*p*	Mean	*p*
Gigante vercelli	18.6	cd	41.8	h	94.8	lm	97.2	h
Carnaroli	16.4	abc	23.8	abc	88.0	hi	95.6	gh
Loto	16.7	bc	49.5	i	89.7	il	94.4	gh
Maratelli	13.6	a	42.0	h	95.8	m	95.8	gh
Vialone Nano	24.4	fg	40.2	gh	84.8	hi	96.2	gh
Volano	25.0	g	36.2	g	83.4	h	94.0	gh
SISR215	21.6	ef	31.2	f	56.2	c	91.6	fgh
Thaibonnet	22.0	ef	30.0	ef	67.4	f	89.8	fg
Baldo	14.6	ab	21.2	a	65.0	ef	85.8	ef
Koral	20.0	de	28.5	def	66.7	ef	80.3	de
Salvo	15.8	abc	22.6	ab	61.6	de	81.2	de
Gladio	14.2	ab	22.2	ab	76.0	g	78.0	d
Venere	26.0	g	32.0	f	52.8	c	69.8	c
Arborio	21.8	ef	30.6	ef	56.6	cd	58.4	b
Asia	16.8	bc	23.6	abc	43.8	b	55.0	b
Augusto	18.2	cd	25.4	bcd	27.8	a	30.6	a
Eurosis	21.6	ef	27.0	cde	29.0	a	35.8	a

Data were taken following 0, 3, 24 and 48 h of PEG treatment. Each percentage value is the mean of 3 plants. For each sampling time, data were subjected to a one-way analysis of variance to compare the different varieties. Different letters in the same column show significant differences based on a Tukey's test ($p \leq 0.001$)

sensitive phenotype for Loto and Vialone Nano in response to osmotic stress.

Based on the obtained physiological results and on the rice grain type classification, Eurosis and Loto (both Long A grain rice cultivars) were selected as tolerant and sensitive genotypes, respectively, for transcriptome analysis, to identify differentially expressed genes (DEGs) and differentially regulated pathways under osmotic stress.

RNA-Seq Data Analysis and Evaluation of DEGs

RNA samples that were isolated from leaves and roots of Eurosis and Loto genotypes at 3 and 24 h of osmotic stress treatment were subjected to whole transcriptome sequencing, to analyze the short-term response to osmotic stress. The corresponding control samples were also analysed. Three biological replicates were performed for each rice genotype and condition (48 samples in total). Raw reads (50 bases, single-end) obtained from Illumina HiSeq sequencing were filtered (Illumina passed-filter call), and the fastQC application was employed to detect sequence contaminants. Contaminant-free, filtered reads (trimmed to 43 bases to discard low-quality 3′ terminal regions) ranging from 17 to 36 million per sample (Additional file 1: Table S1) were mapped with Bowtie 0.12.7 and TopHat 1.4.1 to the rice Nipponbare genome (Ensembl plants release MSU 6.16). Raw read counts were obtained from

BAM alignment files by counting with HTSeq software. An RPKM (Reads per Kilobase per Million) cut-off value of 0.1 was set to declare a locus expressed resulting in 32,862 and 32,778 *loci* above the expression cut-off for Eurosis and Loto varieties, respectively. All of the biological replicates exhibited Pearson correlation coefficients above 0.9, indicating a good level of reproducibility among biological replicates (Fig. 1 and Additional file 2: Table S2).

To identify DEGs, the R package DESeq was employed. Two countset instances were created for Eurosis and Loto. The false discovery rate (FDR) threshold was set to 0.001, and gene dispersion values were calculated using the "maximum" mode. Given the large number of DEGs detected in all of the contrasts, a threshold of 3-fold expression change was further established to focus on a subset of DEGs showing highest modulation.

Gene expression changes for 10 selected genes, that were identified as DEGs during the RNA-Seq experiments and were reported to be involved in the osmotic stress response, were validated using qRT-PCR. The

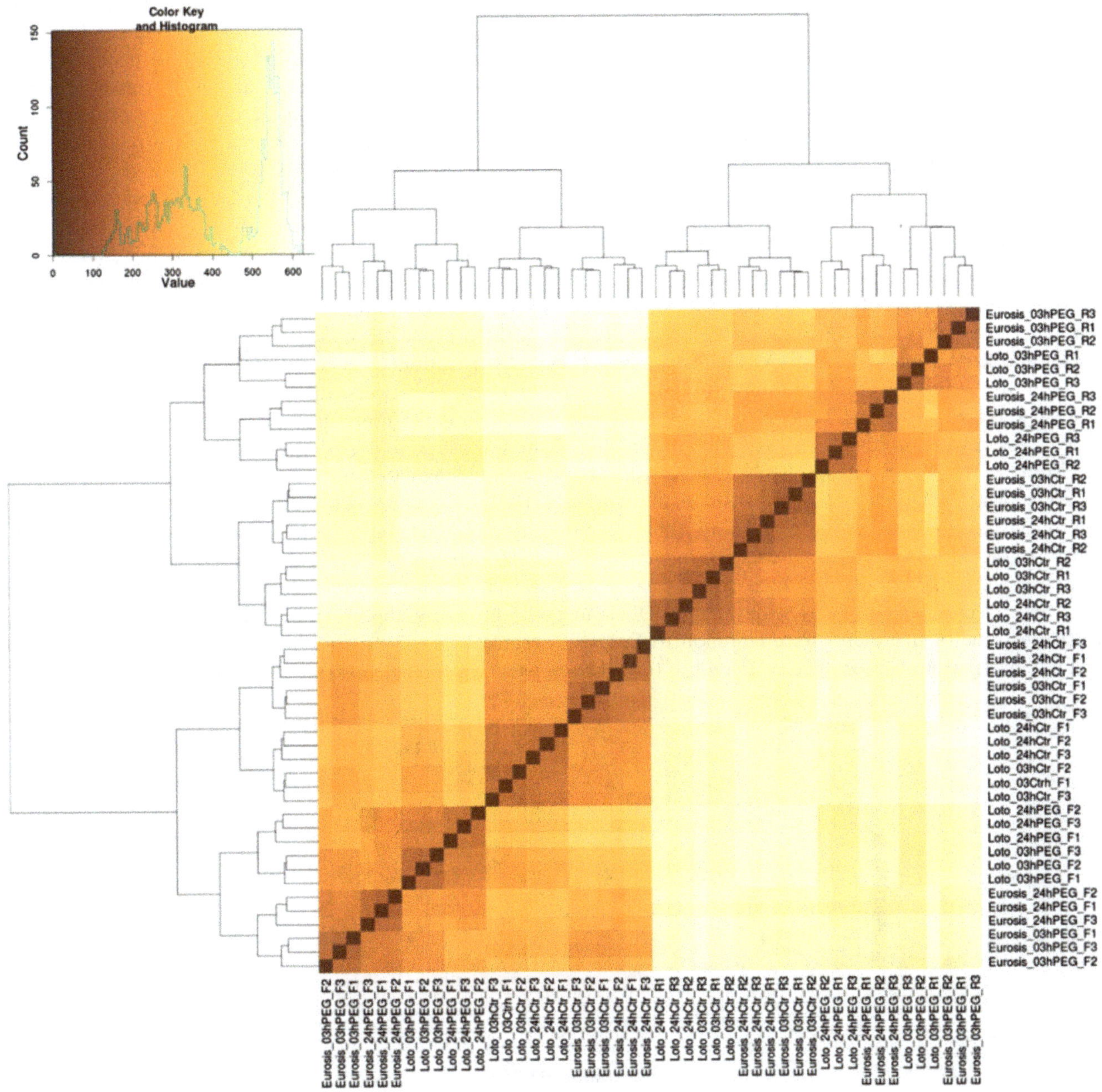

Fig. 1 Clustering and heatmaps of samples. DESeq-normalized count data were transformed with the VST function (DESeq package). Eurosis and Loto (control/Ctr or osmotic/PEG treatment) samples and biological replicates (F1, F2 and F3 for leaves, and R1, R2 and R3 for roots) are shown. The colour scale indicates the degree of correlation (*white-yellow, low correlation; orange-red, strong correlation*)

comparison between RNA-Seq and qRT-PCR fold change data, which was obtained by a regression analysis, revealed a substantial agreement in the extent of the osmotic stress-induced variations in transcript accumulation for the 10 tested genes (Fig. 2 and Additional file 3: Table S3). The transcript levels of 6 of these genes (LOC_Os01g07120, LOC_Os01g66120, LOC_Os02g44870, LOC_Os04g43680, LOC_Os04g45810 and LOC_Os10g33810) have been previously analysed using qRT-PCR in rice samples of Eurosis and Loto plants under the same stress conditions (Baldoni et al. 2013). The comparison between qRT-PCR data presented here and in the previous work revealed comparable stress-induced fold changes, indicating a good reproducibility of the osmotic stress experiments (data not shown).

The RNA-Seq analysis showed that osmotic stress caused significant changes in gene expression in the roots and leaves of both genotypes (Fig. 3, Table 3 and Additional file 4: Table S4). In particular, 3 h of osmotic stress led to a substantial modulation of gene expression in the roots of the tolerant genotype Eurosis (6007 genes), whereas fewer DEGs were called in the roots of the susceptible genotype Loto (3962 genes). At 24 h of treatment, a similar number of DEGs was found in Eurosis and Loto roots (3065 and 3102 genes, respectively). In the leaves, a similar number of DEGs was found in the two genotypes, both at 3 and 24 h of osmotic stress treatment (2977 and 4223 in Eurosis, and 3088 and 4813 in Loto, respectively) (Table 3). The stress response of the two cultivars consisted of a common and a specific component (Fig. 3 and Table 3). In particular, the specific response was prominent in the roots. Indeed, in the roots at 3 h of stress treatment, 2499 DEGs (representing only 34 % of the total DEGs in this tissues/time of treatment) were common to Eurosis and Loto, whereas 66 % were specific to one or the other genotype (47 and 19 % to Eurosis and Loto, respectively). Only 53 DEGs (0.7 %) were oppositely modulated in the two genotypes in this tissue/time of treatment. It is noteworthy that in all of the other contrasts, no oppositely modulated DEGs were observed. In 24 h-treated roots, a proportion of 41 % of the DEGs was common to Eurosis and Loto, whereas 59 % were specific to one or the other genotype. In the leaves, the proportion of common DEGs was higher than in the roots, with 51 and 54 % at 3 and 24 h of stress treatment, respectively (Fig. 3 and Table 3). Regarding up-and down-regulated genes, the mean expression vs. logfold change plots (MA-plots, Fig. 4) illustrated that, among the DEGs, the proportion of up- and down-regulated genes was substantially similar in all the control vs. treated comparisons, with the exception of the 3 h root samples of Eurosis, in which the number of up-regulated genes (61.6 % of all DEGs) was higher than the number of down-regulated genes (38.4 %). In summary, major differences between the responses of the two cultivars were observed at 3 h of osmotic treatment in roots, where a higher number of genes were significantly regulated and where the common response was less represented than in the other tested conditions.

As mentioned above, DEGs showing an opposite modulation between the two genotypes were found only in 3 h-treated roots; that is, 4 DEGs were up-regulated in Eurosis and down-regulated in Loto, and 49 DEGs were up-regulated in Loto and down-regulated in Eurosis (Table 3 and Additional file 5: Table S5). Thus far, no literature data are available on the function of the 4 genes that were up-regulated in Eurosis and down-regulated in Loto; only little information is present in the literature about some of the 49 genes that were up-regulated in Loto and down-regulated in Eurosis. In particular, LOC_Os01g06500 codes for a lectin-like

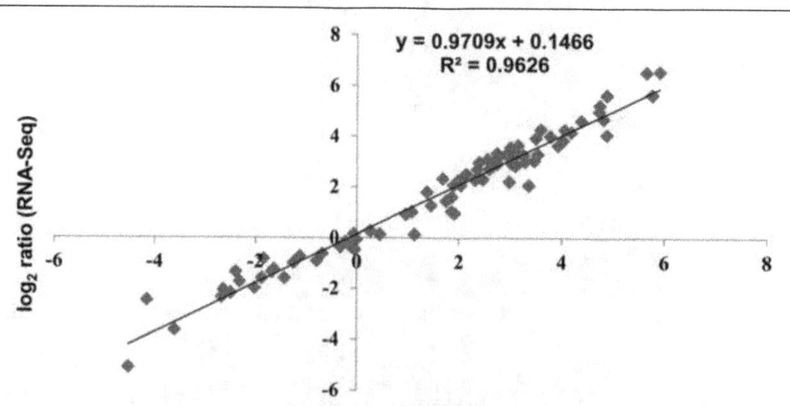

Fig. 2 Validation of the expression of selected genes from RNA-Seq using qRT-PCR. Fold changes in gene expression were transformed to a $\log_2$ scale. The qRT-PCR data $\log_2$-values *(X-axis)* were plotted against the RNA-Seq $\log_2$ values *(Y-axis)*. The function of the regression line and the R^2 value are shown

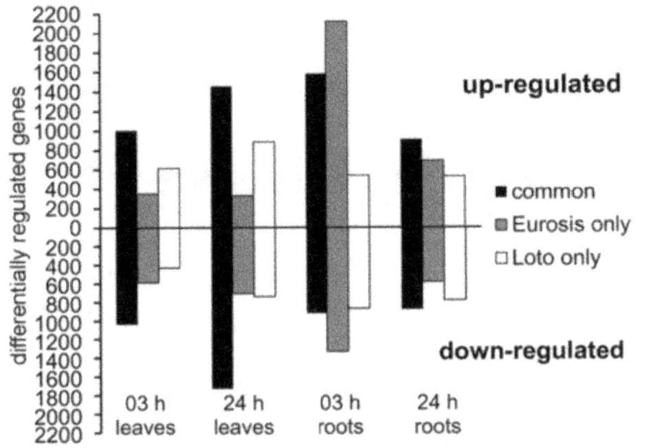

Fig. 3 Up- and down-regulated genes in leaf or root samples at 3 and 24 h of PEG treatment. *Black bars*: common regulated genes in the two cultivars. *Grey bars*: Eurosis specific up- or down-regulated genes. *White bars*: Loto specific up- or down-regulated genes

protein, and its expression correlated with the isoprenoid biosynthetic pathway (Suzuki et al. 2007). In rice, the salt-inducible gene *SalT* encodes a cytoplasmic mannose-binding lectin, whose expression increases after salt or drought stress (de Souza Filho et al. 2003). In general, the expression of lectin-encoding genes may be regulated under abiotic or biotic stress conditions, and a putative role for these genes in plant stress response has been proposed (Rüdiger and Gabius 2001; Van Damme et al. 2004). Moreover, among these 49 genes, some putative TF-encoding genes were present, namely, 1 MYB (LOC_Os02g17190), three putative PHD finger family proteins (LOC_Os03g53630, LOC_Os11g29240 and LOC_Os12g24540) and one putative bZIP (LOC_Os03g20530). These genes may be involved in the upstream regulation of the osmotic stress response and may have a crucial role in establishing a different response to the imposed stress in the two genotypes.

Gene Ontology Enrichment Analysis of RNA-Seq Data

Gene ontology (GO) enrichment analysis of DEGs in the various contrasts was conducted to reveal biological trends differentiating tolerant and susceptible genotypes. The goseq bioconductor package (Robinson and Oshlack 2010) was used to account for the RNA length bias that is typical of RNA-Seq approaches (Oshlack and Wakefield 2009). Both common and genotype-specific enriched terms were identified, confirming the presence of a consistent common response to osmotic stress and genotype-specific responses, as observed in the analysis of DEGs (Fig. 3).

Common Enriched GO Terms

Sixty-one enriched GO terms were shared by the 2 rice genotypes in all of the analysed tissues and treatments (Additional file 6: Table S6). Among the 21 GO terms related to Biological Processes (BP), some were consistent with responses to water/osmotic stress; these included "response to water deprivation" (GO:0009414), "response to water stimulus" (GO:0009415) and "response to stress" (GO:0006950). Moreover, several common GO terms were related to the response to abiotic stresses, such as "response to oxidative stress" (GO:0006979), "response to cold" (GO:0009409), "response to wounding" (GO:0009611), "response to salt stress" (GO:0009651) and "response to freezing" (GO:0050826). This result was not unexpected because the existence of cross-talk among pathways involved in the response to different abiotic stresses is well known (Qin et al. 2011; Nakashima et al. 2014).

Moreover, the GO terms "response to abscisic acid stimulus" (GO:0009737) and "cellular response to abscisic acid stimulus" (GO:0071215) were shared by all of the samples, and this is consistent with the well known pivotal role of abscisic acid (ABA) signalling in the response to osmotic stress (Fujita et al. 2011). Additional GO terms related to hormone signalling were shared by all of the root samples but not by leaf samples, or vice-versa; in the roots, two GO terms were related to salicylic acid and jasmonic acid-mediated responses (GO:0009751 and GO:0009753, respectively), whereas the GO terms "response to ethylene stimulus" (GO:0009723) and "indolebutyric acid metabolic process" (GO:0080024) were shared only by the leaf samples.

The enriched GO terms shared by all of the root samples or by all of the leaf samples were 20 and 70, respectively, confirming a prominent common response in leaves. Water stress highly affects the photosynthetic

Table 3 Summary of the total number of DEGs, DEGs common to both cultivars and cultivar-specific DEGs

Tissue	Contrasts	Total number of DEGs		Common DEGs	Specific DEGs	
		Eurosis	Loto		Eurosis	Loto
Roots	3 h control vs. 3 h stress	6007	3962	2552[a]	3455	1410
Roots	24 h control vs. 24 h stress	3065	3102	1789	1276	1313
Leaves	3 h control vs. 3 h stress	2977	3088	2040	937	1048
Leaves	24 h control vs. 24 h stress	4223	4813	3187	1036	1626

[a]Among the 2552 common DEGs, 53 genes were oppositely regulated between the 2 genotypes, whereas 2499 were modulated in the same manner in the 2 genotypes

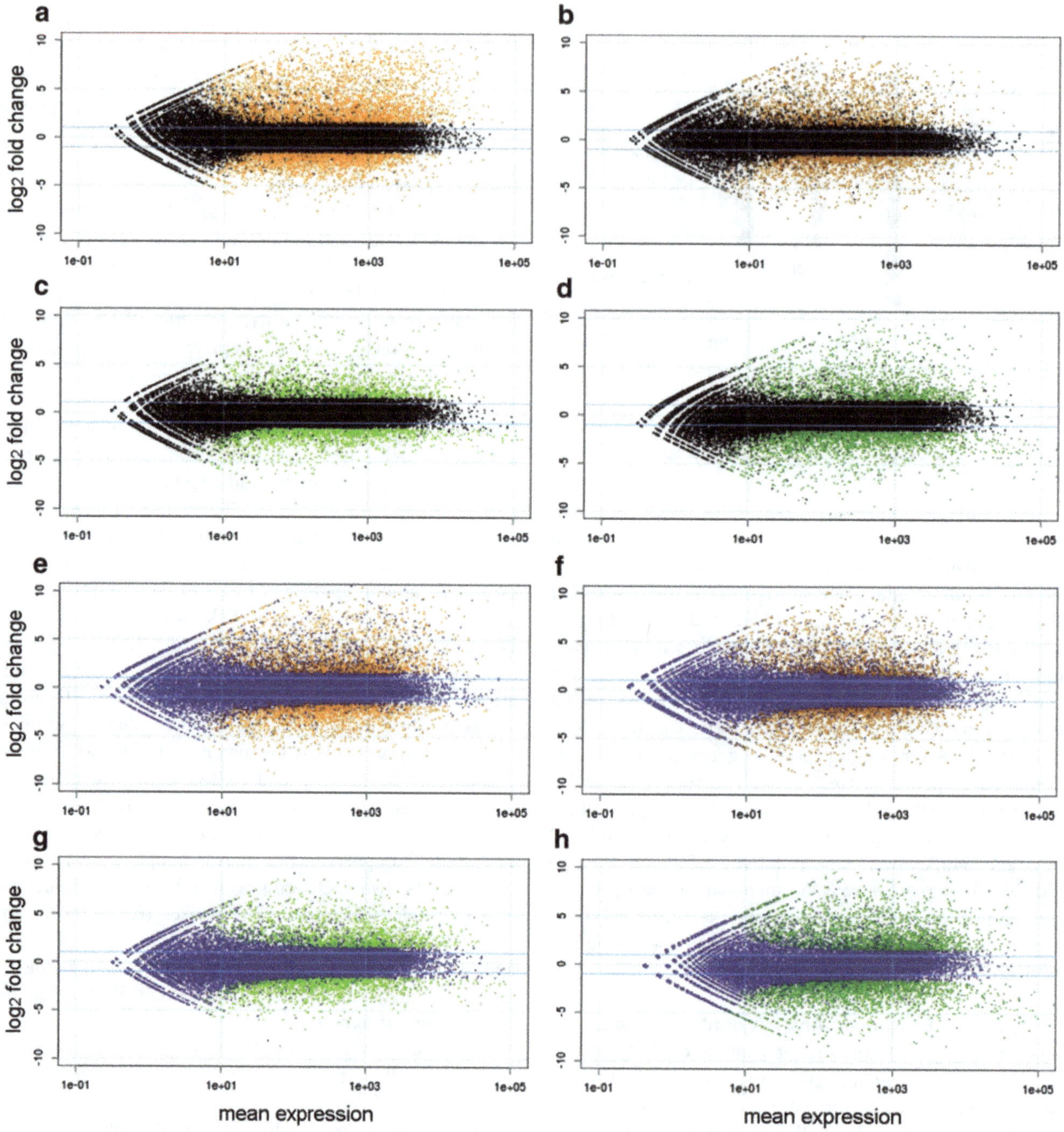

Fig. 4 Mean expression vs. log fold change plots (MA-plots) of Eurosis and Loto samples. Mean expression vs. log fold change plots (MA-plots) were computed for Eurosis (*plotted in black*; **a-d**) and Loto (*plotted in blue*; **e-h**) contrasts. Normalised expression mean values are plotted vs. $\log_2$fold changes. Called DEGs (FDR < 0.001 and fold change > 3) are plotted in orange for root samples (**a**, **b**, **e** and **f**) and in green for leaf samples (**c**, **d**, **g** and **h**)

processes (Chaves et al. 2009). As expected, many enriched GO terms common to leaf samples were related to photosynthesis (Additional file 6: Table S6). Because of the inhibition of photosynthesis during stress, plants must mobilize energy from storage resources, such as carbohydrates, fatty acids and proteins (Shu et al. 2011). Actually, some GO terms related to fatty acid metabolism were shared by leaf samples of the 2 genotypes, namely, "lipid metabolic process" (GO:0006629), "fatty acid biosynthetic process" (GO:0006633) and "lipid biosynthetic process" (GO:0008610) (Additional file 6: Table S6).

Genotype-Specific Enriched GO Terms

Several GO terms were genotype-specific (i.e., enriched GO terms from one genotype/organ/treatment missing in the alternative genotype in the same organ/treatment). Additional file 7: Table S7 reports all of the enriched GO terms for the contrasts with accompanying enrichment *p*-values resulting from goseq analysis. Figures 5 and 6 depict genotype-specific barplots. Because DEG analysis found that the genotype-specific response was prominent in roots, our analysis mainly focused on root samples.

Stress and Hormone-Related GO Terms

Some enriched GO terms specific to either the Eurosis or the Loto genotype were related to osmotic stress

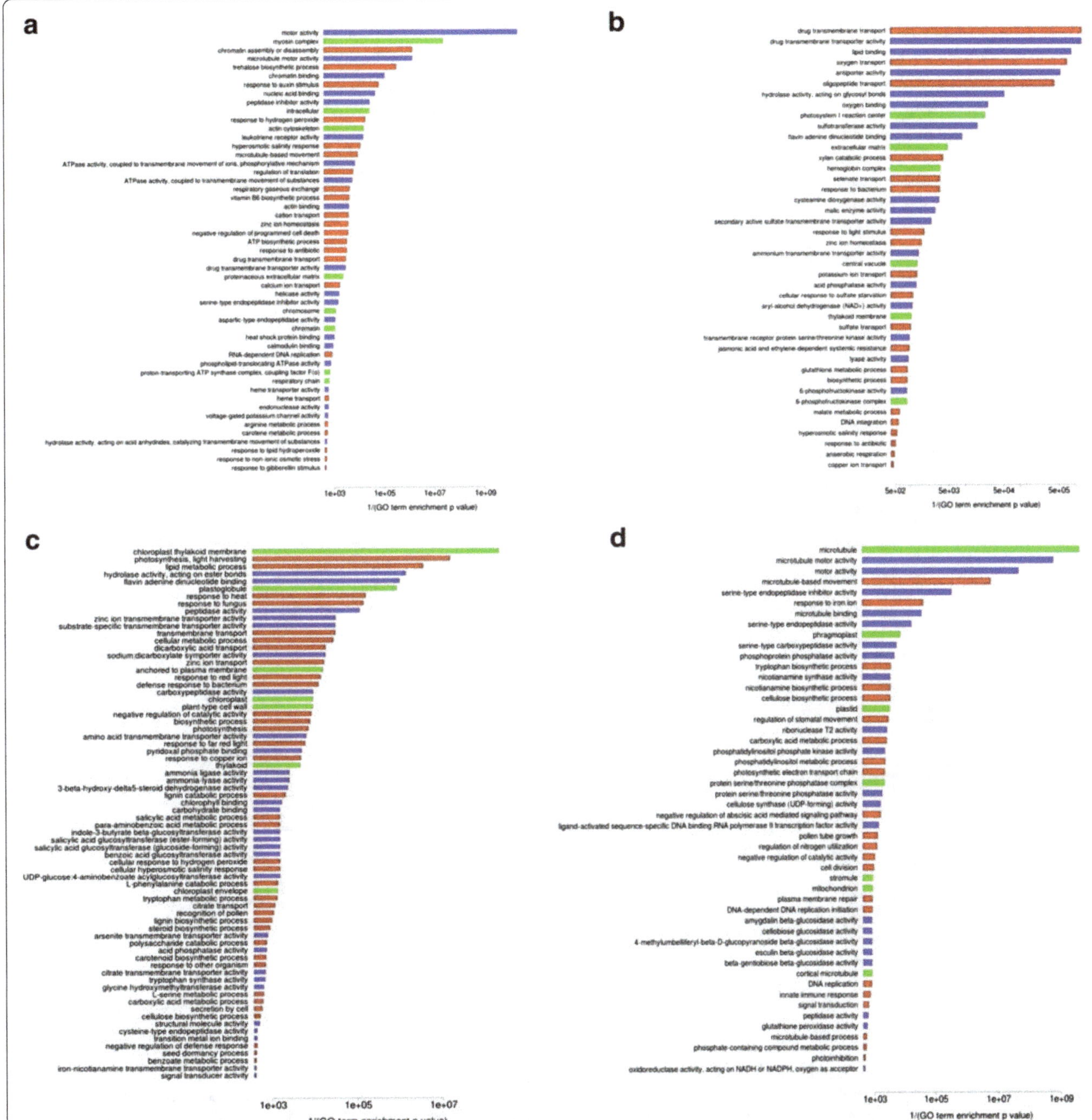

Fig. 5 Barplots of differentially enriched GO terms in the roots. Barplots of differentially enriched GO terms (from one genotype/organ/treatment missing in the alternative genotype in the same organ/treatment) in the roots, as estimated by goseq at an FDR cut-off of 0.05. Reciprocal of enrichment *p*-values are shown on the X axis. Red boxes indicate Biological Process GO terms; blue boxes indicate Molecular Function GO terms; green boxes indicate Cell Compartment GO terms. **a** Eurosis 3 h; **b** Eurosis 24 h; **c** Loto 3 h; **d** Loto 24 h

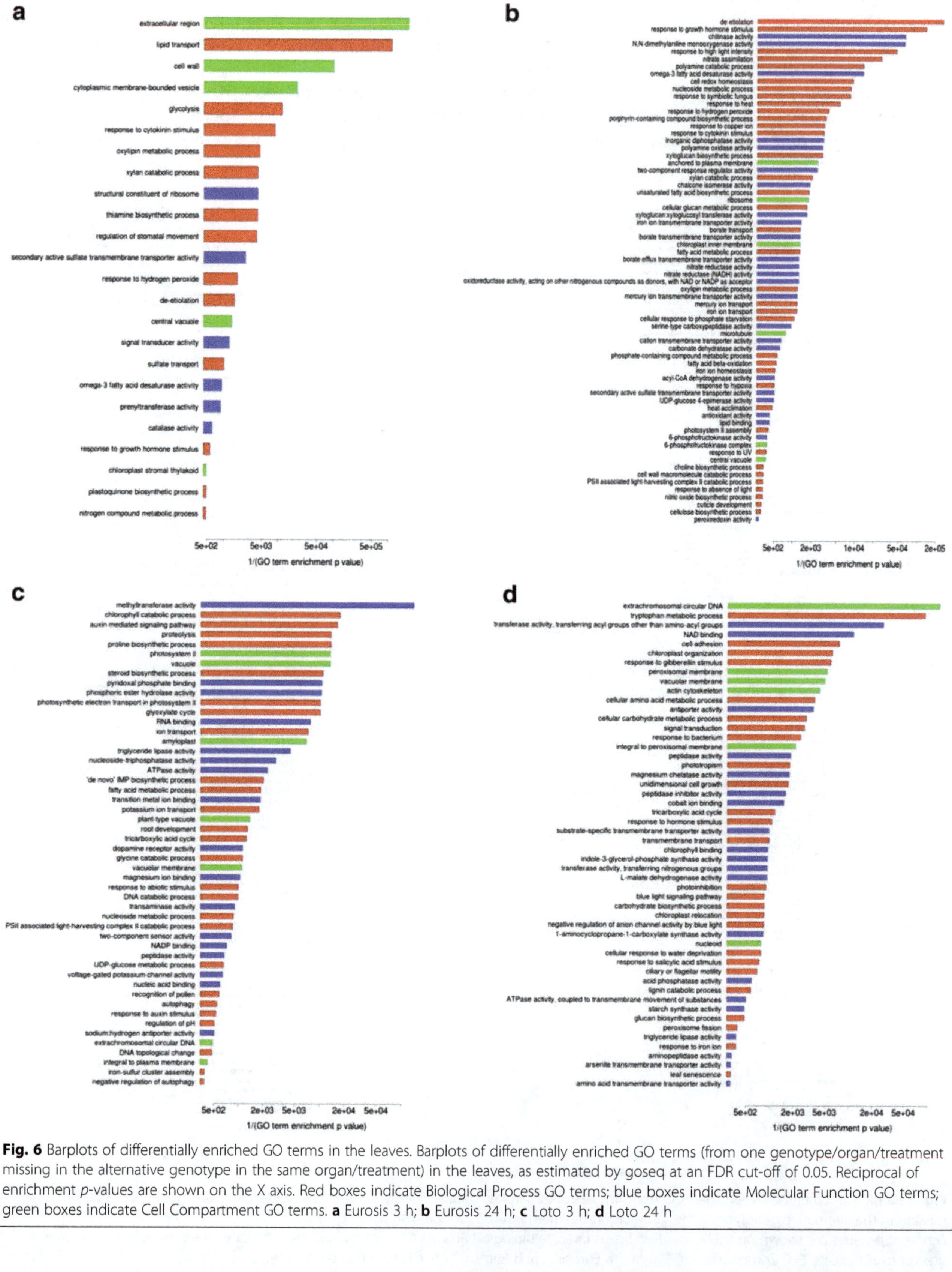

Fig. 6 Barplots of differentially enriched GO terms in the leaves. Barplots of differentially enriched GO terms (from one genotype/organ/treatment missing in the alternative genotype in the same organ/treatment) in the leaves, as estimated by goseq at an FDR cut-off of 0.05. Reciprocal of enrichment *p*-values are shown on the X axis. Red boxes indicate Biological Process GO terms; blue boxes indicate Molecular Function GO terms; green boxes indicate Cell Compartment GO terms. **a** Eurosis 3 h; **b** Eurosis 24 h; **c** Loto 3 h; **d** Loto 24 h

response, i.e., "trehalose biosynthetic process" (GO:0005992), "response to non-ionic osmotic stress" (GO:0010335) and "hyperosmotic salinity response" (GO:0042538) in Eurosis roots (Fig. 5a and b); "cellular hyperosmotic salinity response" (GO:0071475) in Loto roots (Fig. 5c); "choline biosynthetic process" (GO:0042425) in Eurosis leaves (Fig. 6b); "proline biosynthetic process" (GO:0006561), "response to abiotic stimulus" (GO:0009628) and "cellular response to water deprivation" (GO:0042631) in Loto leaves (Fig. 6c and d). Some genes belonging to these GO terms may play a role in the different response of the two genotypes. Among them, LOC_Os03g04570, which was up-regulated only in Eurosis 3 h-treated roots and belonged to the GO term GO:0042538 (Additional file 7: Table S7), codes for a putative peptide transporter (PTR). The PTR family consists of di/tripeptides transporters, and their role in plants is not yet well defined (Saier et al. 2006). This gene is similar to *AtPTR3* (AT5G46050), whose expression is induced by salt stress and mechanical wounding and is regulated by jasmonic acid and salicylic acid (Karim et al. 2005; Karim et al. 2007). Another interesting gene is LOC_Os07g01020, which was up-regulated in Eurosis 3 h- and 24 h-treated roots and in Loto 24 h-treated roots and belonged to the GO terms GO:0010335 and GO:0042538 (Additional file 7: Table S7). This gene codes for a putative SOR/SNZ family protein that has high homology with the pyridoxine synthase gene AT5G01410 (Chen et al. 2014). Interestingly, Jeong et al. (2013) observed that in transgenic rice plants over-expressing *OsNAC5* under the control of a root-specific promoter, LOC_Os07g01020 was up-regulated together with other genes implicated in root growth and development. These plants showed a higher grain yield under drought conditions compared to wild-type plants and an increased root diameter, which could contribute to the enhanced drought tolerance. In our work, the specific induction of LOC_Os07g01020 in Eurosis 3 h-treated roots suggests a possible role for this gene in the early response to osmotic stress of this genotype.

Moreover, several enriched GO terms specific to Eurosis or Loto were related to hormone metabolism (Figs. 5 and 6, Additional file 7: Table S7). Some genes belonging to these GO terms may have a role in the different responses of the two genotypes. For instance, a *MYB* gene (LOC_Os01g34060) belonging to GO terms related to the response to auxin and gibberellin stimulus (GO:0009733 and GO:0009739, respectively) was up-regulated in Eurosis roots after 3 h of treatment and in Loto roots after 24 h, suggesting a delay in the Loto response (Additional file 7: Table S7).

Chromatin-Related GO Terms

Among GO terms specifically enriched in 3 h-treated roots of the tolerant genotype Eurosis, 5 were associated with chromatin (Fig. 5a and Additional file 7: Table S7), namely, "chromatin" (GO:0000785), "chromatin binding" (GO:0003682), "helicase activity" (GO:0004386), "chromosome" (GO:0005694) and "chromatin assembly or disassembly" (GO:0006333). The enrichment of these GO terms was due to the statistically significant modulation of 71 genes (12 up- and 59 down-regulated), which occurred exclusively in Eurosis. Notably, only 1 out of the up-regulated genes and 3 out of the down-regulated genes were listed among the DEGs in the 24 h-treated root samples, suggesting that the transcriptomic response associated with chromatin-related processes is an early and transient response in the tolerant cultivar Eurosis.

Globally, the most represented protein family encoded by DEGs annotated within chromatin-related GO categories was the helicase protein superfamily. Helicases are molecular motors that mainly use the energy derived from ATP hydrolysis to bind, unwind or remodel energetically stable double-stranded DNA (DNA helicases) or local RNA secondary structures (RNA helicases). They are central players in virtually all facets of nucleic acid metabolism, and may play essential roles in the response to stress conditions (Vashisht and Tuteja 2006; Tuteja et al. 2013; Zhu et al. 2015).

In rice, 40 Snf2 proteins, which are ATP-dependent chromatin remodelling factors, have been identified, and some of these proteins were affected by stress treatments (Hu et al. 2013). In our 3 h-treated root dataset, 19 out of the 40 *Snf2* genes were listed among the DEGs. Among them, 3 genes were up-regulated exclusively in Loto, whereas the other 16 were down-regulated (14 only in Eurosis, 1 only in Loto and 1 in both genotypes). One of the genes specifically down-regulated in Eurosis 3 h-treated roots was LOC_Os06g08480, which is annotated as the CHD3-type chromatin-remodelling factor PICKLE. Interestingly, a role of the PICKLE factor in balancing osmotic stress responses during seed germination has been well demonstrated in Arabidopsis (Perruc et al. 2007). Li et al. (2011) found that LOC_Os06g08480 and another down-regulated *Snf2* gene in Eurosis 3 h-treated roots, namely, LOC_Os01g44990, were induced by several abiotic stresses, including PEG-mediated osmotic treatment, in the sensitive cultivar Zhonghua11.

Another represented protein family encoded by DEGs that are annotated within chromatin-related GO categories was the histone protein family. Notably, in our dataset, 7 DEGs annotated as histone proteins were exclusively up-regulated in Eurosis, and 1 was down-regulated in Loto. Recent studies have reported that chromatin regulation, and, in particular, histone modification and DNA methylation, are key elements of the transcriptional response to abiotic stresses, such as water deficit, high-salinity, and temperature shifts. However, it still remains unclear how transcriptional changes and chromatin changes are linked (Kim et al. 2015).

Among the DEGs, which are annotated with chromatin-related GO terms, 2 showed an opposite regulation in the two cultivars, LOC_Os02g04050 and LOC_Os02g50370, both being down-regulated in Eurosis and up-regulated in Loto (Additional file 4: Table S4). The former is annotated in MSU 6.16 as a putative chromosome segregation protein and the latter as a putative helicase domain-containing protein (Additional file 4: Table S4). To our knowledge, no function in response to any abiotic stress has been reported for these two genes.

It is currently understood that the regulation of abiotic stress responsive genes is related to chromatin alterations. Consistently, our results suggest that chromatin remodelling processes occur in response to osmotic stress and may contribute to differentiating the response between the tolerant and the sensitive cultivar.

Cytoskeleton-Related GO Terms

In 3 h-treated root samples, 6 GO categories specifically enriched in the Eurosis genotype were related to the cytoskeleton (Fig. 5a and Additional file 7: Table S7), namely, "motor activity" (GO:0003774), "microtubule motor activity" (GO:0003777), "actin binding" (GO:0003779), "microtubule-based movement" (GO:0007018), "actin cytoskeleton" (GO:0015629) and "myosin complex" (GO:0016459). A total of 57 DEGs (21 up-regulated and 36 down-regulated) were responsible for the enrichment of these GO terms in Eurosis 3 h-treated roots. Their expression did not significantly change in Loto 3 h-treated roots. Among the 36 down-regulated DEGs, several genes are annotated either as kinesins (16 genes) or myosins (9 genes), and few additional genes encode actin regulatory proteins, such as formin, villin, and SCAR-like proteins. Notably, 11 out of these 36 genes were significantly down-regulated in Loto roots after 24 h of treatment, thus suggesting a delay in the response of the sensitive cultivar in comparison to the tolerant one.

Interestingly, some genes associated with these GO terms showed an opposite regulation in the 3 h-treated roots of the two genotypes, all of them exhibiting a down-regulation in Eurosis and an up-regulation in Loto. Among them, two genes belong to the myosin gene family, namely, LOC_Os01g51632 and LOC_Os01g51634. One additional gene, LOC_Os03g06510, was annotated as KIP1. Kip-related proteins (KRPs) play a central role in the regulation of the cell cycle and differentiation through the modulation of cyclin-dependent kinases. In Arabidopsis, the overexpression of a member of the KRP family, KRP2, prevents pericycle activation and reduces the number of lateral roots, suggesting a significant role of KRP2 in the regulation of early lateral root initiation (Himanen et al. 2002). Another KRP protein, ICK3/KRP5, is a positive regulator of both cell growth and endoreduplication in roots (Wen et al. 2013). Some studies analysed the relationship between drought stress and the expression of the *KRP* genes involved in the reprogramming of cell proliferation and cell expansion in leaves (Claeys and Inze 2013; Guan et al. 2014). To our knowledge, no data are available in the literature on a correlation between the expression of *KRP* genes in roots and the tolerance to water stress.

Three additional genes, which showed an opposite regulation in the 3 h-treated roots of the two genotypes, belong to the kinesin gene family, namely, LOC_Os06g36080, LOC_Os07g44400 and LOC_Os09g25380. Kinesins are microtubule-based motor proteins that are ubiquitous in all eukaryotic organisms and use the energy derived from ATP hydrolysis to move along the cytoskeletal elements of microtubules. Kinesins are involved in microtubule organization, organelle and vesicle transport, cellulose microfibril order, and ultimately contribute to cell division, cell growth and the cross-talk of microtubules and actin microfilaments (Li et al. 2012). Recently, a kinesin-like calmodulin-binding protein has been reported to be involved in the signalling network that negatively regulates root growth in Arabidopsis (Humphrey et al. 2015). Considering the kinesin encoding DEGs that were genotype-specific in 3 h-treated roots, 16 were down-regulated in Eurosis and 6 were up-regulated in Loto. Notably, all of the kinesin encoding DEGs showed either a down-regulation in Eurosis or an up-regulation in Loto. This differential regulation of kinesin-encoding genes observed in 3 h-treated roots of Eurosis and Loto suggested that this protein family affects the root growth in a genotype-specific manner, thus contributing to the different osmotic stress phenotypes (tolerance/sensitivity) of the 2 cultivars.

Transmembrane Transport-Related GO Terms

Six enriched GO terms specific to the tolerant genotype Eurosis in the roots at 3 h of stress treatment were related to transmembrane transport (Fig. 5a and Additional file 7: Table S7), namely, "phospholipid-translocating ATPase activity" (GO:0004012), "drug transmembrane transport" (GO:0006855), "drug transmembrane transporter activity" (GO:0015238), "ATPase activity, coupled to transmembrane movement of ions, phosphorylative mechanism" (GO:0015662), "hydrolase activity, acting on acid anhydrides, catalyzing transmembrane movement of substances" (GO:0016820) and "ATPase activity, coupled to transmembrane movement of substances" (GO:0042626). Thirty-three genes were responsible for the enrichment of these GO terms in the tolerant cultivar. Among them, 10 genes were significantly up-regulated only in Eurosis 3 h-treated roots, including 7 genes encoding multidrug and toxic compound extrusion (MATE) family proteins and 2 genes encoding ATP-binding cassette (ABC) transporter family proteins, whereas 23 genes were specifically down-regulated in Eurosis 3 h-treated roots, including 12 genes encoding ABC transporter family proteins. ABC and MATE proteins represent the two largest families of active transporters that have been characterized in plants thus far. ABC proteins are primary transporters that mediate the energy-driven transport of a large and diverse multitude of substrates across biological membranes. ABC transporters play key roles in many physiological processes, such as plant development and plant nutrition, and in stress response (Moons 2008; Matsuda et al. 2012; Jarzyniak and Jasinski 2014; Nguyen et al. 2014; Saha et al. 2015). In the rice genome, 133 genes coding for ABC transporters have been identified (Saha et al. 2015). In our study, in 3 h-treated roots, 15 out of the 133 ABC transporter genes were modulated in the same way in the two cultivars, 4 and 7 ABC transporter genes were specifically up-regulated in Eurosis and Loto, respectively, and 14 and 3 ABC transporter genes were specifically down-regulated in Eurosis and Loto, respectively. Among the 4 Eurosis-specific up-regulated genes, the *OsABCG22* gene (LOC_Os09g29660) is induced by mannitol and by ABA treatments in rice roots (Matsuda et al. 2012). Moreover, Nguyen et al. (2014) found that drought stress induced this gene in the leaves. Our results are consistent with these data, as this gene was also up-regulated in our leaf samples, both at 3 and 24 h of stress treatment.

In addition, the superfamily of MATE proteins represents one of the largest transporter families in plants, with more than 50 MATE genes in the rice genome. MATEs are secondary active transporters; in plants they have many physiological functions, such as the accumulation of secondary metabolites, aluminium detoxification and iron translocation (Takanashi et al. 2014; Jarzyniak and Jasinski 2014). Some MATEs are involved in hormone signalling in Arabidopsis, such as AtEDS5, which is responsible for intracellular salycilic acid transport, and AtDTX50, which functions as an ABA efflux transporter and plays a role in ABA-mediated growth inhibition and responses to drought conditions (Takanashi et al. 2014). In our study, among the 50 genes annotated as MATE efflux family proteins or MATE domain containing proteins, 8 DEGs (7 up- and 1 down-regulated) were specific to Eurosis 3 h-treated roots, and 2 DEGs (both down-regulated) were specific to Loto 3 h-treated roots. To our knowledge, no data are available in the literature on a putative role for any of these MATE efflux genes in water stress response. However, the observed stress-mediated regulation of several MATE genes suggests a possible role for some of them in the response to osmotic stress.

Finally, LOC_Os06g12876, which belongs to the GO terms GO:0016820, was 1 of the 49 genes oppositely modulated (down- and up-regulated in Eurosis and Loto, respectively) in 3 h-treated roots (Table 3). This gene is annotated in RAP-DB as an "ATPase, F1/V1/A1 complex, alpha/beta subunit". Thus far, no data are available on the putative function of this gene.

Oxygen-Related GO Terms

In Eurosis 24 h-treated roots, 5 enriched GO terms were related to oxygen (Fig. 5b), namely, "response to hypoxia" (GO:0001666), "anaerobic respiration" (GO:0009 061), "oxygen transport" (GO:0015671), "oxygen binding" (GO:0019825) and "haemoglobin complex" (GO:0005833). The rice genome contains five genes encoding non-symbiotic haemoglobins (ns-Hbs), namely *OsNSHB1* to *OsNSHB5* (LOC_Os03g13140, LOC_Os03g12510, LOC_Os03g13150, LOC_Os03g13160 and LOC_Os05g44140, respectively) (Garrocho-Villegas et al. 2008), and one gene encoding a putative haemoglobin-like protein HbO (LOC_Os06g39140). Notably, 4 out of the 5 genes coding for ns-Hbs (OsNSHB1-4) and the putative HbO-encoding gene were up-regulated in 24 h-treated roots only in Eurosis, suggesting a possible role of this class of molecules in the differential response of the two genotypes to the treatment. The ns-Hbs are localized in diverse plant organs, their expression is differently regulated under stress conditions, and their apparent function in vivo is to modulate the levels of ATP and nitric oxide (ns-Hbs class 1) or to facilitate the diffusion of O_2 (ns-Hbs class 2) (Garrocho-Villegas et al. 2008), but the precise function of these genes in rice has not yet been revealed.

Photosynthesis-Related GO Terms

Surprisingly, in Loto treated roots, 11 enriched GO-terms were related to photosynthesis (Fig. 5c and d), namely, "chloroplast thylakoid membrane" (GO:0009535), "photosynthesis, light harvesting" (GO:0009765), "plastoglobule" (GO:0010287), "chloroplast" (GO:0009507), "photosynthesis" (GO:0015979), "thylakoid" (GO:0009579), "chlorophyll binding" (GO:0016168), "chloroplast envelope" (GO:0009941), "plastid" (GO:0009536), "photosynthetic electron transport chain" (GO:0009767) and "stromule" (GO:0010319). Nevertheless, this result is consistent with other literature data. For example, in a transcriptomic analysis of leaves and roots of rice seedlings subjected to acute dehydration, many photosynthesis-related GO terms were enriched in roots under dehydration, including GO:0009579, GO:0009765 and GO:0016168 (Minh-Thu et al. 2013). Moreover, a transcriptome analysis revealed a down-regulation of genes encoding Sigma70-like family proteins in rice roots under osmotic stress (Ma et al. 2009). These proteins are involved in the control of chloroplast gene expression. The authors suggested that root genes may regulate enzymes and proteins that are related to photosynthesis.

Cell Wall-Related Enriched GO Terms

Major differences between the leaf samples of the 2 genotypes were found at 24 h of stress treatment, where Eurosis was mainly characterized by GO terms related to cell wall-related processes and Loto by GO terms related to peroxisomes and photoinhibition (Fig. 6b and d). In the cell wall-related GO terms that were enriched in Eurosis, several genes associated with lignin polymerization were present, including many peroxidases at 3 h of treatment and several laccases both at 3 and 24 h of treatment. Increased lignification has been observed in the leaf elongation zone of drought-stressed maize leaves (Vincent et al. 2005) and increased cell wall peroxidase activity has been implicated in the cessation of leaf growth under drought in darnel (Bacon et al. 1997). Moreover, several genes, which were listed in these GO terms, encode xyloglucan-modifying enzymes; this finding is consistent with the observation that more xyloglucan is synthesized during dehydration, promoting the strengthening of the cell wall through cross-linking and tightening (Moore et al. 2008). Similar results were reported in a recent transcriptome profiling study of the leaf elongation zone under drought that compared the gene expression between drought-tolerant and drought-sensitive rice varieties (Cal et al. 2013). Many of the differentially expressed genes were involved in secondary cell wall deposition, lignin polymerization, or coded for glycosyl hydrolases and xyloglucan endotransglucosylase/hydrolases, suggesting that the two genotypes have alternative strategies for the regulation of leaf elongation under drought (Cal et al. 2013). Moreover, drought tolerance involves a restructuring of the cell wall that allows growth processes to occur at a lower water level; thus, cell wall adjustment under water stress is an important phenomenon in plant adaptation. These mechanisms, however, are complex and differ among plant species; many studies are still needed to understand how these processes influence plant stress tolerance (Moore et al. 2008).

MapMan Pathway Analysis of 3 h-Treated Roots

Pathway-based analysis helps to further understand the biological function of genes. We used the MapMan package (http://mapman.gabipd.org/) as a tool to more thoroughly visualize the pathways involved in the osmotic stress response of the two genotypes. Because major differences in the stress response between the 2 cultivars were found in 3 h-treated roots, all of the DEGs specifically modulated in 3 h-treated root samples of Eurosis or Loto (3455 and 1410 genes in Eurosis and Loto, respectively; Table 3, Additional file 8: Table S8) were analysed using MapMan to identify differences in metabolic and regulatory pathways. The analysis showed that some pathways related to cell division, stress response, hormones, regulation, enzyme families and transport were differently regulated between the 2 cultivars, confirming the major results of the GO enrichment analysis (Fig. 7). In particular, the gene sets specific to each genotype included genes that encoded TFs, proteins involved in ethylene signalling and enzymes belonging to several families with a role in the plant stress response (e.g., cytochrome P450; Additional file 9: Table S9).

Among TF-encoding genes, approximately 40 genes were up-regulated only in Eurosis (Additional file 9: Table S9). Among them, LOC_Os02g22020 codes for a GARP TF, which regulates the response to osmotic stress in rice (Mito et al. 2011). This gene is the putative orthologue of *HRS1*, which acts as a negative regulator of ABA signalling during Arabidopsis seed germination. HRS1 may participate in the suppression of ABA signalling in the germinating embryo axis, which in turn promotes the germination of Arabidopsis seeds in either normal or salt stress environments (Wu et al. 2012). Very recently, it has been proposed that HRS1 represses primary root growth in response to phosphorus deficiency, integrating the signalling pathways related to phosphate and nitrate (Medici et al. 2015).

It is well known that MYB TFs have a prominent role in the response to drought in plant species (Baldoni et al. 2015). Among the five *MYB* genes (LOC_Os01g34060, LOC_Os01g65370, LOC_Os05g48010, LOC_Os06g43090 and LOC_Os08g33150) that were up-regulated only in Eurosis (Additional file 9: Table S9), the LOC_Os01g65370 gene is very close to *OsMYB2P-1*, which is associated with Pi starvation signalling and is involved in the regulation of root architecture (Dai et al. 2012). This TF may have a role in the root growth of Eurosis plants.

In addition to the originally identified role in defence signalling, WRKY TFs have a pivotal function in seed germination, flower development, senescence and the abiotic stress response (Tripathi et al. 2014). Three *WRKY* genes (LOC_Os01g09080/*OsWRKY107*, LOC_Os03g55080/*OsWRKY3* and LOC_Os08g29660/*OsWRKY69*) were up-regulated only in Eurosis (Additional file 9: Table S9). Consistently, *OsWRKY69* expression is up-regulated in a drought tolerant cultivar (Douradão) compared to a sensitive cultivar (Primavera) under water shortage conditions, suggesting that its action in the signalling of the drought response may contribute to the tolerant phenotype of rice genotypes (Silveira et al. 2015). Interestingly, *OsWRKY69* expression is regulated by ABL1, which is a rice basic region/leucine zipper motif TF involved in the ABA response, suggesting that OsWRKY69 may be involved in the signalling of the abiotic stress response (Yang et al. 2011).

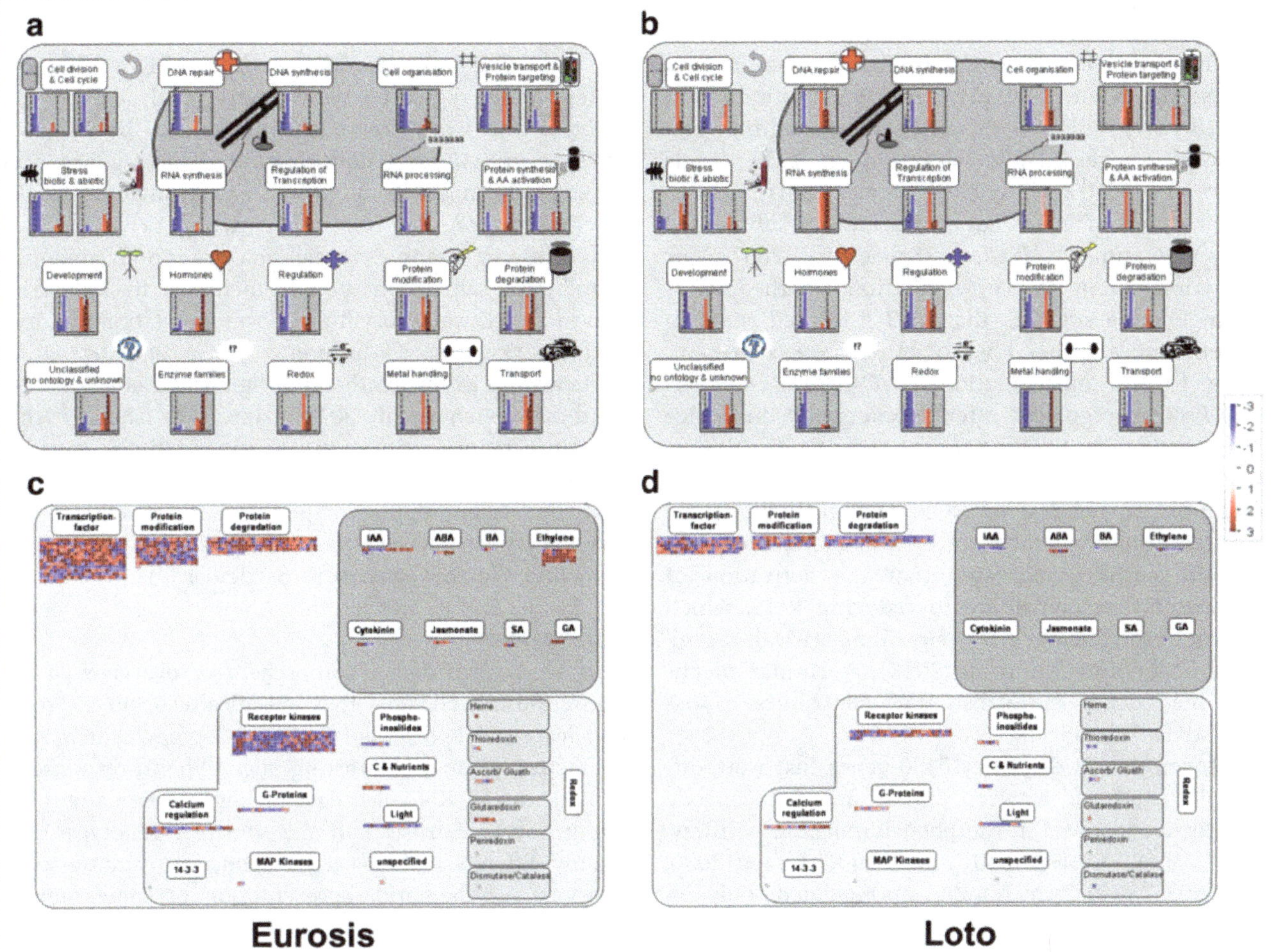

Fig. 7 MapMan analysis of genotype specific up- and down-regulated genes in 3 h-treated roots. MapMan analysis of Eurosis (*left*) or Loto (*right*) DEGs in 3 h-treated roots. **a**, **b** Cell functions overview; **c**, **d** Regulation overview. The colour scale indicates the degree of fold change expression values (*light blue-blue, down-regulation; pink-red, up-regulation*)

In higher plants, ethylene is synthesized via two enzyme-catalysed steps from S-adenosyl-L-methionine (SAM). The 1-aminocyclopropane-1-carboxylic acid (ACC) synthase (ACS) catalyses the cyclization of SAM to ACC and subsequently, the ACC oxidase (ACO) catalyses the oxidative conversion of ACC to ethylene (Yang and Hoffman 1984). ACS and ACO are encoded by medium and small-sized gene families, respectively, and their expression is differentially regulated by various developmental, environmental, and hormone signals (Bleecker and Kende 2000; Chen and McManus 2006; Binnie and McManus 2009). Here, MapMan analysis highlighted that some *ACO* genes were differentially regulated between the two genotypes: LOC_Os03g64280 (*ACO1*) and LOC_Os08g30080 were up-regulated only in Eurosis, and LOC_Os06g37590 was down-regulated only in Eurosis, whereas LOC_Os03g63900 was down-regulated only in Loto (Additional file 9: Table S9). Literature data indicate that *ACO1* is involved in the accumulation of ethylene during submergence, and its activity may contribute to the initiation of adventitious root formation during submergence through the activation of epidermal cell death signalling (Steffens and Sauter 2009; Fukao and Bailey-Serres 2008).

Furthermore, MapMan analysis showed that 6 cytochrome P450 encoding genes (LOC_Os01g41810, LOC_Os02g47470, LOC_Os03g04660, LOC_Os05g25640, LOC_Os05g31740 and LOC_Os10g09110) were up-regulated only in Eurosis, 2 cytochrome P450 encoding genes (LOC_Os01g12750 and LOC_Os11g18570) were up-regulated only in Loto, and 2 cytochrome P450 encoding genes (LOC_Os01g72260 and LOC_Os10g12080) were down-regulated only in Loto (Additional file 9: Table S9). LOC_Os02g47470 and LOC_Os05g31740, which were up-regulated only in Eurosis, seem to be involved in submergence tolerance (Kottapalli et al. 2007; Jung et al. 2010). In particular, LOC_Os02g47470 encodes the ABA 8'-hydroxylase (OsABA8ox1), which catalyses the major regulatory step of the predominant pathway for ABA inactivation via the conversion of ABA to phaseic acid (Kushiro et al. 2004). The involvement of *OsABA8ox1* in the drought response has been previously reported; the induction of *OsABA8ox1* under a water-deficient condition significantly suppressed the elevation of ABA levels (Yazawa et al. 2012). Ethylene partially contributes to the reduction of ABA concentration in submerged rice by activating *OsABA8ox1* (Saika et al. 2007); our observation about the up-regulation of the *ACO1* gene in Eurosis suggests that in 3 h-treated roots of this genotype a higher level of ethylene was present, leading to the up-regulation of *OsABA8ox1*. *OsABA8ox1* is up-regulated after submergence in a rice genotype that is tolerant to a prolonged submergence (Jung et al. 2010). Since restriction of ROS production or removal of ROS during submergence and after desubmergence are critical factors for survival, the authors suggested that the activation of *OsABA8ox1* may contribute to reducing ROS, which are produced during ABA signalling (Kwak et al. 2003; Mittler and Blumwald 2015). A similar mechanism may occur in Eurosis and contribute to osmotic stress tolerance.

Moreover, two of the 10 cytP450 genes that were differentially expressed between Eurosis and Loto encode two enzymes involved in the phenylpropanoid pathway, namely, LOC_Os05g25640 and LOC_Os10g12080. LOC_Os05g25640, which was up-regulated only in Eurosis, codes for the cinnamic acid 4-hydroxylase (C4H), which transforms the cinnamic acid into P-coumaric acid at the beginning of the phenylpropanoid pathway. LOC_Os10g12080, which is also named *CYP98A15p* (Liu et al. 2010), was down-regulated exclusively in Loto and encodes coumarate 3-hydroxylase (C3H), which transforms the P-coumaric acid into caffeic acid at the beginning of the lignin branch. An expression analysis conducted on two rice cultivars under oxidative stress showed a down-regulation of *CYP98A15p* in the more sensitive cultivar, which is consistent with our results (Liu et al. 2010). The authors hypothesized that the repression of lignin synthesis in the sensitive genotype may drive metabolic flux into flavonoids, which leads to a different metabolite flux between the tolerant and the sensitive genotypes (Liu et al. 2010). In a recent study, the expression level of genes involved in suberin and lignin production, including the *C4H* gene, correlates well with the absolute suberin and lignin content in rice roots (Ranathunge et al. 2015). Our observation of the up-regulation of *C4H* only in Eurosis and the down-regulation of *C3H* only in Loto suggests the presence of a higher amount of suberin and lignin in Eurosis roots. Similarly to the *C4H* gene, 4 above mentioned genes (e.g., LOC_Os07g01020 encoding a putative SOR/SNZ family protein; LOC_OsOs03g06510 encoding a KRP; LOC_Os02g22020 encoding a GARP TF and LOC_Os01g65370 encoding a MYB TF) were differentially regulated between the analysed rice genotypes and were previously reported to be involved in root growth and development. These genes may have a role in the root growth and contribute to the tolerant response of Eurosis plants.

Recently, a transcriptomic analysis on Arabidopsis found that salt acclimation is mediated by DEGs involved in cell wall biosynthesis, osmoregulation and oxidative stress, by TF-encoding DEGs and by DEGs participating in the synthesis of lignin and ethylene biosynthesis (Shen et al. 2014). Here, we showed that similar categories of genes characterized the osmotic stress response of a tolerant rice genotype, suggesting that the tolerant phenotype of Eurosis to osmotic stress may be characterized by a similar mechanism involving the root system in particular.

Conclusions

This study provided a comprehensive overview of the transcriptome changes that specifically occur in roots and leaves of two *japonica* rice genotypes, which were characterized by a contrasting phenotype in response to osmotic stress. RNA-Seq data highlighted that the tolerant genotype Eurosis and the sensitive genotype Loto mainly differed in their early response to osmotic stress in roots. A schematic representation of some mechanisms putatively involved in the higher tolerance to osmotic stress of Eurosis is shown in Fig. 8. Differential response included a prompt regulation of genes related to chromatin, cytoskeleton and transmembrane transporters in the tolerant genotype. Moreover, the differential regulation of TFs (Fig. 8a) and hormone-mediate signalling was observed. In particular, some of the hormone-related DEGs may reduce ROS content in the roots of the tolerant genotype (Fig. 8b). Furthermore, genes involved in the biosynthesis of lignin in the roots may determine the tolerant vs. sensitive response (Fig. 8c). The function of these genes in the water stress response and the physiological mechanisms in which they are involved still have to be elucidated. Nonetheless, some of the genes identified in this work may be important players in the tolerance response and could represent good candidate genes for improvement of the rice germplasm.

Methods

Plant Material, Growth Conditions and Osmotic Stress Treatment

Seventeen Italian rice cultivars (*O. sativa* L. ssp *japonica*) were subjected to PEG-mediated osmotic stress, namely

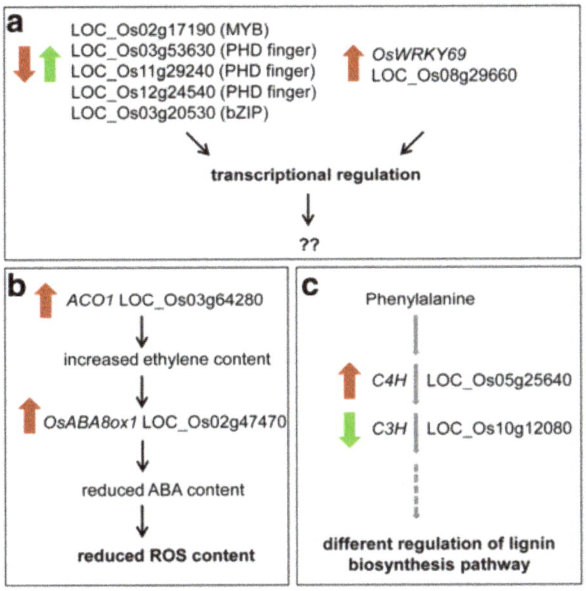

Fig. 8 Schematic representation of some root mechanisms putatively involved in osmotic stress tolerance of Eurosis. *Red and green arrows* indicate gene regulation under osmotic treatment in Eurosis and Loto, respectively. *Grey arrows* indicate enzymatic steps in the lignin biosynthetic pathway

Arborio, Asia, Augusto, Baldo, Carnaroli, Eurosis, Gigante Vercelli, Gladio, Koral, Loto, Maratelli, Salvo, SISR215, Thaibonnet, Venere, Vialone Nano and Volano. The rice seeds were kindly provided by the Rice Research Unit (CREA-RIS, Vercelli, Italy).

All of the hydroponic experiments were performed in a controlled growth room at 28 °C/23 °C under a 14 h light/10 h dark photoperiod with a light intensity of 200 μmol m^{-2} s^{-1} and 60 % relative humidity. Plant growth and osmotic stress treatment were performed as described in Baldoni et al. (2013). After 4 days of germination in Petri dishes, seedlings were grown in hydroponic culture with standard Yoshida nutrient solution (Duchefa Biochemistry, Harlem, The Netherlands). After 20 days of growth (third-leaf stage), half of the plants were transferred to a nutrient solution with 20 % (w/v) PEG6000 (Duchefa Biochemistry, Harlem, The Netherlands) at 11.00 am (3 h after the onset of illumination), to impose an osmotic stress treatment. For RNA purification, leaves (the second and third leaf) and roots from control or treated plants were collected separately in pools of 6 plants after 3 (14:00 pm) and 24 h (11:00 am) of PEG treatment.

Physiological Measurements

Leaf RWC was measured after 0, 3, 24 and 48 h of PEG treatment, as described in Baldoni et al. (2013). Leaf EL was measured after 0, 3, 24 and 48 h of PEG treatment to evaluate cell membrane stability following the method of Shou et al. (2004) with minor modifications. For each sample, 6 leaves from 3 plants (2 leaves per plant) were collected, rinsed with deionized water and cut into 7-mm pieces. Leaf discs were incubated in tubes with 20 ml of deionized water, and the tubes were shaken overnight in a slanted position. The initial conductivity of the incubation (*Ci*) was measured using a conductivity meter (Thermo Orion star Plus, Beverly MA) to estimate the amount of ions released from cells under normal conditions or PEG treatment. Leaf tissue in the incubation solution was then boiled at 100 °C for 30 min to completely disrupt the cell structure. The conductivity of the boiled solution (*C*max) was determined after cooling at room temperature. These 2 measurements were carried out individually for all of the samples from both the control and stressed plants. The percentage of EL was calculated by dividing the *Ci* by the *C*max. For each genotype and condition, 7 biological replicates were measured.

For each sampling time, RWC and EL data were subjected to a one-way analysis of variance (ANOVA) after angular transformation of raw data. Comparison among means was performed using a Tukey's test. Significant differences were accepted at $p < 0.01$.

RNA Isolation and Quantitative RT-PCR Analysis

Total RNA was extracted from the leaves and roots using the TRIzol® RNA Purification Kit (Invitrogen, Carlsbad, CA) following the manufacturer's instructions. RNA purity was checked spectrophotometrically (Nanodrop ND-1000 Spectrophotometer; Celbio, Italy), and only samples with a 260 / 280 nm ratio of absorbance comprised between 1.7 and 2.1 were further used. The integrity of the RNA was verified using the RNA 6000 Nano Labchip Kit on an Agilent 2100 Bioanalyzer (Agilent Technologies, USA), following the manufacturer's protocol. Only samples with a 28S / 18S ratio ≥ 2 were used for further experiments.

The cDNA synthesis and quantitative RT-PCR was performed as described in Baldoni et al. (2013).

Illumina RNA-Sequencing

Four micrograms of total RNA was subjected to library preparation using the TruSeq mRNA Sample Prep Kit from Illumina (Illumina, Inc., CA, USA) following the manufacturer's instructions. Library concentration and size were assayed on a 2100 Bioanalyzer (Agilent). Single-end sequencing (50 bases) was conducted on an Illumina HiSeq2000 with samples run in 6-plex. Illumina sequencing was performed at IGA Technology Services Srl Service Provider (Udine, Italy).

Bioinformatic Methods

Mapping of Illumina Reads

A preliminary read trimming (from 50 to 43 bases) step was performed to discard poor quality bases that were abundantly detected in the 3′ terminal region. The resulting trimmed reads were checked for contaminants, and low quality bases and contaminants were removed using the cutadapt software (Martin 2011). The spliced read mapper TopHat version 1.4.1 (Trapnell et al. 2012) was used to map reads to rice *O. sativa* Nipponbare genome (MSU release 6.16). A minimum and maximum intron length of 40 and 50,000 were used, respectively. Read counts were collected with HTSeq version 0.5.3 (Anders et al. 2014) in the single end and 'union' mode using *O. sativa* MSU 6.16 gtf file as performed for the ensemble plant repository.

Differentially Regulated Gene Calling

The DESeq Bioconductor package version 1.10.1 (Anders and Huber 2010) was used to call DEGs under R release 2.15.2. The cut-off for considering a gene as expressed was set to 0.1 RPKM (Reads per Kilobase per Million). Two distinct DESeq countSet object instances were created for Eurosis and Loto. DESeq parameters for dispersion estimation were the method 'pooled' and sharingMode 'Maximum'. For DEG calling, the FDR threshold was set to 0.001 and the fold change threshold to 3.

Gene Ontology Enrichment Analyses

For goseq analyses, gene lengths were retrieved with BiomaRt queries (*O. sativa* MSU 6.16) out of rice Nipponbare cDNA, and the median length for each rice locus was obtained by parsing with R custom scripts. GO annotations were obtained using BiomaRt queries (*O. sativa* MSU 6.16). An FDR cut-off of 0.05 was used for GO enrichments.

Miscellaneous Bioinformatic Techniques

Heatmaps of the expression and various graphical outputs were generated with custom R scripts based on Bioconductor packages (Gentleman et al. 2004). Sample clustering was performed after data transformation via variance stabilizing transformation function (DESeq package; Anders and Huber 2010), and heatmaps were generated via the heatmap.2 function as available in the 'gplots' Bioconductor package. MapMan figures were generated upon binning of DEG sequences to MapMan bins by the Mercator application (Lohse et al. 2014). Unless otherwise stated, further graphical outputs were generated with custom R and Python scripts.

Additional Files

Additional file 1: Table S1. Read number and alignment statistics. Passed-filter reads refer to preliminary Illumina inbuilt read scoring.

Additional file 2: Table S2. Pearson correlation coefficients of samples.

Additional file 3: Table S3. Comparisons of RNA-Seq and qRT-PCR data. Comparisons of transcript fold changes as detected by RNA-Seq and qRT-PCR expression analyses for 10 selected genes in all of the experimental conditions. Comparisons of the differences between the means of treated and untreated samples of qRT-PCR data were performed using Student's *t*-tests (*p*-value ≤ 0.05 except for the data marked with ^{ns}). *OsCAT-A*: Catalase isozyme A; *OsZCO*: Zeaxanthin cleavage oxygenase.

Additional file 4: Table S4. List of DEGs in all of the analysed samples.

Additional file 5: Table S5. List of DEGs showing contrasting regulation between Eurosis and Loto samples.

Additional file 6: Table S6. List of common GO terms. List of GO terms common in all of the tissues and treatments, in all of the root samples or in all of the leaf samples.

Additional file 7: Table S7. List of enriched GO terms for the contrasts with accompanying enrichment *p*-values as resulting from goseq analysis.

Additional file 8: Table S8. List of DEGs specifically modulated in 3 h-treated root samples of Eurosis or Loto.

Additional file 9: Table S9. List of DEGs belonging to MapMan categories "Ethylene", "Transcription Factor", "Enzyme families - Cytochrome P450."

Abbreviations

ABA: abscisic acid; ABC: ATP-binding cassette; ACC: 1-aminocyclopropane-1-carboxylic acid; ACO: ACC oxidase; ACS: ACC synthase; BP: biological process; C3H: coumarate 3-hydroxylase; C4H: cinnamic acid 4-hydroxylase; CC: cellular component; Ci: initial conductivity; Cmax: absolute conductivity; DEG: differentially expressed gene; EL: electrolyte leakage; FDR: false discovery rate; GO: gene ontology; KRP: Kip-related protein; MATE: multidrug and toxic compound extrusion; MF: molecular function; ns-Hb: non-symbiotic haemoglobin; PEG: polyethylene glycol; RPKM: reads per kilobase per million; RWC: relative water content; SAM: S-adenosyl-L-methionine; TF: transcription factor.

Acknowledgments

This work was supported by Progetto AGER, grant n° 2010–2369 - Integrated Genetic And Genomic Approaches For New Italian Rice Breeding Strategies (RISINNOVA).

Authors' contributions

E.B. and A.G. conceived and designed the research. E.B., F.L. and M.M. performed the stress treatments and the physiological measurements. E.B. performed the RNA isolation and the quantitative RT-PCR analysis. P.B. performed the bioinformatic analysis. E.B. and P.B. analyzed the data. E.B. and A.G. wrote the manuscript. All authors read and approved the final manuscript.

Competing interests

The authors declare that they have no competing interests.

Author details

[1]Institute of Agricultural Biology and Biotechnology - National Research Council, via Bassini 15, 20133 Milan, Italy. [2]Dipartimento di Scienze Agrarie e Ambientali - Produzione, Territorio, Agroenergia, Università degli Studi di Milano, Via Celoria 2, 20133 Milan, Italy. [3]Consiglio per la ricerca in agricoltura e l'analisi dell'economia agraria, Genomics Research Centre, Fiorenzuola d'Arda, Piacenza, Italy.

References

Anders S, Huber W (2010) Differential expression analysis for sequence count data. Genome Biol 11:R106. doi:10.1186/gb-2010-11-10-r106

Anders S, Pyl PT, Huber W (2014) HTSeq - A Python framework to work with high-throughput sequencing data. Bioinformatics 2014:1–4. doi:10.1093/bioinformatics/btu638

Bacon MA, Thompson DS, Davies WJ (1997) Can cell wall peroxidase activity explain the leaf growth response of *Lolium temulentum* L. during drought? J Exp Bot 48:2075–2085. doi:10.1093/jxb/48.12.2075

Baldoni E, Mattana M, Locatelli F, Consonni R, Cagliani LR, Picchi V, Abbruscato P, Genga A (2013) Analysis of transcript and metabolite levels in Italian rice (*Oryza sativa* L.) cultivars subjected to osmotic stress or benzothiadiazole treatment. Plant Physiol Biochem 70:492–503. doi:10.1016/j.plaphy.2013.06.016

Baldoni E, Genga A, Cominelli E (2015) Plant MYB transcription factors: their role in drought response mechanisms. Int J Mol Sci 16:15811–15851. doi:10.3390/ijms160715811

Bates BC, Kundzewicz ZW, Wu S, Palutikof JP (eds) (2008) Climate change and water. Technical paper of the intergovernmental panel on climate change. IPCC Secretariat, Geneva, p 210

Binnie IE, McManus MT (2009) Characterization of the 1-aminocyclopropane-1-carboxylic acid (ACC) oxidase multigene family of *Malus domestica* Borkh. Phytochemistry 70:348–360. doi:10.1016/j.phytochem.2009.01.002

Bleecker AB, Kende H (2000) Ethylene: a gaseous signal molecule in plants. Annu Rev Cell Dev Biol. 1–18. doi:10.1146/annurev.cellbio.16.1.1

Bouman BAM, Humphreys E, Tuong TP, Barker R, Sparks DL (2007) Rice and water. Adv Agron 92:187–237. doi:10.1016/S0065-2113(04)92004-4

Cal AJ, Liu DC, Mauleon R, Hsing YIC, Serraj R (2013) Transcriptome profiling of leaf elongation zone under drought in contrasting rice cultivars. PLoS One 8(1):e54537. doi:10.1371/journal.pone.0054537

Chaves MM, Flexas J, Pinheiro C (2009) Photosynthesis under drought and salt stress: regulation mechanisms from whole plant to cell. Ann Bot London 103:551–560. doi:10.1093/aob/mcn125

Chen BCM, McManus MT (2006) Expression of 1-aminocyclopropane-1-carboxylate (ACC) oxidase genes during the development of vegetative tissues in white clover (*Trifolium repens* L.) is regulated by ontological cues. Plant Mol Biol 60:451–467. doi:10.1007/s11103-005-4813-3

Chen W, Gao YQ, Xie WB, Gong L, Lu K, Wang WS, Li Y, Liu XQ, Zhang HY, Dong HX, Zhang W, Zhang LJ, Yu SB, Wang GW, Lian XM, Luo J (2014) Genome-wide association analyses provide genetic and biochemical insights into natural variation in rice metabolism. Nat Genet 46:714–721. doi:10.1038/ng.3007

Claeys H, Inze D (2013) The agony of choice: how plants balance growth and survival under water-limiting conditions. Plant Physiol 162:1768–1779, http://dx.doi.org/10.1104/pp.113.220921

Dai XY, Wang YY, Yang A, Zhang WH (2012) *OsMYB2P-1*, an R2R3 MYB transcription factor, is involved in the regulation of phosphate-starvation responses and root architecture in rice. Plant Physiol 159:169–183, http://dx.doi.org/10.1104/pp.112.194217

de Souza Filho GA, Ferreira BS, Dias JM, Queiroz KS, Branco AT, Bressan-Smith RE, Oliveira JG, Garcia AB (2003) Accumulation of SALT protein in rice plants as a response to environmental stresses. Plant Sci 164:623–628. doi:10.1016/S0168-9452(03)00014-1

Degenkolbe T, Do PT, Zuther E, Repsilber D, Walther D, Hincha DK, Kohl KI (2009) Expression profiling of rice cultivars differing in their tolerance to long-term drought stress. Plant Mol Biol 69:133–153. doi:10.1007/s11103-008-9412-7

Degenkolbe T, Do PT, Kopka J, Zuther E, Hincha DK, Kohl KI (2013) Identification of drought tolerance markers in a diverse population of rice cultivars by expression and metabolite profiling. PLoS One 8(5):e63637. doi:10.1371/journal.pone.0063637

Fujita Y, Fujita M, Shinozaki K, Yamaguchi-Shinozaki K (2011) ABA-mediated transcriptional regulation in response to osmotic stress in plants. J Plant Res 124:509–525. doi:10.1007/s10265-011-0412-3

Fukao T, Bailey-Serres J (2008) Ethylene - A key regulator of submergence responses in rice. Plant Sci 175:43–51. doi:10.1016/j.plantsci.2007.12.002

Garrocho-Villegas V, Bustos-Rivera G, Gough J, Vinogradov SN, Arredondo-Peter R (2008) Expression and in silico structural analysis of a rice (*Oryza sativa*) hemoglobin 5. Plant Physiol Biochem 46:855–859. doi:10.1016/j.plaphy.2008.05.004

Gentleman RC, Carey VJ, Bates DM, Bolstad B, Dettling M, Dudoit S, Ellis B, Gautier L, Ge YC, Gentry J, Hornik K, Hothorn T, Huber W, Iacus S, Irizarry R, Leisch F, Li C, Maechler M, Rossini AJ, Sawitzki G, Smith C, Smyth G, Tierney L, Yang JYH, Zhang JH (2004) Bioconductor: open software development for computational biology and bioinformatics. Genome Biol 5:R80. doi:10.1186/gb-2004-5-10-r80

Guan C, Ji J, Guan W, Li X, Jin C, Li J, Wang Y, Wang G (2014) LcKRP, a cyclin-dependent kinase inhibitor from *Lycium chinense*, is involved in an ABA-dependent drought stress-signaling pathway. Plant Soil 382:43–59. doi:10.1007/s11104-014-2134-5

Guimaraes PM, Brasileiro ACM, Morgante CV, Martins ACQ, Pappas G, Silva OB, Togawa R, Leal-Bertioli SCM, Araujo ACG, Moretzsohn MC, Bertioli DJ (2012) Global transcriptome analysis of two wild relatives of peanut under drought and fungi infection. BMC Genomics 13:387. doi:10.1186/1471-2164-13-387

Hadiarto T, Tran LSP (2011) Progress studies of drought-responsive genes in rice. Plant Cell Rep 30:297–310. doi:10.1007/s00299-010-0956-z

Himanen K, Boucheron E, Vanneste S, de Almeida Engler J, Inzé D, Beeckman T (2002) Auxin-mediated cell cycle activation during early lateral root initiation. Plant Cell 14:2339–2351, http://dx.doi.org/10.1105/tpc.004960

Hu YF, Zhu N, Wang XM, Yi QP, Zhu DY, Lai Y, Zhao Y (2013) Analysis of rice Snf2 family proteins and their potential roles in epigenetic regulation. Plant Physiol Biochem 70:33–42. doi:10.1016/j.plaphy.2013.05.001

Humphrey TV, Haasen KE, Aldea-Brydges MG, Sun H, Zayed Y, Indriolo E, Goring DR (2015) PERK-KIPK-KCBP signalling negatively regulates root growth in *Arabidopsis thaliana*. J Exp Bot 66:71–83. doi:10.1093/jxb/eru390

Jarzyniak KM, Jasinski M (2014) Membrane transporters and drought resistance - a complex issue. Front Plant Sci 5:687. doi:10.3389/fpls.2014.00687

Jeong JS, Kim YS, Redillas MCFR, Jang G, Jung H, Bang SW, Choi YD, Ha SH, Reuzeau C, Kim JK (2013) *OsNAC5* overexpression enlarges root diameter in rice plants leading to enhanced drought tolerance and increased grain yield in the field. Plant Biotechnol J 11:101–114. doi:10.1111/pbi.12011

Jung KH, Seo YS, Walia H, Cao PJ, Fukao T, Canlas PE, Amonpant F, Bailey-Serres J, Ronald PC (2010) The submergence tolerance regulator *Sub1A* mediates stress-responsive expression of *AP2/ERF* transcription factors. Plant Physiol 152:1674–1692, http://dx.doi.org/10.1104/pp.109.152157

Karim S, Lundh D, Holmstrom KO, Mandal A, Pirhonen M (2005) Structural and functional characterization of AtPTR3, a stress-induced peptide transporter of *Arabidopsis*. J Mol Model 11:226–236. doi:10.1007/s00894-005-0257-6

Karim S, Holmstrom KO, Mandal A, Dahl P, Hohmann S, Brader G, Palva ET, Pirhonen M (2007) AtPTR3, a wound-induced peptide transporter needed for defence against virulent bacterial pathogens in *Arabidopsis*. Planta 225:1431–1445. doi:10.1007/s00425-006-0451-5

Kim JM, Sasaki T, Ueda M, Sako K, Seki M (2015) Chromatin changes in response to drought, salinity, heat, and cold stresses in plants. Front Plant Sci 6:114. doi:10.3389/fpls.2015.00114

Kottapalli KR, Satoh K, Rakwal R, Shibato J, Doi K, Nagata T, Kikuchi S (2007) Combining in silico mapping and arraying: an approach to identifying common candidate genes for submergence tolerance and resistance to bacterial leaf blight in rice. Mol Cells 24:394–408

Krasensky J, Jonak C (2012) Drought, salt, and temperature stress-induced metabolic rearrangements and regulatory networks. J Exp Bot 63:1593–1608. doi:10.1093/jxb/err460

Kumar A, Bernier J, Verulkar S, Lafitte HR, Atlin GN (2008) Breeding for drought tolerance: direct selection for yield, response to selection and use of drought-tolerant donors in upland and lowland-adapted populations. Field Crop Res 107:221–231. doi:10.1016/j.fcr.2008.02.007

Kushiro T, Okamoto M, Nakabayashi K, Yamagishi K, Kitamura S, Asami T, Hirai N, Koshiba T, Kamiya Y, Nambara E (2004) The *Arabidopsis* cytochrome P450 CYP707A encodes ABA 8′-hydroxylases: key enzymes in ABA catabolism. Embo J 23:1647–1656. doi:10.1038/sj.emboj.7600121

Kwak JM, Mori IC, Pei ZM, Leonhardt N, Torres MA, Dangl JL, Bloom RE, Bodde S, Jones JDG, Schroeder JI (2003) NADPH oxidase *AtrbohD* and *AtrbohF* genes function in ROS-dependent ABA signaling in *Arabidopsis*. Embo J 22:2623–2633. doi:10.1093/emboj/cdg277

Lampayan RM, Rejesus RM, Singleton GR, Bouman BAM (2015) Adoption and economics of alternate wetting and drying water management for irrigated lowland rice. Field Crop Res 170:95–108. doi:10.1016/j.fcr.2014.10.013

Lenka SK, Katiyar A, Chinnusamy V, Bansal KC (2011) Comparative analysis of drought-responsive transcriptome in *Indica* rice genotypes with contrasting drought tolerance. Plant Biotechnol J 9:315–327. doi:10.1111/j.1467-7652.2010.00560.x

Li XY, Wang C, Nie PP, Lu XW, Wang M, Liu W, Yao J, Liu YG, Zhang QY (2011) Characterization and expression analysis of the *SNF2* family genes in response to phytohormones and abiotic stresses in rice. Biol Plant 55:625–633. doi:10.1007/s10535-011-0160-1

Li J, Xu YY, Chong K (2012) The novel functions of kinesin motor proteins in plants. Protoplasma 249:95–100. doi:10.1007/s00709-011-0357-3

Liu FX, Xu WY, Wei Q, Zhang ZH, Xing Z, Tan LB, Di C, Yao DX, Wang CC, Tan YJ, Yan H, Ling Y, Sun CQ, Xue YB, Su Z (2010) Gene expression profiles deciphering rice phenotypic variation between Nipponbare (japonica) and 93-11 (indica) during oxidative stress. PLoS One 5(1):e8632. doi:10.1371/journal.pone.0008632

Lohse M, Nagel A, Herter T, May P, Schroda M, Zrenner R, Tohge T, Fernie AR, Stitt M, Usadel B (2014) Mercator: a fast and simple web server for genome scale functional annotation of plant sequence data. Plant Cell Environ 37: 1250–1258. doi:10.1111/pce.12231

Ma TC, Chen RJ, Yu RR, Zeng HL, Zhang DP (2009) Differential global genomic changes in rice root in response to low-, middle-, and high-osmotic stresses. Acta Physiol Plant 31:773–785. doi:10.1007/s11738-009-0291-6

Martin M (2011) Cutadapt removes adapter sequences from high-throughput sequencing reads. EMBnetjournal 17:10–12, http://dx.doi.org/10.14806/ej.17.1.200

Matsuda S, Funabiki A, Furukawa K, Komori N, Koike M, Tokuji Y, Takamure I, Kato K (2012) Genome-wide analysis and expression profiling of half-size ABC protein subgroup G in rice in response to abiotic stress and phytohormone treatments. Mol Genet Genomics 287:819–835. doi:10.1007/s00438-012-0719-3

Medici A, Marshall-Colon A, Ronzier E, Szponarski W, Wang RC, Gojon A, Crawford NM, Ruffel S, Coruzzi GM, Krouk G (2015) AtNIGT1/HRS1 integrates nitrate and phosphate signals at the *Arabidopsis* root tip. Nat Commun 6:6274. doi:10.1038/ncomms7274

Minh-Thu PT, Hwang DJ, Jeon JS, Nahm BH, Kim YK (2013) Transcriptome analysis of leaf and root of rice seedling to acute dehydration. Rice 6:38. doi:10.1186/1939-8433-6-38

Mito T, Seki M, Shinozaki K, Ohme-Takagi M, Matsui K (2011) Generation of chimeric repressors that confer salt tolerance in *Arabidopsis* and rice. Plant Biotechnol J 9:736–746. doi:10.1111/j.1467-7652.2010.00578.x

Mittler R, Blumwald E (2015) The roles of ROS and ABA in systemic acquired acclimation. Plant Cell 27:64–70, http://dx.doi.org/10.1105/tpc.114.133090

Moons A (2008) Transcriptional profiling of the *PDR* gene family in rice roots in response to plant growth regulators, redox perturbations and weak organic acid stresses. Planta 229:53–71. doi:10.1007/s00425-008-0810-5

Moore JP, Vicre-Gibouin M, Farrant JM, Driouich A (2008) Adaptations of higher plant cell walls to water loss: drought *vs* desiccation. Physiol Plant 134:237–245. doi:10.1111/j.1399-3054.2008.01134.x

Moumeni A, Satoh K, Kondoh H, Asano T, Hosaka A, Venuprasad R, Serraj R, Kumar A, Leung H, Kikuchi S (2011) Comparative analysis of root transcriptome profiles of two pairs of drought-tolerant and susceptible rice near-isogenic lines under different drought stress. BMC Plant Biol 11:174. doi:10.1186/1471-2229-11-174

Moumeni A, Satoh K, Venuprasad R, Serraj R, Kumar A, Leung H, Kikuchi S (2015) Transcriptional profiling of the leaves of near-isogenic rice lines with contrasting drought tolerance at the reproductive stage in response to water deficit. BMC Genomics 16:1110. doi:10.1186/s12864-015-2335-1

Nakashima K, Yamaguchi-Shinozaki K, Shinozaki K (2014) The transcriptional regulatory network in the drought response and its crosstalk in abiotic stress responses including drought, cold, and heat. Front Plant Sci 5:170. doi:10.3389/fpls.2014.00170

Nguyen VNT, Moon S, Jung KH (2014) Genome-wide expression analysis of rice ABC transporter family across spatio-temporal samples and in response to abiotic stresses. J Plant Physiol 171:1276–1288. doi:10.1016/j.jplph.2014.05.006

Oshlack A, Wakefield MJ (2009) Transcript length bias in RNA-seq data confounds systems biology. Biol Direct 4:14. doi:10.1186/1745-6150-4-14

Perruc E, Kinoshita N, Lopez-Molina L (2007) The role of chromatin-remodeling factor PKL in balancing osmotic stress responses during Arabidopsis seed germination. Plant J 52:927–936. doi:10.1111/j.1365-313X.2007.03288.x

Qin F, Shinozaki K, Yamaguchi-Shinozaki K (2011) Achievements and challenges in understanding plant abiotic stress responses and tolerance. Plant Cell Physiol 52:1569–1582. doi:10.1093/pcp/pcr106

Ranathunge K, Schreiber L, Bi YM, Rothstein SJ (2015) Ammonium-induced architectural and anatomical changes with altered suberin and lignin levels significantly change water and solute permeabilities of rice (*Oryza sativa* L.) roots. Planta. doi:10.1007/s00425-015-2406-1

Richards RA (1996) Defining selection criteria to improve yield under drought. Plant Growth Regul 20:157–166. doi:10.1007/BF00024012

Robinson MD, Oshlack A (2010) A scaling normalization method for differential expression analysis of RNA-seq data. Genome Biol 11:R25. doi:10.1186/gb-2010-11-3-r25

Rüdiger H, Gabius HJ (2001) Plant lectins: occurrence, biochemistry, functions and applications. Glycoconjugate J 18:589–613

Saha J, Sengupta A, Gupta K, Gupta B (2015) Molecular phylogenetic study and expression analysis of ATP-binding cassette transporter gene family in *Oryza sativa* in response to salt stress. Comput Biol Chem 54:18–32. doi:10.1016/j.compbiolchem.2014.11.005

Saier MH, Tran CV, Barabote RD (2006) TCDB: the Transporter Classification Database for membrane transport protein analyses and information. Nucleic Acids Res 34:D181–D186. doi:10.1093/nar/gkj001

Saika H, Okamoto M, Kushiro T, Shinoda S, Jikumaru Y, Fujimoto M, Miyoshi K, Arikawa T, Arimura SI, Kamiya Y, Tsutsumi N, Nambara E, Nakazono M (2007) Ethylene promotes submergence-induced expression of *OsABA8oxl*, a gene that encodes ABA 8′-hydroxylase in rice. Plant Cell Physiol 48:287–298. doi:10.1093/pcp/pcm003

Shen XY, Wang ZL, Song XF, Xu JJ, Jiang CY, Zhao YX, Ma CL, Zhang H (2014) Transcriptomic profiling revealed an important role of cell wall remodeling and ethylene signaling pathway during salt acclimation in Arabidopsis. Plant Mol Biol 86:303–317. doi:10.1007/s11103-014-0230-9

Shinozaki K, Yamaguchi-Shinozaki K (2007) Gene networks involved in drought stress response and tolerance. J Exp Bot 58:221–227. doi:10.1093/jxb/erl164

Shou HX, Bordallo P, Fan JB, Yeakley JM, Bibikova M, Sheen J, Wang K (2004) Expression of an active tobacco mitogen-activated protein kinase kinase kinase enhances freezing tolerance in transgenic maize. Proc Natl Acad Sci U S A 101:3298–3303. doi:10.1073/pnas.0308095100

Shu LB, Lou QJ, Ma CF, Ding W, Zhou J, Wu JH, Feng FJ, Lu X, Luo LJ, Xu GW, Mei HW (2011) Genetic, proteomic and metabolic analysis of the regulation of energy storage in rice seedlings in response to drought. Proteomics 11: 4122–4138. doi:10.1002/pmic.201000485

Silveira RDD, Abreu FRM, Mamidi S, McClean PE, Vianello RP, Lanna AC, Carneiro NP, Brondani C (2015) Expression of drought tolerance genes in tropical upland rice cultivars (*Oryza sativa*). Genet Mol Res 14(3):8181–8200. doi:10.4238/2015July.27.6

Steffens B, Sauter M (2009) Epidermal cell death in rice is confined to cells with a distinct molecular identity and is mediated by ethylene and H_2O_2 through an autoamplified signal pathway. Plant Cell 21:184–196, http://dx.doi.org/10.1105/tpc.108.061887

Suzuki S, Yamamura M, Hattori T, Nakatsubo T, Umezawa T (2007) The subunit composition of hinokiresinol synthase controls geometrical selectivity in norlignan formation. Proc Natl Acad Sci U S A 104:21008–21013. doi:10.1073/pnas.0710357105

Takanashi K, Shitan N, Yazaki K (2014) The multidrug and toxic compound extrusion (MATE) family in plants. Plant Biotechnol 31:417–430. doi:10.5511/plantbiotechnology.14.0904a

Trapnell C, Roberts A, Goff L, Pertea G, Kim D, Kelley DR, Pimentel H, Salzberg SL, Rinn JL, Pachter L (2012) Differential gene and transcript expression analysis of RNA-seq experiments with TopHat and Cufflinks. Nat Protoc 7:562–578. doi:10.1038/nprot.2012.016

Tripathi P, Rabara RC, Rushton PJ (2014) A systems biology perspective on the role of WRKY transcription factors in drought responses in plants. Planta 239: 255–266. doi:10.1007/s00425-013-1985-y

Tuteja N, Sahoo RK, Garg B, Tuteja R (2013) OsSUV3 dual helicase functions in salinity stress tolerance by maintaining photosynthesis and antioxidant machinery in rice (*Oryza sativa* L. cv. IR64). Plant J 76:115–127. doi:10.1111/tpj.12277

Utsumi Y, Tanaka M, Morosawa T, Kurotani A, Yoshida T, Mochida K, Matsui A, Umemura Y, Ishitani M, Shinozaki K, Sakurai T, Seki M (2012) Transcriptome analysis using a high-density oligomicroarray under drought stress in various genotypes of cassava: an important tropical crop. DNA Res 19:335–345. doi:10.1093/dnares/dss016

Valliyodan B, Nguyen HT (2006) Understanding regulatory networks and engineering for enhanced drought tolerance in plants. Curr Opin Plant Biol 9:189–195. doi:10.1016/j.pbi.2006.01.019

Van Damme EJM, Barre A, Rouge P, Peumans WJ (2004) Cytoplasmic/nuclear plant lectins: a new story. Trends Plant Sci 9:484–489. doi:10.1016/j.tplants.2004.08.003

Vashisht AA, Tuteja N (2006) Stress responsive DEAD-box helicases: a new pathway to engineer plant stress tolerance. J Photoch Photobio B 84:150–160. doi:10.1016/j.jphotobiol.2006.02.010

Vincent D, Lapierre C, Pollet B, Cornic G, Negroni L, Zivy M (2005) Water deficits affect caffeate O-methyltransferase, lignification, and related enzymes in maize leaves. A proteomic investigation. Plant Physiol 137:949–960, http://dx.doi.org/10.1104/pp.104.050815

Wen B, Nieuwland J, Murray JAH (2013) The *Arabidopsis* CDK inhibitor ICK3/KRP5 is rate limiting for primary root growth and promotes growth through cell elongation and endoreduplication. J Exp Bot 64:1135–1144. doi:10.1093/jxb/ert009

Wu CM, Feng JJ, Wang R, Liu H, Yang HX, Rodriguez PL, Qin HJ, Liu X, Wang DW (2012) *HRS1* acts as a negative regulator of abscisic acid signaling to promote timely germination of *Arabidopsis* seeds. PLoS One 7(4):e35764. doi:10.1371/journal.pone.0035764

Yang SF, Hoffman NE (1984) Ethylene biosynthesis and its regulation in higher plants. Ann Rev Plant Physiol 35:155–189. doi:10.1146/annurev.pp.35.060184.001103

Yang X, Yang YN, Xue LJ, Zou MJ, Liu JY, Chen F, Xue HW (2011) Rice ABI5-Like1 regulates abscisic acid and auxin responses by affecting the expression of ABRE-containing genes. Plant Physiol 156:1397–1409, http://dx.doi.org/10.1104/pp.111.173427

Yazawa K, Jiang CJ, Kojima M, Sakakibara H, Takatsuji H (2012) Reduction of abscisic acid levels or inhibition of abscisic acid signaling in rice during the early phase of *Magnaporthe oryzae* infection decreases its susceptibility to the fungus. Physiol Mol Plant Pathol 78:1–7. doi:10.1016/j.pmpp.2011.12.003

Yoshida T, Mogami J, Yamaguchi-Shinozaki K (2014) ABA-dependent and ABA-independent signaling in response to osmotic stress in plants. Curr Opin Plant Biol 21:133–139. doi:10.1016/j.pbi.2014.07.009

Zhu MK, Chen GP, Dong TT, Wang LL, Zhang JL, Zhao ZP, Hu ZL (2015) *SlDEAD31*, a putative DEAD-Box RNA helicase gene, regulates salt and drought tolerance and stress-related genes in tomato. PLoS One 10(8): e0133849. doi:10.1371/journal.pone.0133849doi:10.1371/journal.pone.0133849

7

Mapping QTLs using a novel source of salinity tolerance from Hasawi and their interaction with environments in rice

M. Akhlasur Rahman[1,2], Isaac Kofi Bimpong[3], J. B. Bizimana[4], Evangeline D. Pascual[5], Marydee Arceta[1], B. P. Mallikarjuna Swamy[1], Faty Diaw[3], M. Sazzadur Rahman[2] and R. K. Singh[1*]

Abstract

Background: Salinity is one of the most severe and widespread abiotic stresses that affect rice production. The identification of major-effect quantitative trait loci (QTLs) for traits related to salinity tolerance and understanding of QTL × environment interactions (QEIs) can help in more precise and faster development of salinity-tolerant rice varieties through marker-assisted breeding. Recombinant inbred lines (RILs) derived from IR29/Hasawi (a novel source of salinity) were screened for salinity tolerance in the IRRI phytotron in the Philippines (E1) and in two other diverse environments in Senegal (E2) and Tanzania (E3). QTLs were mapped for traits related to salinity tolerance at the seedling stage.

Results: The RILs were genotyped using 194 polymorphic SNPs (single nucleotide polymorphisms). After removing segregation distortion markers (SDM), a total of 145 and 135 SNPs were used to construct a genetic linkage map with a length of 1655 and 1662 cM, with an average marker density of 11.4 cM in E1 and 12.3 cM in E2 and E3, respectively. A total of 34 QTLs were identified on 10 chromosomes for five traits using ICIM-ADD and segregation distortion locus (SDL) mapping (IM-ADD) under salinity stress across environments. Eight major genomic regions on chromosome 1 between 170 and 175 cM (*qSES1.3, qSES1.4, qSL1.2, qSL1.3, qRL1.1, qRL1.2, qFWsht1.2, qDWsht1.2*), chromosome 4 at 32 cM (*qSES4.1, qFWsht4.2, qDWsht4.2*), chromosome 6 at 115 cM (*qFWsht6.1, qDWsht6.1*), chromosome 8 at 105 cM (*qFWsht8.1, qDWsht8.1*), and chromosome 12 at 78 cM (*qFWsht12.1, qDWsht12.1*) have co-localized QTLs for the multiple traits that might be governing seedling stage salinity tolerance through multiple traits in different phenotyping environments, thus suggesting these as hot spots for tolerance of salinity. Forty-nine and 30 significant pair-wise epistatic interactions were detected between QTL-linked and QTL-unlinked regions using single-environment and multi-environment analyses.

Conclusions: The identification of genomic regions for salinity tolerance in the RILs showed that Hasawi possesses alleles that are novel for salinity tolerance. The common regions for the multiple QTLs across environments as co-localized regions on chromosomes 1, 4, 6, 8, and 12 could be due to linkage or pleiotropic effect, which might be helpful for multiple QTL introgression for marker-assisted breeding programs to improve the salinity tolerance of adaptive and popular but otherwise salinity-sensitive rice varieties.

Keywords: Hasawi-*aus* rice landrace, Single nucleotide polymorphism (SNP), Novel QTLs, QTL × environment interactions, Seedling-stage salinity tolerance

* Correspondence: r.k.singh@irri.org
[1]International Rice Research Institute (IRRI), DAPO Box 7777, Metro Manila, Philippines
Full list of author information is available at the end of the article

Background

Unfavorable environmental conditions such as salinity, drought, heat, and submergence pose a huge threat to agricultural production and productivity and challenge future food security. Abiotic stresses cause crop yield losses of more than 50% and this is expected to worsen further because of climate change, so there is an urgent need to develop climate-smart crop varieties to counteract abiotic stresses and to sustain food production (Zeigler and Barclay 2008; Kumari et al. 2009; Khan et al. 2016). Salinity is one of the widest-spread and most severe abiotic stresses that affect rice production and productivity worldwide (Flowers 2004).

Rice (*Oryza sativa* L.) is the major staple food for almost one-half of the world's population, so sustained rice production and productivity are essential for food security. However, the rice crop is sensitive to salinity stress during different stages of its growth and development, and this stress at the seedling stage is most severe and can sometimes cause complete crop failure (Munns and Tester 2008; Singh et al. 2010; Hossain et al. 2015; Munns and Gilliham 2015). The prevalence of higher sodium ions (Na^+) in saline conditions is harmful to the growth and development of rice plants because of the negative effect on photosynthesis that leads to a reduction in plant growth, chlorophyll content, and metabolic processes (Qados 2011; Munns and Gilliham 2015; Rahman et al. 2016). Transplanting cost is one of the major resource-consuming activities and it could be reduced using recent techniques such as direct-seeded rice (DSR). But, DSR is not feasible in salt-affected areas as rice seedlings are very sensitive to salinity stress; hence, the recommendation for salt-affected soils is to plant seedlings older than the normal 21-day-old seedlings. Thus, salinity tolerance at the seedling stage is crucial for good crop establishment, especially in coastal areas. Various mechanisms such as preferential uptake of potassium ions (K^+), sodium exclusion from roots, and its restricted transport to shoots have been reported to confer salinity tolerance in rice (Kader et al. 2006; Wu et al. 2009; Singh and Flowers 2010; Rahman et al. 2016). However, salinity tolerance is a complex trait governed by genetic factors such as multiple QTLs and their interactions (epistasis), and is also significantly influenced by environmental factors (Wurschum et al. 2013; Roy et al. 2014; Khan et al. 2016; Liu et al. 2016).

Recent advances in molecular marker technology have enabled the dissection of the genetic basis of salinity tolerance to identify major-effect QTLs and their use in marker-assisted breeding to develop salinity-tolerant rice varieties (Munns 2005; Tuberosa and Salvi 2006; Passioura et al. 2007; Thomson et al. 2010, 2012; Thomson 2014; Hossain et al. 2015). Several studies have reported QTLs for traits related to salinity tolerance.

Precise and rapid exploitation of rice germplasm by identifying useful alleles and introgressing them into elite rice varieties is a key to successful breeding programs (Negrao et al. 2008). A new source of salinity tolerance, "Hasawi," an *aus* landrace from Saudi Arabia, is found to have higher Na^+ exclusion and early seedling vigor (Thomson et al. 2010; Al-Mssallem et al. 2011; Rahman et al. 2016; Bizimana et al. 2017). Salinity tolerance in Pokkali is due to its capacity to maintain a low Na^+-K^+ ratio in the shoot tissue (ion-homeostasis) and its faster growth rate under saline conditions (Walia et al. 2005; Ismail et al. 2007; Singh et al. 2010; Thomson et al. 2010), and in Nona Bokra to maintaining higher shoot K^+ content under salt stress (Ren et al. 2005). Bimpong et al. (2014) have reported four grain yield-enhancing QTLs (*qPH8*, *qDTF8*, *qTN8*, and *qTN8*) from Hasawi even for the reproductive stage under saline conditions.

Even though Hasawi is a highly salt-tolerant genotype, it does not have the same tolerance allele as Pokkali and Nona Bokra at *Saltol* and *SKC1* (*OsHKT1;5*), which is a major QTL/gene for salinity tolerance. It has been recently reported that Hasawi is a new source of alleles for salinity tolerance (Bimpong et al. 2014; Bizimana et al. 2017).

The main objectives of our study were to screen IR29/Hasawi-derived RIL populations for seedling-stage salinity tolerance to identify large-effect novel QTLs for traits related to salinity tolerance, to identify epistatic QTLs, to understand the effect of QTL × environment interactions on salinity tolerance, and to identify the RILs with higher salinity tolerance. The novel QTLs identified can be used in marker-assisted QTL pyramiding with other known QTLs to enhance the degree of salinity tolerance in rice.

Methods

In an attempt to identify robust QTLs for salinity tolerance at the seedling stage using a novel donor, the following activities were carried out.

Plant materials

A population (about 600) of recombinant inbred lines was developed from a cross between IR29 (salt sensitive) and Hasawi (salt tolerant; IRGC acc. no. 16817) at the International Rice Research Institute (IRRI), Philippines. A set of 300 RILs from this cross was used for phenotyping under controlled environment in the IRRI phytotron in the Philippines (SE Asia – E1). Another set of 300 RILs, different from the one phenotyped in the Philippines, was sent to Africa Rice Center's Sahel regional station in Senegal (West Africa – E2) and IRRI's Eastern and Southern Africa office in Tanzania (East Africa – E3) for phenotyping, but under uncontrolled natural environment except for rain protection. Both the E2 and E3 sites received the identical set of RILs.

Each RIL was advanced to constitute the phenotyping population ($F_{5:6}$), in which F_5 plants were used for genotyping and F_5-derived F_6 seedlings/plants were phenotyped at the seedling stage for salinity tolerance at all three phenotyping sites. We followed selective mapping for the study, which is a well-established and robust method for QTL mapping if resources are limited (Lander and Botstein 1989). A subset of 142 common RILs evaluated in both Senegal and Tanzania was genotyped at IRRI, Philippines. The RILs used for genotyping from the set phenotyped at IRRI were completely different (except for three RILs) from those in the set phenotyped in Senegal and Tanzania. This different RIL set used for phenotyping at IRRI consisted of 155 individuals.

Environmental classification

This IR29/Hasawi RIL population was evaluated in three different countries using the nutrient solution culture technique following modified Yoshida nutrient solution (Singh and Flowers 2010) in the IRRI phytotron, Philippines (Southeast Asia), and in open screenhouse conditions in Tanzania (East Africa) and the AfricaRice Sahel regional station in Senegal (West Africa).

Evaluation of $F_{5:6}$ RILs for salt tolerance

Screening of Hasawi, IR29, and the 300 $F_{5:6}$ RILs for salinity tolerance was carried out in a hydroponic system following IRRI standard protocol (Gregorio et al. 1997). Seeds were heat-treated for 5 days in a convection oven set at 48 °C to break seed dormancy, and, after that, the seeds were placed in petri dishes with two layers of paper towels, moistened with distilled water during 48 h for uniform germination. The germinated seeds were sown one seed per hole on a styrofoam sheet with 96 holes, attached to a nylon net bottom, and the sheet was floated on modified Yoshida nutrient solution (Singh et al. 2010). The seedlings were salinized after 5 days using 6 dS m^{-1} salt (NaCl) concentration (equivalent to about 60 mM NaCl). This concentration was increased to 12 dS m^{-1} (~120 mM) after 2 days of 6 dS m^{-1} treatments to reduce the immediate shock. Each genotype was represented by five seedlings per row of styrofoam and replicated thrice in the experiment. The experiment in the Philippines was conducted in a controlled phytotron with 29/21 °C day and night temperature with 70 ± 10% relative humidity. The screenhouse temperature recorded during the experiments in Senegal ranged from 17 to 28 °C in the morning and from 36 to 44 °C in the afternoon. The mean relative humidity varied between 35% and 95%. However, the experiment in Tanzania was conducted in a screenhouse covered by plastic on top only, with a minimum temperature of 24 °C and maximum of 37 °C. The minimum relative humidity was 51% and the maximum 84%, with natural daylight of about 14 h. The pH of the solution was adjusted and maintained at 5.0 to 5.1 every day with acid (1 N HCl) or base (1 N NaOH). The nutrient solution was renewed once every week to limit the effect of algae and to replenish the nutrients. Scoring as per the standard evaluation system (SES) was used and recorded 12 and 25 days after the imposition of salinity stress to finally score the genotype for overall degree of tolerance. All RILs were monitored and scored based on visual symptoms of salt stress injury. The following IRRI modified standard evaluation system (SES) for rice was used (IRRI 2007).

Score	Symptom/observation	Degree of tolerance
1	Normal growth, only the old leaves show white tips while no symptoms on young leaves	Highly tolerant
3	Near normal growth, but only leaf tips burn, few older leaves become whitish partially and rolled	Tolerant
5	Growth severely retarded; most leaves severely injured, few young leaves elongating	Moderately tolerant
7	Complete cessation of growth; most leaves dried; only a few young leaves still green	Sensitive
9	Almost all plants dead or dying	Highly sensitive

Other phenotypic parameters such as root and shoot length (RL and SL), shoot fresh weight (FWSht), and shoot dry weight (DWSht) were measured after 25 days.

Genotyping

DNA extraction, quantification, and quality control

We followed the approach of selective genotyping (Lander and Botstein 1989) to increase the efficiency of QTL mapping (Lin and Ritland 1996). Seedling SES injury score (final) was used as the parameter to select the genotypes from extremes and also randomly. Out of 300 genotypes, we picked 142 RILs that comprised the 7 most tolerant genotypes (SES score 1–3) and 34 most sensitive ones (SES score 9), and the rest (101) were random genotypes with intermediate to sensitive SES score (5–7). However, in the Philippines, an almost equal number of tolerant, intermediate, and sensitive genotypes was used for genotyping. Genomic DNA was isolated from young leaves using the CTAB (cetyl trimethyl ammonium bromide) mini-preparation method (Murray and Thompson 1980).

Scoring of SNPs and analysis of polymorphism

A chip (indica × indica) comprising 384 SNP markers spread throughout 12 chromosomes of the rice genome was used for the parental polymorphic survey between the two parents (Hasawi and IR29). For each OPA (Oligo Pool All, reagent) run, the final DNA concentration was normalized to 50 ng/μL. For the SNP analysis, the (Illumina 2008) GoldenGate assay (Fan et al. 2003) was performed using VeraCode technology on a BeadXpress Reader according to the manufacturer's protocol. Briefly, about 250 ng of genomic DNA was used to make biotinylated genomic DNA, which then underwent oligonucleotide hybridization to bind the samples to paramagnetic particles, followed by allele-specific extension and ligation, PCR, hybridization to the VeraCode Bead Plate, and scanning on the BeadXpress Reader. The analysis employed the VC0011439-OPA set of 384 SNP markers designed to be informative across *indica* and *aus* germplasm (Thomson et al. 2012) and was run in the Genotyping Services Laboratory at IRRI (Thomson 2014; http://gsl.irri.org). Raw hybridization intensity data processing was performed using the genotyping module in the BeadStudio package (Illumina, San Diego, CA, USA), followed by allele calling using ALCHEMY software (Wright et al. 2010). Graphical genotyping of both IR29 and Hasawi was performed using Flapjack (https://ics.hutton.ac.uk/flapjack/) software developed by the Scottish Crop Research Institute to check polymorphisms. After the polymorphism survey and filtering for low call rates, we found 194 SNPs as polymorphic (which is about 50%) for analysis.

SNP linkage map and QTL analysis

Of the 384 SNPs used for the parental polymorphism survey, 194 that were polymorphic were selected for QTL analysis but we had to drop a large number of SNPs for the construction of a linkage map due to segregation distortion (SD).

We used ICIMapping ver. 4.0.1 software (www.isbreeding.net) for the genetic linkage map construction and QTL analysis. Then, a genetic linkage group was constructed based on recombination frequency and SNP ordering was done using the ordering algorithm of RECORD (without imposing marker order) coming from REcombination Counting and ORDering, proposed by van Os et al. (2005a, 2005b). RECORD was developed to produce accurate marker orders in a relatively short time by employing the total number of observable recombination events between adjacent markers as a target function (Wang et al. 2012), although RECORD is not capable of handling populations with high heterozygous loci (Liu et al. 2014). However, it (RECORD) can deal with BC_1, F_2, F_3, and RIL (in fact, any generation obtained by repeated selfing of a hybrid between homozygous parents) mapping populations (van Os et al. 2005a, 2005b). We chose the best ordering algorithm and rippling criteria for fine tuning of the linkage map, which was not used as the input method. COUNT (number of recombination events) algorithm was used for rippling. Rippling was used for fine-tuning of the ordered chromosomes. The Chi-square (χ^2) test was performed using whole data on a 1:1 basis. The highly distorted markers with <5% probability of either allele for IR29 or Hasawi were discarded; while less or non-distorted markers were included for the linkage map construction and QTL analysis. Segregation distortion markers (SDM) were removed to increase statistical power. The inclusive composite interval mapping (ICIM-ADD) method was used to identify more precise QTLs. For additive mapping, ICIM-ADD retains all advantages of CIM over IM, and avoids the possible increase of sampling variance and the complicated background marker selection process in CIM. Extensive simulations using different genomes and various genetic models indicate that ICIM has increased detection power, reduced false detection rate, and resulted in less biased estimates of QTL effects than CIM in additive (and dominance) mapping. Extensive simulations also show that ICIM is an efficient method for epistasis mapping, and QTL epistatic networks can be identified no matter whether the two QTL have any additive effects (Wang et al. 2014; Xu 2008). The minimal LOD value required to declare a significant QTL was obtained empirically from 1000 permutation tests (Churchill and Doerge 1994). The proportion of the total phenotypic variance explained (PVE) by each QTL was calculated as R^2 value (R^2 = PVE). The QTLs were named based on the nomenclature procedure suggested by McCouch et al. (1997) and McCouch and CGSNL (2008).

The digenic (epistatic) interactions between marker loci were determined and single environment (SE) and multi-environment (ME) joint analyses were performed using the multi-environment trials (MET) program in QTL IciMapping ver. 4.0.1 (Wang et al. 2014) to detect QEIs with LOD thresholds of 3.0. We followed the approach of Zhang et al. (2010) in which only reliable QTLs detected by both single and multi-environment analysis were reported.

Results

The QTLs and QEIs identified by ICIM in the IR29/Hasawi RIL population were based on phenotypic evaluation in three different environments.

Correlation analysis between traits in the F_6 RIL population

Significant negative correlations were observed between the SES and all other parameters related to salt tolerance such as SL, RL, FWsht, and DWsht (Table 1), which is expected because a lower SES score indicates higher tolerance, which is based on seedling survival and vigor. In addition, significant positive correlations were observed among traits other than SES score (except between RL and SL in E1), suggesting the importance of these parameters in mechanisms associated with tolerance of

Table 1 Correlation coefficients among different traits in an F_5 (RIL) population of a cross between IR29 (salt sensitive) and Hasawi (salt tolerant) at seedling stage under three different environments

Environment	Trait	SES	RL	SL	FWsht
Philippines	RL	−0.45[a]			
	SL	−0.18[b]	0.12[ns]		
	FWsht	−0.76[a]	0.35[a]	0.43[a]	
	DWsht	−0.76[a]	0.34[a]	0.41[a]	0.98[a]
Tanzania	RL	−0.75[a]			
	SL	−0.64[a]	0.70[a]		
	FWsht	−0.79[a]	0.67[a]	0.64[a]	
	DWsht	−0.86[a]	0.75[a]	0.74[a]	0.84[a]
Senegal	RL	−0.44[a]			
	SL	−0.46[a]	0.33[a]		
	FWsht	−0.50[a]	0.40[a]	0.46[a]	
	DWsht	−0.34[a]	0.19[b]	0.33[a]	0.45[a]

SES Standard evaluation system score based on salt stress symptoms, *SL* Shoot length, *RL* Root length, *FWsht* Shoot fresh weight, *DWsht* Shoot dry weight, *ns* non-significant

[a] and [b] indicate significance at the 1% and 5% level, respectively

salt stress at the seedling stage as indicated by the higher values for these parameters (Table 1).

Genotypic analysis

We used single nucleotide polymorphic markers to determine the polymorphism between the two parents (Hasawi and IR29). In all, 194 SNPs out of 384 (50.52%) showed polymorphism between the parents. After removing distorted markers, 145 SNP markers were used in E1 and 135 markers in E2 and E3 to perform genetic linkage analysis. They were distributed throughout the rice genome and covered a total length of 1655 cM using E1 genotyping data and 1662 cM using E2 and E3 genotyping data, with an average interval of 11.4 and 12.3 cM between markers, respectively. The highest marker density was found on chromosome 1 (23), with an average interval of 10 cM (Additional file 1: Table S1).

Marker segregation distortion analysis

The expected genotypic ratio of 1:1 in the F5:6 RIL population for homozygous IR29:homozygous Hasawi allele varied with three categories. First, with no segregation distortion that accounted for 40 SNPs (in E1) and 37 SNPs (E2 and E3); second, very highly distorted markers (<5% probability of either allele) varied from Mendelian segregation ratio for this RIL population, and less distorted markers (105 for E1 and 98 for E2 and E3). Only non-distorted and less distorted markers were used for mapping studies. When selective mapping is followed, some of the segregation distortions happen due to a sampling effect as well. The whole population was subjected to χ^2 significance (P = 0.05) before analyzing the data. This was done to avoid false linkages from the expected Mendelian segregation ratio. Indeed, little SD for the specific markers is expected to have the effect of the allele through marker-trait association; otherwise, this association will not be reflected as a significant QTL. Hackett and Broadfoot (2003) suggested that segregation distortion causes very little effect on both marker order and map length.

QTL analysis

Inclusive composite interval mapping (ICIM-ADD) was employed to identify putative QTLs. However, the QTLs reported here were identified after constructing the genetic linkage map using ordering algorithm (RECORD) instead of the input method. A total of 34 different QTLs were identified in three diverse saline environments (Table 2). The QTLs conferring tolerance of salinity at the seedling stage were identified on 10 chromosomes: chromosome 1 (*qSES1.1, qSES1.2, qSES1.3, qSES1.4, qSES1.5, qSL1.1, qSL1.2, qSL1.3, qSL1.4, qRL1.1, qRL1.2, qFWsht1.1, qFWsht1.2, qFWsht1.3, qDWsht1.1, qDWsht1.2, qDWsht1.3*); chromosome 2 (*qFWsht2.1*); chromosome 3 (*qRL3.1*); chromosome 4 (*qSES4.1, qFWsht4.1, qFWsht4.2, qDWsht4.1, qDWsht4.2*); chromosome 5 (*qDWsht5.1*); chromosome 6 (*qFWsht6.1, qDWsht6.1*); chromosome 7 (*qDWsht7.1*); chromosome 8 (*qFWsht8.1, qDWsht8.1*); chromosome 11 (*qRL11.1*); and chromosome 12 (*qSL12.1, qFWsh12.1, qDWsht12.1*). One QTL (*qFWsht 6.1*) was identified in two different environments (E2 and E3) within the same chromosomal location. The details of the QTLs are presented in Table 2. The QTLs with a large effect are also illustrated on the molecular linkage map (Fig. 1 based on E1 and Fig. 2 based on E2 and E3).

QTLs for salinity tolerance at seedling stage

Overall phenotypic performance (SES)

Six QTLs (*qSES1.1, qSES1.2, qSES1.3, qSES1.4, qSES1.5,* and *qSES4.1*) were evaluated by the SES in controlled (evaluated for salinity tolerance under phytotron conditions) Environment-1 (Philippines) with significant LOD value (3.2–20.6) and PVE ranging from 5.4% to 42.3%. The Hasawi allele increased the overall phenotypic performance and reduced SES visual scores at five loci except *qSES1.5*, where the allelic effect comes from IR29 (Table 2). The QTLs (*qSES1.3* and *qSES1.4*) on the long arm of chromosome 1 are located in a similar region as *qSL1.2* and *qSL1.3* in E1. *qSES1.4*, located between markers id1023892 and id1017885 on chromosome 1, had the largest PVE (42.3%) and was additive in nature.

Shoot length

Four QTL regions (*qSL1.1, qSL1.2, qSL1.3, qSL1.4*) with significant LOD value (3.3–6.5) were identified for SL on the long arm of chromosome 1 and one QTL (*qSL12.1*)

Table 2 QTLs for traits related to salt tolerance during seedling stage in an IR29/Hasawi RIL population in three environments

Trait	QTL	Chr	Position (cM)	Marker interval	Allele	LOD	PVE (%)	Add	E	Method
SES	*qSES1.1*	1	110	ud1000711-id1004348	Hasawi	3.2	10.7	−0.31	E1	ICIM
	qSES1.2	1	128	id1004348-id1015258	Hasawi	4.9	8.8	−0.27	E1	ICIM
	qSES1.3	1	170	id1024972-id1023892	Hasawi	17.5	39.9	−0.58	E1	IM, ICIM
	qSES1.4	1	175	id1023892-id1017885	Hasawi	20.6	42.3	−0.60	E1	IM, –ICIM
	qSES1.5	1	194	id1003559-id1002308	IR29	3.8	5.4	0.24	E1	-IM, ICIM
	qSES4.1	4	32	id4008522-id4008092	Hasawi	3.8	5.8	−0.23	E1	IM, ICIM
SL	*qSL1.1*	1	59	id1024836-id1025983	Hasawi	4.4	18.5	−2.13	E2	IM, ICIM
	qSL1.2	1	169	id1024972-id1023892	Hasawi	6.5	17.9	−3.25	E1	IM, ICIM
	qSL1.3	1	177	id1023892-id1017885	Hasawi	5.7	19.5	−3.40	E1	IM, ICIM
	qSL1.4	1	221	id1024836-id1016633	IR29	5.6	15.3	3.36	E1	IM, ICIM
	qSL12.1	12	58	id12007988-id12005823	IR29	3.3	7.2	2.23	E1	IM, ICIM
RL	*qRL1.1*	1	168	id1024972- id1023892	Hasawi	4.8	14.3	−1.09	E1	IM, ICIM
	qRL1.2	1	175	id1023892-id1017885	Hasawi	4.2	13.5	−1.06	E1	IM, ICIM
	qRL3.1	3	104	id3200001-id3010345	Hasawi	3.2	21.9	−1.53	E1	IM
	qRL11.1	11	22	id11007488-id11008862	IR29	5.0	14.7	2.11	E1	IM, ICIM
FWsht	*qFWsht1.1*	1	0	id1002899-id1016436	Hasawi	5.5	10.2	−0.09	E1	IM, ICIM
	qFWsht1.2	1	175	id1023892-id1017885	Hasawi	3.6	21.1	−0.12	E1	IM
	qFWsht1.3	1	194	id1003559-id1002308	Hasawi	4.5	13.8	−0.11	E1	IM
	qFWsht2.1	2	55	id2007526-fd12	Hasawi	3.1	27.2	−0.12	E2	IM, ICIM
	qFWsht4.1	4	12	id4003259-id4007105	IR29	6.1	28.9	0.14	E1	IM, ICIM
	qFWsht4.2	4	32	id4008522-id4008092	IR29	4.3	8.8	0.08	E1	IM, ICIM
	qFWsht6.1	6	115	id6016941-id6001397	IR29	3.2	37.8	0.22	E2	IM, ICIM
	qFWsht6.1	6	115	id6016941-id6001397	IR29	3.3	47.1	0.56	E3	IM, ICIM
	qFWsht8.1	8	105	id8007301-id8000240	IR29	3.5	47.1	0.55	E3	IM, ICIM
	qFWsht12.1	12	78	id12003019-id12005205	IR29	4.4	8.0	0.09	E1	ICIM
DWsht	*qDWsht1.1*	1	0	id1002899-id1016436	Hasawi	3.4	5.5	−0.02	E1	IM, ICIM
	qDWsht1.2	1	175	id1023892-id1017885	Hasawi	3.4	19.6	−0.03	E1	IM
	qDWsht1.3	1	194	id1003559-id1002308	Hasawi	3.9	12.0	−0.02	E1	IM
	qDWsht4.1	4	13	id4003259-id4007105	IR29	8.3	30.1	0.03	E1	IM, ICIM
	qDWsht4.2	4	32	id4008522-id4008092	IR29	3.2	5.9	0.02	E1	IM, ICIM
	qDWsht5.1	5	38	id5007714-id5014589	IR29	3.2	46.9	0.12	E3	IM, ICIM
	qDWsht6.1	6	115	id6016941-id6001397	IR29	3.7	48.4	0.12	E3	IM, ICIM
	qDWsht7.1	7	107	ud7000066-id7000461	IR29	3.3	5.7	0.02	E1	IM, ICIM
	qDWsht8.1	8	105	id8007301-id8000240	IR29	3.8	47.2	0.12	E3	IM, ICIM
	qDWsht12.1	12	78	id12003019-id12005205	IR29	4.6	7.5	0.02	E1	IM, ICIM

Chr Chromosome, *LOD* Log of odds, *Add* Additive effect, *IM* Interval mapping, *ICIM* Inclusive composite interval mapping, *E* Environment, *E1* Philippines, *E2* Senegal, *E3* Tanzania, *MI*: Marker interval, *PVE* Phenotypic variation of the rice RIL population explained by each QTL

located at 58 cM on chromosome 12 in E1 had PVE ranging from 7.2% to 19.5%. *qSL1.2* is located between SNP markers id1024972 and id1023892 and co-located with two other QTLs (*qSES1.3* and *qRL1.1*). *qSL1.3* is detected between id1023892 and id1017885 and shared a common genomic region with four other large-effect QTLs (*qSES1.4, qRL1.2, qFWsht1.2,* and *qDWsht1.2*) on the long arm of chromosome 1 (Table 2).

Root length

Four root-length QTLs (*qRL1.1, qRL1.2, qRL3.1,* and *qRL11.1*) with significant LOD value (3.2–5.0) were detected on chromosomes 1, 3, and 11 in the Philippines. These QTLs had PVE of 14.3%, 13.5%, 21.9%, and 14.7%, respectively. Hasawi alleles contributed to longer root length. Two QTLs (*qRL1.1* and *qRL1.2*) located between id1024972-id1023892 and id1023892-id1017885 were additive.

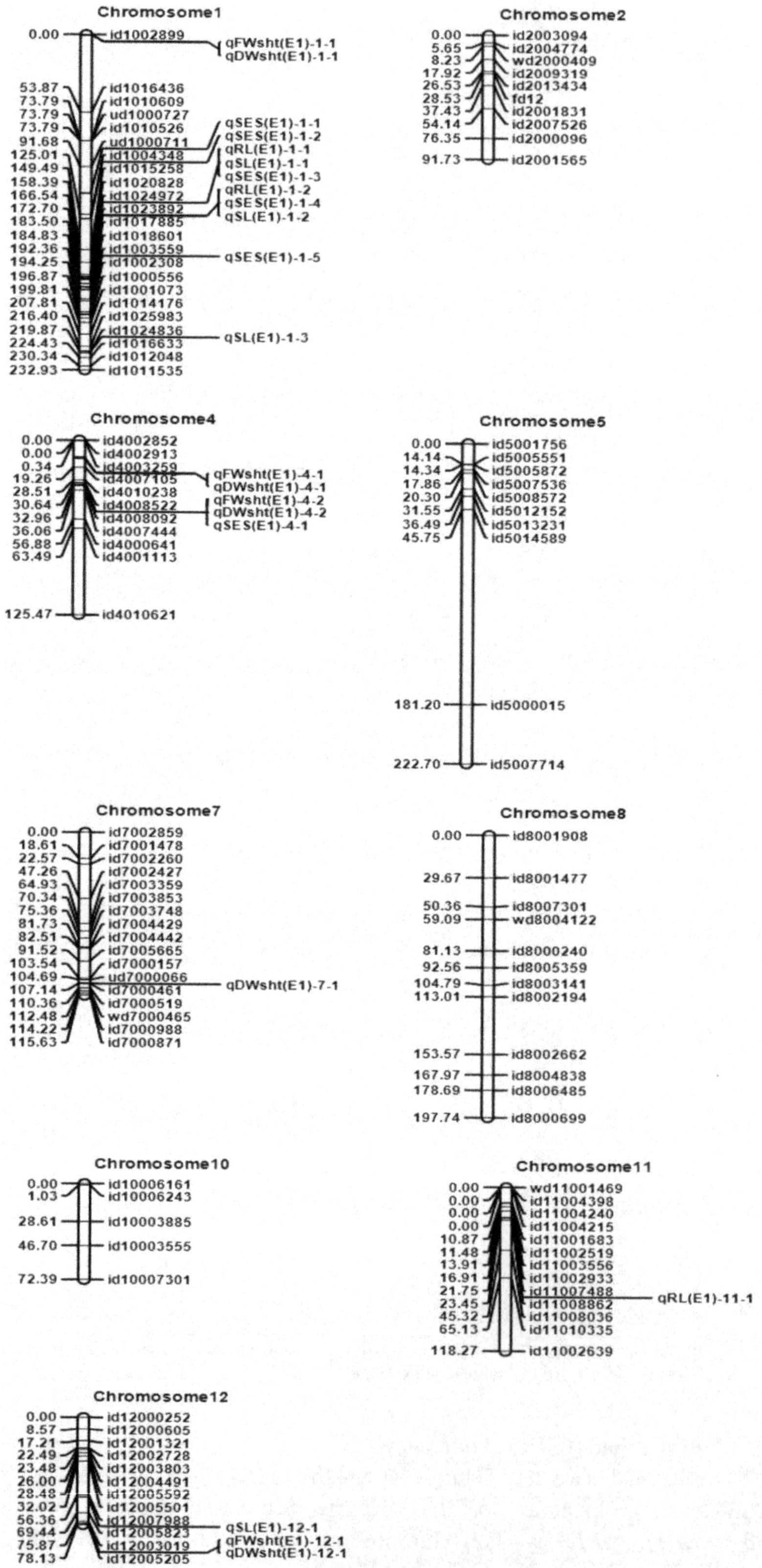

Fig. 1 Genetic linkage maps of the 12 chromosomes constructed using ordering algorithm (RECORD) based on an IR29/Hasawi RIL population phenotyped in the phytotron at IRRI (E1). The names of the SNP markers with position are listed to the right and the approximate locations of the QTLs detected for salinity tolerance are shown to the left of the chromosomes

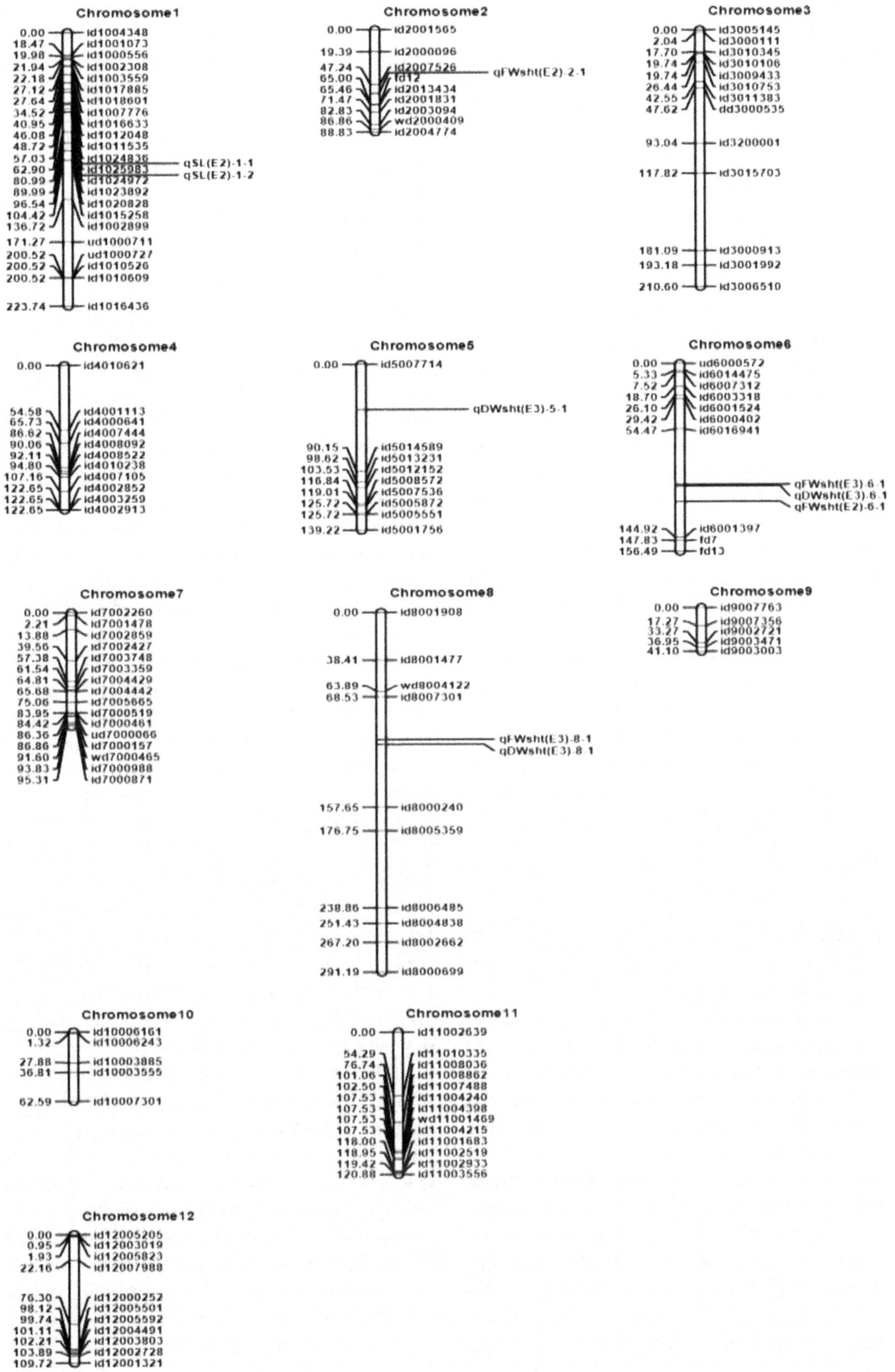

Fig. 2 Genetic linkage maps of the 12 chromosomes constructed using ordering algorithm (RECORD) based on an IR29/Hasawi RIL population phenotyped in Senegal (E2) and Tanzania (E3). The names of the SNP markers with position are listed to the right and the approximate locations of the QTLs detected for salinity tolerance are shown to the left of the chromosomes

Shoot fresh weight

Nine genomic regions were identified for shoot fresh weight with significant LOD of 3.1 to 6.1. Three FWsht QTLs (*qFWsht1.1, qFWsht1.2, qFWsht1.3*) were located on chromosome 1 with LOD value of 3.6–5.5 and they accounted for PVE of 10.2%, 21.1%, and 13.8%, respectively, and Hasawi had positive effects on shoot fresh weight in the Philippines. This QTL (*qFWsht1.2*) was co-located with four other important genomic regions (*qSES1.4, qRL1.2, qSL1.3,* and *qDWsht1.2*) on the long arm of chromosome 1. One QTL (*qFWsht6.1*) was identified in both E2 and E3 with positive effects from IR29. *qFWsht12.1* shared a common location with *qDWsht12.1* at 78 cM on chromosome 12 (Table 2).

Shoot dry weight

In the Philippines, ICIM-ADD detected 10 significant QTLs (*qDWsht1.1, qDWsht1.2, qDWsht1.3, qDWsht4.1, qDWsht4.2, qDWsht5.1, qDWsht6.1, qDWsht7.1, qDWsht8.1,* and *qDWsht12.1*) with LOD value ranging from 3.2 to 8.3 for shoot dry weight on seven chromosomes (1, 4, 5, 6, 7, 8, and 12) near markers id1002899, id1023892, id1003559, id4003259, id4008522, id5007714, id6016941, ud7000066, id8007301, and id12003019, which explained PVE ranging from 5.5% to 48.4%. Out of seven QTLs, three located on chromosome 1 (*qDWsht1.1* at zero (0) cM; *qDWsht1.2* at 175 cM; *qDWsht1.3* at 194 cM), two on chromosome 4 (*qDWsht4.1* at 12 cM and *qDWsht4.2* at 32 cM), one on chromosome 8 (*qDWsht8.1* at 105 cM), and one on chromosome 12 (*qDWsht12.1* at 78 cM) were observed to share common genomic regions with QTLs conferring for shoot fresh weight and SES score. *qDWsht6.1* (E3) is co-localized with QTLs *qFWsht6.1* (E2) and *qFWsht6.1* (E3) and has PVE of 48.4% (Table 2).

As per the multi-locus analysis, QTLs identified on different chromosomes are considered independent of each other, and their effects were generally additive in nature. For example, the major QTLs for SES score on chromosome 1 (flanked by markers id1024972-id1023892, with PVE of 39.9%, and another located between id1023892 and id1017885, with PVE of 42.3%) and chromosome 4 (flanked by markers id4008522-id4008092, with PVE of 5.8%) together accounted for 88.0% of the total PVE observed in the study from the Philippines. Likewise, the combined effect of three QTLs (*qDWsht1.2*, R^2 = 19.6%; *qDWsht1.3*, R^2 = 12.0%; *qDWsht4.1*, R^2 = 30.1%) explained 61.7% of the total phenotypic variance for overall shoot dry weight, which is a part of the phenotypic performance that affects SES score in the Philippines. SES score is based on plant vigor and higher plant vigor comes from vigorous shoot growth; hence, a lower SES score and more shoot biomass will eventually give more shoot dry weight and this clearly indicates that these are traits indirectly linked to each other.

QTL interactions for seedling-stage salinity tolerance

(i) Epistatic interactions A two-way test to detect epistatic interactions between marker loci was performed for single environment (SE) and multi-environment (ME) analyses with stringent threshold LOD of 5.0 using the ICIM-EPI method of ICIM version 4.0.1 software for all traits. The SE analysis detected 49 significant digenic interactions (2 in E1, 31 in E2, and 16 in E3) located across 12 different chromosomes for SES score, SL, FWsht, and DWsht, whose PVE ranged from 13.6% to 44.9% in E1, from 16.0% to 75.9% in E2, and from 21.9% to 74.9% in E3 (Additional file 1: Table S2). Out of 49 digenic interactions, we identified 13 interactions between QTLs and background loci, 36 interactions between complementary loci, and no interaction between QTLs observed (Additional file 1: Table S2). The ME analysis identified a total of 30 significant interactions consisting of one interaction for SL, two marker loci intervals (MI) for SES score, three MI for FWsht, and 24 intervals for DWsht that were spread across 10 different chromosomes (1, 2, 3, 4, 5, 6, 7, 8, 11, and 12) (Table 3). Two types of digenic interactions were identified (Table 3; Fig. 3): (I) interaction between the QTL (marker interval fd12-id2013434; *qFWsht2.1*) on chromosome 2 for FWsht and background loci (such as marker interval id1007776-id1016633) on chromosome 1 with LOD of 7.1 for shoot fresh weight (Fig. 3a); and (II) interaction between complementary locus 1 (35 cM; MI: id1007776-id1016633) on chromosome 1 and 95 cM; MI: id4010238-id4007105 on chromosome 4 for shoot dry weight (Fig. 3b) and interaction between background loci at MI: id6001397-fd7 (145 cM) on chromosome 6 and MI: id11002639- id11010335) on chromosome 11 (0 cM) with LOD of 6.3 and PVE of 7.5% for SES score (Table 3; Fig. 3c). Interaction between QTLs is not found in this study. Four marker pairs had a significant effect on the final phenotype through the interaction between the QTL and background loci and 26 significant interactions between complementary loci, thus indicating strong interaction effects. The interaction component study on shoot dry weight revealed that one of the marker intervals (id1004348-id1001073) on chromosome 1 hosted a main-effect QTL (*qDWsht1.1*) that interacted with background loci on chromosome 8 at marker interval id8000240-id8005359 to express the phenotype with 4.0% of explainable variation due to total interaction components. Out of 26 type II interactions (the interaction between complementary loci), the marker intervals (ud6000572-id6014475) on chromosome 6 interacted with other background loci on chromosome 11 at marker interval id11002933-id11003556 to express the phenotype with a very high LOD for both epistatic interaction (LOD: 53.7; the highest PVE of 14.1%) and QTL × E interaction (LOD = 31.0, PVE of 8.1%), suggesting the higher probability of occurrence for digenic and QTL × E interactions (Table 3). More epistatic interactions seem to be present for shoot dry weight than

Table 3 Epistasis for traits related to salinity tolerance in an IR29/Hasawi RIL population in two different environments (E2 and E3)

Chr 1	Position 1 (cM)	Marker interval at position 1	Chr 2	Position 2 (cM)	Marker interval at position 2	LOD			PVE (%)			AA × E2	AA × E3	TI
						Epi	AA	QEI	Epi	AA	QEI			
SL: Shoot length														
2	20	id2000096-id2007526	8	35	id8002194-id8001477	7.0	2.4	3.5	9.3	4.6	4.7	0.87	−0.87	C
SES														
5	0	id5007714-id5014589	11	0	id11002639-id11010335	6.3	4.3	2.0	7.8	7.2	0.6	−0.72	0.72	C
6	145	id6001397-fd7	11	0	id11002639-id11010335	6.3	4.1	2.1	7.5	7.1	0.4	−0.72	0.72	C
FWsht: Shoot fresh weight														
1	40	id1007776-id1016633	2	65	fd12-id2013434	7.1	4.1	3.0	5.4	3.3	2.1	0.05	−0.05	B
3	90	dd3000535-id3200001	6	155	fd7-fd13	7.0	3.2	3.7	8.6	3.8	4.8	−0.16	0.16	C
3	115	id3200001-id3015703	7	65	id7004429-id7004442	5.8	3.1	2.6	7.8	4.0	3.7	0.12	−0.12	C
DWsht: Shoot dry weight														
1	35	id1007776-id1016633	2	80	id2001831-id2003094	6.8	1.0	5.8	3.0	0.4	2.6	−0.04	0.04	C
2	65	fd12-id2013434	4	90	id4007444-id4008092	6.9	3.7	3.2	2.1	2.1	0.0	−0.04	0.04	B
1	35	id1007776-id1016633	4	95	id4010238-id4007105	16.5	4.8	11.7	7.1	2.5	4.7	−0.07	0.07	C
1	220	id1010609-id1016436	5	120	id5007536-id5005872	15.9	3.8	12.0	1.2	1.5	0.0	−0.07	0.07	B
1	50	id1011535-id1024836	6	5	ud6000572-id6014475	55.2	29.1	26.1	5.6	5.6	0.0	−0.14	0.14	C
2	80	id2001831-id2003094	6	5	ud6000572-id6014475	12.5	2.3	10.2	8.0	1.1	6.8	−0.06	0.06	C
4	90	id4007444-id4008092	6	5	ud6000572-id6014475	14.5	3.3	11.1	8.1	1.6	6.5	−0.06	0.06	C
5	120	id5007536-id5005872	6	5	ud6000572-id6014475	68.8	39.1	29.7	7.9	7.9	0.1	−0.16	0.16	C
6	5	ud6000572-id6014475	6	20	id6003318-id6001524	19.1	5.2	13.8	4.8	3.9	0.9	−0.06	0.06	C
1	145	id1002899-ud1000711	7	0	id7002260-id7001478	5.0	1.6	3.3	1.0	0.8	0.2	0.03	−0.03	C
5	125	id5007536-id5005872	7	5	id7001478-id7002859	9.1	0.0	9.1	2.6	2.7	0.0	0.06	−0.06	C
4	90	id4007444-id4008092	7	15	id7002859-id7002427	9.2	2.9	6.2	1.0	1.0	0.0	0.05	−0.05	C
6	5	ud6000572-id6014475	7	60	id7003748-id7003359	6.6	3.8	2.7	5.1	1.6	3.5	−0.05	0.05	C
4	90	id4007444-id4008092	8	70	id8007301-id8000240	7.9	2.0	5.9	1.2	1.0	0.2	0.04	−0.04	C
6	5	ud6000572-id6014475	8	90	id8007301-id8000240	9.0	0.0	9.5	2.5	1.0	1.5	0.05	−0.05	C
1	0	id1004348-id1001073	8	175	id8000240-id8005359	6.3	4.3	2.0	4.0	4.0	0.1	0.02	−0.02	B
2	80	id2001831-id2003094	8	175	id8000240-id8005359	5.0	0.4	4.6	3.5	0.2	3.3	0.03	−0.03	C
7	60	id7003748-id7003359	8	175	id8000240-id8005359	5.9	1.7	4.1	3.9	1.1	2.8	0.03	−0.03	C
5	125	id5007536-id5005872	8	180	id8005359-id8006485	20.3	4.9	15.4	5.8	3.5	2.3	0.06	−0.06	C
8	175	id8000240-id8005359	11	75	id11010335-id11008036	20.7	4.9	15.7	2.8	2.8	0.0	0.08	−0.08	C
1	35	id1007776-id1016633	11	115	id11004215-id11001683	14.0	2.6	11.4	4.2	1.3	2.9	−0.06	0.06	C
6	5	ud6000572-id6014475	11	120	id11002933-id11003556	53.7	22.7	31.0	14.1	5.9	8.1	−0.12	0.12	C
7	10	id7001478-id7002859	11	120	id11002933-id11003556	8.0	0.0	8.6	5.0	1.1	3.9	−0.05	0.05	C
6	5	ud6000572-id6014475	12	105	id12002728-id12001321	25.0	7.2	17.8	3.7	2.8	0.9	−0.09	0.09	C

AA by E represents the estimated additive effect of epistatic QTLs and QTL × environment interactions (QEI) under two environments, respectively; its positive value indicates that IR29 has the positive allele and is just the opposite for negative values; its positive value indicates that two-loci genotypes are the same as those in the parent IR29 having positive effects, whereas the two-loci recombinants have negative effects (from Hasawi); PVE (epistatic QTLs), PVE (AA), and PVE (QEI) represent the PVE by the epistatic QTLs, additive/main-effect QTLs, and QTL × environment interactions, respectively. *Epi* Epistatic, *TI*: Type of interaction, *A* interaction between QTLs, *B* interaction between QTLs and background loci, *C* interaction between complementary loci

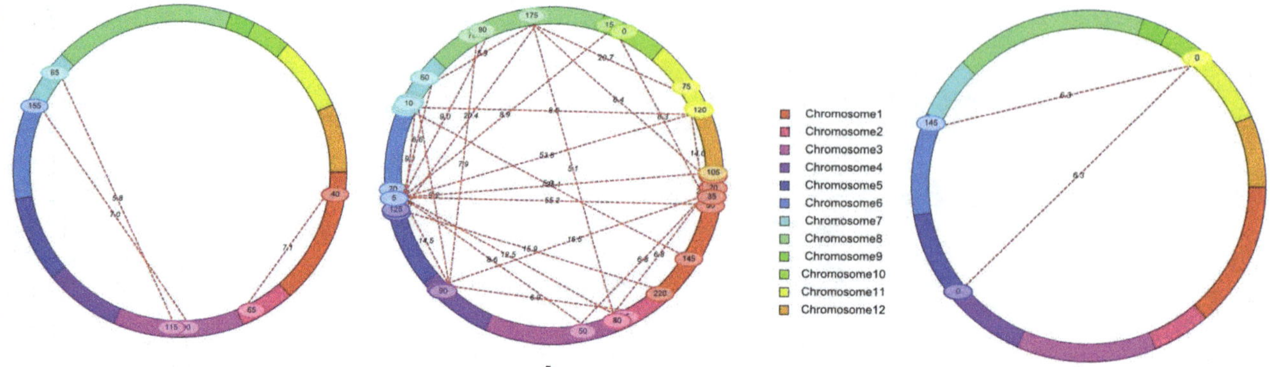

a b c

Fig. 3 Cyclic illustrations of epistatic QTLs for salt tolerance indices: (**a**) shoot fresh weight, (**b**) shoot dry weight, and (**c**) SES score with QTL × environment interactions on linkage groups by ICIM. The dotted lines indicate the interacting marker pairs located on the same or different chromosomes with corresponding LOD score due to epistatic effect. Chromosomes 1 (40 cM) and 2 (*qFWsht2.1*, 65 cM) showed interaction between QTL and background loci for shoot fresh weight (**b**), whereas chromosomes 1 (35 cM) and 4 (95 cM) showed interaction between complementary loci for DWsht (see Table 3)

for others as indicated by the multiple dotted lines connecting the chromosomes.

(ii) QTL × environment interactions (QEI) Out of 30 digenic interactions (1 for SL, 2 for SES score, 3 for FWsht, and 24 for DWsht) identified by the combined analysis of the multi-environment phenotypic values under two environmental conditions, 18 significant QEI were identified with LOD of 5.8 to 31.0 and PVE of 0.2% to 8.1% for DWsht (Table 3; Fig. 3). Three QTLs (*qSES4.1, qSL1.1,* and *qDWsht8.1*) were identified through combined analysis of both E2 and E3 (Additional file 1: Table S3). Importantly, *qSES4.1, qSL1.1,* and *qDWsht8.1* corresponded to main-effect QTLs such as *qSES4.1, qSL1.1,* and *qDWsht8.1*, respectively (Table 2).

Identification of tolerant RILs and use in breeding

There were significant differences in SL, RL, FWsht, DWsht, and SES score between the two parents under salinity (Additional file 1: Table S4). Hasawi had higher SL, RL, FWsht, and DWsht than the sensitive IR29. Transgressive segregation was observed for these traits and the heritabilities of SL, RL, FWsht, DWsht, and SES score ranging from 50% to 74% are high (Additional file 1: Table S4), indicating the repeatability/precision of the trials. Thus, four RILs (IR91477–13–1-1-1, IR91477–64–1-1, IR91477–76–1-1, and IR91477–250–1-1) with high salt tolerance and that had multiple introgression of the related QTLs were selected for evaluation and could be used as potential donors (pre-breeding materials) for further breeding programs (Table 4).

Discussion

Most studies on seedling-stage salinity tolerance have been carried out using common donors such as Pokkali and Nona Bokra. There have been fewer attempts to identify new donors for salinity tolerance, to identify novel QTLs, and to understand epistatic interaction effects among the QTLs and other loci, and also the effect of environmental interaction on QTL expression for salinity tolerance. In this study, we used Hasawi, a landrace from Saudi Arabia found to have strong adaptability to salinity and drought stress environments (Al-Mssallem et al. 2011; Bizimana et al. 2017). We made crosses using Hasawi, a tolerant donor, and IR29, the recipient parent. IR29 is an internationally recognized salinity-sensitive rice check variety. We developed $F_{5:6}$ RILs and screened them under three diverse environments across Asia and Africa.

The major findings of this study were five QTL hotspot (co-localized) regions identified on five different chromosomes for multiple QTLs such as chromosome 1 (*qSES1.3, qSES1.4, qSL1.2, qSL1.3, qRL1.1, qRL1.2, qFWsht1.2, qDWsht1.2*), chromosome 4 (*qSES4.1, qFWsht4.1, qFWsht4.2, qDWsht4.1, qDWsht4.2*), chromosome 6 (*qFWsht6.1, qDWsht6.1*), chromosome 8 (*qFWsht8.1, qDWsht8.1*), and chromosome 12 (*qSL12.1, qFWsht12.1, qDWsht12.1*); five salinity-tolerant RILs identified with multiple QTL introgression and that could be used directly in breeding programs; novel additive robust QTLs on the long arm of chromosome 1 for SES score, SL, RL, FWsht, and DWsht in a similar region; and one QTL (*qFWsht6.1*) detected in E2 and E3 for FWsht. Thirty-four QTLs identified through ICIM-ADD and IM-ADD mapping are additive because of the use of a permanent mapping population RIL with no or few heterozygotes; consequently, a dominant effect was absent.

SES score is a visual parameter for assessing the tolerance of seedlings under salinity stress. The lower the SES score (1 or 3), the higher the tolerance, whereas a higher SES score (7 or 9) suggests sensitivity. An SES score of 5 indicates moderate tolerance. Significant negative correlations were observed for SES score with all the other four parameters (SL, RL, FWsht, and DWsht) (Table 1). This is obvious because seedlings can be scored with low SES only if they have attained higher shoot length and high vigor, which means long root length, although higher SES scores could be given to plants with poor vigor and growth. This suggests that all four growth-related parameters directly relate to visually based SES scores. The negative associations observed between SES score and plant growth attributes clearly demonstrate the significance and the detrimental effects of high Na^+ accumulation in plant tissue under saline conditions. The most common salt injury symptoms in rice are leaf tip burning, early senescence, and complete necrosis, particularly among sensitive varieties such as IR29. The detrimental effects of salt stress on the growth and yield of rice genotypes are well documented in several earlier reports (Flowers and Yeo 1981, 1995; Gregorio and Senadhira 1993; Ashraf et al. 1999; Ismail et al. 2007; Munns and Tester 2008; Ding et al. 2010; Singh et al. 2010; Bimpong et al. 2016). The mostly positive significant correlations among SL, RL, FWsht, and DWsht suggest that these traits ultimately contribute to seedling-stage salinity tolerance.

This finding is in agreement with earlier findings that wide genetic variation exists for the traits RL and DWsht in rice under saline conditions at the seedling stage (Maiti et al. 2006; Al-Amin et al. 2013; Bimpong et al. 2014). It has also been reported that salinity stress leads to negative root growth and development (Roy et al. 2002; Rodrigues et al. 2002).

Stable genomic regions for salinity tolerance in multi-environments

Genetic linkage maps, based on pair-wise distance estimates, have emerged as pivotal tools for locating genes or QTLs. The analysis of recombination events from marker segregation data is especially helpful when a large number of markers segregate in a single mapping population. But, mapping larger numbers of markers also exponentially increases the potential orders of these loci on a chromosome. Hence, advanced and efficient algorithms are required to achieve near-perfect ordering of large numbers of loci. REcombination Counting and ORDering (RECORD) (van Os et al. 2005a, 2005b) is a faster, more accurate method for ordering of loci on genetic linkage maps and it performs especially well in regions of maps with high marker density. Therefore, RECORD was used to identify the best marker order in each linkage group and to generate a linkage map of rice. A total of 34 QTL regions were identified for five traits in our study across three environments. Eight QTLs for SES score, SL, RL, FWsht, and DWsht (*qSES1.3, qSES1.4, qSL1.2, qSL1.3, qRL1.1, qRL1.2, qFWsht1.2,* and *qDWsht1.2*) were identified within 170–175 cM (Table 2) on the long arm of chromosome 1. Although they are not at an identical position, it looks like an important region. *qSES1.4,* which is an additive QTL and has a high LOD value with PVE of 42.3%, is also identified in the same region. The identified QTLs (*qSES1.3, qSES1.4, qSL1.2, qSL1.3, qRL1.1, qRL1.2, qFWsht1.2,* and *qDWsht1.2*) on chromosome 1 in our study are located in different regions (170–175 cM) than the earlier reported QTLs on chromosome 1 such as *Saltol* (10.6–11.5 Mb) and *SKC1* (11.46 Mb) (Bonilla et al. 2002; Ren et al. 2005; Kim et al. 2009). It is interesting to note that most of the QTLs controlling seedling-stage salinity tolerance are reported on chromosome 1 in different studies and in the current study also we identified QTLs for all the studied traits except DWsht. A large number of QTLs for salinity tolerance have already been reported on the short arm of chromosome 1 (Claes et al. 1990; Flowers et al. 2000; Takehisa et al. 2004; Lin et al. 2004; Ren et al. 2005; Zhang et al. 2008; Sabouri et al. 2009; Ammar et al. 2009; Thomson et al. 2010; Bimpong et al. 2014).

The novel QTLs identified on chromosome 1 (*qSES1.3, qSES1.4, qSL1.2, qSL1.3, qRL1.1, qRL1.2*) from Hasawi needs further investigation to identify the candidate genes and to understand the molecular mechanisms conferring early seedling-stage salinity tolerance in rice, which could be through similar or different physiological pathways.

The other common QTL detected in two environments is *qFWsht6.1* (E2 and E3), which is co-localized with *qDWsht6.1* (E3) (Table 2). *qDWsht8.1* for DWsht shared a common genomic region (105 cM) with *qFWsht8.1* on chromosome 8. Indeed, both the traits are very closely associated. Three QTLs (*qSES4.1, qFWsht4.2,* and *qDWsht4.2*) are located in a similar position (32 cM) on chromosome 4 as the cluster at the marker interval of id4008522-id4008092. They may be working as QTL clusters of large QTLs due to co-localization. This region needs to be saturated with more markers for further fine mapping.

It was observed in multi-locus analysis that, if the pooled effects of some major QTLs such as those for SES score on different chromosomes (the pooled PVE of three SES score QTLs is 88.0%, with *qSES1.3,* R^2 = 39.9%; *qSES1.4,* R^2 = 42.3%; *qSES4.1,* R^2 = 5.8%) and for DWsht (the pooled PVE of three DWsht QTLs is 61.7%, with *qDWsht1.2,* R^2 = 19.6%; *qDWsht1.3,* R^2 = 12.0%; *qDWsht4.1,* R^2 = 30.1%) are considered as for one trait, then the overall phenotypic manifestation of their individual effects seems to be not as strong as expected (Table 2). This observation might be due to (i) co-localization of the QTLs, (ii) less-than-additive epistatic interactions among QTLs as suggested by Eshed and Zamir (1996), and (iii) some of the QTLs may be affecting the common traits through similar developmental processes.

Importantly, out of a total of 34 QTLs, 9 main-effect QTLs identified for five different traits such as 2 QTLs for SES score (*qSES1.3, qSES1.4*), 2 for shoot length (*qSL1.2, qSL1.3*), 2 for root length (*qRL1.1, qRL1.2*), 1 for shoot fresh weight (*qFWsht6.1*), and 2 for shoot dry weight (*qDWsht1.2, qDWsht6.1*) could be focused on for further details and fine mapping for the important traits linked with salt tolerance.

Digenic interaction

Epistasis is a major factor underlying quantitative traits (Caicedo et al. 2004). In our study, potential epistatic interactions between marker loci identified on 10 different chromosomes revealed 49 significant digenic interactions (2 interactions for E1, 31 for E2, and 16 for E3) through single environment analysis for SES score, shoot length, root length, shoot fresh weight, and shoot dry weight, with quite a wide range of PVE. But, we are not elaborating on this as our focus is on QTL interactions with environments; hence, we worked with multi-environment analysis. The ME analysis illustrated 17 marker intervals resulting in 30 two-way interactions (Table 3; Fig. 3); these interactions significantly influenced the traits, suggesting that epistasis is an important component of the genetic basis for complex traits, including tolerance of salt stress. Individually, all the complementary/background markers reported in Table 3 had no significant effect on the trait alone (otherwise, they would have mapped as reliable large-effect QTLs), but resulted in an enhanced effect when combined and interacting with QTLs and other markers. However, only one significant digenic interaction each was identified for SL, two interactions for SES score, and three interactions for FWsht, as several large-effect QTLs were detected for these traits. Some researchers have suggested the presence of significant epistatic interactions among QTL-linked or -unlinked markers (Cocherham and Zeng 1996; Eshed and Zamir 1996; Li et al. 1997; Hossain et al. 2015). So, there is a need to assess the importance of epistatic gene interactions as this complicates the genotype-phenotype relationships; in addition, different computing models used in analyzing epistasis vary and give different results (Malmberg et al. 2005). The interaction effect may enhance or reduce the expected manifestation of overall QTL effect depending upon the degree and direction of the interaction. Type II interactions (between complementary loci) had relatively higher PVE than type I interactions, probably because the trait manifests itself only through interaction between two complementary loci as no main-effect QTL is involved. The influence of epistatic QTL interactions alone explained the trait variation, ranging from 2.1% to 14.1%, which could be crucial when the threshold limits for salinity tolerance of a variety are to be enhanced to withstand environmental stress.

Studies in *Arabidopsis* and rice (Malmberg et al. 2005; Mei et al. 2005; Wang et al. 2011) have suggested that epistatic QTL effects are more important than additive QTL effects for fitness traits, for example, the loss of effect of the *Saltol* region in the introgression process to develop near-isogenic lines (NILs) (Thomson et al. 2010). The clean NILs were susceptible to salinity stress compared with NILs having interaction between *Saltol* and other background loci from the donor parent Pokkali. By contrast, studies designed to explicitly model epistatic interactions in other crops such as maize revealed that epistasis is of little or only moderate importance for quantitative traits (Schon et al. 2004; Mihaljevic et al. 2005; Blanc et al. 2006). These contrasting results might be due in part to the relative importance of epistatic effects in predominantly inbreeding or predominantly outcrossing species, and in part to differences in modeling procedures.

Several chromosomal regions were associated with more than one trait, indicating linkage or pleiotropic effects. For instance, three QTLs (*qSES1.3, qSL1.2,* and *qRL1.1*) located at 170 cM linked with MI: id1024972-id1023892 and five other QTLs such as *qSES1.4, qSL1.3, qRL1.2, qFWsht1.2,* and *qDWsht1.2* are closely associated with MI: id1023892-id1017885 at 175 cM. One SNP (id1023892) is found to be common in both marker intervals, which looks like a major common SNP in the region located on the long arm of chromosome 1, conferring salt tolerance. It is important to note that the same QTL might contribute to several traits associated with a specific phenotype because of closely associated traits. Hence, epistatic effects and pleiotropy can play a notable role in the interaction and function of a QTL; the presence of a QTL with a very small effect may have a large effect on a regulatory pathway (Koyama et al. 2001). Further characterization of this region by fine mapping and the identification of the genes underlying its tolerance will shed more light on whether the same set of genes or an entirely different set of linked genes governs these phenotypes.

QTL × environment interaction (QEI) for salinity tolerance

QTL × environment interaction plays an important role in adaptation to changing environments. QEI is particularly high in self-pollinated plants such as rice (Jain and Marshall 1967; Wang et al. 2014) and the complex epistatic interactions and QTL × environment interaction effects are important in controlling salt tolerance (Wang et al. 2011). In comparison to main-effect QTLs whose LOD threshold was kept at 3.0, the threshold was kept at >5.0 as the threshold for two-way interactions to identify only very significant QTLs. All the interactions were significant when LOD (epistasis) and LOD (QEI) ranged from 5.8 to 68.8 and 5.8 to 31.0, respectively. However, the LOD (AA) of 12 interactions is non-significant (LOD = 0.0 to 2.9) (Table 3). This suggests that these 12 interactions have less additive effect than QTL × environment interaction effects to express the phenotype. Combined LOD partitioned into PVE for additive-effect QTL and QTL × E interactions suggests potential enhancement of stress tolerance by a genotype for a specific environment through the sum of overall manifestation effects of QEI. Thus, QEI has a huge influence on salinity tolerance as the degree of salinity is dependent on environmental factors (temperature, humidity, rainfall), crop season, and crop growth stage. Besides this, the negative AA × E2 value (Table 3) indicates that interaction with Hasawi alleles rather than IR29 alleles may be one of the reasons for making the seedlings more tolerant of salinity stress. The relatively lower contribution of QTL × E interaction through additive components does not eliminate the possibility and importance of dominance or epistatic QTLs or interactions between the QTLs and the environment. E1 was the controlled environment whereas E2 and E3 were uncontrolled natural environments except for rain protection, and this was expected to have more QTL × E interaction, but the experiment infers that there is not a high order of interaction component but enough for affecting and elevating the threshold tolerance limits. Nine additive major QTLs that were identified (seven additive QTLs in the Philippines, one in both Tanzania and Senegal, and one in Senegal) and 30 epistatic QTLs that were identified by joint analyses suggest that epistatic QTLs and QTL × environment interactions are important components for FWsht, DWsht, SES score, and SL. Our investigation revealed a significant combined effect of epistatic interaction and QTL × E interaction with high PVE, but, on further dissection, we found more epistatic interaction than QTL × E interaction probably because of the higher heritable trait (Additional file 1: Table S2; Tables 3 and 4). Thus, major QTL effects, QEI, and epistatic interactions need to be considered together to improve selection efficiency using genomic-assisted breeding (Zhao et al. 2010; Liu et al. 2014).

Comparison of the new QTL loci with previously mapped QTLs

Results of comparative analysis of the QTL positions identified in the study compared with the QTL positions identified in earlier studies as being associated with salinity tolerance at various growth stages are shown. Rice cultivars grown in saline environments are most sensitive at both the vegetative and reproductive stages. However, the relationship between tolerances at the two stages is poor, suggesting that they are regulated by different processes and genes (Singh and Flowers 2010; Hossain et al. 2015; Rahman et al. 2016; Ahmadizadeh et al. 2016). The major QTL *Saltol*, derived from salt-tolerant landrace Pokkali, has been mapped on chromosome 1. This QTL confers salt tolerance at the vegetative stage and explains between 39.2% and 43.9% of the PVE in the original RIL population (Bonilla et al. 2002), but further studies found that *Saltol* alone does not work as a robust QTL (Thomson et al. 2010). A gene for salt tolerance at the vegetative stage, *SKC1*, has been identified in the same region from Nona Bokra and positionally cloned (Ren et al. 2005). *SKC1* maintains K^+ homoeostasis in the salt-tolerant cultivar under salt stress, and the gene encodes a member of HKT-type transporters. This gene turns out to be a protein in the HKT family that exclusively mediates K^+ and Na^+ translocation between roots and shoots, thereby regulating K^+/Na^+ homeostasis in the shoots, resulting in improved salt tolerance (Ren et al. 2005). The eight novel QTLs (*qSES1.3, qSES1.4, qSL1.2, qSL1.3, qRL1.1, qRL1.2, qFWsht1.2,* and *qDWsht1.2*) responsible for seedling-stage salinity tolerance on the long arm of chromosome 1 as reported in our study were found to be very different from *SKC1* and *Saltol*. These eight novel QTLs span a region of 170 to 175 cM. There is a need to further test the stability of the identified QTLs being expressed before drawing a conclusion.

Koyama et al. (2001) identified 10 QTLs for five shoot traits related to salt tolerance: Na^+ concentration (one QTL) at 74 cM on chromosome 1; K^+ concentration (three QTLs) on chromosomes 4, 6, and 9; Na^+ concentration

Table 4 Phenotypic values of traits for salt tolerance among selected RILs and parents common in Senegal and Tanzania with their related QTLs

Selected RILs	SL (cm)	RL (cm)	FWsht (g)	DWsht (g)	SES score	QTLs
IR91477-13-1-1-1	42.5	16.7	0.98	0.24	**3.0**	*qSES1.4, qSES4.1, qSL1.3*
IR91477-16-1-1-1	43.5	20.9	1.07	0.25	4.3	*qSES1.4, qSL1.3, qFWsht6.1*
IR91477-64-1-1	35.8	20.8	1.18	0.26	4.0	*qSES1.4, qSL1.3, qRL1.1, qFWsht6.1*
IR91477-76-1-1	36.4	22.2	0.79	0.20	4.3	*qSES1.4, qSES4.1, qDWsht6.1, qFWsht4.2*
IR91477-250-1-1	37.3	17.5	1.02	0.27	3.5	*qSES1.4, qSL1.3, qFWsht4.2, qFWsht6.1*
IR29 (sensitive check)	20.5	11.3	0.61	0.13	8.0	
Hasawi (tolerant check)	40.0	22.9	1.30	0.28	3.0	
Heritability (%)	72.5	78.5	55.0	59.3	68.6	

(two QTLs) on chromosomes 4 and 6; K^+ concentration (two QTLs) on chromosomes 1 (at 56 cM) and 4; and Na^+-K^+ ratio in shoots (two QTLs) on chromosomes 1 (at 74 cM) and 4. In our study, eight QTLs were identified for SES score and shoot and root traits and are located on chromosome 1 but in a different position (~175.0 cM) on the long arm versus the short arm of the chromosomes in the previous studies. Lin et al. (2004) detected five QTLs for four traits associated with salt tolerance in roots and three QTLs for three shoot traits associated with salt tolerance, but none of these QTLs have the same map locations as any of the QTLs identified here across environments, suggesting that most of these QTLs are novel and could be important for breeding. Wang et al. (2012) reported five additive QTLs, for Na^+ in shoots (*qSNC9*), K^+ in shoots (*qSKC1* and *qSKC9*), K^+ in roots (*qRKC4*), and for salt tolerance rating (*qSTR7*), as new salt tolerance loci. However, *qSES4.1, qFWsht4.1, qFWsht4.2, qDWsht1.1, qDWsht1.2,* and *qDWsht7.1* in E1 were identified that shared similar chromosomal positions in our study.

The QTLs that co-localize in a similar region (such as id1023892 marker) probably indicate some functional relatedness among them. This major QTL cluster might also have pleiotropic effects on other traits. The cluster of QTLs on chromosome 1 for different traits, such as SES score, SL, and RL (Fig. 1), was also supported by the strong correlations observed among these traits (Table 1). This clustering of loci and correlation of effects can be attributed to different linked QTLs occurring on the same segment or pleiotropic effects of a single QTL. High-resolution mapping is required to determine whether pleiotropic effects are present.

The QTLs identified in this study that overlap with others mapped previously fall into two categories: (i) QTLs that share a similar map position and are mapped to the same trait, and (ii) QTLs that share a similar map position but are mapped to a different trait. However, we found a tight cluster of QTLs localizing around 170–175.0 cM on chromosome 1, and three QTLs (*qSES4.1, qFWsht4.2,* and *qDWsht4.2*) on chromosome 4 at 32.0 cM and two QTLs (*qFWsht6.1, qDWsht6.1*) on chromosome 6 at 115.0 cM, which may be considered as novel loci. Several studies have indicated that many genomic loci controlling important rice traits are clustered in the same chromosome regions (Cai and Morishima 2002; Angaji 2008; Hossain et al. 2015). These major loci should be targeted for pyramiding through MABC (Singh et al. 2007; Thomson et al. 2010). Many QTLs for SES score, shoot length, root length, shoot fresh weight, and shoot dry weight were identified. The ones accounting for higher LOD and PVE could subsequently be used for QTL pyramiding after the development of low-cost diagnostic molecular markers linked to them.

Responses of the RILs and their parental lines to salt stress

The most crucial step of QTL mapping for salt tolerance in rice is the evaluation of salt tolerance (Wang et al. 2011). We selected five salt-tolerant RILs based on visual phenotypic score (SES score) that are commonly tolerant in two environments (Senegal and Tanzania). Salinity has large effects on crop growth, yield, and productivity (Tester and Davenport 2003; Munns and Tester 2008; Munns and Gilliham 2015). Initial vigor that directly relates to higher shoot and root length and fresh and dry weight of shoots through faster growth at early seedling stage could reduce the Na^+ concentration in plant tissues probably because of a dilution effect besides other salt-tolerance mechanisms operating in plants.

Conclusions

Salinity tolerance is a complex quantitative trait and previous studies established its strong association with visual symptoms (as indicated by the SES score) and other traits such as shoot length, root length, shoot fresh weight, and shoot dry weight. Overall phenotypic performance reflected by SES scores is determined by these key traits. We identified genomic regions on chromosomes 1, 2, 3, 4, 5, 6, 7, 8, 11, and 12 that are associated with salinity tolerance at the seedling stage by affecting SES scores, shoot length, root length, shoot fresh weight, and shoot dry weight. Thirty-four QTLs for the five traits were detected on 10 chromosomes. The QTLs (*qSES1.3, qSES1.4, qSL1.2, qSL1.3, qRL1, qRL1.2, qFWsht1.2,* and *qDWsht1.2*) detected on chromosome 1 could be of much interest and termed as novel QTLs as, unlike earlier reported *Saltol* and *SKC1*, they are on the long arm of chromosome 1. The study also detected 30 digenic two-way interactions through ME analysis that are quite important for gene expression, especially for complex traits such as salinity tolerance. Significant QTL × environment interaction for FWsht, DWsht, SES score, and SL indicated as high as a 8.1% contribution for phenotypic manifestation through interaction between QTLs and background loci, or complementary loci. The robust QTLs, digenic interactions, and QEI could be good targets for more detailed QTL studies, fine mapping, and subsequent pyramiding to develop highly tolerant varieties that could lead to the development of improved rice varieties for salt-affected areas where salt stress is a major impediment to rice production.

Additional files

Additional file 1: Table S1. Linkage map information. **Table S2.** Digenic interactions/epistatic QTLs (LOD >5.0) using ICIM-EPI method in single environments E1, E2, and E3. **Table S3.** Significant QTLs detected in two environments (E2 and E3) for traits related to salt tolerance in an IR29/Hasawi RIL population by inclusive composite interval mapping (ICIM) through combined analysis. **Table S4.** Responses of different salt-tolerant RIL genotypes of rice and their parents under salt stress at seedling stage in (a) the Philippines, (b) Senegal, and (c) Tanzania.

Abbreviations
DWsht: Dry weight of shoot; FWsht: Fresh weight of shoot; PVE: Phenotypic variance explained; QTLs: Quantitative trait loci; RL: Root length; SDM: Segregation distortion markers; SES score: Standard evaluation system (SES) score; SL: Shoot length

Acknowledgments
We thank the Genotyping Services Laboratory for technical assistance for genotyping. This work was funded in part by the Global Rice Science Partnership (GRiSP), Japan Rice Breeding Project (JRBP), and Bill & Melinda Gates Foundation (BMGF).

Authors' contributions
RKS, MAR, and BPMS conceived and designed the research. IKB, JPB, and EDP carried out the phenotyping experiment in Senegal, Tanzania, and the Philippines, respectively. MAR, JPB, MA, and EDP processed the genotyping data. MAR and MSR analyzed the data. MAR, IKB, BPMS, and RKS wrote the paper. All authors read and approved the final manuscript.

Competing interests
The authors declare that they have no competing interests.

Author details

[1]International Rice Research Institute (IRRI), DAPO Box 7777, Metro Manila, Philippines. [2]Bangladesh Rice Research Institute, Gazipur 1701, Bangladesh. [3]Africa Rice Center, Sahel Regional Station, BP 96 St Louis, Senegal. [4]IRRI-ESA Office, Bujumbura, Burundi. [5]Institute of Biological Sciences, University of the Philippines at Los Baños, Laguna, Philippines.

References

Ahmadizadeh M, Vispo NA, Calapit-Palao CDO, Pangaan I, Dela Viña C, Singh RK (2016) Reproductive stage salinity tolerance in rice: a complex trait to phenotype. IJPP – Springer. https://doi.org/10.1007/s40502-016-0268-6

Al-Amin M, Islam MM, Begum SN, Alam MS, Moniruzzaman M, Patwary MAK (2013) Evaluation of rice germplasm under salt stress at the seedling stage through SSR markers. Int J Agric Res Innov Technol 3(1):52–59

Al-Mssallem MQ, Hampton SM, Frost GS, Brown JE (2011) A study of Hasawi rice (*Oryza sativa* L.) in terms of its carbohydrate hydrolysis (*in vitro*) and glycaemic and insulinaemic indices (*in vivo*). Eur J Clin Nutr 65:627–634

Ammar MHM, Pandit A, Singh RK, Sameena S, Chauhan MS et al (2009) Mapping of QTLs controlling Na^+, K^+ and Cl^- ion concentrations in salt tolerant *indica* rice variety CSR27. J Plant Biochem Biotechnol 18:139–150

Angaji SA (2008) Mapping QTLs for submergence tolerance during germination in rice. Afr J Biotechnol 7(15):2551–2558

Ashraf A, Khalid A, Ali K (1999) Effect of seedling age and density on growth and yield of rice in saline soil. Pak J Biol Sci 2(3):860–862

Bimpong IK, Manneh B, Diop B, Ghislain K, Sow A, Amoah NKA, Gregorio G, Singh RK, Ortiz R, Wopereis M (2014) New quantitative trait loci for enhancing adaptation to salinity in rice from Hasawi, a Saudi landrace, into three African cultivars at the reproductive stage. Euphytica 200:45–60

Bimpong IK, Manneh B, Sock M, Diaw F, Amoah NKA, Ismail AM, Gregorio GB, Singh RK, Wopereis M (2016) Improving salt tolerance of lowland rice cultivar 'Rassi' through marker-aided backcross breeding in West Africa. Plant Sci 242:288–299

Bizimana JB, Luzi-kihupi A, Murori RW, Singh RK (2017) Identification of quantitative trait loci for salinity tolerance in rice (*Oryza sativa* L.) using IR29/Hasawi mapping population. J Genet 96(4):571–582. https://doi.org/10.1007/s12041-017-0803-x

Blanc G, Charcosset A, Mangin B, Gallais A, Moreau L (2006) Connected populations for detecting quantitative trait loci and testing for epistasis: an application in maize. Theor Appl Genet 113:206–224

Bonilla PS, Dvorak J, Mackill D, Deal K, Gregorio G (2002) RFLP and SSLP mapping of salinity tolerance genes in chromosome 1 of rice (*Oryza sativa* L.) using recombinant inbred lines. Philipp Agric Sci 85(1):64–74

Cai HW, Morishima H (2002) QTL clusters reflect character associations in wild and cultivated rice. Theor Appl Genet 104:1217–1228

Caicedo AL, Stinchcombe JR, Olsen KM, Schmitt J, Purugganan MD (2004) Epistatic interaction between *Arabidopsis* FRI and FLC flowering time genes generates a latitudinal cline in a life history trait. Proc Natl Acad Sci U S A 101:15670–15675

Churchill GA, Doerge RW (1994) Empirical threshold values for quantitative trait mapping. Genetics 138:963–971

Claes B, Dekeyser R, Villarroel R, den Bulcke MV, Bauw G, Montagu MV, Caplan A (1990) Characterization of a rice gene showing organ-specific expression in response to salt stress and drought. Plant Cell 2:19–27

Cocherham CC, Zeng ZB (1996) Design III with marker loci. Genetics 143:1437–1456

Ding M, Hou P, Shen X, Wang M, Deng S, Sun J et al (2010) Salt-induced expression of genes related to Na^+/K^+ and ROS homeostasis in leaves of salt-resistant and salt-sensitive poplar species. Plant Mol Biol 73:251–269

Eshed Y, Zamir D (1996) Less-than-additive epistatic interactions of quantitative trait loci in tomato. Genetics 143:1807–1817

Fan JB, Oliphant A, Shen R et al (2003) Highly parallel SNP genotyping. Cold Spring Harb Symp Quant Biol 68:69–78

Flowers TJ (2004) Improving crop salt tolerance. J Exp Bot 55(396):307–319

Flowers TJ, Koyama ML, Flowers SA, Sudhakar C, Singh KP, Yeo AR (2000) QTL: their place in engineering salt tolerance of rice to salinity. J Exp Bot 51:99–106

Flowers TJ, Yeo AR (1981) Variability in the resistance of sodium chloride salinity within rice (*Oryza sativa* L.) varieties. New Phytol 88:363–373

Flowers TJ, Yeo AR (1995) Breeding for salinity resistance in crop plants. Where next? Aust J Plant Physiol 22:875–884

Gregorio GB, Senadhira D (1993) Genetics analysis of salinity tolerance in rice. Theor Appl Genet 86:333–338

Gregorio GB, Senadhira D, Mendoza RD (1997) Screening rice for salinity tolerance. In: IRRI discussion paper series N°. 22, IRRI, Metro Manila. pp 3–19

Hackett CA, Broadfoot LB (2003) Effects of genotyping errors, missing values and segregation distortion in molecular marker data on the construction of linkage maps. Heredity 90:33–38. https://doi.org/10.1038/sj.hdy.6800173

Hossain H, Rahman MA, Alam MS, Singh RK (2015) Mapping of quantitative trait loci associated with reproductive-stage salt tolerance in rice. J Agro Crop Sci 201:17–31. https://doi.org/10.1111/jac.12086

Illumina (2008) GoldenGate genotyping assay for VeraCode manual protocol. In: Catalog # VC-901-1001, part # 11275211 rev. a

IRRI (2007) Standard evaluation system for rice (SES), 4th edn. International Rice Research Institute, Los Baños (Philippines), p 52

Ismail AM, Heuer S, Thomson MJ, Wissuwa M (2007) Genetic and genomic approaches to develop rice germplasm for problem soils. Plant Mol Biol 65:547–570

Jain S, Marshall D (1967) (1967) Population studies in predominantly self-pollinating species. X. Variation in natural populations of *Avena fatua* and *A. barbata*. Am Nat 101:19–33

Kader MA, Seidel T, Golldack D, Lindberg S (2006) Expressions of OsHKT1, OsHKT2, and OsVHA are differentially regulated under NaCl stress in salt-sensitive and salt-tolerant rice (*Oryza sativa* L.) cultivars. J Exp Bot 57:4257–4268

Khan S, Javed MA, Jahan N, Manan FA (2016) A short review on the development of salt tolerant cultivars in rice. Int J Public Health Sci (IJPHS) 5(2):201–212

Kim S-H, Bhat PR, Cui X, Walia H, Xu J, Wanamaker S et al (2009) Detection and validation of single feature polymorphisms using RNA expression data from a rice genome array. BMC Plant Biol 9:65. https://doi.org/10.1186/1471-2229-9-65

Koyama ML, Levesley A, Koebner RMD, Flowers TJ, Yeo AR (2001) Quantitative trait loci for component physiological traits determining salt tolerance in rice. Plant Physiol 125:406–422

Kumari S, Panjabinee Sabharwal V, Kushwaha HR, Sopory SK, Singla-Pareek SL, Pareek A (2009) Transcriptome map for seedling stage specific salinity stress response indicates a specific set of genes as candidate for saline tolerance in *Oryza sativa* L. Funct Integr Genomics 9:109–123

Lander ES, Botstein D (1989) Mapping Mendelian factors underlying quantitative traits using RFLP linkage maps. Genetics 121:185–199

Li ZK, Pinson SRM, Park WD, Paterson AH, Stansel JW (1997) Epistasis for three grain-yield components in rice (*Oryza sativa* L.) Genetics 145:453–465

Lin HX, Zhu MZ, Yano M, Gao JP, Liang ZW, Su WA, Hu XU, Ren ZH, Chao DY (2004) QTLs for Na^+ and K^+ uptake of the shoots and roots controlling rice salt tolerance. Theor Appl Genet 108:253–260

Lin JZ, Ritland K (1996) The effects of selective genotyping on estimates of proportion of recombination between linked quantitative trait loci. Theor Appl Genet 93:1261–1266

Liu D, Ma C, Hong W, Huang L, Liu M et al (2014) Construction and analysis of high-density linkage map using high-throughput sequencing data. PLoS One 9(6):e98855. https://doi.org/10.1371/journal.pone.0098855

Liu Y, Chen L, Fu D, Lou Q, Mei H, Xiong L, Li M, Xu X, Mei X, Luo L (2014) Dissection of additive, epistatic effect and QTL × environment interaction of quantitative trait loci for sheath blight resistance in rice. Hereditas 151:28–37

Liu Y, Li C, Shi X, Feng H, Wang Y (2016) Identification of QTLs with additive, epistatic, and QTL × environment interaction effects for the bolting trait in *Brassica rapa* L. Euphytica 210:427–439

Maiti RK, Vidyasagar P, Banerjee PP (2006) Salinity tolerance in rice (*Oryza sativa* L.) hybrids and their parents at emergence and seedling stage. Crop Res Hisar 31(3):427–433

Malmberg RL, Held S, Waits A, Mauricio R (2005) Epistasis for fitness related quantitative traits in *Arabidopsis thaliana* grown in the field and in the greenhouse. Genetics 171:2013–2027

McCouch SR, CGSNL (Committee on Gene Symbolization, Nomenclature and Linkage, Rice Genetics Cooperative) (2008) Gene nomenclature system for rice. Rice 1:72–84

McCouch SR, Cho YG, Yano M et al (1997) Report on QTL nomenclature. Rice Genet Newsl 14:11

Mei HW, Li ZK, Shu QY, Guo LB, Wang YP, Yu XQ et al (2005) Gene actions of QTLs affecting several agronomic traits resolved in a recombinant inbred rice population and two backcross populations. Theor Appl Genet 110:649–659

Mihaljevic R, Utz HF, Melchinger AE (2005) No evidence for epistasis in hybrid and *per se* performance of elite European flint maize inbreds from generation means and QTL analyses. Crop Sci 45:2605–2613

Munns R (2005) Genes and salt tolerance: bringing them together. New Phytol 167:645–663

Munns R, Gilliham M (2015) Salinity tolerance of crops – what is the cost? New Phytol 208:668–673. https://doi.org/10.1111/nph.13519

Munns R, Tester M (2008) Mechanisms of salinity tolerance. Annu Rev Plant Biol 59:651–681

Murray MG, Thompson WF (1980) Rapid isolation of high molecular weight plant DNA. Nucl Acids Res 8:4321–4325

Negrao S, Oliveira MM, Jena KK, Mackill D (2008) Integration of genomic tools to assist breeding in the *japonica* subspecies of rice. Mol Breed 22:159–168

Passioura J, Spielmeyer W, Bonnett DG (2007) Requirements for success in Marker-Assisted breeding for Drought-Prone Environments. In: Jenks MA, Hasegawa PM, Jain SM. (eds) Advances in Molecular-Breeding Toward Drought and Salt Tolerant Crops. Springer, Dordrecht. doi:https://doi.org/10.1007/978-1-4020-5578-2_19

Qados AMSA (2011) Effect of salt stress on plant growth and metabolism of bean plant *Vicia faba* (L.) J Saudi Soc Agric Sci 10:7–15

Rahman MA, Thomson MJ, Alam MS, De Ocampo M, Egdane J, Ismail AM (2016) Exploring novel genetic sources of salinity tolerance in rice through molecular and physiological characterization. Ann Bot 117:1083–1097. https://doi.org/10.1093/aob/mcw030

Ren ZH, Gao JP, Li LG, Cai XL, Huang W, Chao DY, Zhu MZ, Wang ZY, Luan S, Lin HX (2005) A rice quantitative trait locus for salt tolerance encodes a sodium transporter. Nat Genet 37:1141–1146

Rodrigues LN, Fernandes PD, Gheyi HR, Viana SBA (2002) Germination and formation of rice seedlings under saline stress. Esc Agrot Fed Castanhal 6(3):397–403

Roy SJ, Negrao S, Tester M (2014) Salt resistant crop plants. Curr Opin Biotechnol 26:115–124

Roy SK, Patra SK, Sarkar KK (2002) Studies on the effect of salinity stress on rice (*Oryza sativa* L.) at seedling stage. J Interacademicia 6(3):254–259

Sabouri H, Reizai AM, Mouomeni A, Kavoussi M, Sabouri A (2009) QTL mapping of physiological traits related to salt tolerance in young rice seedlings. Biol Plant 53:657–662

Schon CC, Utz HF, Groh S, Truberg B, Openshaw S, Melchinger AE (2004) Quantitative trait locus mapping based on resampling in a vast maize testcross experiment and its relevance to quantitative genetics for complex traits. Genetics 167:485–498

Singh RK, Flowers TJ (2010) The physiology and molecular biology of the effects of salinity on rice. In: Pessarakli M (ed) Third edition of "handbook of plant and crop stress". Taylor and Francis, Florida, USA, pp 899–939

Singh RK, Gregorio GB, Jain RK (2007) QTL mapping for salinity tolerance. Physiol Mol Biol Plants 13:87–99

Singh RK, Redoña E, Refuerzo L (2010) Varietal improvement for abiotic stress tolerance in crop plants: special reference to salinity in rice. In: Pareek A, Sopory SK, Bohnert HJ, Govindjee N (eds) Abiotic stress adaptation in plants: physiological, molecular and genomic foundation. Springer, Dordrecht, Netherlands, pp 387–415

Takehisa H, Shimodate T, Fukuta Y, Ueda T, Yano M, Yamaya T, Kameya T, Sato T (2004) Identification of quantitative trait loci for plant growth of rice in paddy field flooded with salt water. Field Crops Res 89:85–95

Tester M, Davenport R (2003) Na^+ tolerance and Na^+ transport in higher plants. Ann Bot 91:503–527

Thomson MJ (2014) High-throughput SNP genotyping to accelerate crop improvement. Plant Breed Biotechnol 2:195–212

Thomson MJ, de Ocampo M, Egdane J, Rahman MA, Sajise AG, Adorada DL, et al. (2010) Characterizing the *Saltol* quantitative trait locus for salinity tolerance in rice. Rice 3:148–160

Thomson MJ, Zhao K, Wright M et al (2012) High-throughput single nucleotide polymorphism genotyping for breeding applications in rice using the BeadXpress platform. Mol Breed 29:875–886

Tuberosa R, Salvi S (2006) Genomics approaches to improve drought tolerance in crops. Trends Plant Sci 11:412–415

van Os H, Stam P, Visser RG, Van Eck HJ (2005a) RECORD: a novel method for ordering loci on a genetic linkage map. Theor Appl Genet 112(1):30–40

van Os H, Stam P, Visser RG, van Eck HJ (2005b) SMOOTH: a statistical method for successful removal of genotyping errors from high-density genetic linkage data. Theor Appl Genet 112:187–194

Walia H, Wilson C, Condamine P, Liu X, Ismail AM, Zeng L et al (2005) Comparative transcriptional profiling of two contrasting rice genotypes under salinity stress during the vegetative growth stage. Plant Physiol 139: 822–835

Wang J, Li H, Zhang L, Meng L (2014) Users' manual of QTL IciMapping. The Quantitative Genetics Group, Institute of Crop Science, Chinese Academy of Agricultural Sciences (CAAS), Beijing 100081, China, and Genetic Resources Program, International Maize and Wheat Improvement Center (CIMMYT), Apdo. Postal 6–641, 06600 Mexico, D.F., Mexico

Wang X, Pang Y, Zhang J, Zhang Q, Tao Y, Feng B, Zheng T, Xu J, Li Z (2014) Genetic background effects on QTL and QTL × environment interaction for yield and its component traits as revealed by reciprocal introgression lines in rice. Crop J 2:345–357

Wang ZF, Wang JF, Bao YM, Wu YY, Zhang HS (2011) Quantitative trait loci controlling rice seed germination under salt stress. Euphytica 178:297–307

Wang Z, Chen Z, Cheng J, Lai Y, Wang J, et al. (2012) QTL analysis of Na+ and K+ concentrations in roots and shoots under different levels of NaCl stress in rice (Oryza sativa L.). PLoS ONE 7(12):e51202. doi:10.1371/journal.pone.0051202.

Wright MH, Tung CW, Zhao K, Reynolds A, McCouch SR, Bustamante CD (2010) ALCHEMY: a reliable method for automated SNP genotype calling for small batch sizes and highly homozygous populations. Bioinformatics 26:2952–2960

Wu Y, Hu Y, Xu G (2009) Interactive effects of potassium and sodium on root growth and expression of K/Na transporter genes in rice. Plant Growth Regul 57:271–280

Wurschum T, Maurer HP, Dreyer F, Reif JC (2013) Effect of inter- and intragenic epistasis on the heritability of oil content in rapeseed (*Brassica napus* L.) Theor Appl Genet 126(2):435–441

Xu S (2008) Quantitative trait locus mapping can benefit from segregation distortion. Genetics 180:2201–2208

Zeigler R, Barclay A (2008) The relevance of rice. Rice 1:3–10. https://doi.org/10.1007/s12284-008-9001-z

Zhang L, Li H, Li Z, Wang J (2008) Interactions between markers can be caused by the dominance effect of quantitative trait loci. Genetics 180:1177–1190

Zhang L, Wang S, Li H, Deng Q, Zheng A, Li S, Li P, Li Z, Wang J (2010) Effects of missing marker and segregation distortion on QTL mapping in F2 populations. Theor Appl Genet (2010) 121:1071–1082. https://doi.org/10.1007/s00122-010-1372-z

Zhao X, Qin Y, Jia B, Kim SM, Lee H-S, Eun M-Y, Kim K-M, Sohn J-K (2010) Comparison and analysis of main effects, epistatic effects, and QTL × environment interactions of QTLs for agronomic traits using DH and RIL populations in rice. J Crop Sci Biotechnol 13(4):235–241. https://doi.org/10.1007/s12892-010-0076-x

8

Exploring traditional aus-type rice for metabolites conferring drought tolerance

Alberto Casartelli[1], David Riewe[2], Hans Michael Hubberten[3], Thomas Altmann[2], Rainer Hoefgen[3] and Sigrid Heuer[1,4*]

Abstract

Background: Traditional varieties and landraces belonging to the aus-type group of rice (*Oryza sativa* L.) are known to be highly tolerant to environmental stresses, such as drought and heat, and are therefore recognized as a valuable genetic resource for crop improvement. Using two aus-type (Dular, N22) and two drought intolerant irrigated varieties (IR64, IR74) an untargeted metabolomics analysis was conducted to identify drought-responsive metabolites associated with tolerance.

Results: The superior drought tolerance of Dular and N22 compared with the irrigated varieties was confirmed by phenotyping plants grown to maturity after imposing severe drought stress in a dry-down treatment. Dular and N22 did not show a significant reduction in grain yield compared to well-watered control plants, whereas the intolerant varieties showed a significant reduction in both, total spikelet number and grain yield. The metabolomics analysis was conducted with shoot and root samples of plants at the tillering stage at the end of the dry-down treatment. The data revealed an overall higher accumulation of N-rich metabolites (amino acids and nucleotide-related metabolites allantoin and uridine) in shoots of the tolerant varieties. In roots, the aus-type varieties were characterised by a higher reduction of metabolites representative of glycolysis and the TCA cycle, such as malate, glyceric acid and glyceric acid-3-phosphate. On the other hand, the oligosaccharide raffinose showed a higher fold increase in both, shoots and roots of the sensitive genotypes. The data further showed that, for certain drought-responsive metabolites, differences between the contrasting rice varieties were already evident under well-watered control conditions.

Conclusions: The drought tolerance-related metabolites identified in the aus-type varieties provide a valuable set of protective compounds and an entry point for assessing genetic diversity in the underlying pathways for developing drought tolerant rice and other crops.

Keywords: Rice, Aus-type landraces, Metabolites, Allantoin, Drought tolerance, Roots, Spikelets, Genetic diversity

Background

To meet the increasing demand for food due to an increasing world population, future agricultural systems need to become more productive and, at the same time, more resource-use efficient and sustainable. Rice (*Oryza sativa* L.) is currently the main source of calories for more than half of the world's population and considerable breeding efforts are undertaken globally to increase the yield and yield potential of rice (for a recent review see Khan et al. 2015). However, there is also considerable scope for increasing yield by closing yield gaps, i.e., reducing the difference between the actual yield and the attainable yield, which is generally determined by light intensity and temperature, as well as nutrient- and water-availability. Modelling of yield gaps caused by water and nutrient limitations showed that closing the yield gap in maize, wheat and rice to 75% of the attainable yield would equal a 29% increase in global production (Mueller et al. 2012).

For closing such yield gaps it will be important to improve water and farm management, but equally important to develop crops that maintain high yield under adverse conditions, such as heat and drought or submergence, and increasing pest and disease pressure. The enhancement of drought tolerance in rice is one of the key challenges due

* Correspondence: sigrid.heuer@rothamsted.ac.uk
[1]School of Agriculture, Food and Wine, Waite Campus, The University of Adelaide, Adelaide, SA, Australia
[4]Rothamsted Research, Harpenden, UK
Full list of author information is available at the end of the article

to more frequent and more severe drought events caused by climate change (Porter et al. 2014; Lesk et al. 2016) and the need to reduce water consumption of rice production. First drought-tolerant rice varieties developed by marker-assisted selection (MAS) are now being released conferring yield advantages under drought across different environments of about 11% on average compared with the control (Swamy et al. 2013). Likewise, submergence tolerant Sub1-rice varieties had a yield advantage of more than 50% in submergence-prone regions across India (Mackill et al. 2012). This shows the potential impact of breeding for stress tolerance.

In recent years, a specific group of rice, so called aus-type rice, has been discovered as a valuable source of stress tolerance. Aus-type rice is closest related to indica-type rice but constitutes a distinct genetic group (McNally et al. 2009). These landraces have evolved and are still cultivated under environmental stress conditions in India and Bangladesh (Londo et al. 2006) and therefore have developed and preserved tolerance mechanisms for a diversity of stresses. For example, the submergence tolerance gene *OsSUB1A* mentioned above and the phosphorus (P) -starvation tolerance gene *OsPSTOL1* have both been identified from aus-type rice varieties (Xu et al. 2006; Gamuyao et al. 2012) and the aus-type variety N22 has been described as one of the most heat-tolerant rice cultivars currently known (Li et al. 2015; González-Schain et al. 2016). The variety Dular showed the highest P uptake under low-P field conditions (Wissuwa and Ae 2001) and consistently ranked highest in a drought study showing the least yield reduction over multiple seasons compared to other genotypes (Henry et al. 2011).

Aus-type rice is therefore highly valuable for breeding applications as a source of novel tolerance traits but also for gene discovery research. With the availability of an N22 *de-novo* reference genome (https://pag.confex.com/pag/xxiv/webprogram/Paper21395.html) and "omics" technologies it is now possible to assess aus-type rice at the molecular level and more easily gain access to genes and pathways that are specific to this group of rice. This will be important since so far molecular studies on stress tolerance have been predominantly carried out using the japonica type variety Nipponbare, for which a reference genome and genetic resources are available. However, Nipponbare is a modern irrigated variety and as such intolerant to drought and other abiotic stresses. Genes and pathways that are stress responsive in Nipponbare are therefore representative of an intolerant response and might be distinct from those in tolerant genotypes. In fact, important genes such as *OsSUB1A* and *OsPSTOL1*, or *Deep Root 1* (DRO1) and the *SNORKEL* deep water rice genes are not present in Nipponbare (Xu et al. 2006; Hattori et al. 2009; Gamuyao et al. 2012; Uga et al. 2013) and a comparative genome analysis of the aus-type variety DJ123 with Nipponbare and an indica genome (IR64) identified more than 600 genes that were specific to the aus-type variety (Schatz et al. 2014).

Metabolomics is regarded as the most transversal among the "*omics*" technologies mainly because it is not dependent on the availability of reference genomes and because it is untargeted and as such comprehensive, high throughput and facilitates the discovery of novel biomarkers (Beckles and Roessner 2012). Metabolites provide a direct readout of the physiological status of plants, reflecting the end products of the effect of environmental factors and the genetically determined, physiological and developmental responses of plants regulated by highly complex signalling and posttranslational processes. Therefore, metabolomics is closer to the phenotype than transcriptomics or proteomics alone (Beckles and Roessner 2012). Mapping of metabolites has already been applied to identify quantitative trait loci (QTL e.g. Matsuda et al. 2012; Hill et al. 2015) and potentially metabolites associated with a given trait of interest can be used as a screening and phenotyping tool in breeding programs, such as for quality traits in rice (Redestig et al. 2011; Matsuda et al. 2012). A comparative study of metabolomics and whole-genome SNP markers in maize has furthermore shown that metabolite profiles can predict the heterotic potential and yield of adult hybrid plants (Riedelsheimer et al. 2012).

In this study, we have conducted an untargeted, factorial metabolome analysis to compare the drought response of two tolerant aus-type varieties (Dular and N22) with two modern irrigated rice varieties (IR74 and IR64) grown under well-watered and dry-down conditions in soil. The main objective was to identify metabolites and their underlying pathways that are associated with drought tolerance, i.e., metabolites that show a distinct drought response in the aus-type rice varieties.

Methods

Plant material and growth conditions

Seeds of the two aus-type varieties (N22: IRGC19379; Dular: IRGC32561) and two indica-type irrigated varieties (IR64: IRGC66970; IR74: IRGC76331) used in this study were derived from IRRI's International Rice Genebank Collection (IRGC) in the Philippines.

Plants were grown in a glasshouse at IRRI (Los Banos, Laguna, Philippines) under the natural tropical conditions from September to December. Pots were filled with 6 kg of sifted local soil (anthraquic Gleysols) with basal fertilizer application equivalent to 45-30-20 kg ha^{-1} N-P-K. In total, 48 plants were grown for each genotype with two plants in each pot.

All pots were kept well-watered (WW) until 18 days after sowing (DAS), when water was withheld from half of the pots for the dry-down drought (D) treatment. Leaf rolling in D stressed plants occurred at 32 DAS in all genotypes and roots and shoots of 16 plants (8 pots) for each genotype were harvested. Roots and shoots of the same number of plants were harvested from WW controls at 33 DAS. For each genotype and treatment, 4 pots were harvested during the morning and four in the afternoon to account for time-of-day variation in metabolite compositions. Because the D-treated soil was very hard, pots had to be soaked in water for about 30 min before plants could be removed from the soil without damaging the root system. Roots, still attached to the shoot, were then washed with tap water on a sieve and rinsed twice with de-ionized water. Root and shoot length was measured with a ruler before shoots and roots were separated and frozen in liquid nitrogen. Samples were stored at -80 °C until they were further processed for metabolite analysis.

The remaining plants (8 plants in 4 pots for each genotype and treatment) were grown to maturity to assess phenotypic differences in the effect of the D treatment among the selected rice genotypes. D stressed plants were re-watered at 32 DAS for 2 days and water was withheld for a second dry-down until leaf rolling, which occurred at 42 DAS. From then on, the soil was kept flooded until plant maturity with a second fertilizer application (same as above) at 46 DAS. Pots with the WW control plants were kept flooded at all times.

In summary, the sample sets for the metabolite analysis and the phenotyping at maturity consisted of 15–16 biological replicates for each the two treatments (WW and D) and tissues (shoot and root) for each of the four genotypes (N22, Dular, IR64 and IR74).

GC-MS and IC analysis and data acquisition

Metabolites were extracted from 50 ± 5 mg fresh weight, measured and processed as described in Riewe et al. (2012 and 2016) using a LECO Pegasus HT mass spectrometer (LECO, St. Joseph, MI, USA) hyphenated with an Agilent 7890 gas chromatograph (Agilent, Santa Clara, CA, USA) and a Gerstel MPS2-XL autosampler (Gerstel, Muelheim/Ruhr, Germany). Eighty-nine known and 226 unknown metabolites were quantified in splitless mode. Lactate, malate, fructose, glucose and sucrose were quantified using split injections (1:50). Data were normalized regarding sample weight, measurement day and median of the respective metabolite per analysed batch. Outliers (more or less than replicate median +/−2xSD) were removed. For the WW shoots samples, a cluster of 28 samples was also excluded from the analysis due to technical problems during the sample preparation and the number of replicates for these samples was therefore reduced to 8–11. The full list of annotated metabolite peaks is provided as Additional file 1: Table S1.

Ion chromatography analysis

A subfraction of the polar phase containing polar metabolites and inorganic ions was filtered using an Ultrafree MC 5000 MC NMWL filter unit (Millipore). Subsequently, anions and cations were analyzed by high-performance anion- and cation-exchange chromatography with conductivity detection facilitated by a Dionex ICS-3000 system as described in detail in Schmidt et al. (2013).

Data analysis

For the identification of tolerant-specific metabolites, the normalised data was log10 transformed to improve normality and analysed by a two factorial ANOVA with interaction, where the factors were treatment (WW or D) and genotype. For this purpose, the tolerant aus-type varieties Dular and N22 were combined into a tolerant group and IR64 and IR74 into a sensitive group. A Bonferroni correction (Broadhurst and Kell 2006) was applied to account for multiple testing.

For the pathway map, Student's *t*-test analysis coupled with Bonferroni correction was performed between treatment and control values to highlight individual metabolite trends and metabolite networks were constructed using KEGG pathway maps web tool (http://www.genome.jp/kegg/).

Results

Phenotypic effect of drought stress in tolerant and intolerant rice genotypes

For this study we have chosen two representative aus-type varieties (Dular and N22) and two irrigated indica-type rice varieties (IR64 and IR74) to assess differential responses to water deficit at the metabolite level. Plants were grown in soil-filled pots under well-watered (WW) and dry-down drought (D) stress conditions until leaf rolling. A parallel set of WW plants and plants re-watered after drought stress was grown to maturity for yield component analysis.

All genotypes responded to the applied drought conditions by significantly increasing length of the longest root by 40% to almost 80% (Fig. 1). At the same time, root fresh weight (FW) was decreased by about 50%. Of the four genotypes, N22 showed the least root elongation and the highest reduction in root FW (Fig. 1). No differences in root length (longest root) were observed at the time of sampling for the metabolite analysis and the aus-type varieties had only slightly longer roots on average (Dular: 26.3 cm; N22: 27.5 cm) compared with the irrigated varieties (IR64: 25.7 cm; IR74: 23.6 cm) (data not shown).

Plant height was generally not affected by the treatment in any genotype but a significant reduction in tiller number by about 20% was observed in N22, IR64 and IR74 (Fig. 1), which corresponded to a reduction by

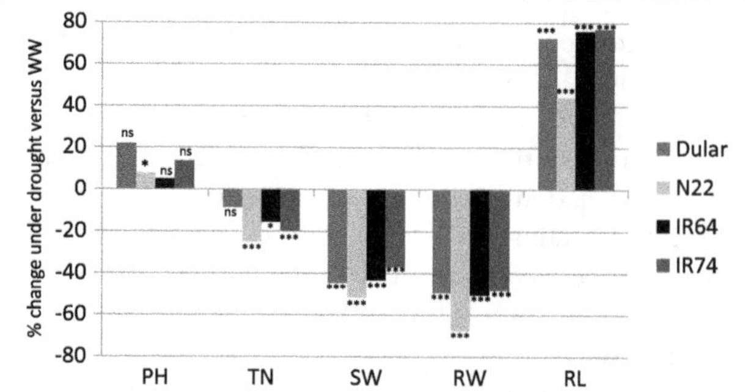

Fig. 1 Phenotypic data at the vegetative stage of four rice genotypes used for metabolite profiling. Four rice genotypes representing stress-tolerant aus-type varieties (Dular, N22) and intolerant indica-type irrigated rice varieties (IR64, IR74) were grown in soil-filled pots under well-watered and dry-down conditions. Root and shoot samples for metabolite analysis were harvest at 33 days after seeding. PH = plant height; TN = tiller number; SW = shoot fresh weight; RW = root fresh weight; RL = length of longest root. Asterisks indicate significant differences between WW and drought conditions: * = $p < 0.05$; ** = $p < 0.01$; *** = $p < 0.005$; ns = not significant

about 2 tillers (data not shown). All genotypes accordingly showed a significant reduction in shoot FW by about 40% (Fig. 1).

In plants grown to maturity, likewise no effect of the treatment on plant height was observed and tiller number of re-watered plants remained about 20% lower compared with WW controls in all genotypes (Additional file 2: Figure S1). In contrast, shoot DW recovered in all genotypes, except in IR64 which showed 20% lower shoot DW compared to the WW control. Likewise, reduction in root DW in re-watered plants remained highest in IR64 (> 40%) compared to about 20% reduced root DW in the other genotypes (Additional file 2: Figure S1).

Recording of the flowering time showed that the D treatment delayed flowering by four (IR64) to 7 days (Dular) and up to 10 days (N22, IR74). However, spikelet fertility and grain yield clearly differentiated between tolerant and intolerant genotypes. Whereas the aus-type varieties Dular and N22 did not show a significant reduction in the number of filled spikelets (grain) compared to the WW controls, the D treatment significantly reduced the total number of spikelets and the number of grains per panicle in IR64 and IR74 (Additional file 2: Figure S2). Interestingly, total spikelet number per panicle increased significantly in Dular, which compensated for the low spikelet fertility (62%) in the re-watered plants. Spikelet fertility in the other genotypes was largely unaffected by the D treatment, i.e., the reduced yield in IR64 and IR74 is due to a reduced spikelet number not due to reduced fertility (Additional file 2: Figure S2). This long-term negative effect of vegetative drought on yield appeared to be independent of phenology since IR74 flowered much later (91 DAS) than IR64 (63 DAS), which was closer to the aus-type varieties (Dular: 68 DAS; N22: 58 DAS).

Genotype and treatment effect on metabolite profiles

Root and shoot material of plants harvested from WW and D plants was analysed by ion chromatography-conductivity and gas chromatography-mass spectrometry (GC-MS). The combined dataset comprised a total of 328 metabolites and inorganic ions, of which 102 could be annotated. The metabolite data sets derived from root and shoot samples of the four genotypes were subjected to principal component analysis (PCA) to score for general trends (Roessner et al. 2001). This analysis showed that the first two principal components are sufficient for a clear separation of the genotypes according to treatment (PC1) and drought response (PC2) (Fig. 2a, b; Additional file 1: Table S1 for PC loadings). For both, roots and shoots, PC1 separates metabolome data sets of WW from drought stressed plants and in both cases samples derived from WW plants were less variable than those from the dry-down plants, suggesting less homogenous growth conditions as can be expected during dry-down of large pots. For the roots, PC2 separated the tolerant varieties Dular and N22 from the drought-intolerant indica varieties (IR64 and IR74) at comparable levels in both, the WW and the dry-down samples (Fig. 2a). PC2 also separated the aus-type and the indica varieties in the shoot samples (Fig. 2b), however, due to specific changes in the sensitive cultivars IR64 and IR74, separation of the drought samples in shoots is higher. Hierarchical clustering of all metabolites confirmed the PC analysis and clearly separated between genotypes and treatment in both, shoots and roots (Additional file 2: Figures S3 and S4). Other parameters, such as the spatial distribution of pots in the greenhouse, time and day of harvest and measurements was tested but did not show an effect (data not shown). The PC analysis therefore provided sufficient evidence that the metabolome data are indicative of the genetic differences between the genotypes and of the treatment effect and hence suitable for scoring specific diagnostic metabolite signatures.

Metabolic response of the primary metabolism of shoot and root subjected to drought

Metabolic and ion profiling allowed the identification and annotation of 102 molecules including amino acids, sugars, organic acids and nutrient ions. The results are summarised in the pathway map shown in Fig. 3. The relative abundance of each metabolite in the individual genotypes and tissues is colour coded and presented as the log_2 value of the fold change (log2FC) between stress and control conditions (see M & M for details). Overall, a large number (83) of metabolites showed a response to the drought treatment by either increasing (positive

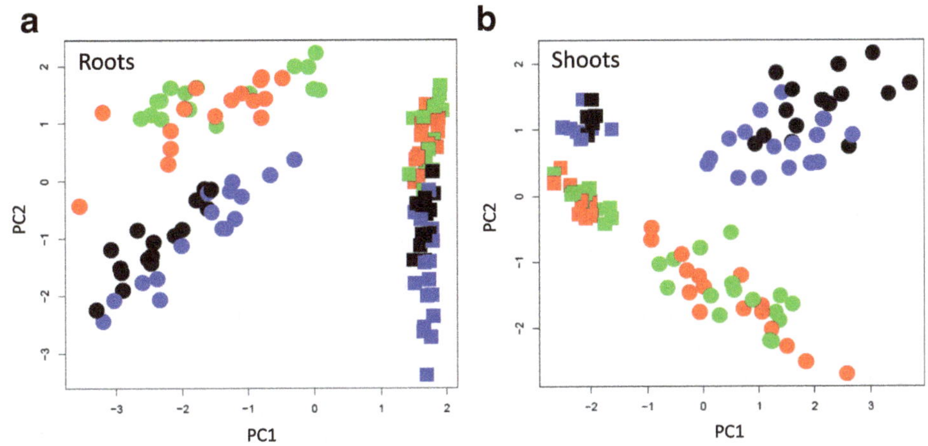

Fig. 2 PCA analysis of metabolites in roots and shoots. The first two principal components of root (**a**) and shoot (**b**) samples of four rice genotypes and two treatments (well-watered and dry down) are shown. Squares = samples from well-watered plants; circles = samples from plants exposed to a dry-down treatment; black = Dular; blue = N22; red = IR64; green = IR74

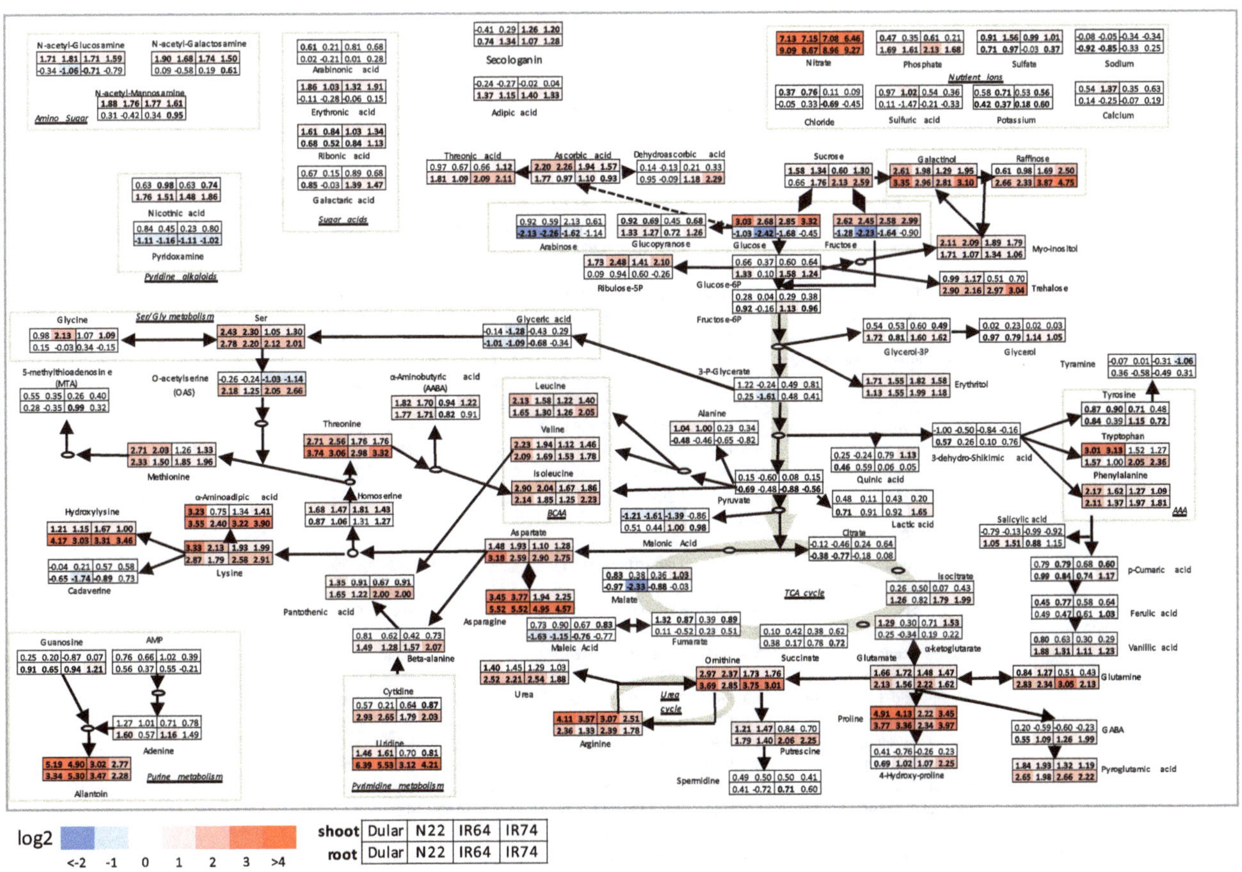

Fig. 3 Metabolic changes in the primary metabolism in shoots and roots in response to drought. Variation in metabolite levels in shoots and roots are presented as a $\log_2$ ratio per variety, pairwise by stress versus control conditions. Bold values indicate significant differences between drought and well-watered conditions (Student t-test with Bonferroni correction, $P < 0{,}05$). Colour coding indicates significantly different metabolites with a log2FC higher or lower than one (two-fold in linear scale)

log2FC) or reducing (negative log2FC) the levels under drought (Fig. 3).

To reduce complexity, only those up- and down-regulated metabolites presenting a log2FC larger or smaller than 1 ($P < 0.05$) were considered as significantly altered under the applied experimental conditions. This was further supported by an estimation of the theoretical possible accumulation of metabolites simply caused by the reduction in FW under drought, which was 0.86-log2FC in shoots and 1.14-log2FC in roots. The median accumulation over all the metabolite features identified was 0.74-log2FC in shoots and 0.93-log2FC roots. While this is close to the theoretical enrichment, most of the identified metabolites were well above that threshold as shown in Figs. 4 and 5.

Of all analysed metabolites, nitrate (NO_3^-) showed the highest accumulation in the shoots (7.2-log2FC) as well as in the roots (9.3-log2FC). In contrast, phosphate (PO_4^{-3}) also increased in roots (up to 2.1-log2FC) but did not show any significant changes in shoots. The remaining nutrient ions (Cl^-, SO_4^{-2}, H_2SO_4, Na^+, K^+, Ca^{++}) did not show major significant differences between genotypes or treatment response.

A range of sugars were affected by the treatment with the majority showing a positive log2FC in both, roots and shoots (Fig. 3). Sucrose, which is thought to play a key role in osmotic adjustment (Lemoine et al. 2013), increased only slightly in shoots but up to 2.6-log2FC in roots, whilst glucose and fructose positively accumulated in shoots (up to 3.3-log2FC) but were reduced in roots (up to −2.4-log2FC). Sugars belonging to the raffinose family oligosaccharides (RFO), known to have a protective role under stress (Nishizawa et al. 2008), also increased significantly (raffinose: 4.8-log2FC; galactinol: 3.4-log2FC; myo-inositol: 2.1-log2FC) together with the polyol erythritol (2-log2FC). Interestingly, other sugars were responsive specifically in roots but not in shoots, such as trehalose (3-log2FC) and arabinose (−2.3-log2FC). In contrast, the phospho-sugar ribulose-5-phosphate showed a positive accumulation (2.5-log2FC) specifically in shoots.

Metabolite levels of intermediates of the glycolysis and the tricarboxylic acid cycle (TCA) were not strongly affected by the treatment (Fig. 3). Although in roots, reduction under drought was observed for 3-P-glycerate (−1.6-log2FC, N22-specific), malate (−2.33-log2FC,

	Two-way ANOVA analysis			Fold-change (D/WW)				Response
	Treatment	Genotype	Interaction	Dular	N22	IR64	IR74	type
Shoot								
Allantoin	1.29E-39	3.21E-10	3.77E-10	5.19	4.90	3.02	2.77	T
Proline	1.11E-24	1.95E-02	1.28E-02	4.91	4.13	2.22	3.45	T
Arginine	7.57E-38	1.94E-02	2.07E-02	4.11	3.57	3.07	2.51	T
Asparagine	6.44E-26	6.86E+00	1.95E-03	3.45	3.77	1.94	2.25	T
Ornithine	8.29E-23	1.32E-01	1.53E-03	2.97	2.37	1.73	1.76	T
Threonine	1.09E-33	6.93E-01	2.97E-02	2.71	2.56	1.76	1.76	T
Methionine	5.10E-24	8.60E-02	8.32E-04	2.71	2.03	1.26	1.33	T
Serine	6.48E-32	2.64E+01	1.11E-07	2.43	2.30	1.05	1.30	T
2-amino-butanoic acid (AABA)	4.79E-26	9.28E-02	4.36E-02	1.82	1.70	0.94	1.22	T
Uridine	1.77E-18	2.10E+02	3.08E-02	1.46	1.61	0.70	0.81	T
Chloride	1.29E-08	9.28E-01	5.39E-04	0.37	0.76	0.11	0.09	T
Raffinose	4.54E-20	6.85E-01	8.23E-04	0.61	0.98	1.69	2.50	S
Secologanin	1.57E-09	6.39E-11	1.35E-07	-0.41	0.29	1.26	1.20	S
Quinic acid	2.53E-03	7.05E-02	3.68E-03	0.25	-0.24	0.79	1.13	S
O-acetyl-serine (OAS)	3.83E-09	1.30E+01	2.92E-02	-0.26	-0.24	-1.03	-1.14	S
Root								
Uridine	2.22E-63	2.05E+01	5.31E-11	6.39	5.53	3.12	4.21	T
2-amino-butanoic acid (AABA)	2.06E-28	1.44E+00	5.34E-06	1.77	1.71	0.82	0.91	T
Sulphate	5.77E-16	5.96E+00	2.68E-04	0.71	0.97	-0.03	0.37	T
Malate	9.35E-19	1.88E-13	2.18E-05	-0.97	-2.33	-0.88	-0.03	T
Glyceric acid	1.84E-16	1.03E-04	2.07E-03	-1.01	-1.09	-0.68	-0.34	T
Sodium	2.14E-04	2.66E+02	1.07E-02	-0.92	-0.85	-0.33	0.25	T
Glyceric acid-3-phosphate	1.07E+02	2.11E-06	1.06E-02	0.25	-1.61	0.48	0.41	T
Citrate	3.58E-06	6.79E+01	1.27E-02	-0.38	-0.77	-0.18	0.08	T
Raffinose	2.57E-60	7.13E+01	2.29E-14	2.66	2.33	3.87	4.75	S
Dehydroascorbic acid	2.02E-10	4.13E-12	4.47E-02	0.95	-0.09	1.18	2.29	S
5-methylthio-adenosine (MTA)	1.60E+00	2.18E-02	8.94E-04	0.28	-0.35	0.99	0.32	S
N-acetyl-mannosamine	1.59E-01	1.98E-12	3.43E-02	0.31	-0.42	0.34	0.95	S
N-acetyl-galactosamine	2.65E+01	6.76E-11	4.48E-02	0.09	-0.58	0.19	0.61	S
Chloride	1.17E-03	3.51E+01	3.63E-08	-0.05	0.33	-0.69	-0.45	S

Fig. 4 Drought-responsive metabolites with differential accumulation in tolerant and sensitive rice genotypes. Metabolites with significant differences in shoots and roots between the analysed rice genotypes based on a Two-Way ANOVA are listed. Red and blue color shading highlight the degree of postive and negative-fold change in response to the drought treatment. Metabolites with higher fold-change in the tolerant or intolerant genotypes, respectively, were classidified as tolerant-specific (T) or sensitive-specific (S)

	Two-way ANOVA analysis			Fold-change (D/WW)				Response
	Treatment	Genotype	Interaction	Dular	N22	IR64	IR74	type
Shoot								
Unknown MST 94	1.27E-47	3.88E-04	2.41E-02	4.90	5.38	4.17	3.88	T
Unknown MST 89	1.09E-38	1.64E-09	2.01E-05	4.47	5.32	3.33	2.82	T
Unknown MST 57	9.53E-28	7.20E+01	2.52E-03	4.80	4.25	3.07	2.68	T
Unknown MST 37	1.06E-26	3.66E+01	1.83E-02	3.84	4.65	2.66	2.23	T
Unknown MST 191	8.38E-47	1.20E-13	1.09E-04	3.87	4.00	2.88	2.66	T
Unknown MST 52	2.21E-46	1.14E-07	3.51E-06	4.10	3.52	2.78	2.47	T
Unknown MST 151	8.08E-34	1.08E-08	3.44E-06	3.53	3.96	2.36	2.02	T
Unknown MST 160	4.24E-30	1.41E-18	8.99E-04	3.07	2.52	1.25	1.45	T
Unknown MST 48	4.16E-25	1.93E+02	2.14E-05	2.41	2.70	1.00	1.41	T
Unknown MST 72	2.49E-17	2.30E+02	4.18E-03	2.14	2.40	1.37	1.08	T
Unknown MST 175	1.95E-24	1.53E+02	1.96E-02	2.15	1.61	0.96	1.08	T
Unknown MST 205	4.67E-09	8.99E-29	3.76E-05	1.80	1.89	-0.04	0.19	T
Unknown MST 107	1.09E-13	4.89E-02	4.59E-02	2.12	1.27	0.80	1.06	T
Unknown MST 79	2.48E-02	2.95E-12	1.11E-10	1.85	0.85	-0.18	-0.35	T
Unknown MST 177	2.20E+02	3.01E+02	1.71E-02	0.89	1.68	-0.09	-1.18	T
Unknown MST 61	1.11E+02	2.04E-22	1.20E-04	-0.28	-0.90	0.08	0.62	T
Unknown MST 165	6.67E+00	2.06E+02	7.82E-03	-0.22	0.90	-0.32	-1.92	S
Root								
Unknown MST 37	9.31E-62	1.72E+02	3.10E-06	4.16	4.75	3.41	3.28	T
Unknown MST 44	2.92E-43	4.88E+01	3.69E-02	4.28	4.05	3.41	2.67	T
Unknown MST 89	6.01E-26	1.39E-03	9.01E-03	2.33	4.85	2.67	1.42	T
Unknown MST 123	2.09E-42	1.45E+02	2.64E-03	3.26	2.54	2.01	2.02	T
Unknown MST 199	6.04E-54	4.48E-01	3.86E-03	3.19	2.27	2.01	2.01	T
Unknown MST 198	6.52E-55	3.77E+00	1.74E-04	2.97	2.09	1.93	1.89	T
Unknown MST 103	6.44E-51	6.99E-03	1.47E-02	2.35	2.08	1.71	1.51	T
Unknown MST 45	1.35E-08	2.70E-01	3.23E-03	0.79	1.00	0.17	0.23	T
Unknown MST 205	3.87E-48	1.19E-17	1.35E-12	-3.52	-4.07	-1.94	-1.90	T
Unknown MST 87	2.97E-35	1.03E-09	2.99E-05	-2.09	-2.78	-0.93	-1.58	T
Unknown MST 42	3.78E-20	5.03E-05	9.31E-04	-0.71	-1.27	-0.71	-0.16	T
Unknown MST 22	7.61E-10	2.21E-09	4.90E-02	-0.80	-1.15	0.24	-0.88	T
Unknown MST 92	1.23E-02	2.44E-02	1.50E-02	-0.48	-1.37	-0.31	0.09	T
Unknown MST 120	3.18E-04	5.63E-13	3.48E-02	-0.33	-1.18	-0.21	-0.31	T
Unknown MST 61	2.64E+02	9.21E-15	1.23E-05	-0.28	-0.88	-0.32	1.26	T
Unknown MST 197	1.21E-59	3.39E-68	7.23E-25	2.28	2.97	5.86	5.99	S
Unknown MST 193	7.88E-64	6.61E-67	3.08E-21	1.94	2.84	4.53	4.89	S
Unknown MST 206	1.85E-57	1.67E+00	2.50E-12	2.78	2.22	3.91	5.07	S
Unknown MST 194	2.20E-52	2.60E-81	3.41E-18	1.23	2.23	4.17	3.81	S
Unknown MST 121	8.18E-63	2.03E-60	4.55E-10	2.42	2.57	3.69	4.17	S
Unknown MST 187	1.32E-52	5.97E-63	2.54E-09	1.84	2.71	3.92	3.83	S
Unknown MST 136	1.36E-65	1.24E-74	5.93E-08	3.09	2.10	3.34	4.27	S
Unknown MST 211	3.82E-46	3.42E-01	2.93E-04	2.77	1.60	3.17	3.89	S
Unknown MST 209	1.04E-38	1.91E+02	2.57E-02	2.33	1.40	2.41	3.39	S
Unknown MST 160	7.63E-51	5.05E-07	4.26E-04	1.92	1.37	2.18	2.40	S
Unknown MST 208	1.55E-08	6.19E+00	4.28E-05	0.20	0.16	1.76	2.77	S
Unknown MST 196	1.55E-07	2.84E+01	4.50E-02	0.83	0.82	2.72	1.01	S
Unknown MST 216	3.93E-04	1.24E-21	5.68E-11	0.05	-0.46	1.20	0.97	S
Unknown MST 202	7.57E+00	1.28E-25	2.64E-06	-0.28	-0.40	0.90	0.48	S
Unknown MST 214	3.09E+02	7.83E-21	1.36E-10	-0.56	-0.59	0.80	0.43	S
Unknown MST 96	1.80E+00	2.08E-23	7.72E-05	0.01	-0.43	0.32	0.75	S
Unknown MST 157	2.00E+02	7.15E+00	1.13E-02	-0.03	-0.37	-0.04	0.78	S

Fig. 5 Unknown drought-responsive metabolites in tolerant and sensitive rice genotypes. Metabolites without annotation but significant differences in shoots and roots between the analysed rice genotypes based on a Two-Way ANOVA are listed. Red and blue color shading highlight the degree of postive and negative-fold change in response to the drought treatment. Metabolites with higher fold-change in the tolerant or intolerant genotypes, respectively, were classidified as tolerant-specific (T) or sensitive-specific (S)

N22-specific), glyceric acid (−1-log2FC, Dular and N22-specific), whilst accumulation for isocitrate (up to 2-log2FC) was observed in Dular, IR64 and IR74.

Amino acids (AA) were the class of primary metabolites that presented the most widespread response, which is in agreement with other studies conducted on abiotic stresses (e.g. Bowne et al. 2011; Planchet et al. 2011; Witt et al. 2012) and an important role of proline for drought tolerance in rye grass was already suggested 60 years ago (Kemble and Macpherson 1954).

In our study, AA belonging to the glutamate family showed a strong positive accumulation in response to drought in both, shoots and roots, with the highest log2FC observed for proline (4.9-log2FC) followed by arginine (4.1-log2FC), the intermediate ornithine (3.8-log2FC) and glutamate (2.2-log2FC). In contrast to these AA, glutamine showed a positive accumulation (3.1-log2FC) mainly in roots (Fig. 3). Increased levels were also observed for other metabolites related to the family, such as GABA (2-log2FC) and urea (2.6-log2FC). AAs belonging to the aspartate family overall increased under drought in both, shoots and roots, with asparagine increasing the most (5.5-log2FC), followed by threonine (3.7-log2FC), lysine (3.3-log2FC), aspartate (3.2-log2FC) and methionine (2.7-log2FC). Interestingly, aspartate, asparagine and the lysine-related metabolite hydroxylysine showed a relatively higher accumulation in roots than in shoots (Fig. 3). The latter, for example, accumulated up to 4.2-log2FC in roots, while it was only up to 1.7-log2FC in shoots. The branched-chain amino acids (BCAA) presented similar positive accumulations in both tissues under stress (leucine: 2.1-log2FC, valine: 2.2-log2FC, isoleucine: 2.9-log2FC). Within the serine family, serine showed a positive (2.8-log2FC) accumulation in shoots and roots, while the intermediate O-acetyl serine (OAS) increased in roots (2.2-log2FC) but decreased in shoots (−1.4-log2FC), especially in the intolerant genotypes IR64 and IR74 (Fig. 3). Among the aromatic AA, only phenylalanine showed a consistent accumulation (2.2-log2FC) under drought in all genotypes and tissues, while changes in tryptophan were genotype specific with high accumulation in shoots of the tolerant genotypes (>3-log2FC) but in roots of the intolerant genotypes (>2-log2FC). Tissue-specific accumulation under stress was also observed for three amino sugars (N-acetyl-glucosamine, N-acetyl-galactosamine and N-acetyl-mannosamine) which increased specifically in drought stressed shoots (up to 1.9-log2FC). In contrast, the pyridine alkaloids nicotinic acid increased (1.9-log2FC) and pyridoxamine decreased (−1.2-log2FC) specifically in roots.

Metabolites related to nucleotide metabolisms were strongly affected by the drought treatment. Allantoin, indicative of the purine catabolic pathway, increased in roots and shoots about 5-log2FC in the tolerant genotypes and about 3-log2FC in the intolerant genotypes (Fig. 3). Allantoin accumulation was recently reported to increase tolerance under various stress conditions in Arabidopsis (Watanabe et al. 2014; Irani and Todd 2016; Lescano et al. 2016). Pyrimidine-related molecules, such as cytidine and uridine, increased 2.9- and 6.4-log2FC, respectively, but preferentially accumulated in drought-stressed roots.

Identification of tolerant- and sensitive-specific metabolites

The principal aim of this study was to identify metabolites that are associated with drought tolerance and are therefore specifically responsive in the aus-type varieties N22 and Dular. We have therefore conducted a two-way ANOVA analysis using drought treatment as one factor and tolerance group as the second factor. For the latter, Dular and N22 were combined into a tolerant group and the IR64 and IR74 into a sensitive group (Additional file 3: Table S2) which is also justified by the fact that in the PCA analysis (Fig. 2) no separation within the group of aus-type and indica type varieties could be detected. Overall, in shoots, 79 metabolites were significantly ($P < 0.05$) changed because of the drought treatment, 26 because of the genotype and 15 because of the interaction of the two factors (Additional file 3: Table S2). The latter represent metabolites that accumulated under drought differentially in tolerant and sensitive genotypes and the data are shown in Fig. 4 as $\log_2$ ratios. Similarly, in roots, 85 metabolites were significantly changed because of the stress, 30 because of the genotype, and 14 as a result of the interaction of the two factors (Fig. 4 and Additional file 3: Table S2). Metabolites were further classified depending on whether the log2FC was higher in the tolerant group or in the drought sensitive group.

A general comparison between the identified metabolites in roots and shoots revealed major differences between the two tissue types. In shoots of the tolerant group, several AA (serine, methionine, asparagine, proline, threonine, arginine and its derivate ornithine) specifically accumulated in response to drought, whereas no AA showed significant interaction between treatment and genotype in roots. Conversely, several organic acids (glyceric acid, glyceric acid 3-phoshate, malic acid and citric acid), which are all components of glycolysis and the TCA cycle, were significant for roots of the tolerant group but not in shoots. In contrast, two metabolites were identified in the tolerant group in both, shoots and roots, namely uridine and 2-amino-butanoic acid (AABA), whereas raffinose was specific to the sensitive groups in shoots and roots (Fig. 4). Chloride was associated with the tolerant genotypes in shoots but with the sensitive genotypes in roots.

Amongst the identified metabolites different drought response pattern were apparent as can be seen in the box plots of representative metabolites from shoots and roots shown in Fig. 6. Generally, three different response types can be distinguished as (i) the magnitude of the response between control and stress (either positive or negative) is higher in the tolerant or sensitive genotypes; (ii) metabolite abundances are different between tolerant and sensitive genotypes under WW conditions rather than under stress; and (iii) the metabolite is responsive to the treatment in only one of the two groups.

Methionine and raffinose are examples of category one, as they showed a higher magnitude of positive change under drought in shoots of the tolerant genotypes and roots of the sensitive genotypes, respectively (Fig. 6). Allantoin is a representative of category two

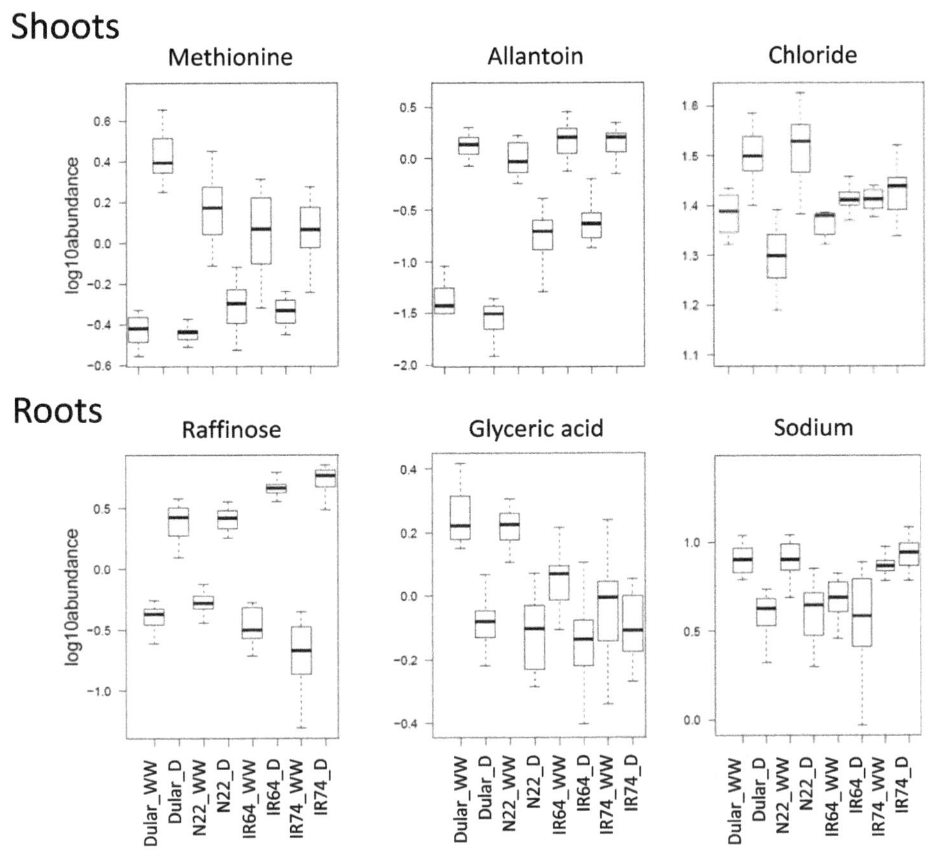

Fig. 6 Representative drought responsive metabolites in tolerant and sensitive rice genotypes. Box plots showing the log10 abundance of the metabolites indicated in shoot and root samples of four rice genotypes grown under well-watered (WW) conditions or exposed to a dry-down (D) treatment

since it accumulated to about the same level in drought-stressed shoots in all genotypes but was less abundant in the tolerant genotypes under well-watered conditions, thus, explaining the higher log2FC in Dular and N22 compared with the sensitive genotypes. The opposite is true for glyceric acid, which was more abundant under WW conditions in roots of the tolerant genotypes compared with IR64 and IR74 but showed no differences under drought. Chloride and sodium are representative of category three since they were not responsive to stress in intolerant genotypes but increased and decreased in N22 and Dular shoots and roots, respectively.

Although metabolomics is becoming an increasingly popular technique for studying plant stress responses, one of its main current limitations is the large amount of unknown metabolites that are identified during the analysis. For example, in our experiment 226 out of 328 of the registered metabolic features were annotated as unknown. However, despite being unable to place these metabolites in a pathway map, unknowns are still indicative of the genetic diversity between genotypes and their contrasting stress responses. Applying the same analysis as conducted for the annotated metabolites, we identified 17 metabolites in shoots with significant interaction between treatment and genotype, of which 16 were classified as tolerant-specific and one as sensitive-specific (Fig. 5). Similarly, in roots 32 metabolites presented significant interaction of which 15 and 17 were associated with tolerant and sensitive genotypes, respectively. As was the case for the annotated metabolites, shoots and roots showed a contrasting stress response and only five unknown metabolites were significantly stress responsive in both tissues (Unknowns 37, 61, 89, 160 and 205). The identified unknown metabolites could be assigned to the same three categories as described above for the annotated metabolites and representative examples are shown in Fig. 7.

Discussion

The aim of this study was to assess differences in the response to drought at the metabolite level in tolerant and intolerant rice genotypes. For this purpose, we have selected two traditional varieties (Dular and N22) that belong to the aus-type group and are known for their tolerance to drought, as well as P deficiency, heat and other stresses. For the intolerant genotypes, we have

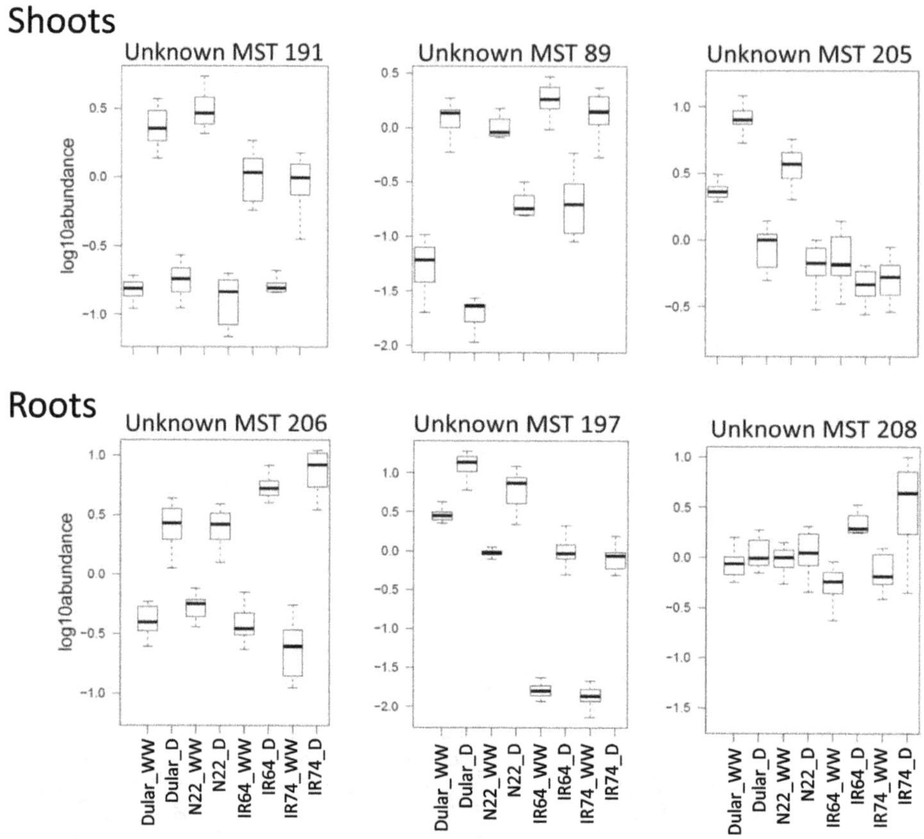

Fig. 7 Representative unknown metabolites associated with drought tolerance. Box plots showing the $\log_{10}$ abundance of the unknown metabolites indicated in shoot and root samples of four rice genotypes grown under well-watered (WW) conditions or exposed to a dry-down (D) treatment

selected IR64 and IR74, which are well adapted to the irrigated paddy rice system and overall represent typical modern, semi-dwarf varieties.

Changes in balances of carbon supply and plant growth under drought

The comparison of phenotypic data of plants at the vegetative stage and at plant maturity showed that all four rice varieties selected for this study responded similarly to the drought treatment by reducing shoot and root biomass and tiller numbers (Fig. 1; Additional file 2: Figure S1). Drought inhibits gas exchange and photosynthesis affecting the balance between carbon supply and plant growth as shown in Arabidopsis (Meyer et al. 2007; Sulpice et al. 2009), with starch as the major determinant of growth, in conjunction with other metabolites, such as sucrose and amino acids. In our study, drought induced the accumulation of compounds related to the central carbon metabolism (Fig. 3) suggesting that carbon utilisation for biomass formation and growth was impaired under stress. It is interesting to note that accumulation of monosaccharides in shoots was paralleled by their reduction in roots, in particular glucose, fructose and arabinose (Fig. 3). Roots are heterotrophic in nature and rely on photo-assimilates provided by photosynthesising leaves via the phloem. Reduction of these sugars and reduction in root weight, as observed in all genotypes, thus suggests that roots suffered from limited shoot C supply. This is in agreement with the finding that N22 presented the highest reduction of key metabolites (glucose, fructose, 3-P-glycerate, malate) under drought and a comparably higher reduction in root DW and lower increase in root length (Fig. 1). However, overall tolerant and intolerant genotypes all showed a similar root response to drought (Fig. 1; Additional file 2: Figure S1) showing that root plasticity and the ability to forage for deep water under drought is an important trait and highly conserved even in paddy rice.

Despite these similarities, differences in drought tolerance were obvious at plant maturity and both Dular and N22 out-yielded the irrigated varieties (Additional file 2: Figure S2). The lower grain yield in IR64 and IR74 was related to a reduced total spikelet number, rather than reduced fertility or grain size, indicative of a smaller inflorescence meristem and reduced number of floret primordia, and thus yield potential. The development of inflorescence meristems is regulated, at least in part, by cytokinin and depends on invertases and sugar supply during development (Ashikari et al. 2005; for a review see Jameson and Song 2016). However, in contrast to roots, sucrose, glucose and fructose accumulated under drought in leaves in all genotypes and sugar starvation can therefore not explain the reduced spikelet number in IR64 and IR74. It will be interesting to investigate this further and determine hormone levels and the long-term effect of drought-induced metabolites on inflorescence meristem development under drought.

Free AA and allantoin are the main metabolites increased under drought in shoots

The metabolite analysis allowed the identification of aus-type specific metabolites that might be indicative of the pathways specifically implicated with tolerance. In fact, in shoots the most represented class of metabolites of primary metabolism were AA which accumulated to a higher level in the tolerant genotypes. Accumulation of AA is a common and well documented response to abiotic stresses (Rai 2002; Planchet and Limami 2015) although it is still a matter of debate whether this is due to increased protein degradation, decreased protein synthesis, or to enhanced AA synthesis or interconversion. However, decades of molecular studies on e.g. proline have shown that AA accumulation may have a functional role in tolerance, as first shown in drought tolerant barley by Singh et al. (1972). Since then, several studies have described proline as being an osmolyte (Yoshiba et al. 1995), a regulator of redox potential (Hare and Cress 1997), a molecular chaperone (Verbruggen and Hermans 2008; Szabados and Savoure 2010), a ROS scavenger (Mohanty and Matysik 2001) and a signalling molecule (Khedr et al. 2003). A recent field study using two contrasting rice genotypes further showed that the tolerant genotype accumulated significantly higher levels of proline under drought in roots, however, proline levels in leaves were higher in the intolerant genotype (Raorane et al. 2015).

Our data are in support of a positive role for proline under drought since it showed the second highest log2FC in shoots of all rice genotypes analysed, but a higher fold increase (5-log2FC) in the tolerant aus-type varieties compared with IR64 and IR74 (about 3-log2FC) (Fig. 4). In agreement with that, glutamate, the direct precursor of proline, was increased under drought, but this occurred in all genotypes and to about the same extent (Fig. 3). In contrast, ornithine and arginine, that can offer an alternative route for proline biosynthesis (Delauney et al. 1993; Verslues and Sharma 2010), showed a greater enrichment in the tolerant genotypes and might contribute to the observed higher log2FC-change of proline in N22 and Dular.

Aus-type genotypes also showed a greater accumulation of AA belonging to the aspartate family. Asparagine is well known to be involved in long-distance transport of nitrogen (N) and acts as a reserve of reduced N (Lea et al. 2007). Thus, asparagine may be used by plants as a reserve of N and C during stress and/or as AA storage to be used during recovery. Other components of this family associated with aus-type rice were threonine and methionine, the precursor of S-adenosyl methionine (SAM) and thus of polyamines and ethylene. Ethylene is a well-known stress hormone and any changes to this pathway that affect ethylene levels might be directly relevant for stress tolerance (for recent reviews see Müller and Munné-Bosch 2015; Salazar et al. 2015).

As the carbon backbone of methionine is derived from aspartate and the whole aspartate family of AA is tightly co-regulated (Galili et al. 2005) it is conceivable that we see parallel increases of lysine, threonine and isoleucine as well as increases of the branched chain amino acids isoleucine, valine and leucine under stress conditions. How the need for an increased flux from aspartate into the other members of the aspartate family is signalled is not clear, however, it is likely that this involves aspartate kinase which produces the common precursor aspartylphosphate (Galili et al. 2005).

In addition, the non-proteinogenic AA α-amino butyric acid (AABA), a derivate of threonine to isoleucine biosynthesis, also significantly increased in tolerant genotypes. Exogenous application of AABA to tomato plants was shown to induce the accumulation of the phytohormone ethylene (Cohen et al. 1994), suggesting that AA intermediates may indeed have important roles.

Serine and its acetylated form O-acetylserine (OAS), that are also involved in methionine biosynthesis, showed a significant differential accumulation in the contrasting rice genotypes included in this study. Interestingly, serine levels showed a higher positive log2FC-change in aus-type rice, while OAS levels were unaltered in tolerant genotypes but decreased in sensitive genotypes (Fig. 4). A recent study in *Vitis vinifera* showed no changes of OAS concentration between WW and droughted plants, while it showed altered expression of serine acetyltransferases genes (*VvSERAT1;2* up-regulated; *VvSERAT3;1* down-regulated) that convert serine to OAS (Tavares et al. 2015). OAS is then converted to cysteine, which is the precursor of glutathione (GSH), a major antioxidant, for which accumulation under drought, cold and heat shock has been well documented (Nieto-Sotelo and Ho 1986; Dhindsa 1991; Kocsy et al. 1996). Since, cysteine and GSH and methionine-downstream metabolites (SAM, spermidine, spermine, and ethylene) were not included in our study, further analyses are required to support their putative role in drought tolerance.

Interestingly, allantoin was the metabolite that presented the most significant interaction between treatment and genotypes, and also had the highest magnitude of log2FC-change in shoots (Fig. 4). Allantoin is an intermediate of purine catabolism that allows the plant to recycle N present in the purine ring. Allantoin has recently been shown to positively activate ABA and jasmonic acid in Arabidopsis, both important hormones in stress signalling (Watanabe et al. 2014; Takagi et al. 2016). It was also suggested that allantoin reduces accumulation of reactive oxygen species (ROS) under stress conditions (Brychkova et al. 2008; Watanabe et al. 2010; Irani and Todd 2016) though the exact mechanisms are unclear since allantoin did not show antioxidant activity in-vitro (Wang et al. 2012). In agreement with our data, recent findings showed that allantoin accumulates under different abiotic stresses, especially in drought tolerant genotypes in rice (Degenkolbe et al. 2013) and wheat (Bowne et al. 2011), as well as in resurrection plants (Oliver et al. 2011; Yobi et al. 2013). In contrast, allantoin was found to accumulate under drought in a sensitive barley cultivar (Chmielewska et al. 2016), suggesting that differences among plant species may exist. In the rice study by Degenkolbe et al. (2013), including twenty-one genotypes mainly originating from a Vietnamese drought breeding program, a positive correlation was revealed between allantoin levels under drought and physiological traits associated with tolerance. However, a negative correlation between levels of asparagine, serine and threonine was also reported, which is in contrast to our data but might be explained by the different drought treatments (dry-down versus 18 d drought) applied. However, the fact that allantoin was associated with tolerance independently in the present study and by Degenkolbe et al. (2013) suggests that it might indeed be a robust metabolic marker for drought tolerance justifying more in-depth studies of the pathway and its underlying genes.

Metabolites in roots mainly show negative log2FC

Overall, the drought responsive metabolites identified in roots represent a more diverse set compared with shoots and the data were more variable, with several metabolites significant in only one of the two tolerant or intolerant genotypes and those metabolites can therefore be considered less robust (Figs. 3 and 4). That these metabolites were identified by the ANOVA is because average values for the tolerant and intolerant genotypes, respectively, were used for the analysis. However, the variability of the data also reflects the complexity of root systems and heterogeneity of soil in large pots but might also be caused by the soaking of the soil before harvest, which was inevitable for extracting intact root systems from natural soil.

Nevertheless, some metabolites showed consistent changes, especially raffinose and uridine. Raffinose has also been identified in shoots and in both tissues, intolerant genotypes showed the higher log2FC (Fig. 4). Raffinose is a soluble carbohydrate, synthesised from sucrose and galactinol (Peterbauer and Richter 2001) and the raffinose family oligosaccharides (RFOs) and biosynthetic genes are well known to differentially accumulate upon abiotic stress treatments (for a review see Sengupta et al. 2015). They are thought to play important roles in stabilizing membranes, stress signalling and as antioxidants. In Arabidopsis it has been shown that overexpression of galactinol synthase, the key enzyme for RFO synthesis, increased the concentration of galactinol and raffinose, and tolerance to ROS, salinity and chilling stress (Nishizawa et al. 2008). That, in our study, raffinose showed the higher log2FC in the intolerant genotypes (up to 4.8- versus 2.7-fold) is somewhat surprising but might in fact be indicative of the higher stress level experienced by the indica rice varieties due to the absence of protective mechanism that are present in roots of the aus-type varieties. Galactinol was identified as highly drought responsive also in this study, however, there were no significant differences between the aus-type and the irrigated varieties (Fig. 3) and galactinol is therefore not a tolerant-specific metabolite in rice.

In N22 and Dular roots, uridine was the metabolite with the highest log2FC and most significant interaction of treatment and genotype, and it was also showing a higher log2FC in tolerant shoots (Fig. 4). Uridine is a RNA-specific nucleoside containing the pyrimidine base uracil and the pentose sugar ribose. Cytidine, another pyrimidine nucleoside, as well as the products of pyrimidine catabolism (beta-alanine and its conjugate with pantoate, pantothenic acid) were also detected in this study and showed positive accumulation, although without significant genotypic differences (Fig. 3). Interestingly, the use of exogenous uridine and cytidine has been subject of a recent commercial patent, as it was shown that this enhanced plants growth under control and heat stress (45 °C) in cucumber (*Cucumis sativus*) (Cansev et al. 2014). It still remains unclear how these metabolites influence plant growth; perhaps via increasing levels of uridine-diphosphate-glucose (UDPG), which is a key metabolite involved in cell wall synthesis, glycosylation of proteins and lipids, secondary metabolism and lipid sulfonylation (for a review see Kleczkowski et al. 2010).

As seen for uridine, AABA was responsive to drought in roots and in shoots but showed a higher log2FC in the aus-types, reinforcing the notion that threonine and isoleucine might be important for tolerance (see above). The other metabolites identified in roots were less consistent and showed, with the exception of dehydroascorbic acid, negative fold changes (Fig. 4).

Interestingly, in roots, monosaccharides and metabolites representative of glycolysis and the TCA cycle, all showed negative log2FC in the aus-type varieties, notably in N22 (Fig. 4). Glycolysis and TCA cycle are involved in energy generation and are strictly connected with AA metabolism as they provide carbon skeletons required for their synthesis. Therefore, the reduction of these metabolites in roots and accumulation of AA in shoots of tolerant genotypes might suggest a controlled process to enable AA accumulation in shoots for drought protection. That this might be at the expense of root growth is indicated by the relatively lower root response in N22 compared with the other genotypes as discussed above (Fig. 1).

It is also noteworthy that the reduction of central carbon metabolites in roots was accompanied by an accumulation of the di-saccharides sucrose, galactinol, and trehalose. However, this was a general response observed in all genotypes and is therefore not tolerant-specific. Nevertheless, the importance of galactinol under drought was already mentioned above and trehalose is widely recognized for its importance in stress tolerance in different plant species (e.g. Delorge et al. 2014 and references therein). Importantly, the signalling molecule trehalose-6-phosphate (T-6-P) has recently been shown to significantly increase yield and recovery from drought in Arabidopsis, maize and wheat and is now being tested as an agro-chemical (Nuccio et al. 2015; Griffiths et al. 2016).

Many unknown metabolites are highly drought responsive

About two-third of the metabolites that we identified as responsive to drought and associated with tolerance were unknowns (Fig. 5). Identifying a large number of unknowns is not surprising and it is estimated that plants have up to 1 million metabolites (Saito and Matsuda 2010) while commercial libraries generally include only a few thousand. It will therefore require a major effort and investment to reveal the nature of these molecules and the underlying pathways.

For now, our data may therefore serve as another example for the untapped potential of molecules that accumulate under drought and for the genetic diversity within rice, and other crops. Virtually all unknown metabolites that were significant in shoots based on the two-way ANOVA showed a higher magnitude of change in the aus-types (Fig. 5), as was also observed for the known metabolites, especially allantoin and AA as discussed above (Fig. 4). Similarly, a greater number of both, known and unknowns, showed a negative log2FC in roots, with the known metabolites mainly associated with glycolysis and TCA. However, in contrast to the known metabolites, many unknowns showed a higher-fold positive change in the intolerant genotypes, including MST 197, which is the most significant metabolite identified in the two-way ANOVA across the entire experiment (Fig. 5). Interestingly, the main differences between tolerant and intolerant genotypes are under WW conditions where the aus-types had about 100 times the levels of MST 197 compared to the intolerant lines, while under drought the difference was only 10 times (Fig. 7).

Metabolites show different drought response patterns

Our results showed that for the majority of significant metabolites the determinant of the difference between tolerant and sensitive genotypes was the magnitude of log2FC, rather than the abundance (Fig. 6). For some metabolites, such as allantoin and the unknowns MST 89 and 197, the higher log2FC and association with tolerant genotypes is indeed determined at WW control conditions, rather than under stress (Figs. 6 and 7). This was also indicated by the PCA analysis (Fig. 2) which showed some separation of the tolerant and intolerant genotypes under WW conditions, especially in shoots. To determine whether or not metabolites have a role in conferring tolerance it therefore seems important to assess the magnitude of change and the absolute concentration. It has for example been shown that, in a drought tolerant rice variety (TKM-1), proline levels under drought increased from 250 to 1350 μg g^{-1} DW (5.4-fold), while in a sensitive variety (Sabarmati) it increased from 755 to 900 μg g DW^{-1} (1.2-fold). Despite the fact that proline levels under WW conditions were about 3-times higher in the sensitive variety, the authors proposed a correlation of proline accumulation and drought tolerance based on the differences in fold-change (Mali and Mehta 1977).

Metabolites correlating with beneficial traits have the potential to be used in breeding (see Matros et al. 2017 and reference therein). However, in contrast to DNA-based markers, metabolites are much less robust since they are responsive to the environment, and might be tissue-specific and developmentally regulated. Rather than using metabolite markers, it will therefore be important to develop DNA-based markers targeting the genes underlying the differential response of a given metabolite. This is, however, only possible if the genes and regulators of the pathway are well known and allelic variation between tolerant and intolerant genotypes exist. However, metabolites can also be used for statistical association with genomic regions, i.e. mQTL mapping, which opens opportunities to employ unknown metabolites for breeding as well as known metabolites (Fernie and Schauer 2009; Matsuda et al. 2012). As more and higher quality *de-novo* assembled genome sequences become available (Huang et al. 2012; Schatz et al. 2014; Du et al. 2017), mQTL mapping will also facilitate the identification of the underlying genes and pathways, which might be genotype specific as is being shown for an increasing number of agronomically important genes (Xu et al. 2006; Hattori et al. 2009; Gamuyao et al. 2012).

Metabolites that show differences under WW control conditions, such as allantoin or MST 197, might be generally more eligible for high-throughput screens in breeding programs because they do not require stress treatment but could be predictive of the stress tolerance capacity. Curiously, the example from proline and our data on allantoin suggest that in some cases, plants with a low concentration of a given metabolite should be selected as a prerequisite for the required high magnitude of change desirable under stress.

Conclusions

The comparison of traditional aus-type rice with irrigated varieties allowed us to identify tolerant-specific metabolites that accumulate in shoots and/or roots under drought. These metabolites have protective roles as osmolytes (proline), N storage molecule (asparagine), stress signalling (allantoin) and growth enhancer (uridine). Therefore, together with the underlying genes and pathways, they are interesting targets for in-depth studies on their role in drought tolerance. Our data suggest that the protective function of certain metabolites under stress may depend on the magnitude of the accumulation upon stress rather than on abundance (e.g. allantoin and unknown MST 197). In addition, for certain tolerance-related metabolites we show that the difference among genotypes is already pre-determined under control conditions. If a causal relationship to stress tolerance can be demonstrated, these metabolites may be suitable candidates for high throughput screens in breeding programs, avoiding the need to screen under stress. Mapping of both, known and unknown metabolites in conjunction with the availability of *de-novo* genome sequences of tolerant genotypes will enable us to gain access to the underlying genes and pathways and devise strategies for crop improvement.

Acknowledgements
This work was supported by the International Rice Research Institute (IRRI; Philippines), the Max Planck Institute of Molecular Plant Physiology (Golm, Germany), and the Leibniz-Institute of Plant Genetics and Crop Plant Research (IPK; Germany). We would like to thank A. Cruz and E. Ramos and for their technical assistance in growing and sampling the plants and Andrea Apelt for her technical assistance in the GC-MS analysis. Rothamsted Research is supported by the Biotechnology and Biological Sciences Research Council (BBSRC, UK).

Authors' contributions
SH and RH conceptualized and managed the project. DR and HMH conducted the metabolite analyses supervised by TA. AC and DR analysed the data and AC and SH wrote the manuscript with inputs from RH. All authors read and approved the final manuscript.

Competing interests
The authors declare that they have no competing interests.

Author details
[1]School of Agriculture, Food and Wine, Waite Campus, The University of Adelaide, Adelaide, SA, Australia. [2]Julius Kühn-Institute (JKI), Federal Research Centre for Cultivated Plants, Institute for Ecological Chemistry, Plant Analysis and Stored Product Protection, Berlin, Germany. [3]Max Planck Institute of Molecular Plant Physiology, Potsdam-Golm, Germany. [4]Rothamsted Research, Harpenden, UK.

References
Ashikari M, Sakakibara H, Lin S, Yamamoto T, Takashi T, Nishimura A, Angeles ER, Qian Q, Kitano H, Matsuoka M (2005) Cytokinin oxidase regulates rice grain production. Science 309:741–745

Beckles DM, Roessner U (2012) 5 - plant metabolomics: applications and opportunities for agricultural biotechnology. In: Altman A, Hasegawa PM (eds) Plant biotechnology and agriculture. Elsevier Inc., San Diego

Bowne JB, Erwin TA, Juttner J, Schnurbusch T, Langridge P, Bacic A, Roessner U (2011) Drought responses of leaf tissues from wheat cultivars of differing drought tolerance at the metabolite level. Mol Plant 5:418–429

Broadhurst DI, Kell DB (2006) Statistical strategies for avoiding false discoveries in metabolomics and related experiments. Metabolomics 2:171–196

Brychkova G, Alikulov Z, Fluhr R, Sagi M (2008) A critical role for ureides in dark and senescence-induced purine remobilization is unmasked in the *Atxdh1* Arabidopsis mutant. Plant J 54:496–509

Cansev A, Gülen H, Kesicizengin M, Ergin S, Cansev M, Kumral NA (2014) Use of pyrimidines in stimulation of plant growth and development and enhancement of stress tolerance. US patent 14,769,652, 28 august 2014

Chmielewska K, Rodziewicz P, Swarcewicz B, Sawikowska A, Krajewski P, Marczak Ł, Ciesiołka D, Kuczyńska A, Mikołajczak K, Ogrodowicz P, Krystkowiak K, Surma M, Adamski T, Bednarek P, Stobiecki M (2016) Analysis of drought-induced proteomic and metabolomic changes in barley (*Hordeum vulgare* L.) leaves and roots unravels some aspects of biochemical mechanisms involved in drought tolerance. Front Plant Sci 7:1108

Cohen Y, Niderman T, Mosinger E, Fluhr R (1994) β-Aminobutyric acid induces the accumulation of pathogenesis-related proteins in tomato (*Lycopersicon esculentum* L.) plants and resistance to late blight infection caused by *Phytophthora infestans*. Plant Physiol 104:59–66

Degenkolbe T, Do PT, Kopka J, Zuther E, Hincha DK, Köhl KI (2013) Identification of drought tolerance markers in a diverse population of rice cultivars by expression and metabolite profiling. PLoS One 8(5):e63637

Delauney A, Hu C, Kishor P, Verma D (1993) Cloning of ornithine δ-aminotransferase cDNA from *Vigna aconitifolia* by *trans*-complementation in *Escherichia coli* and regulation of proline biosynthesis. J Biol Chem 268:18673–18678

Delorge I, Janiak M, Carpentier S, Van Dijck P (2014) Fine tuning of trehalose biosynthesis and hydrolysis as novel tools for the generation of abiotic stress tolerant plants. Front Plant Sci 5:147

Dhindsa RS (1991) Drought stress, enzymes of glutathione metabolism, oxidation injury, and protein synthesis in *Tortula ruralis*. Plant Physiol 95:648–651

Du H, Yu Y, Ma Y, Gao Q, Cao Y, Chen Z, Ma B, Qi M, Li Y, Zhao X, Wang J, Liu K, Qin P, Yang X, Zhu L, Li S, Liang C (2017) Sequencing and *de novo* assembly of a near complete *indica* rice genome. Nat Commun 8:15324

Fernie AR, Schauer N (2009) Metabolomics-assisted breeding: a viable option for crop improvement? Trends Genet 25:39–48

Galili G, Amir R, Hoefgen R, Hesse H (2005) Improving the levels of essential amino acids and sulfur metabolites in plants. J Biol Chem 386:817–831

Gamuyao R, Chin JH, Pariasca-Tanaka J, Pesaresi P, Catausan S, Dalid C, Slamet-Loedin I, Tecson-Mendoza EM, Wissuwa M, Heuer S (2012) The protein kinase Pstol1 from traditional rice confers tolerance of phosphorus deficiency. Nature 488:535

González-Schain N, Dreni L, Lawas LMF, Galbiati M, Colombo L, Heuer S, Jagadish KSV, Kater MM (2016) Genome-wide transcriptome analysis during anthesis reveals new insights into the molecular basis of heat stress responses in tolerant and sensitive rice varieties. Plant Cell Physiol 57:57–68

Griffiths CA, Sagar R, Geng Y, Primavesi LF, Patel MK, Passarelli MK, Gilmore IS, Steven RT, Bunch J, Paul MJ, Davis BG (2016) Chemical intervention in plant sugar signalling increases yield and resilience. Nature 540:574–578

Hare P, Cress W (1997) Metabolic implications of stress-induced proline accumulation in plants. Plant Growth Regul 21:79–102

Hattori Y, Nagai K, Furukawa S, Song XJ, Kawano R, Sakakibara H, Wu J, Matsumoto T, Yoshimura A, Kitano H, Matsuoka M, Mori H, Ashikari M (2009) The ethylene response factors SNORKEL1 and SNORKEL2 allow rice to adapt to deep water. Nature 460:1026–1030

Henry A, Gowda VRP, Torres RO, McNally KL, Serraj R (2011) Variation in root system architecture and drought response in rice (*Oryza sativa*): Phenotyping of the OryzaSNP panel in rainfed lowland fields. Field Crop Res 120:205–214

Hill CB, Taylor JD, Edwards J, Mather D, Langridge P, Bacic A, Roessner U (2015) Detection of QTL for metabolic and agronomic traits in wheat with adjustments for variation at genetic loci that affect plant phenology. Plant Sci 233:143–154

Huang X, Kurata N, Wei X, Wang Z-X, Wang A, Zhao Q, Zhao Y, Liu K, Lu H, Li W, Guo Y, Lu Y, Zhou C, Fan D, Weng Q, Zhu C, Huang T, Zhang L, Wang Y, Feng L, Furuumi H, Kubo T, Miyabayashi T, Yuan X, Xu Q, Dong G, Zhan Q, Li C, Fujiyama A, Toyoda A, Lu T, Feng Q, Qian Q, Li J, Han B (2012) A map of rice genome variation reveals the origin of cultivated rice. Nature 490:497–501

Irani S, Todd CD (2016) Ureide metabolism under abiotic stress in *Arabidopsis thaliana*. J Plant Physiol 199:87–95

Jameson PE, Song J (2016) Cytokinin: a key driver of seed yield. J Exp Bot 67:593–606

Kemble AR, Macpherson HT (1954) Liberation of amino acids in perennial rye grass during wilting. Biochem J 58:46–49

Khan MH, Dar ZA, Dar SA (2015) Breeding strategies for improving rice yield—a review. Agric Sci 06:467–478

Khedr AHA, Abbas MA, Wahid AAA, Quick WP, Abogadallah GM (2003) Proline induces the expression of salt-stress-responsive proteins and may improve the adaptation of *Pancratium maritimum* L. to salt-stress. J Exp Bot 54:2553–2562

Kleczkowski LA, Kunz S, Wilczynska M (2010) Mechanisms of UDP-glucose synthesis in plants. Crit Rev Plant Sci 29:191–203

Kocsy G, Brunner M, Rüegsegger A, Stamp P, Brunold C (1996) Glutathione synthesis in maize genotypes with different sensitivities to chilling. Planta 198:365–370

Lea PJ, Sodek L, Parry MA, Shewry PR, Halford NG (2007) Asparagine in plants. Ann Appl Biol 150:1–26

Lemoine R, La Camera S, Atanassova R, Dédaldéchamp F, Allario T, Pourtau N, Bonnemain J-L, Laloi M, Coutos-Thévenot P, Maurousset L, Faucher M, Girousse C, Lemonnier P, Parrilla J, Durand M (2013) Source-to-sink transport of sugar and regulation by environmental factors. Front Plant Sci 4:272

Lescano C, Martini C, González C, Desimone M (2016) Allantoin accumulation mediated by allantoinase downregulation and transport by Ureide Permease 5 confers salt stress tolerance to *Arabidopsis* plants. Plant Mol Biol 91:581–595

Lesk C, Rowhani P, Ramankutty N (2016) Influence of extreme weather disasters on global crop production. Nature 529:84–87

Li X, Lawas LMF, Malo R, Glaubitz U, Erban A, Mauleon R, Heuer S, Zuther E, Kopka J, Hincha DK, Jagadish KSV (2015) Metabolic and transcriptomic signatures of rice floral organs reveal sugar starvation as a factor in reproductive failure under heat and drought stress. Plant Cell Environ 38: 2171–2192

Londo JP, Yu-Chung C, Kuo-Hsiang H, Tzen-Yuh C, Barbara AS (2006) Phylogeography of Asian wild rice, *Oryza rufipogon*, reveals multiple independent domestications of cultivated rice, *Oryza sativa*. PNAS 103:9578

Mackill D, Ismail A, Singh U, Labios R, Paris T (2012) 6 development and rapid adoption of submergence-tolerant (Sub1) rice varieties. Adv Agron 115:299

Mali PC, Mehta SL (1977) Effect of drought on enzymes and free proline in rice varieties. Phytochemistry 16:1355–1357

Matros A, Liu G, Hartmann A, Jiang Y, Zhao Y, Wang H, Ebmeyer E, Korzun V, Schachschneider R, Kazman E (2017) Genome–metabolite associations revealed low heritability, high genetic complexity, and causal relations for leaf metabolites in winter wheat (*Triticum aestivum*). J Exp Bot 68:415–428

Matsuda F, Okazaki Y, Oikawa A, Kusano M, Nakabayashi R, Kikuchi J, Yonemaru J, Ebana K, Yano M, Saito K (2012) Dissection of genotype-phenotype associations in rice grains using metabolome quantitative trait loci analysis. Plant J 70:624–636

McNally KL, Childs KL, Bohnert R, Davidson RM, Zhao K, Ulat VJ, Zeller G, Clark RM, Hoen DR, Bureau TE, Stokowski R, Ballinger DG, Frazer KA, Cox DR, Padhukasahasram B, Bustamante CD, Weigel D, Mackill DJ, Bruskiewich RM, Rätsch G, Buell CR, Leung H, Leach JE (2009) Genomewide SNP variation reveals relationships among landraces and modern varieties of rice. PNAS 106:12273–12278

Meyer RC, Steinfath M, Lisec J et al (2007) The metabolic signature related to high plant growth rate in Arabidopsis Thaliana. Proc Natl Acad Sci U S A 104:4759–4764

Mohanty P, Matysik J (2001) Effect of proline on the production of singlet oxygen. Amino Acids 21:195–200

Mueller ND, Gerber JS, Johnston M, Ray DK, Ramankutty N, Foley JA (2012) Closing yield gaps through nutrient and water management. Nature 490:254–257

Müller M, Munné-Bosch S (2015) Ethylene response factors: a key regulatory hub in hormone and stress signaling. Plant Physiol 169:32–41

Nieto-Sotelo J, Ho T-HD (1986) Effect of heat shock on the metabolism of glutathione in maize roots. Plant Physiol 82:1031–1035

Nishizawa A, Yabuta Y, Shigeoka S (2008) Galactinol and Raffinose constitute a novel function to protect plants from oxidative damage. Plant Physiol 147:1251–1263

Nuccio ML, Wu J, Mowers R, Zhou H-P, Meghji M, Primavesi LF, Paul MJ, Chen X, Gao Y, Haque E (2015) Expression of trehalose-6-phosphate phosphatase in maize ears improves yield in well-watered and drought conditions. Nature. Biotechnol 33:862–869

Oliver MJ, Guo L, Alexander DC, Ryals JA, Wone BW, Cushman JC (2011) A sister group contrast using untargeted global metabolomic analysis delineates the biochemical regulation underlying desiccation tolerance in *Sporobolus stapfianus*. Plant Cell 23:1231–1248

Peterbauer T, Richter A (2001) Biochemistry and physiology of raffinose family oligosaccharides and galactosyl cyclitols in seeds. Seed Sci Res 11:185–197

Planchet E, Limami AM (2015) Amino acid synthesis under Abiotic stress. In: D'Mello JPF (ed) Amino acids in higher plants. CABI Publishers, Oxfordshire and Boston, pp 262–276

Planchet E, Rannou O, Ricoult C, Boutet-Mercey S, Maia-Grondard A, Limami AM (2011) Nitrogen metabolism responses to water deficit act through both abscisic acid (ABA)-dependent and independent pathways in *Medicago truncatula* during post-germination. J Exp Bot 62:605–615

Porter JR, Xie L, Challinor AJ, Cochrane K, Howden SM, Iqbal MM, Lobell DB, Travasso MI (2014) Food security and food production systems. In: Field CB, Barros VR, Dokken DJ, Mach KJ, Mastrandrea MD, Bilir TE, Chatterjee M, Ebi KL, Estrada YO, Genova RC, Girma B, Kissel ES, Levy AN, MacCracken S, Mastrandrea PR, White LL (eds) Climate change 2014: impacts, adaptation, and vulnerability. Part a: global and Sectoral aspects. Contribution of working group II to the fifth assessment report of the intergovernmental panel of climate change. Cambridge University Press, Cambridge and New York, pp 485–533

Rai VK (2002) Role of amino acids in plant responses to stresses. Biol Plantarum 45:481–487

Raorane ML, Pabuayon IM, Miro B, Kalladan R, Reza-Hajirezai M, Oane RH, Kumar A, Sreenivasulu N, Henry A, Kohli A (2015) Variation in primary metabolites in parental and near-isogenic lines of the QTL qDTY12. 1: altered roots and flag leaves but similar spikelets of rice under drought. Mol Breed 35:138

Redestig H, Kusano M, Ebana K, Kobayashi M, Oikawa A, Okazaki Y, Matsuda F, Arita M, Fujita N, Saito K (2011) Exploring molecular backgrounds of quality traits in rice by predictive models based on high-coverage metabolomics. BMC Syst Biol 5:176

Riedelsheimer C, Grieder C, Melchinger AE, Lisec J, Willmitzer L, Czedik-Eysenberg A, Sulpice R, Flis A, Stitt M, Altmann T (2012) Genome-wide association mapping of leaf metabolic profiles for dissecting complex traits in maize. PNAS 109:8872–8877

Riewe D, Jeon HJ, Lisec J, Heuermann MC, Schmeichel J, Seyfarth M, Meyer RC, Willmitzer L, Altmann T (2016) A naturally occurring promoter polymorphism of the Arabidopsis *FUM2* gene causes expression variation, and is associated with metabolic and growth traits. Plant J 88:826–838

Riewe D, Koohi M, Lisec J, Pfeiffer M, Lippmann R, Schmeichel J, Willmitzer L, Altmann T (2012) A tyrosine aminotransferase involved in tocopherol synthesis in Arabidopsis. Plant J 71:850–859

Roessner U, Luedemann A, Brust D, Fiehn O, Linke T, Willmitzer L, Fernie A (2001) Metabolic profiling allows comprehensive phenotyping of genetically or environmentally modified plant systems. Plant Cell 13:11–29

Saito K, Matsuda F (2010) Metabolomics for functional genomics, systems biology, and biotechnology. Annu Rev Plant Biol 61:463–489

Salazar C, Hernández C, Pino MT (2015) Plant water stress: associations between ethylene and abscisic acid response. Chil J Agr Res 75:71–79

Schatz MC, Maron LG, Stein JC, Wences AH, Gurtowski J, Biggers E, Lee H, Kramer M, Antoniou E, Ghiban E, Wright MH, Chia J-m, Ware D, McCouch SR, McCombie WR (2014) Whole genome *de novo* assemblies of three divergent strains of rice, *Oryza sativa*, document novel gene space of *aus* and *indica*. Genome Biol 15:506

Schmidt R, Mieulet D, Hubberten H-M, Obata T, Hoefgen R, Fernie AR, Fisahn J, San Segundo B, Guiderdoni E, Schippers JHM, Mueller-Roeber B (2013) SALT-RESPONSIVE ERF1 regulates reactive oxygen species-dependent signaling during the initial response to salt stress in Rice. Plant Cell 25:2115–2131

Sengupta S, Mukherjee S, Basak P, Majumder AL (2015) Significance of galactinol and raffinose family oligosaccharide synthesis in plants. Front Plant Sci 6:656

Singh TN, Aspinall D, Paleg LG (1972) Proline accumulation and varietal adaptability to drought in barley: a potential metabolic measure of drought resistance. Nature. New Biol 236:188–190

Sulpice R, Pyl E-T, Ishihara H, Trenkamp S, Steinfath M, Witucka-Wall H, Gibon Y, Usadel B, Poree F, Piques MC (2009) Starch as a major integrator in the regulation of plant growth. Proc Natl Acad Sci 106:10348–10353

Swamy BPM, Ahmed HU, Henry A, Mauleon R, Dixit S, Vikram P, Tilatto R, Verulkar SB, Perraju P, Mandal NP, Variar MSR, Chandrababu R, Singh ON, Dwivedi JL, Das SP, Mishra KK, Yadaw RB, Aditya TL, Karmakar B, Satoh K, Moumeni A, Kikuchi S, Leung H, Kumar A (2013) Genetic, physiological, and gene expression analyses reveal that multiple QTL enhance yield of Rice mega-variety IR64 under drought. PLoS One 8:e62795

Szabados L, Savoure A (2010) Proline: a multifunctional amino acid. Trends Plant Sci 15:89–97

Takagi H, Ishiga Y, Watanabe S, Konishi T, Egusa M, Akiyoshi N, Matsuura T, Mori IC, Hirayama T, Kaminaka H, Shimada H, Sakamoto A (2016) Allantoin, a stress-related purine metabolite, can activate jasmonate signaling in a MYC2-regulated and abscisic acid-dependent manner. J Exp Bot 67:2519

Tavares S, Wirtz M, Beier MP, Bogs J, Hell R, Amâncio S (2015) Characterization of the serine acetyltransferase gene family of *Vitis vinifera* uncovers differences in regulation of OAS synthesis in woody plants. Front Plant Sci 6:74

Uga Y, Sugimoto K, Ogawa S, Rane J, Ishitani M, Hara N, Kitomi Y, Inukai Y, Ono K, Kanno N, Inoue H, Takehisa H, Motoyama R, Nagamura Y, Wu J, Matsumoto T, Takai T, Okuno K, Yano M (2013) Control of root system architecture by DEEPER ROOTING 1 increases rice yield under drought conditions. Nature Genet 45:1097–1102

Verbruggen N, Hermans C (2008) Proline accumulation in plants: a review. Amino Acids 35:753–759

Verslues PE, Sharma S (2010) Proline metabolism and its implications for plant-environment interaction. Arabidopsis Book 8:e0140

Wang P, Kong CH, Sun B, XH X (2012) Distribution and function of allantoin (5-ureidohydantoin) in rice grains. J Agr Food Chem 60:2793–2798

Watanabe S, Matsumoto M, Hakomori Y, Takagi H, Shimada H, Sakamoto A (2014) The purine metabolite allantoin enhances abiotic stress tolerance through synergistic activation of abscisic acid metabolism. Plant Cell Environ 37:1022–1036

Watanabe S, Nakagawa A, Izumi S, Shimada H, Sakamoto A (2010) RNA interference-mediated suppression of xanthine dehydrogenase reveals the role of purine metabolism in drought tolerance in Arabidopsis. FEBS Lett 584: 1181–1186

Wissuwa M, Ae N (2001) Genotypic variation for tolerance to phosphorus deficiency in rice and the potential for its exploitation in rice improvement. Plant Breed 120:43–48

Witt S, Galicia L, Lisec J, Cairns J, Tiessen A, Araus JL, Palacios-Rojas N, Fernie AR (2012) Metabolic and phenotypic responses of greenhouse-grown maize hybrids to experimentally controlled drought stress. Mol Plant 5:401–417

Xu K, Xu X, Fukao T, Canlas P, Maghirang-Rodriguez R, Heuer S, Ismail AM, Bailey-Serres J, Ronald PC, Mackill DJ (2006) *Sub1A* is an ethylene-response-factor-like gene that confers submergence tolerance to rice. Nature 442:705–708

Yobi A, Wone BW, Xu W, Alexander DC, Guo L, Ryals JA, Oliver MJ, Cushman JC (2013) Metabolomic profiling in *Selaginella lepidophylla* at various hydration states provides new insights into the mechanistic basis of desiccation tolerance. Mol Plant 6:369–385

Yoshiba Y, Kiyosue T, Katagiri T, Ueda H, Mizoguchi T, Yamaguchi-Shinozaki K, Wada K, Harada Y, Shinozaki K (1995) Correlation between the induction of a gene for δ1-pyrroline-5-carboxylate synthetase and the accumulation of proline in *Arabidopsis thaliana* under osmotic stress. Plant J 7:751–760

9

Inoculation with the endophyte *Piriformospora indica* significantly affects mechanisms involved in osmotic stress in rice

Muhammad Abu Bakar Saddique[1,3], Zulfiqar Ali[1,4*], Abdus Salam Khan[1], Iqrar Ahmad Rana[2] and Imran Haider Shamsi[3*]

Abstract

Background: Rice is a drought susceptible crop. A symbiotic association between rice and mycorrhizal fungi could effectively protect the plant against sudden or frequent episodes of drought. Due to its extensive network of hyphae, the endophyte is able to deeply explore the soil and transfer water and minerals to the plant, some of them playing an important role in mitigating the effects of drought stress. Moreover, the endophyte could modify the expression of drought responsive genes and regulate antioxidants.

Results: Three rice genotypes, WC-297 (drought tolerant), Caawa (moderately drought tolerant) and IR-64 (drought susceptible) were inoculated with *Piriformospora indica* (*P. indica*), a dynamic endophyte. After 20 days of co-cultivation with the fungus, rice seedlings were subjected to 15% polyethylene glycol-6000 induced osmotic stress. *P. indica* improved the growth of rice seedlings. It alleviated the destructive effects of the applied osmotic stress. This symbiotic association increased seedling biomass, the uptake of phosphorus and zinc, which are functional elements for rice growth under drought stress. It boosted the chlorophyll fluorescence, increased the production of proline and improved the total antioxidant capacity in leaves. The association with the endophyte also up regulated the activity of the Pyrroline-5-carboxylate synthase (*P5CS*), which is critical for the synthesis of proline.

Conclusion: A mycorrhizal association between *P. indica* and rice seedlings provided a multifaceted protection to rice plants under osmotic stress (– 0.295 MPa).

Keywords: *Oryza sativa*, *Piriformospora indica*, Symbiosis, Osmotic stress, Fv/Fm, Phosphorus, Zinc, Proline, Total antioxidant capacity, Pyrroline-5-carboxylate synthase

Background

Rice is staple food for nearly half of the world's seven billion people. Almost 754.6 million tons of paddy are annually produced (FAO 2017). Considering the increasing population and the associated demand of rice there is a need to increase this crop's yield. Because of its semiaquatic ancestors, rice is very susceptible to drought stress, which drastically affects its growth and grain yield (Yue et al. 2006; Fahad et al. 2017). Drought is one of the major cause of yield reduction in rice belts of different countries (FAO 2017). Drought interferes with and damages morphological, physiological and molecular features that are responsible for growth and development (Farooq et al. 2009a). In the case of rice, low water availability reduces germination, plant biomass, number of tillers, plant height and modifies root angle (Ji et al. 2012; Sokoto and Muhammad 2014; Uga et al. 2015). It decreases transpiration, stomatal conductance, water use efficiency, relative water content, chlorophyll content, photosynthesis, photosystem II activity

* Correspondence: zulfiqarpbg@hotmail.com; drimran@zju.edu.cn
[1]Department of Plant Breeding and Genetics, University of Agriculture, 38040, Faisalabad, Pakistan
[3]Department of Agronomy, College of Agriculture and Biotechnology, Zhejiang University, 310058, Hangzhou, People's Republic of China

and affects membrane stability and abscisic acid content (Ding et al. 2014; Yang et al. 2014). Drought induces the accumulation of osmoprotectants like proline, sugars, polyamines and antioxidants (Li et al. 2011; Fahramand et al. 2014) as well as changes in the expression of genes which encode transcription factors and defense related proteins (Nakashima et al. 2014).

Various strategies can help maintaining rice grain yield under drought prone conditions like, exploitation of diversified germplasm (Liu et al. 2015), effective management practices (Haefele et al. 2016) and by exploiting the association of rice with endophytes (Gill et al. 2016). Some progress has been obtained by genetically improving drought tolerance in rice (Serraj et al. 2011). The management of water during rice field preparation and throughout complete crop growth period can help to save water and make it available during drought episodes (Haefele et al. 2016). Moreover, seed priming, foliar application of growth regulators as well as osmoprotectants and the proper application of phosphorus (P), zinc (Zn) and silicon (Si) can mitigate drought effects under limited water environment (Hattori et al. 2005; Kaya et al. 2006; Taiz and Zeiger 2006; Ashraf and Foolad 2007; Tariq et al. 2017; Gill and Tuteja 2010). Finally, symbiotic association with endophytes can help maintain grain production under stress conditions. Hyphae of the mycorrhizal fungi are able to deeply explore the rhizosphere and transport water and minerals to plant roots and keep roots moist even when there is less water availability (Varma et al. 2013).

Among many endophytes, *P. indica* forms symbiotic association with almost every cultivable crop. It extends its mutualistic link with pteridophytes, bryophytes, angiosperms and gymnosperms (Varma et al. 2001). *P. indica* was found to have a positive effect on host plants growth under saline environment, water stress, temperature shocks as well as biotic stresses. Under osmotic stresses, *P. indica* increases the cellular osmolarity of plant and maintains turgor (Gill et al. 2016). Along with many drought responsive genes *P. indica* up regulates an important proline synthesizing gene, *P5CS* (Abo-Doma et al. 2016). Proline is highly soluble and zwitterionic in nature so its higher accumulation do not damage plants. Instead, it stabilizes cellular structures, acts as water substitute through hydrophilic interactions and hydrogen bonding. It also induces drought tolerance by scavenging ROS and by being utilized as energy source after the release of stress (Verslues and Sharma 2010). Under abiotic stress *P. indica* was also found to stabilize chlorophyll in rice leaves (Abadi and Sepehri 2016). By consequence, it increases the maximum quantum yield of PSII (Fv/Fm) (Vahabi et al. 2016; Shahabivand et al. 2017).

Finally, *P. indica* enhances the uptake of P and Zn. High P uptake is responsible for maintaining optimum leaf relative water contents, efficiency of photosystem II and net photosynthetic rate. High phosphorus level decreases malondialdehyde content and increases osmolytes and nitrogenous compounds concentration (Tariq et al. 2017). Zinc is part of antioxidant complexes (Cu/Zn-SOD) and is also important for scavenging ROS that ultimately reduces the damage to cellular membranes (Gill and Tuteja 2010; Ngwene et al. 2016). In response to these valuable services, the plant could devote up to 15–20% of the produced photosynthates to this fungus but the extent of benefits generally offsets this cost (Varma et al. 2001). In the present study, rice plants have been inoculated with *P. indica* under polyethylene glycol-6000 (PEG-6000) induced water stress in hydroponic conditions. The effects of the symbiotic association under osmotic stress were analyzed on seedling biomass, root and shoot length, P and Zn uptake, expression of *P5CS* gene, integrity of grana in chloroplasts, level of Fv/Fm, proline content and total antioxidant capacity (TAC). Whereas, in most of the previous research articles, the interaction between rice roots and *P. indica* had been studied under salt and heavy metal stress. It had been reported that this symbiotic fungus enhances the seedling biomass, length of root and shoot, chlorophyll content and proline concentration under salt stress (Jogawat et al. 2013; Bagheri et al. 2013). Previously, the interaction between rice roots and *P. indica* under osmotic stress was not fully explored. We have tried to highlight this interaction and concluded that *P. indica* induces drought tolerance in rice seedlings by improving plant morphological, physiological and molecular aspects.

Methods

Growth and cultivation conditions

P. indica was grown in Kafer liquid media (Hill and Kafer 2001). The pH of the media was adjusted at 6.5 and temperature at 28 °C. Growing culture was continuously shaken at 130 rpm. Three rice genotypes were used in this study, WC-297 (drought tolerant), Caawa (moderately drought tolerant) and IR-64 (drought susceptible).

The method of Sarma et al. (2011) was used for the effective coating of fungus on rice seeds. Seeds of the genotypes were surface sterilized and coated with a paste of *P. indica* in vermiculite, used as a carrier to adhere spores on seed surface. Seeds without fungus association were treated with only vermiculite. After 10 days of co-cultivation with *P. indica*, seedlings were transferred to a Yoshida nutrition media (Yoshida et al. 1976). After 10 days of acclimatization in this media, a 15% solution of PEG-6000 was added to induce osmotic stress in half of the inoculated and half of the non-inoculated plants. Whereas, 15% PEG 6000 induces −0.295 MPa of osmotic pressure at 25 °C (Michel and Kaufmann 1973).

Colonization of fungus: Staining of fungal hyphae and spores

To monitor root colonization, roots were analyzed after 4 days of germination and at the end of the experiment. Roots were thoroughly washed with deionized water, cut into 1 cm length segments which were placed overnight in a 10% KOH solution at room temperature. Roots were then washed 5 times with sterilized H_2O and incubated in 1% HCl solution for 3 min. Later on, roots were mounted in 0.05% trypan blue in lactophenol for microscopic examination (Michal Johnson et al. 2011).

Morphological analysis

Root and leaf samples were collected 20 days after the beginning of stress exposure. Under control and osmotic stress the gain in fresh root weight (FRW), fresh shoot weight (FSW), dry root weight (DRW), dry shoot weight (DRW), root length (RL) and shoot length (SL) was determined both from non-inoculated and inoculated plants.

Elemental analysis of leaves and roots

Root and leaf samples of the rice plants were collected 20 days after the beginning of stress exposure. Samples were dried in a hot air oven, initially for 3 h at 105 °C and then for another 24 h at 80 °C. Approximately 0.1 g of dried leaves and roots was used to determine the concentration of P and Zn. Samples were digested in 5 ml of HNO_3 at 120 °C for 2 h and then at 80 °C for another 2 h in microwave digester (Microwave 3000; Antoon Paar). After digestion, they were diluted to 20 ml with Mili-Q water, and the elemental contents were determined using an inductively coupled plasma-optical emission spectrometer (ICP-OES;. Optima 8000DV; PerkinElmer).

Determining the chlorophyll fluorescence

Drought induced change in Fv/Fm was determined 20 days after the beginning of stress exposure, using a fluorometer (IMAG-MAXI, Heinz Walz, Effeltrich, Germany). Plants were placed in the dark for 30 min, then 20 cm long leaves were cut and immediately analyzed.

Electron microscopy observation of chloroplasts

Chloroplasts in rice leaves were examined 20 days after the beginning of stress exposure, using a transmission electron microscope (Hitachi Model H-7650 TEM). Samples were first fixed in 2.5% glutaraldehyde in phosphate buffer (0.1 M, pH 7.0) for more than 4 h and washed three times with phosphate buffer (0.1 M, pH 7. 0) for 15 min. They were fixed with 1% OsO_4 in phosphate buffer (0.1 M, pH 7.0) for 2 h, and washed three times in phosphate buffer (0.1 M, pH 7.0) for 15 min. They were then dehydrated stepwise using graded ethanol (30%, 50%, 70%, 80%, 90%, 95% and 100%) for 15 min at each step and transferred to absolute acetone for 20 min. Infiltration of the samples was done by placing them in a mixture of absolute acetone and the final spurr resin mixture (1:1) for 1 h at room temperature. Then samples were transferred to a mixture of absolute acetone and final resin mixture (1:3) for 3 h and to final spurr resin mixture for overnight. The samples were finally placed in eppendorf tubes containing spurr resin and heated at 70 °C for more than 9 h. They were sectioned with a LEICA EM UC7 ultratome and sections were stained with uranyl acetate and alkaline lead citrate for 5 and 10 min respectively and then observed in transmission electron microscope (Hitachi Model H-7650).

Expression pattern of *P5CS* gene

The expression pattern of *P5CS* (proline synthesizing gene) was observed 24 h after the beginning of stress exposure. For primers designing, coding DNA sequences (CDS) were retrieved from The Rice Annotation Project Database (http://rapdb.dna.affrc.go.jp/). Ubiquitin (*UBQ*) was used as a reference gene. Forward and reverse primers of these two genes are given in the Additional file 1: Table S1. TRIzol® reagent (FAVORGEN BIOTECH CORP) was used for the extraction of RNA. Plant leaves were grind in liquid nitrogen and then transferred immediately in microfuge tube holding 700 µl Trizol. This mixture was centrifuged at 12000 rpm for 2 min. Supernatant was shifted in another tube and 300 µl chloroform was added and mixed gently by inversion. The solution was centrifuged to get phase separation at 12000 rpm for 10 min. The upper phase was collected in another microfuge tube, an equal amount of iso-propanol was added followed by an incubation on ice for 10 min. Again centrifuged at 12000 rpm for 10 min, supernatant was discarded and pellet of RNA was collected. Pellet was washed with 75% ethanol. Air dried for 2–3 min and diluted in 25 µl nuclease free water. All centrifugations were done at 4 °C. High quality RNA was used to synthesize cDNA following the protocol of Thermo Fisher Scientific. Gel bands of the PCR products were analyzed with IMAGE. J software. Numerical values showing gel band strength were used to calculate the percentage of change in the expression of *P5CS*.

Proline analysis

For the determination of proline, the method of Bates et al. (1973) was used. Leaves (0.5 g) collected 20 days after the beginning of stress exposure, were homogenized in 5 mL of 3% aqueous sulphosalicylic acid. The homogenate was filtered through Whatman No.2 filter paper. One milliliter of filtrate was taken and mixed with 1 mL of acid ninhydrin and 1 mL of glacial acetic acid in a test tube. The mixture was briefly vortexed and heated at

100 °C in a water bath for 1 h and terminated the reaction in an ice bath. Four milliliter of toluene were added to the solution and vortexed for 15–20 s. The chromophore containing free proline was aspirated from the aqueous phase in a test tube and warmed to room temperature. The absorbance was measured at 520 nm using a spectrophotometer.

Estimation of total antioxidant capacity

The method of Prieto et al. (1999) was used for the determination of total antioxidant capacity (TAC). Leaf samples (0.4 g) collected 20 days after the beginning of stress exposure, were homogenized in 50% methanol. The assay mixture containing 900 μl reagent (0.6 M H_2SO_4, 28 mM sodium phosphate and 4 mM ammonium molybdate) and 100 μl of enzyme extract was boiled at 95 °C for 30 min. Spectrophotometer absorbance was measured at 695 nm, a higher absorbance (nm) indicating a higher TAC (Prasad et al. 2009).

Statistical analysis

Experiment was performed by following factorial under completely randomized design. It had three replications (each replication had one plant per pot), three genotypes and four treatments (no osmotic stress, no osmotic stress + *P. indica* inoculation, osmotic stress and osmotic stress + *P. indica* inoculation). Data were analyzed using Analysis of Variance method and least significant difference ($LSD_{0.05}$) (Statistix 8, version 8.1).

Results

Establishment of symbiotic association

P. indica successfully colonized rice roots. A strong association of root and fungus was observed after 4 days of germination. Fungus was found to develop spores and hyphae within emerging roots (Fig. 1a). This association was maintained until the end of experiment (Fig. 1b).

Morphological analysis

Seedling biomass and length of root and shoot were significantly improved in inoculated plants ($p \leq 0.01$, Table 1). In the absence of osmotic stress, the average FRW of the non-inoculated plants of WC-297, Caawa and IR-64 was 1.73, 1.82 and 1.72 g, respectively. When plants were inoculated with *P. indica* the average FRW was increased to 1.86, 2.06 and 2.24 g in the three genotypes, respectively. Under osmotic stress average FRW in WC-297, Caawa and IR-64 was 0.75, 0.61 and 0.59 g in non-inoculated plants and 1.02, 0.82 and 0.94 g in inoculated plants (Fig. 2a). In the absence of osmotic stress, the average FSW of non-inoculated plants were 3.44, 3.38 and 3.01 g in WC-297, Caawa and IR-64, respectively. In inoculated plants it was 3.79, 3.82 and 3.49 g in WC-297, Caawa and IR-64, respectively. Under osmotic stress the average FSW for the non-inoculated plants of these three genotypes was 1.61, 1.20 and 1.09 g and in inoculated plants it was 1.82, 1.37 and 1.53 g in WC-297, Caawa and IR-64, respectively (Fig. 2b). In the absence of osmotic stress, the average DRW of the non-inoculated plants of WC-297, Caawa and IR-64 was 0.21, 0.26 and 0.22 g, respectively. When plants were inoculated with *P. indica* the average DRW was increased to 0.28, 0.29 and 0.31 g in the three genotypes, respectively. Under osmotic stress average DRW in WC-297, Caawa and IR-64 was 0.12, 0.09 and 0.11 g in non-inoculated plants and 0.18, 0.10 and 0.15 g in inoculated plants, respectively (Fig. 2c). In the absence of osmotic stress, the average DSW of non-inoculated plants were 0.49, 0.52 and 0.49 g in WC-297, Caawa and IR-64, respectively. In inoculated plants it was 0.64, 0.53 and 0.49 g in WC-297, Caawa and IR-64, respectively. Under osmotic stress

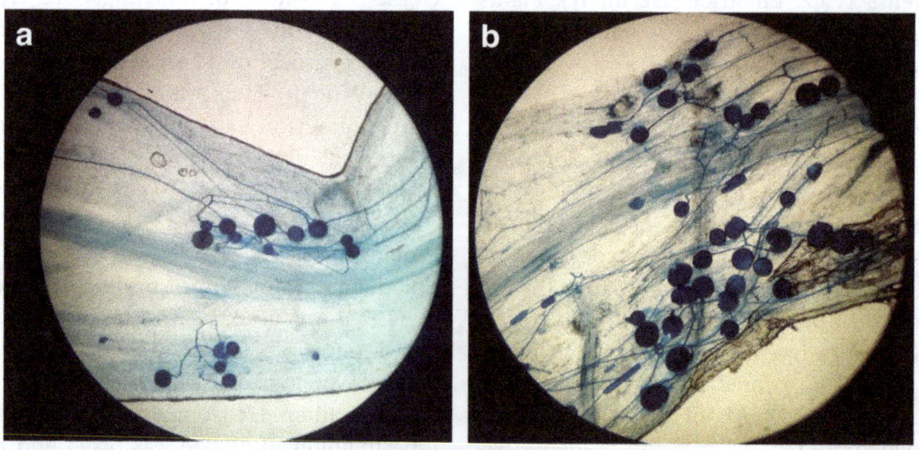

Fig. 1 Association of *P. indica* with rice roots. **a** Established symbiosis after 4 days of germination, a network of hyphae and spores is visible within rice roots. **b** Network of hyphae along with attached spores, observed at the end of experiment

Table 1 Mean squares of absolute values for various seedling traits of the three rice genotypes grown in control, control + *P. indica*, drought and drought + *P. indica*

Residual	Genotype × Treatment	Treatment	Genotype	Source of variation
24	6	3	2	df
0.0018	0.055**	3.9**	0.0066**	Fresh root weight
0.0035	0.067**	13**	0.45*	Fresh shoot weight
0.000024	0.0027**	0.068**	0.00051**	Dry root weight
0.00011	0.0076**	0.21**	0.0035**	Dry shoot weight
1.7	17**	306**	61**	Root length
3.7	14*	202**	7.2^{NS}	Shoot length
0.0000073	0.00066**	0.0050**	0.0070**	Fv/Fm
5.7	229**	1333**	1088**	Proline
0.00045	0.0056**	0.042**	0.0085**	Total antioxidant capacity
0.0086	0.68**	22**	0.015^{NS}	Phosphorus in root
0.018	0.056*	20**	0.14**	Phosphorus in shoot
0.0000012	0.000058**	0.00046**	0.000034**	Zinc in root
0.0000029	0.00018**	0.0020**	0.0000087^{NS}	Zinc in shoot

*, ** and NS indicates significant differences at $p \leq 0.05$, $p \leq 0.01$ and non-significant ($p > 0.05$), respectively

the DSW for the non-inoculated plants of these three genotypes was 0.21, 0.25 and 0.21 g and in inoculated plants it was 0.33, 0.27 and 0.35 g in WC-297, Caawa and IR-64, respectively (Fig. 2d). In the absence of osmotic stress, the average RL of the non-inoculated plants of WC-297, Caawa and IR-64 was 39.33, 36.33 and 35. 34 cm, respectively. When plants were inoculated with *P. indica* the average RL was increased to 46.0, 45.0 and 43.0 cm in the three genotypes, respectively. Under osmotic stress average RL in WC-297, Caawa and IR-64 was 44.6, 30.7 and 34.6 cm in non-inoculated plants and 48.0, 47.6 and 46.0 cm in inoculated plants, respectively (Fig. 2e). In the absence of osmotic stress, the average SL of non-inoculated plants were 37.0, 34.6 and 37.7 cm in WC-297, Caawa and IR-64, respectively. In inoculated plants it was 41.3, 42.3 and 41.0 cm in WC-297, Caawa and IR-64, respectively. Under osmotic stress the average SL for the non-inoculated plants of these three genotypes was 37.6, 31.0 and 33.0 cm and in inoculated plants it was 43.6, 45.0 and 43.0 cm in WC-297, Caawa and IR-64, respectively (Fig. 2f).

Elemental analysis

The concentration of P and Zn were significantly improved in the presence of root symbiosis ($p \leq 0.01$, Table 1).The uptake of P was almost double in inoculated than in non-inoculated plants. In the absence of osmotic stress, average P concentration in leaves of the non-inoculated plants of WC-297, Caawa and IR-64 was 1.85, 1.88 and 1.95 mg.g^{-1}, respectively. When plants were inoculated with *P. indica*, P concentration increased to 4.86, 4.38 and 4.84 mg.g^{-1} in the three genotypes, respectively. Under osmotic stress, P leaf concentration in WC-297, Caawa and IR-64 was 1.3, 1. 13 and 1.16 mg.g^{-1} in non-inoculated plants, and 2.19, 2.16 and 2.42 mg.g^{-1} in inoculated plants (Fig. 3a). Almost similar change in P level was noticed when roots were analyzed. In the absence of osmotic stress, P concentration in roots of non-inoculated plants were 1.71, 1.65 and 1. 81 mg.g^{-1} in WC-297, Caawa and IR-64, respectively. In inoculated plants, P concentration increased to 4.48, 5.51 and 4.56 mg.g^{-1} in WC-297, Caawa and IR-64, respectively. Under osmotic stress P concentration of the three genotypes was 1.26, 1.14 and 1.49 mg.g^{-1} in non-inoculated plants and 2.90, 1.76 and 2.35 mg.g^{-1} in inoculated plants in WC-297, Caawa and IR-64, respectively (Fig. 3b).

The symbiotic association of the rice plant with *P. indica* also increased leaves and roots Zn concentration. In the absence of osmotic stress, leaves Zn concentration of non-inoculated plants of WC-297, Caawa and IR-64 was 0.042, 0.041 and 0.048 mg.g^{-1}, respectively. In inoculated plants, Zn concentration increased to 0.068, 0.084 and 0.067 mg.g^{-1} in the three genotypes, respectively. Under osmotic stress, the average leaf Zn concentration of non-inoculated plants of WC-297, Caawa and IR-64 were 0.040, 0.034 and 0.041 mg.g^{-1}, respectively. In inoculated plants, leaf Zn concentration increased to 0.06, 0.045 and 0.055 mg.g^{-1} in the three genotypes, respectively (Fig. 3c). In the absence of osmotic stress, root Zn concentration was 0.043, 0.045 and 0.051 mg.g^{-1}, in non-inoculated plants and 0.061, 0. 05 and 0.054 mg.g^{-1} in inoculated plants of WC-297, Caawa and IR-64, respectively. Under osmotic stress, root Zn concentration was 0.039, 0.034 and 0.039 mg.g^{-1} in non-inoculated plants and 0.042, 0.047 and 0. 044 mg.g^{-1} in inoculated plants of WC-297, Caawa and IR-64, respectively (Fig. 3d).

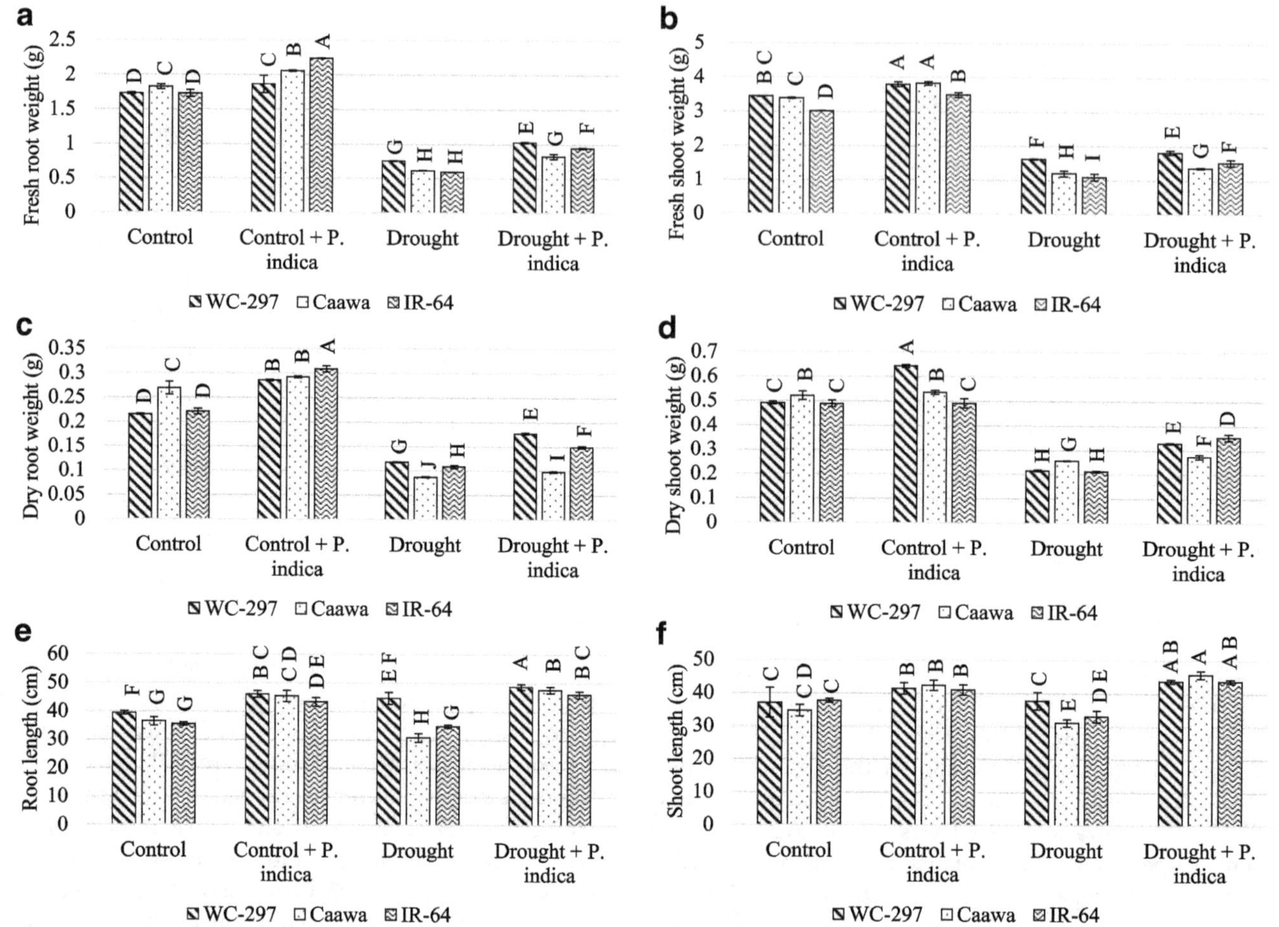

Fig. 2 Gain in biomass and length of root and shoot of rice seedlings. The inoculation with *P. indica* increased the plant biomass and the length of root and shoot. **a** Gain in fresh root weight. **b** fresh shoot weight **c** dry root weight **d** dry shoot weight **e** root length and **f** shoot length. Alphabetic on the top of each bar are showing the $LSD_{0.05}$ difference

Expression pattern of *P5CS* gene

The expression of *P5CS* gene was influenced by the presence of *P. indica*. The change in the expression of *P5CS* was obvious both in the presence and absence of osmotic stress (Fig. 4). In the absence of osmotic stress, the inoculation by *P. indica* increased the expression of this gene up to 102 and 90% in WC-297 and Caawa, respectively. In IR-64 a decrease of 1.5% was noted in the expression of *P5CS*. Under osmotic stress, symbiotic association also increased the expression of *P5CS* gene by 11, 46 and 64% in WC-297, Caawa and IR-64, respectively. Osmotic stress induced an increase of the expression of *P5CS* in non-inoculated plants of WC-297 and Caawa (41% and 36%, respectively) and decrease of this expression (43%) in IR-64.

Proline concentration and total antioxidant capacity

The concentration of proline and the level of TAC were significantly improved in the presence of root symbiosis ($p \leq 0.01$, Table 1). Osmotic stress increased the level of both proline and TAC in rice leaves, while the inoculation of the plant with *P. indica* boosted the concentration of these beneficial compounds. In the absence of osmotic stress, proline concentration in non-inoculated plants was 19.96, 17.92 and 18.83 ppm in WC-297, Caawa and IR-64, respectively. In inoculated plants, proline concentration was increased to 33.94, 23.94 and 25.18 ppm. Under osmotic stress proline concentration was 41.31, 21.74 and 20.23 ppm in non-inoculated plants and 70.23, 42.81 and 29.96 ppm in inoculated plants, in WC-297, Caawa and IR-64, respectively (Fig. 5a). Plant analysis for TAC also shown similar results. In the absence of osmotic stress, TAC in non-inoculated plants was 0.34, 0.29 and 0.35 nm in WC-297, Caawa and IR-64, respectively. In inoculated plants TAC was 0.49, 0.43 and 0.36 nm in WC-297, Caawa and IR-64, respectively. In the presence of osmotic stress TAC was 0.33, 0.31 and 0.37 nm in non-inoculated plants and 0.51, 0.46 and 0.42 nm in inoculated plants, in WC-297, Caawa and IR-64, respectively (Fig. 5b).

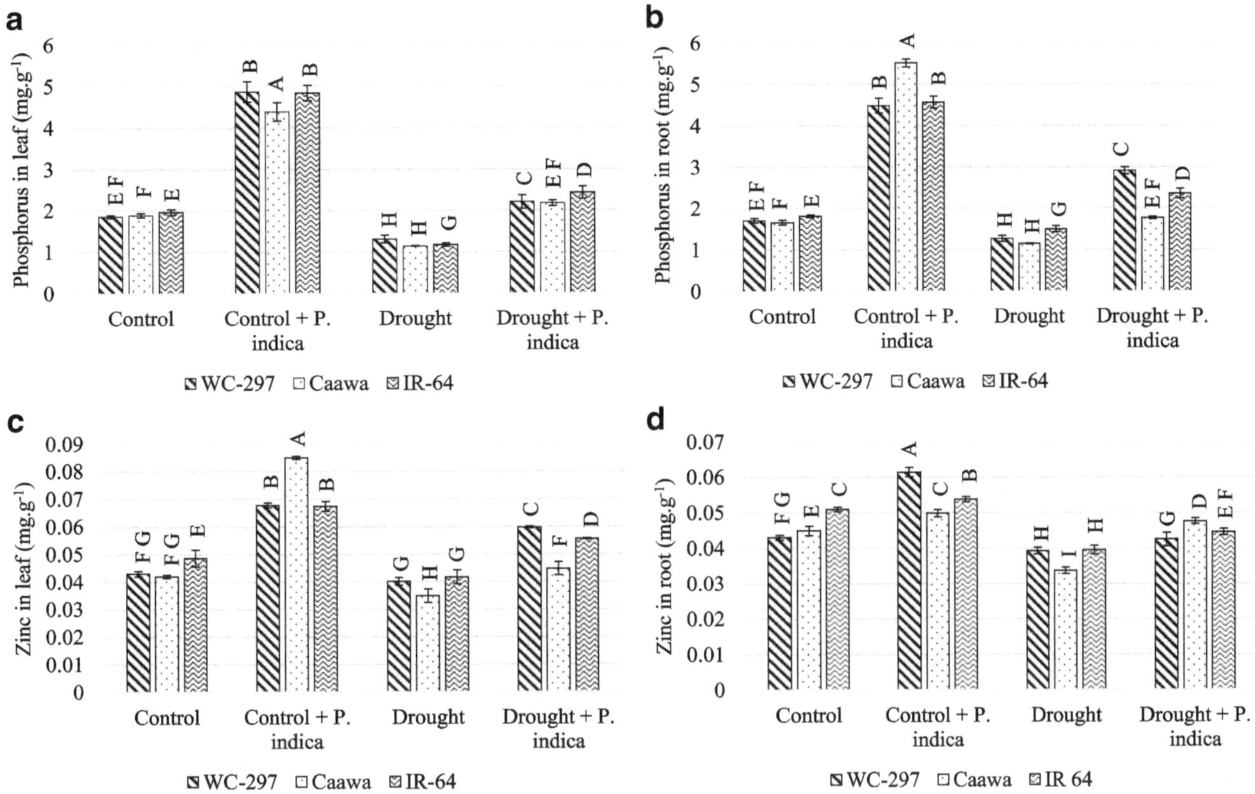

Fig. 3 Phosphorus and zinc concentration in rice leaf and root. The inoculation with *P. indica* increased the level of P and Zn in leaf and root. **a** Increase in the concentration of P in leaves. **b** P concentration in roots. **c** Zn concentration in leaves. **d** Zn concentration in roots. Alphabetic on the top of each bar are showing the $LSD_{0.05}$ difference

Chlorophyll fluorescence (Fv/Fm)

Fv/Fm was significantly improved in the presence of root symbiosis ($p \leq 0.01$, Table 1). Water limited environment affected the photosynthetic machinery and so, reduced Fv/Fm. This decrease was higher in IR-64 than in Caawa and WC-297. The symbiotic association increased Fv/Fm both in the presence and absence of osmotic stress. In the absence of osmotic stress, the value of Fv/Fm in non-inoculated plants of WC-297, Caawa and IR-64 was 0.79, 0.81 and 0.81, respectively. In inoculated plants, it was 0. 82, 0.82 and 0.81 in the three genotypes. In the presence of osmotic stress, Fv/Fm was 0.79, 0.75 and 0.74 in non-inoculated plants and 0.80, 0.80 and 0.78 in inoculated plants, in WC-297, Caawa and IR-64, respectively (Fig. 6).

Electron microscopy

The inoculation with *P. indica* also stabilized the stacking of grana in chloroplasts in both the absence and presence of osmotic stress. In the absence of osmotic stress the grana were more organized when plants were inoculated with *P. indica* (Fig. 7). Similarly, under osmotic stress the grans were healthy when plants were inoculated with *P. indica* (Fig. 8). In non-inoculated plants, under osmotic stress, the highest damage to chloroplasts was noted in IR-64. Some damage was also observed in Caawa. In WC-297 the chloroplasts were healthy even under drought stress. In inoculated plants, the grana in chloroplast were well organized under osmotic stress.

Discussion

The effects of osmotic stress have been frequently reported in rice. Osmotic stress hinders mineral uptake, transport and distribution (Zain et al. 2014), reduces chlorophyll content, disintegrate the grana in chloroplasts and disturbs photosynthetic efficiency (Farooq et al. 2009b; Zain et al. 2014; Swapna and Shylaraj 2017; Korres et al. 2017). Several research papers has been published during the last two decades which emphasize the supportive role of *P. indica* in mitigating the osmotic stress in different host plants. It has been reported that *P. indica* serves as a bioregulator, biofertilizer and bioprotector for many host plants. Regardless of the stress, *P. indica* has evolved to protect its food source-the plant. It activates different signaling, transport, metabolic and developmental programs in plants (Ngwene et al. 2016; Gill et al. 2016; Bakshi et al. 2017). We demonstrate that *P. indica* increased the rice seedling biomass,

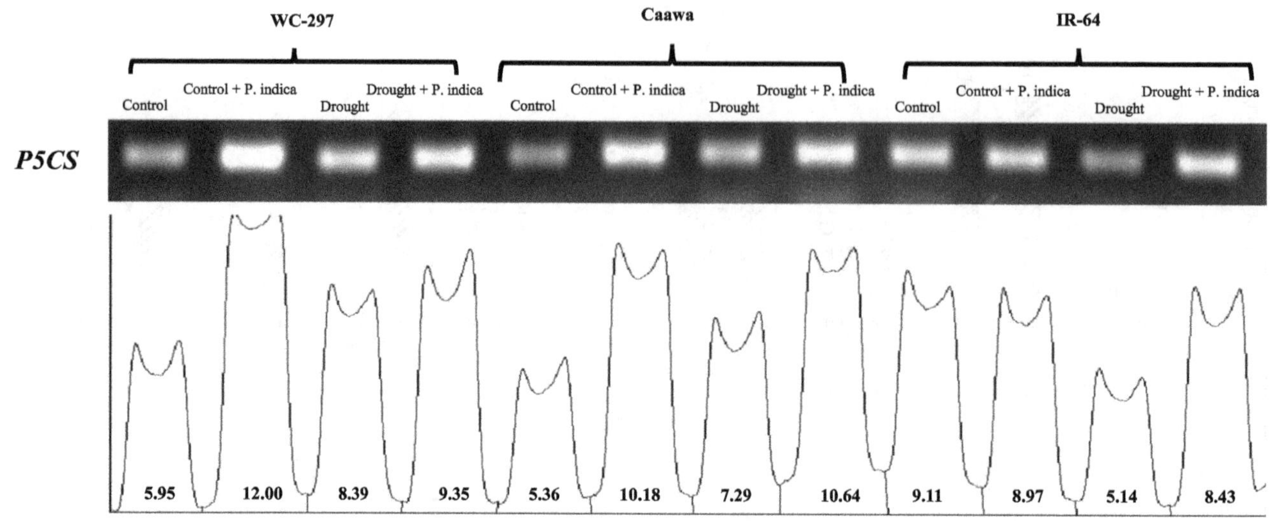

Fig. 4 Change in the expression pattern of *P5CS*. *P. indica* increased the expression of *P5CS* both in the presence and absence of osmotic stress. With a little exception for IR-64, where without inoculation, a little decrease in expression was recorded under osmotic stress

length of root and shoot, P and Zn concentration in root and shoot, proline concentration, TAC, Fv/Fm and the expression of *P5CS*. Most of the previous research articles elaborated the mechanisms that how *P. indica* ameliorates the drastic effects of salt, heavy metal and some biotic stresses in rice plant. It had been reported that this symbiotic fungus increase the concentration of proline, antioxidants and decreases the concentration of malondialdehyde under these stresses (Bagheri et al. 2014; Nassimi and Taheri 2017; Mohd et al. 2017). Most of the yield damage in rice is due to the water shortage. There is very less literature available explaining the rice and *P. indica* symbiosis under osmotic stress. So, there is a need to investigate the role of *P. indica* for inducing drought tolerance in rice. This research article highlighted the *P. indica* induced improvement in some of the important features of rice plants under osmotic stress. We determined that *P. indica* significantly increased the rice plant biomass and the length of root and shoot under control and osmotic stress. Similar growth improvement was reported in Arabidopsis and many other field crops (Franken 2012; Im et al. 2002). In our study, the analysis of mineral profile showed that symbiotic association almost doubled the concentration of both P and Zn in leaves and roots of the rice plant. This increase in mineral acquisition was observed in all three genotypes both in the presence and absence of osmotic stress. The increase in the availability of both macro and micronutrients in the rhizosphere due to the association with *P. indica* has been reported in many crops (Franken 2012). Present study, focused on the uptake of P and Zn as these two elements have been

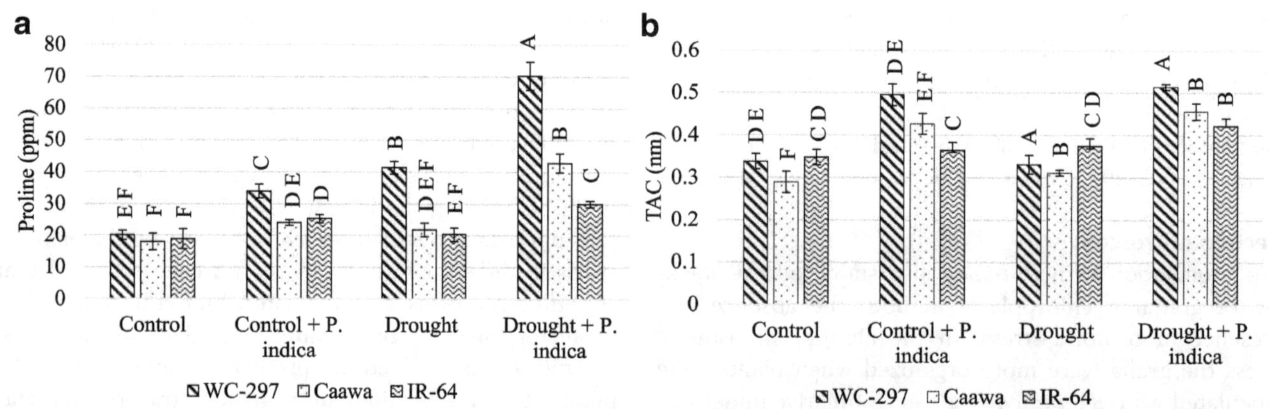

Fig. 5 Change in the concentration of proline and TAC. The inoculation with *P. indica* increased the concentration of proline and total antioxidant capacity (TAC) in both the absence and presence of osmotic stress. **a** Increase in proline concentration. **b** Increase in TAC. Alphabetic on the top of each bar are showing the $LSD_{0.05}$ difference

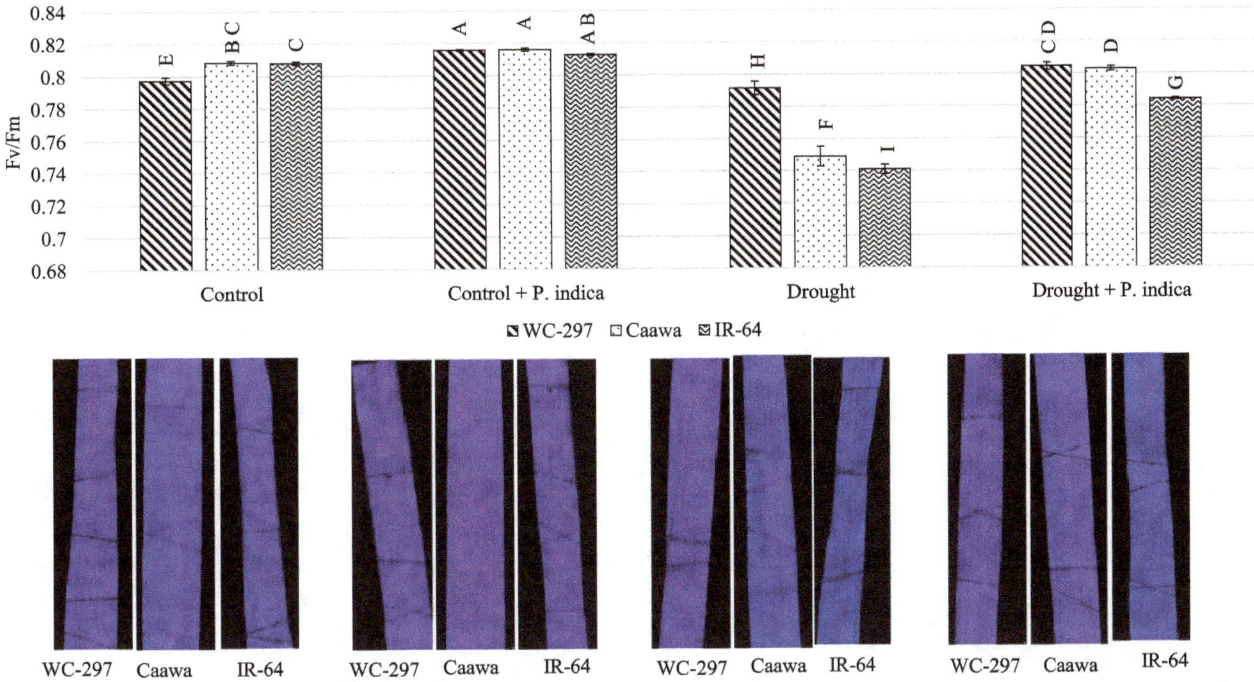

Fig. 6 Change in Fv/Fm. The inoculation with *P. indica* increased Fv/Fm in both the absence and presence of osmotic stress. Under each bar of the graph there is a picture of the leaf section showing Fv/Fm activity. Alphabetic on the top of each bar are showing the $LSD_{0.05}$ difference

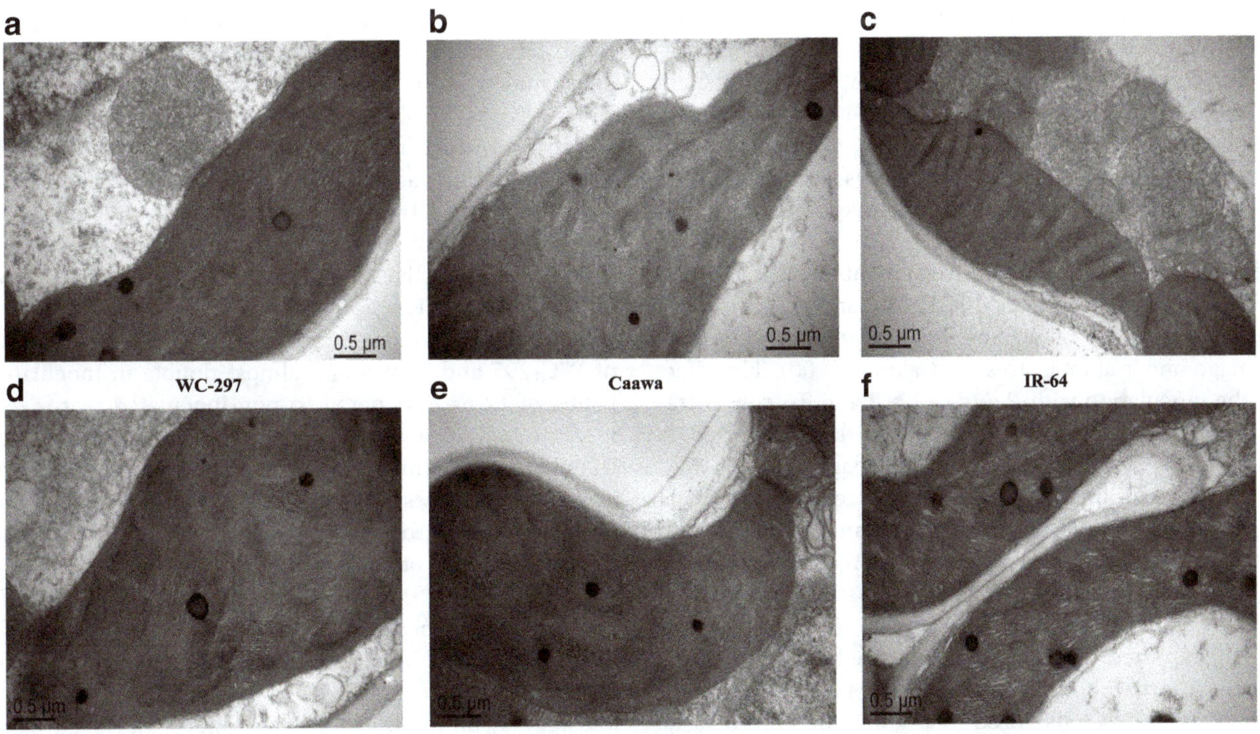

Fig. 7 Chloroplast of rice genotypes in the absence of osmotic stress. **a-c** Chloroplasts of the non-inoculated plants of WC-297, Caawa and IR-64, respectively. **d-f** Chloroplasts of the inoculated plants of WC-297, Caawa and IR-64, respectively. Grana in chloroplasts are better organized in rice plants inoculated with *P. indica*

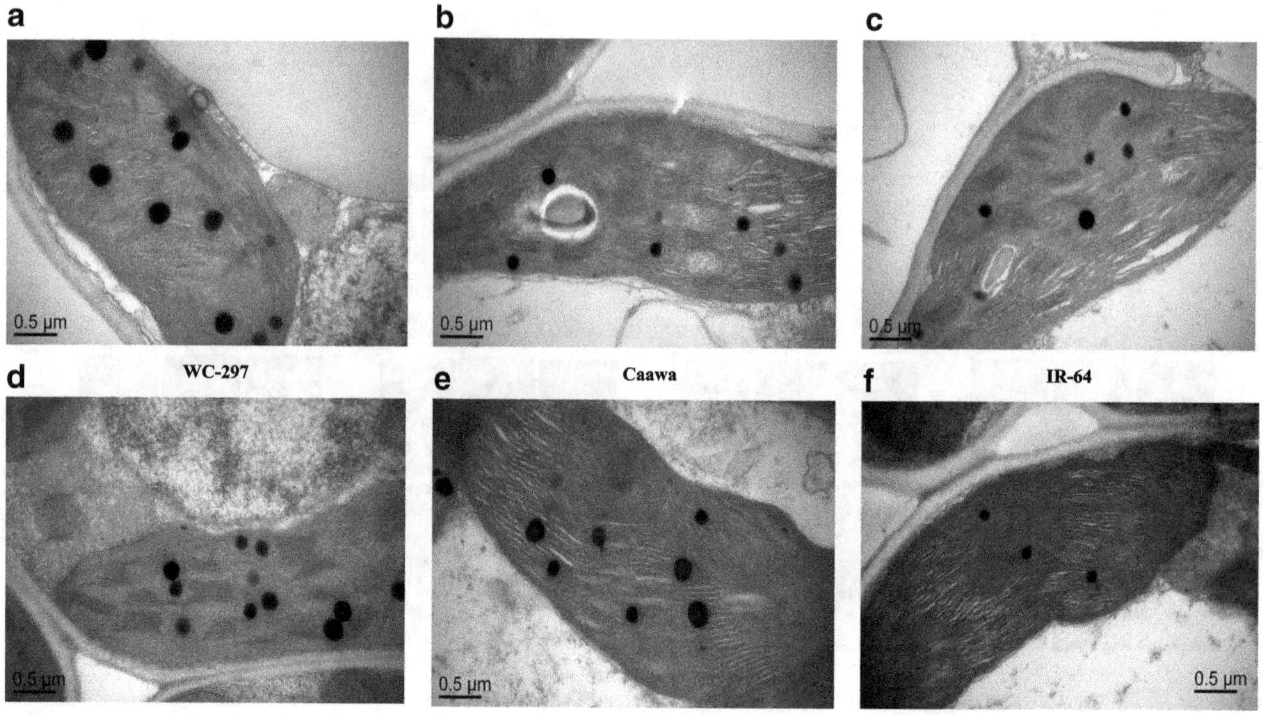

Fig. 8 Chloroplast of rice genotypes in the presence of osmotic stress. **a-c** Chloroplasts of the non-inoculated plants of WC-297, Caawa and IR-64, respectively. **d-f** Chloroplasts of the inoculated plants of WC-297, Caawa and IR-64, respectively. Grana in chloroplasts are better organized in rice plants inoculated with *P. indica*

described to have an important role in maintaining plant growth and development under water-limited environments (Ngwene et al. 2016; Tariq et al. 2017). Sufficient supply of P increases drought tolerance by inducing deeper rooting and higher inorganic phosphorus (Pi) supply for carbon assimilation in leaves. It is also responsible for maintaining optimum leaf relative water contents (Tariq et al. 2017). Similarly, Zn is critical for the growth of rice plants under stress environments as it is the part of antioxidative enzymes like SOD and CAT. Proper concentration of Zn in soil helps plant avoiding ROS damage under abiotic stresses (Cakmak 2000). The effect of the inoculation with *P. indica* on P and Zn concentration, observed in this study, has also been reported in other crops by Franken (2012) and Padash et al. (2016).

Fv/Fm is an important marker to determine the effect of stress on the photosystem II in plants (Murchie and Lawson 2013). In present research, *P. indica* increased Fv/Fm both in control and osmotic stressed plants of rice. Previously Sherameti et al. (2008) and Bakshi et al. (2017) reported that *P. indica* increased the level of Fv/Fm in *Arabidopsis thaliana* under abiotic stress. Similarly, it also stabilized the grana in chloroplasts. Under osmotic stress it helped maintaining grana in proper stacked arrangement, thus protecting the photosynthetic system. .

The association with *P. indica* also altered the expression of *P5CS*. *P. indica* increased the expression of this gene under both control and stressed environments. This gene is involved in the synthesis of proline, an amino acid which is critical for plant growth and development. Proline acts as both an osmotic agent and a radical scavenger (Kishor et al. 2014; Kishor et al. 2015). The supportive role of *P. indica* in increasing the synthesis of this protective compound was also evidenced by our experiment. The accumulation of proline due to osmotic stress was more than double in inoculated plants of WC-297 and Caawa and almost double in inoculated plants of IR-64, compared to non-inoculated plants. An increase in TAC was also associated with the presence of the endophyte. This increase in TAC was significant in all three genotypes. Increase in proline and TAC accumulation due to the inoculation with *P. indica* has also been reported in other crops by Waller et al. (2005) and Prasad et al. (2013). Most of the rice inoculation with *P. indica* had been performed to mitigate the salinity, heavy metal and some biotic stresses. We have tested the association of *P. indica* with rice plant under osmotic stress and it was summarized that this fungus improved plant performance under water deficient environment as it is already reported for inducing tolerance in rice under many other stresses.

Conclusion

P. indica is known to help maintaining growth and yield in many field crops under drought stress, but the underlying mechanisms are not fully elucidated. In the present study, *P. indica* was found to improve morphological traits, mineral uptake and protect the photosynthetic machinery in rice plants submitted to osmotic stress. Moreover, the inoculation with *P. indica* increased Fv/Fm, up regulated the expression of *P5CS*, raised the accumulation of proline and TAC. These results confirmed that the endophyte has a significant role in protecting the rice plant against osmotic stress. Further investigation is needed, however, to validate these effects of the inoculation in rice fields under various drought scenarios and use the inoculation with *P. indica* as an additional agronomical tool to improve rice grain yield under drought-prone conditions.

Abbreviations

Caawa: Moderately drought tolerant genotype; CDS: Coding DNA sequence; Fv/Fm: Maximum quantum yield of PSII; IR-64: Drought susceptible genotypes; P: Phosphorus; *P. indica*: *Piriformospora indica*; P5CS: Pyrroline-5-carboxylate synthase; PEG-6000: Polyethylene glycol-6000; ROS: Reactive oxygen species; TAC: Total antioxidant capacity; TEM: Transmission electron microscope; WC-297: Drought tolerant genotypes; Zn: Zinc

Acknowledgments

Research in the author's laboratories was financed by the Higher Education Commission of Pakistan, 973 National Project (517102-N51501/001), National Natural Science Foundation of China Project, International (Regional) Cooperation and Exchange Program (517102-N11808ZJ), The Fundamental Research Funds for the Central Universities (517102*172210172) and Jiangsu Collaborative Innovation Center for Modern Crop Production (JCIC-MCP). P. indica was provided by Dr. Alga Zuccaro, Max Planck Institute for Terrestrial Microbiology, Germany and Dr. Bing-Gan Lou, Institute of Biotechnology, Zhejiang University, Hangzhou, People's Republic of China.

Funding

This work was funded by the Higher Education Commission of Pakistan. Project: Indigenous 5000 PhD Fellowship, Pin No.: 213–66438-2Av2–077, International Research Support Initiative Program (IRSIP), HEC project # 2297 and 973 national project (517102-N51501/001), National Natural Science Foundation of China Project, International (Regional) Cooperation and Exchange Program (517102-N11808ZJ), the Fundamental Research Funds for the Central Universities (517102*172210172) and Jiangsu Collaborative Innovation Center for Modern Crop Production (JCIC-MCP).

Authors' contributions

Conceived and designed the experiments: MABS ZA ASK. Performed the experiments: MABS. Analyzed the data: MABS ZA IHS. Contributed reagents/materials/analysis tools: ZA IHS IAR. Wrote the paper: MABS. Critical review and editing: ZA ASK IAR. All authors read and approved the final manuscript.

Competing interests

The authors declare that they have no competing interests.

Author details

[1]Department of Plant Breeding and Genetics, University of Agriculture, 38040, Faisalabad, Pakistan. [2]Centre of Agricultural Biochemistry and Biotechnology, University of Agriculture, 38040, Faisalabad, Pakistan. [3]Department of Agronomy, College of Agriculture and Biotechnology, Zhejiang University, 310058, Hangzhou, People's Republic of China. [4]Department of Plant Breeding and Genetics, Muhammad Nawaz Shareef University of Agriculture, 60000, Multan, Pakistan.

References

Abadi VAJM, Sepehri M (2016) Effect of *Piriformospora indica* and Azotobacter chroococcum on mitigation of zinc deficiency stress in wheat (Triticum aestivum L.). Symbiosis 69(1):9–19

Abo-Doma A, Edrees S, Abdel-Aziz S (2016) The effect of mycorrhiza growth and expression of some genes in barley. Egypt J Genet Cytol 40(2):301–313

Ashraf M, Foolad M (2007) Roles of glycine betaine and proline in improving plant abiotic stress resistance. Environ Exp Bot 59(2):206–216

Bagheri A, Saadatmand S, Niknam V, Nejadsatari T, Babaeizad V (2014) Effects of Piriformospora indica on biochemical parameters of Oryza sativa under salt stress. Int J Biosci 4(4):24–32

Bagheri AA, Saadatmand S, Niknam V, Nejadsatari T, Babaeizad V (2013) Effect of endophytic fungus, Piriformospora indica, on growth and activity of antioxidant enzymes of rice (Oryza sativa L.) under salinity stress. Int J Adv Biol Biomed Res 1(11):1337–1350

Bakshi M, Sherameti I, Meichsner D, Thürich J, Varma A, Johri AK, Yeh K-W, Oelmüller R (2017) *Piriformospora indica* reprograms gene expression in Arabidopsis phosphate metabolism mutants but does not compensate for phosphate limitation. Front Microbiol 8:1262

Bates L, Waldren R, Teare I (1973) Rapid determination of free proline for water-stress studies. Plant Soil 39(1):205–207

Cakmak I (2000) Tansley review no. 111 possible roles of zinc in protecting plant cells from damage by reactive oxygen species. New Phytol 146(2):185–205

Ding L, Li Y, Li Y, Shen Q, Guo S (2014) Effects of drought stress on photosynthesis and water status of rice leaves. Chin J Rice Sci 28(1):65–70

Fahad S, Bajwa AA, Nazir U, Anjum SA, Farooq A, Zohaib A, Sadia S, Nasim W, Adkins S, Saud S (2017) Crop production under drought and heat stress: plant responses and management options. Front Plant Sci 8:1147

Fahramand M, Mahmoody M, Keykha A, Noori M, Rigi K (2014) Influence of abiotic stress on proline, photosynthetic enzymes and growth. Int Res J Appl Basic Sci 8(3):257–265

Farooq M, Wahid A, Kobayashi N, Fujita D, Basra S (2009a) Plant drought stress: effects, mechanisms and management. Agron Sustain Dev 29(1):185–212

Farooq M, Wahid A, Lee D-J, Ito O, Siddique KH (2009b) Advances in drought resistance of rice. Crit Rev Plant Sci 28(4):199–217

Food and Agriculture Organization of the United Nations. http://www.fao.org/3/I8317EN/I8317EN.pdf. Accessed 25 Dec 2017.

Franken P (2012) The plant strengthening root endophyte *Piriformospora indica*: potential application and the biology behind. Appl Microbiol Biotechnol 96(6):1455–1464

Gill SS, Gill R, Trivedi DK, Anjum NA, Sharma KK, Ansari MW, Ansari AA, Johri AK, Prasad R, Pereira E (2016) *Piriformospora indica*: potential and significance in plant stress tolerance. Front Microbiol 7:1–20

Gill SS, Tuteja N (2010) Reactive oxygen species and antioxidant machinery in abiotic stress tolerance in crop plants. Plant Physiol Biochem 48(12):909–930

Haefele SM, Kato Y, Singh S (2016) Climate ready rice: augmenting drought tolerance with best management practices. Field Crop Res 190:60–69

Hattori T, Inanaga S, Araki H, An P, Morita S, Luxova M, Lux A (2005) Application of silicon enhanced drought tolerance in Sorghum bicolor. Physiol Plant 123(4):459–466

Hill TW, Kafer E (2001) Improved protocols for aspergillus minimal medium: trace element and minimal medium salt stock solutions. Fungal Genet Rep 48(1):20–21

Im YJ, Han O, Chung GC, Cho BH (2002) Antisense expression of an Arabidopsis omega-3 fatty acid desaturase gene reduces salt/drought tolerance in transgenic tobacco plants. Mol Cells 13(2):264–271

Ji K, Wang Y, Sun W, Lou Q, Mei H, Shen S, Chen H (2012) Drought-responsive mechanisms in rice genotypes with contrasting drought tolerance during reproductive stage. J Plant Physiol 169(4):336–344

Jogawat A, Saha S, Bakshi M, Dayaman V, Kumar M, Dua M, Varma A, Oelmüller R, Tuteja N, Johri AK (2013) Piriformospora indica rescues growth diminution of rice seedlings during high salt stress. Plant Signal Behav 8(10):e26891

Kaya MD, Okçu G, Atak M, Çıkılı Y, Kolsarıcı Ö (2006) Seed treatments to overcome salt and drought stress during germination in sunflower (Helianthus annuus L.). Eur J Agron 24(4):291–295

Kishor K, Polavarapu B, Sreenivasulu N (2014) Is proline accumulation per se correlated with stress tolerance or is proline homeostasis a more critical issue? Plant Cell Environ 37(2):300–311

Kishor PBK, Kumari PH, Sunita M, Sreenivasulu N (2015) Role of proline in cell wall synthesis and plant development and its implications in plant ontogeny. Front Plant Sci 6:1–17

Korres N, Norsworthy J, Burgos N, Oosterhuis D (2017) Temperature and drought impacts on rice production: an agronomic perspective regarding short-and long-term adaptation measures. Water Resour Rural Dev 9:12–27

Li X, Zhang L, Li Y (2011) Preconditioning alters antioxidative enzyme responses in rice seedlings to water stress. Procedia Environ Sci 11:1346–1351

Liu R, Zhang H, Chen Z, Shahid M, Fu X, Liu X (2015) Drought-tolerant rice germplasm developed from an Oryza officinalis transformation-competent artificial chromosome clone. Genet Mol Res 14(4):13667–13678

Michal Johnson J, Sherameti I, Ludwig A, Nongbri PL, Sun C, Lou B, Varma A, Oelmüller R (2011) Protocols for Arabidopsis thaliana and *Piriformospora indica* co-cultivation–a model system to study plant beneficial traits. Endocytobiosis Cell Res 21:101–113

Michel BE, Kaufmann MR (1973) The osmotic potential of polyethylene glycol 6000. Plant Physiol 51(5):914–916

Mohd S, Shukla J, Kushwaha AS, Mandrah K, Shankar J, Arjaria N, Saxena PN, Narayan R, Roy SK, Kumar M (2017) Endophytic Fungi Piriformospora indica mediated protection of host from arsenic toxicity. Front Microbiol 8:754

Murchie EH, Lawson T (2013) Chlorophyll fluorescence analysis: a guide to good practice and understanding some new applications. J Exp Bot 64(13):3983–3998

Nakashima K, Yamaguchi-Shinozaki K, Shinozaki K (2014) The transcriptional regulatory network in the drought response and its crosstalk in abiotic stress responses including drought, cold, and heat. Front Plant Sci 5:1–7

Nassimi Z, Taheri P (2017) Endophytic fungus Piriformospora indica induced systemic resistance against rice sheath blight via affecting hydrogen peroxide and antioxidants. Biocontrol Sci Technol 27(2):252–267

Ngwene B, Boukail S, Söllner L, Franken P, Andrade-Linares D (2016) Phosphate utilization by the fungal root endophyte *Piriformospora indica*. Plant Soil 405(1–2):231–241

Padash A, Shahabivand S, Behtash F, Aghaee A (2016) A practicable method for zinc enrichment in lettuce leaves by the endophyte fungus *Piriformospora indica* under increasing zinc supply. Sci Hortic 213:367–372

Prasad KN, Yang B, Yang S, Chen Y, Zhao M, Ashraf M, Jiang Y (2009) Identification of phenolic compounds and appraisal of antioxidant and antityrosinase activities from litchi (Litchi sinensis Sonn.) seeds. Food Chem 116(1):1–7

Prasad R, Kamal S, Sharma PK, Oelmüller R, Varma A (2013) Root endophyte Piriformospora indica DSM 11827 alters plant morphology, enhances biomass and antioxidant activity of medicinal plant Bacopa monniera. J Basic Microbiol 53(12):1016–1024

Prieto P, Pineda M, Aguilar M (1999) Spectrophotometric quantitation of antioxidant capacity through the formation of a phosphomolybdenum complex: specific application to the determination of vitamin E. Anal Biochem 269(2):337–341

Sarma M, Kumar V, Saharan K, Srivastava R, Sharma A, Prakash A, Sahai V, Bisaria V (2011) Application of inorganic carrier-based formulations of fluorescent pseudomonads and *Piriformospora indica* on tomato plants and evaluation of their efficacy. J Appl Microbiol 111(2):456–466

Serraj R, McNally KL, Slamet-Loedin I, Kohli A, Haefele SM, Atlin G, Kumar A (2011) Drought resistance improvement in rice: an integrated genetic and resource management strategy. Plant Prod Sci 14(1):1–14

Shahabivand S, Parvaneh A, Aliloo AA (2017) Root endophytic fungus *Piriformospora indica* affected growth, cadmium partitioning and chlorophyll fluorescence of sunflower under cadmium toxicity. Ecotoxicol Environ Saf 145:496–502

Sherameti I, Tripathi S, Varma A, Oelmüller R (2008) The root-colonizing endophyte Pirifomospora indica confers drought tolerance in Arabidopsis by stimulating the expression of drought stress–related genes in leaves. Mol Plant-Microbe Interact 21(6):799–807

Sokoto M, Muhammad A (2014) Response of rice varieties to water stress in Sokoto, Sudan savannah, Nigeria. J Biosci Med 2(01):68

Swapna S, Shylaraj KS (2017) Screening for osmotic stress responses in Rice varieties under drought condition. Rice Sci 24(5):253–263

Taiz L, Zeiger E (2006) Plant physiology. 4th. Sinauer associate, Sunderland, mass, EUA

Tariq A, Pan K, Olatunji OA, Graciano C, Li Z, Sun F, Sun X, Song D, Chen W, Zhang A (2017) Phosphorous application improves drought tolerance of Phoebe zhennan. Front Plant Sci 8:1561

Uga Y, Kitomi Y, Yamamoto E, Kanno N, Kawai S, Mizubayashi T, Fukuoka S (2015) A QTL for root growth angle on rice chromosome 7 is involved in the genetic pathway of DEEPER ROOTING 1. Rice 8(1):8

Vahabi K, Dorcheh SK, Monajembashi S, Westermann M, Reichelt M, Falkenberg D, Hemmerich P, Sherameti I, Oelmüller R (2016) Stress promotes Arabidopsis-*Piriformospora indica* interaction. Plant Signal Behav 11(5): e1136763

Varma A, Chordia P, Bakshi M, Oelmüller R (2013) Introduction to Sebacinales. In: Piriformospora indica: Springer; pp 33:3–24

Varma A, Singh A, Sahay NS, Sharma J, Roy A, Kumari M, Rana D, Thakran S, Deka D, Bharti K (2001) *Piriformospora indica*: an axenically culturable mycorrhiza-like endosymbiotic fungus. In: Fungal associations: Springer; pp 9:125–150

Verslues PE, Sharma S (2010) Proline metabolism and its implications for plant-environment interaction. The arabidopsis book:e0140. https://doi.org/10.1199/tab.0140

Waller F, Achatz B, Baltruschat H, Fodor J, Becker K, Fischer M, Heier T, Hückelhoven R, Neumann C, von Wettstein D (2005) The endophytic fungus *Piriformospora indica* reprograms barley to salt-stress tolerance, disease resistance, and higher yield. Proc Natl Acad Sci U S A 102(38):13386–13391

Yang P-M, Huang Q-C, Qin G-Y, Zhao S-P, Zhou J-G (2014) Different drought-stress responses in photosynthesis and reactive oxygen metabolism between autotetraploid and diploid rice. Photosynthetica 52(2):193–202

Yoshida, S (1976) Routine procedure for growing rice plants in culture solution. Laboratory manual for physiological studies of rice. 61-66

Yue B, Xue W, Xiong L, Yu X, Luo L, Cui K, Jin D, Xing Y, Zhang Q (2006) Genetic basis of drought resistance at reproductive stage in rice: separation of drought tolerance from drought avoidance. Genetics 172(2):1213–1228

Zain NAM, Ismail MR, Mahmood M, Puteh A, Ibrahim MH (2014) Alleviation of water stress effects on MR220 rice by application of periodical water stress and potassium fertilization. Molecules 19(2):1795–1819

10

Suppression of *OsMDHAR4* enhances heat tolerance by mediating H_2O_2-induced stomatal closure in rice plants

Jianping Liu[1], Xinjiao Sun[1], Feiyun Xu[1], Yingjiao Zhang[1], Qian Zhang[1], Rui Miao[1], Jianhua Zhang[2*], Jiansheng Liang[3] and Weifeng Xu[1*]

Abstract

Background: Monodehydroascorbate reductase (MDAR or MDHAR), which is responsible for growth, development and stress response in plants, is a key enzyme in the maintenance of the ascorbate acid (AsA) pool through the AsA–glutathione (AsA–GSH) cycle. High temperature affects a broad spectrum of cellular components and metabolism including AsA–GSH cycle in plants. In rice, however, the detailed roles of *OsMDHAR4* in resistance against heat stress remains unclear.

Results: Here, we report that OsMDHAR4 protein was localized to the chloroplasts. *OsMDHAR4* expression was detected in all tissues surveyed and peaked in leaf blade. *OsMDHAR4* was responsive to multiple stresses and was relatively strongly induced by heat treatment. In comparison with wild type, the *osmdhar4* mutant exhibited improved tolerance to heat stress, whereas *OsMDHAR4* overexpression lines exhibited enhanced sensitivity to heat stress. Moreover, we found that suppression of *OsMDHAR4* promoted stomatal closure and hydrogen peroxide accumulation, and overexpression of *OsMDHAR4* increased stomatal opening and decreased hydrogen peroxide content in rice leaves.

Conclusions: Taken together, these results indicated that *OsMDHAR4* negatively regulates tolerance to heat stress by mediating H_2O_2-induced stomatal closure in rice.

Keywords: Rice, Monodehydroascorbate reductase, Heat tolerance, *OsMDHAR4*, Stomata, Hydrogen peroxide

Background

Extreme temperature is one of the serious threats affecting crop production and distribution worldwide (Yokotani et al., 2008). In plants, a transient increase of 10–15 °C above ambient, is generally considered as heat shock or heat stress, which negatively affects plant growth, seed germination, photosynthesis, respiration, water relation, and membrane stability in plants (Wahid et al., 2007), and is often accompanied by the generation of reactive oxygen species (ROS), such as hydrogen peroxide (H_2O_2), hydroxyl radical, superoxide anion radicals, and singlet oxygen (Liu and Huang, 2000; Mittler, 2002; Apel and Hirt, 2004; Song et al., 2014). The understanding of plant responses to heat stress in physiology, genetics, and molecular biology will be greatly helpful in improving the heat tolerance of plants through genetic engineering.

Compared with other ROS, H_2O_2 is more stable, more diffusive, and have a long half-life (approximately 1 ms) and high permeability across membranes (Levine et al., 1994), so it can readily escape from the organelle where it was produced to the cytosol. Previous research has shown that H_2O_2 plays a dual role in plants: at low concentrations, it acts as a secondary messenger involved in triggering tolerance to various biotic and abiotic stresses; but at high concentrations, it leads to programmed cell death (Quan et al., 2008). H_2O_2 also acts as a key regulator in a broad range of physiological processes,

* Correspondence: jhzhang@cuhk.edu.hk; wfxu@fafu.edu.cn
[2]Department of Biology, Hong Kong Baptist University, Hong Kong, China
[1]Center for Plant Water-use and Nutrition Regulation and College of Life Sciences, Joint International Research Laboratory of Water and Nutrient in Crop, Fujian Agriculture and Forestry University, Jinshan Fuzhou 350002, China
Full list of author information is available at the end of the article

such as photorespiration and photosynthesis (Noctor and Foyer, 1998), growth and development (Foreman et al., 2003), the cell cycle (Mittler et al., 2004), senescence (Peng et al., 2005), and stomatal movement (Bright et al., 2006).

Stomatal pores that are located in the epidermis of plant leaves control the uptake of CO_2 for photosynthesis and the water loss during transpiration, and play a crucial role in abiotic stress tolerance (Schroeder et al., 2001; Hetherington and Woodward, 2003). In the process of ABA-dependent stomatal closure, H_2O_2 plays a vital role as a signal molecule by elevating calcium levels in guard cells through the activation of plasma membrane calcium channels (Pei et al., 2000; Wang and Song, 2008). OsHTAS, a RING finger ubiquitin E3 ligase, functions in rice heat tolerance through H_2O_2-induced stomatal closure and is mainly involved in the ABA-dependent pathway (Liu et al., 2016). The ABA-activated SnRK2 protein kinase OPEN STOMATA1 phosphorylates NADP (NADPH) oxidase (AtrbohF), which functions to produce ABA-induced ROS in guard cells (Sirichandra et al., 2009). There's another way of ABA-independent stomatal closure. DROUGHT AND SALT TOLERANCE (DST), a zinc finger transcription factor, negatively regulates H_2O_2-induced stomatal closure by directly regulating the expression of genes related to H_2O_2 scavenging (Huang et al., 2009). *OsSRO1c* suppresses DST to positively regulate H_2O_2-induced stomatal closure (You et al., 2013).

ROS is normally scavenged by an enzymatic anti-oxidative system containing catalase (CAT), ascorbate peroxidase (APX), glutathione peroxidase (GPX) and superoxide dismutase (SOD), and by a non-enzymatic anti-oxidative system including ascorbic acid (AsA), glutathione (GSH), tocopherols (TOCs) and phenolic compounds to protect plant cells. The monodehydroascorbate reductase (MDAR or MDHAR), well-known as flavin adenine dinucleotide (FAD) enzyme, is involved in the ascorbate–glutathione cycle, and plays an important role in directly reducing monodehydroascorbate (oxidized ascorbate) to ascorbate using NAD(P)H as an electron donor (Apel and Hirt 2004). MDARs were found in many eukaryotes, including cucumbers, potatoes, soybean root nodules, and rot fungus (Gill and Tuteja, 2010), and were localized in chloroplasts, mitochondria, peroxisomes, and the cytosol in plants (Omoto et al., 2013), and in microsomes, mitochondria, the Golgi apparatus, and erythrocytes in animals (Sakihama et al., 2000).

The roles of MDAR have been extensively reported under abiotic and biotic stress. Overexpression of *AtMDAR1* in tobacco confers enhanced tolerance to ozone, salt and polyethylene glycol stresses (Eltayeb et al., 2007). Overexpression of *LeMDAR* from tomato (*Lycopersicon esculentum* Mill.) enhanced tolerance to temperature and methylviologen-mediated oxidative stresses (Li et al., 2010). *OsMDHAR*-expressing yeast cells displayed enhanced tolerance to H_2O_2 (Kim et al., 2016). Silencing *OsMDHAR3* gene increased salt sensitivity in rice (Kim et al., 2017). Knockdown of *TaMDHAR4* and *TaMDAR6* through virus-induced gene silencing (VIGS) enhanced the wheat resistance to *Puccinia striiformis* f. sp. *tritici* (*Pst*) (Feng et al., 2014; Abou-Attia et al., 2016). However, the roles of rice MDAR in heat stress response remain unclear.

In this study, a new gene encoding the chloroplastic MDHAR (*OsMDHAR4*) from the rice plant was cloned, and its function was analyzed. The *osmdhar4* mutant displayed significantly improved heat tolerance with increased stomatal closure and reduced water loss by promoting H_2O_2 accumulation. *OsMDHAR4*-overexpressing transgenic plants showed obviously enhanced sensitivity to heat stress with decreased stomatal closure and accelerated water loss speed. Our results suggest that the *OsMDHAR4* gene negatively regulates resistance to heat stress in rice.

Results

Amino acid sequence alignment and subcellular localization of OsMDHAR4

The coding sequence (CDS) of *OsMDHAR4* gene spans 1431 bp and encodes a protein of 476 amino acids and has a predicted molecular weight of 51.86 kDa. The prediction of protein domains using NCBI and InterProScan databases indicated that OsMDHAR4 protein contains pyridine nucleotide-disulfide oxidoreductase domains (Pyr_redox_2 and NAD(P)-binding domain) and a FAD/NAD-linked reductase, dimerisation domain. The multiple amino acid sequence alignment of OsMDHAR4 and other plant MDHARs was performed using the BLAST tool at the NCBI database. OsMDHAR4 showed 87, 88, 86, 43, 69, 43, 70 and 53% identity and 92, 93, 93, 61, 80, 60, 82 and 69% similarity to ZmMDAR (*Zea mays*), BdMDAR (*Brachypodium distachyon*), TaMDHAR4, TaMDAR6 (*Triticum aestivum*), AtMDAR4, AtMDAR6 (*Arabidopsis thaliana*), GmMDAR (*Glycine max*), and SlMDAR (*Solanum lycopersicum*), respectively (Additional file 1: Figure S1).

To study the subcellular localization of OsMDHAR4 protein, we fused the CDS of *OsMDHAR4* to the N-terminal of yellow fluorescent protein (YFP) driven by the cauliflower mosaic virus 35S promoter. The empty and recombinant vectors were transiently transformed into the epidermis of *Nicotiana benthamiana* by agroinfiltration. Fluorescence microscopic analysis revealed that OsMDHAR4-YFP was mainly detected in chloroplast, partially colocalized with the red autofluorescence of chlorophyll. On the contrary, 35S-YFP was widely distributed in the nucleus and cytoplasm and was not colocalized with the red autofluorescence of chloroplasts (Fig. 1). This

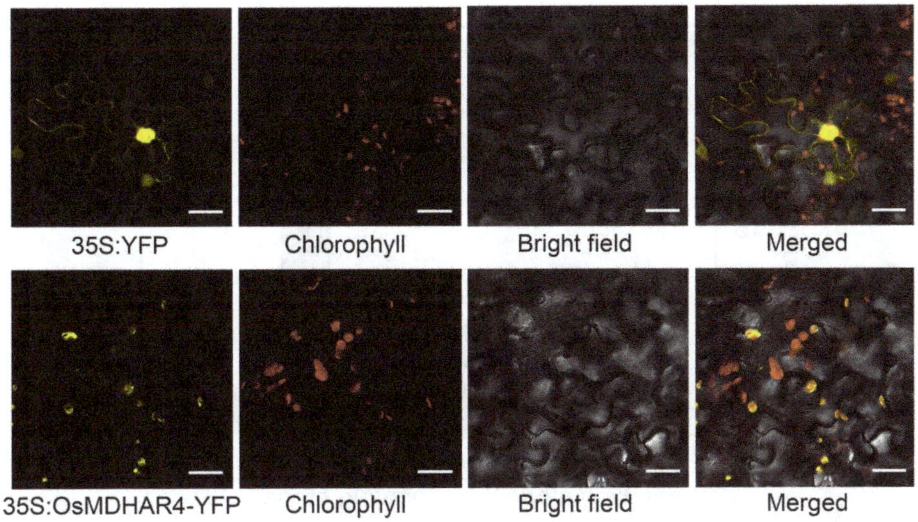

Fig. 1 Subcellular localization of OsMDHAR4 in *Nicotiana benthamiana* by confocal fluorescent microscopy. Red color represents the autofluorescence of chlorophyll; yellow fluorescence shows the localization of OsMDHAR4-YFP. Scale bars = 20 μm

result indicated that OsMDHAR4 protein was localized to the chloroplasts.

Expression pattern of *OsMDHAR4*

A total of 6 representative tissues, including root, shoot, stem, sheath and blade of flag leaf, and panicle, were prepared for spatio-temporal expression analysis of *OsMDHAR4* in rice Zhonghua11 (ZH11). The mRNA abundance of *OsMDHAR4* was detected using quantitative RT-PCR (qRT-PCR). *OsMDHAR4* transcripts were detected in all of the tissues surveyed, and the expression level was relatively lowest in root and peaked in the blade of the flag leaf (Fig. 2a).

To investigate the physiological and functional relevance of the *OsMDHAR4* gene, we further examined the expression level of *OsMDHAR4* under heat stress and other treatments, including polyethylene glycol (PEG), abscisic acid (ABA), H_2O_2, salt and cold, at the seedling stage. It was found that, after heat treatment, the *OsMDHAR4* transcripts increased rapidly at 0.5 h (1.8-fold), after which a slight decrease occurred at 3 h, and then they continued to rise and peaked at 24 h (2.8-fold). Under the PEG treatment, the transcript abundance of *OsMDHAR4* was nearly unchanged at most time points except for 3 h and 24 h, where two induced peaks (1.7 and 2.2-fold) appeared. In the ABA treatment, *OsMDHAR4* was induced quickly at 0.5 h and 3 h, after which a dramatic decrease appeared at 6 h and 12 h, and recovered to initial level until 24 h. When treated with H_2O_2, the *OsMDHAR4* expression showed a significant trend of rising followed by falling. The *OsMDHAR4* transcript level was slightly suppressed by salt and cold treatments at 6 h and 12 h (Fig. 2b). These results demonstrated that *OsMDHAR4* is responsive to multiple stresses.

Disruption of *OsMDHAR4* increased heat tolerance

As shown in Fig. 2b, the expression of *OsMDHAR4* was significantly induced by heat treatment, and the positive roles of MDHAR have been extensively reported under abiotic stress such as ozone, salt, PEG (Sharma and Davis 1997; Eltayeb et al.,2007) and drought (Sharma and Dubey 2005). However, the role of the *MDHAR* gene in rice remains largely unknown. To evaluate the function of *OsMDHAR4* in heat stress response, we firstly searched for the Rice Mutant Database (http://rmd.ncpgr.cn/). Then, the *osmdhar4* mutant (RMD_03Z11DA36), a transfer DNA (T-DNA) insertion line in the japonica rice ZH11 background was obtained (Wu et al., 2003; Zhang et al., 2006). DNA sequencing and genotyping revealed that the T-DNA was inserted in the first intron of the *OsMDHAR4* gene, 254 bp downstream of the translation initiation site (Fig. 3a and Additional file 2: Figure S2). A qRT-PCR assay showed that *OsMDHAR4* was expressed at a very low level in the *osmdhar4* plants (Fig. 3b). Under normal growth conditions, the *osmdhar4* mutant had no obvious differences compared with the wild type (WT, ZH11) (Fig. 3c, left). *osmdhar4* and WT seedlings (5.5- to 6.5-leaf stage) were treated at high temperature (45 °C) for 72 h and subsequently returned to 26 °C for recovery. Survival rates were used as a measure of heat tolerance. After the heat treatment and recovery, 80.5% of the *osmdhar4* seedlings survived compared with 45.6% of the WT seedlings (Fig. 3c, d). Subsequently, we measured the water loss rates of detached leaves from *osmdhar4* and WT. The results showed that the detached leaves of *osmdhar4* lost water more slowly than the WT leaves (Fig. 3e). This decreased water loss of the *osmdhar4* mutant might lead to an increased tolerance to high temperature.

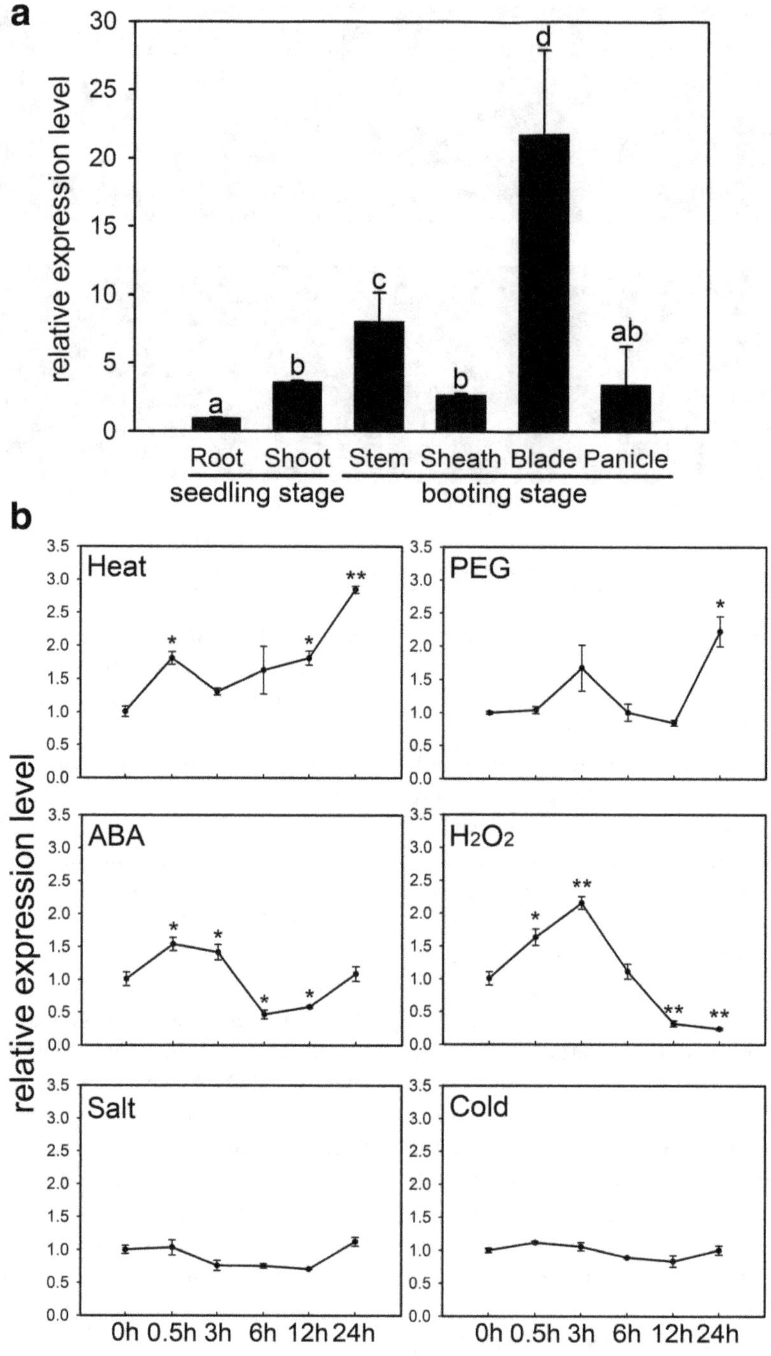

Fig. 2 Expression pattern of *OsMDHAR4*. **a** Expression analysis of *OsMDHAR4* in different tissues containing root, shoot, stem, sheath and blade of flag leaf, and panicle by qRT-PCR. Error bars indicate the SE based on three biological replicates. Statistical differences among the samples are labeled with different letters according to the LSD test ($P < 0.05$, one-way ANOVA). **b** Expression of *OsMDHAR4* under abiotic stress conditions. Seedlings (3.5- to 4.5-leaf stage) were subjected to heat (45 °C), PEG (20% [*w/v*] PEG 6000), ABA (100 μM), H_2O_2 (100 mM), salt (250 mM NaCl), and cold (4 °C). Relative expression levels of *OsMDHAR4* were examined by qRT-PCR. The rice *Actin1* gene was used as the internal control. Error bars indicate the SE based on three biological replicates. *, $P < 0.05$, **, $P < 0.01$, by Student's *t*-test

Overexpression of *OsMDHAR4* decreased heat tolerance

To further validate the functions of *OsMDHAR4* in rice, we constructed an overexpression vector of *OsMDHAR4* driven by 35S promoter. The *OsMDHAR4*-overexpressing vector was transformed into rice ZH11. PCR analysis using *OsMDHAR4*-specific primers (*OsMDHAR4*-OE-F and R) and vector-specific primers (Hyg-F and R) confirmed the presence of the transgene in the two

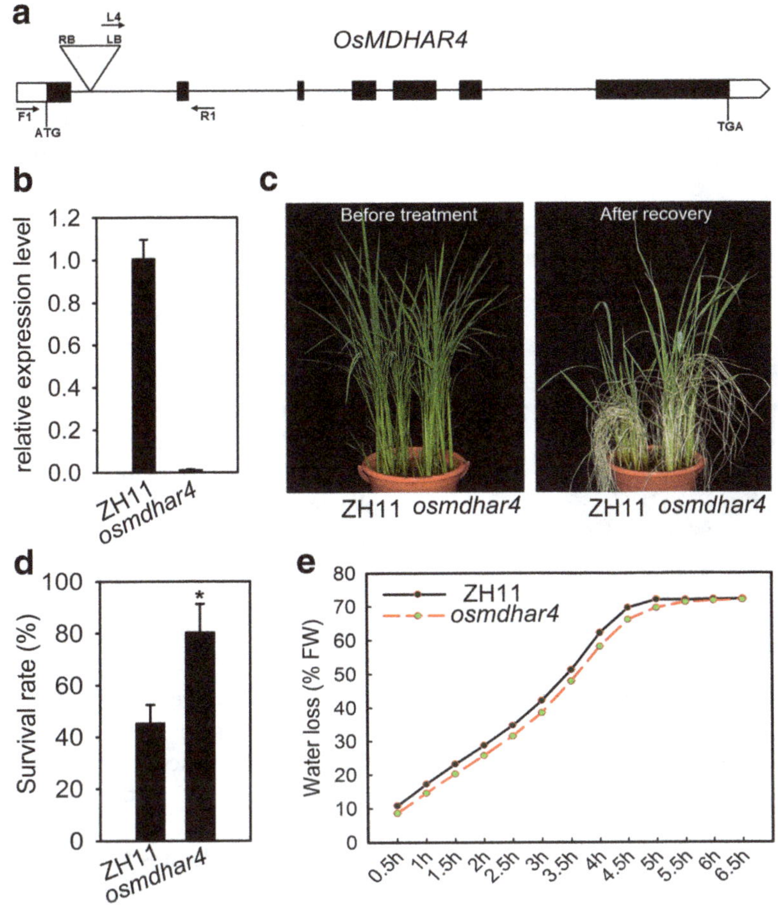

Fig. 3 Increased heat tolerance of the *osmdhar4* mutant at seedling stage. **a** Schematic diagram of the *OsMDHAR4* gene. In the genomic DNA, exons, introns, and untranslated regions are indicated by black boxes, lines between boxes, and white boxes, respectively. The T-DNA insertion site is located in the first intron, 254 bp downstream of the start codon ATG. LB, left border of the T-DNA; RB, right border of the T-DNA. Arrows F1, R1, and L4 represent the primers used in the genotyping of the *osmdhar4* mutant. **b** Expression analysis of *OsMDHAR4* in the *osmdhar4* mutant detected by qRT-PCR. The rice *Actin1* gene was used as the internal control. Error bars indicate the SE based on three technical replicates. **c** 5.5- to 6.5-leaf stage WT and the *osmdhar4* mutant plants were growing in barrels filled with well-mixed soil. The left and right half of each barrel was planted with WT and mutant plants, respectively. Plants were treated with high temperature (45 °C) for 72 h, followed by culturing under normal conditions for 7 d. The photograph was taken before heat stress and after recovery. **d** Comparison of survival rates of WT and the *osmdhar4* mutant after heat stress. Error bars indicated the SE based on three biological replicates. *, $P < 0.05$, by Student's *t*-test. **e** Water loss from detached leaves of WT and the *osmdhar4* mutant at indicated time points. FW, Fresh Weight

independent overexpression lines *OsMDHAR4*-OE-1 and *OsMDHAR4*-OE-5 (Additional file 3: Figure S3). Both the two transgenic lines, in which the *OsMDHAR4* expression level was significantly increased approximately triple and twice of that in WT (Fig. 4a), were selected for heat tolerance testing (45 °C for 48 h at the 5.5- to 6.5-leaf stage). Under normal conditions, we did not observe any phenotypic differences between these two overexpression lines and the WT (Fig. 4b, top). Under the heat stress treatment, both *OsMDHAR4*-OE-1 and OE-5 were more sensitive than the WT, which was consistent with the behavior of the *osmdhar4* mutant in heat treatment. The survival rates ranged from 46.9 to 60.0% for the *OsMDHAR4*-overexpressing lines, and 90.0 to 90.6% for the WT plants after recovery, but there was no significant difference between *OsMDHAR4*-OE-1 and OE-5 (Fig. 4b, c). Consistent with this result, detached leaves of *OsMDHAR4*-overexpressing lines lost water more quickly than the WT leaves (Fig. 4d). Taken together, all of these results suggested that *OsMDHAR4* plays a negative role in heat tolerance in rice.

OsMDHAR4 alters the stomatal aperture status of rice seedlings

The function of *OsMDHAR4* in the control of water loss prompted us to investigate the stomatal aperture status, the major factor affecting the water-holding capacity in rice leaves. The stomatal apertures of the *osmdhar4* mutant, *OsMDHAR4*-OE-1 and the WT plants were checked by scanning electron microscopy. No significant differences in the percentages of the three types of

Fig. 4 Overexpression of *OsMDHAR4* showed increased sensitivity to heat stress. **a** Transcript level of *OsMDHAR4* was measured by qRT-PCR in two *OsMDHAR4*-overexpresing lines (*OsMDHAR4*-OE-1 and *OsMDHAR4*-OE-5, ZH11 background). The rice *Actin1* gene was used as the internal control. Error bars indicate the SE based on three biological replicates.*, $P < 0.05$, by Student's *t*-test. **b** 5.5- to 6.5-leaf stage ZH11 and *OsMDHAR4*-overexpressing plants were subjected to 45 °C high temperature environment for 48 h, followed by growing in normal conditions for 7 d. Surviving seedlings were photographed. **c** Comparison of the survival rates of ZH11 and *OsMDHAR4*-overexpressing plants after heat stress. Error bars indicated the SE based on three biological replicates. *, $P < 0.05$, by Student's *t*-test. **d** Water loss from detached leaves of ZH11and *OsMDHAR4*-overexpressing plants at indicated time points. FW, Fresh Weight

stomata and stomatal conductance between the *osmdhar4* mutant and the WT were observed before heat stress. However, after heat treatment (45 °C for 24 h at 5.5- to 6.5-leaf stage), the results showed that 56.8% of stomata were completely closed in the *osmdhar4* mutant, whereas only 25.0% were completely closed in the WT plants. On the other hand, only 10.5% of stomata were completely open in the *osmdhar4* mutant, but 46% were completely open in WT. The percentage of partially open stomata didn't differ significantly from each other (Fig. 5a, b). Furthermore, the stomatal conductance was dramatically decreased in the *osmdhar4* mutant compared the WT plants after heat treatment (Fig. 5c). These results were in agreement with the slower water loss of detached leaves from the *osmdhar4* mutant. Either before or after heat treatment, *OsMDHAR4*-OE-1 plants had lower percentages of completely closed stomata than WT (before, 32.2% for *OsMDHAR4*-OE-1, 40.4% for the WT; after, 17.2% for *OsMDHAR4*-OE-1, 30.4% for the WT), while the percentages of completely open were higher than the WT (before, 24.6% for *OsMDHAR4*-OE-1, 20.1% for the WT; after, 36.6% for *OsMDHAR4*-OE-1, 25.1% for the WT; Fig. 5d). Moreover, compared with the WT, *OsMDHAR4*-OE-1 had significantly higher stomatal conductance after heat treatment (Fig. 5e). These results were also in agreement with the faster water loss of the *OsMDHAR4*-OE-1 plants.

OsMDHAR4 modulates H_2O_2 homeostasis of rice seedlings

ROS are continuously produced as the byproducts of various metabolic pathways and are scavenged by different antioxidative defense components in plants. The equilibrium between ROS production and scavenging may be perturbed by adverse environmental factors, such as heat stress (Apel and Hirt, 2004). As H_2O_2 is a very stable ROS with a long half-life (Levine et al., 1994) and a signal molecular that induces stomatal closure (McAinsh et al., 1996), so we measured the H_2O_2 content in the rice seedling leaves. A higher level of H_2O_2 accumulation was detected in the *osmdhar4* mutant either before or after heat treatment (45 °C for 24 h at 5.5- to 6.5-leaf stage, Fig. 6a). The DAB staining results also support the quantitative analysis (Fig. 6b). In addition, more H_2O_2 accumulation was detected in the guard cells of the *osmdhar4* mutant under normal conditions, as indicated by the ROS indicator, H2DCFDA (Additional file 4: Figure S4). Thus, ROS homeostasis in *osmdhar4* seedlings was perturbed. These results indicated that the increased stomatal closure in the *osmdhar4* mutant was probably due to accumulation of H_2O_2 in guard cells. As expected, in the *OsMDHAR4*-OE-1 seedling leaves, visibly less H_2O_2 accumulation was detected under normal conditions or after heat stress through the quantitative analysis and DAB staining (Fig. 6c, d). Taken together, our results indicated that *OsMDHAR4* may negatively regulate H_2O_2-induced stomatal closure.

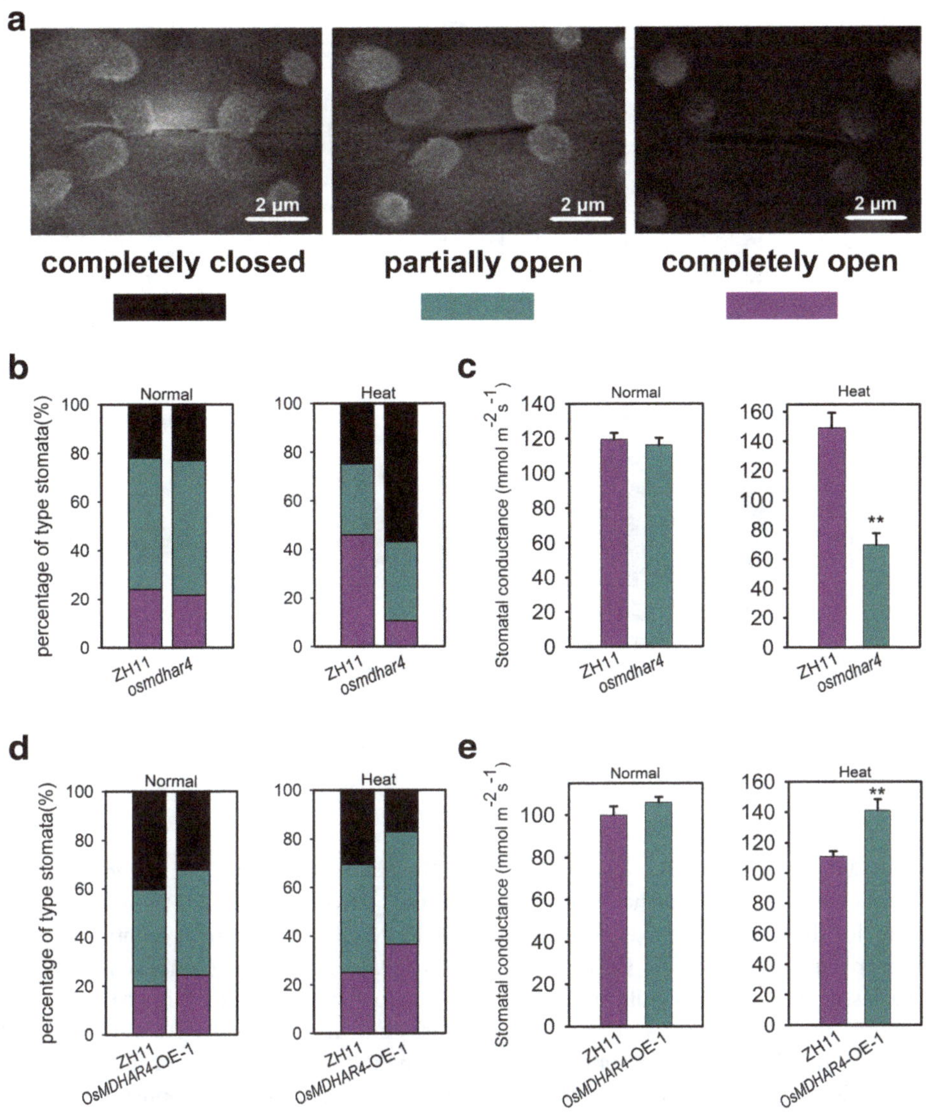

Fig. 5 The stomatal aperture status in the *osmdhar4* mutant and *OsMDHAR4*-overexpressing plants. **a** Scanning electron microscopy images of three levels of stomatal opening. Scale bars = 2 μm. **b** Percentages of three levels of stomatal opening in WT and the *osmdhar4* mutant under normal conditions (48 stomata for WT, 40 stomata for the *osmdhar4* mutant) and heat stress conditions (28 stomata for WT, 25 stomata for the *osmdhar4* mutant). **c** Stomatal conductance of WT and *osmdhar4* mutant under normal conditions and heat stress conditions (3 repeats, 12 plants in each repeat). **d** Percentages of three levels of stomatal opening in WT and *OsMDHAR4*-OE-1 plants under normal conditions (70 stomata for WT, 90 stomata for *OsMDHAR4*-OE-1) and heat stress conditions (60 stomata for WT, 76 stomata for *OsMDHAR4*-OE-1). **e** Stomatal conductance of the WT and *OsMDHAR4*-OE-1 under normal conditions and heat stress conditions (3 repeats, 12 plants in each repeat). FW, Fresh Weight. Error bars indicate the SE based on three biological replicates. *, $P < 0.05$, and **, $P < 0.01$, by Student's *t*-test

Discussion

MDAR has been reported to be distributed widely across kingdoms pointing towards a universal role. In plants, it has been found to localize in chloroplasts, glyoxysomes, mitochondria, and cytosol. Chloroplastic MDAR localizes on the stroma of the chloroplast (Sano et al., 2005) and catalyzes the conversion of monodehydroascorbate (MDA) to AsA. AsA is a major antioxidant molecule that directly neutralizes ROS. Previous research has indicated that overexpression of chloroplastic MDAR from tomato enhanced tolerance to temperature and methyl viologen-mediated oxidative stress (Li et al., 2010). However, the physiological role of the chloroplastic MDAR in rice in response to heat stress is not reported.

In this study, *OsMDHAR4*, one new gene encoding monodehydroascorbate reductase was isolated from rice, and its amino acid sequence showed high identity and similarity to the MDAR sequences from several other plant species (Additional file 1: Figure S1). Bioinformatics studies predicted that ZmMDAR and BdMDAR, the

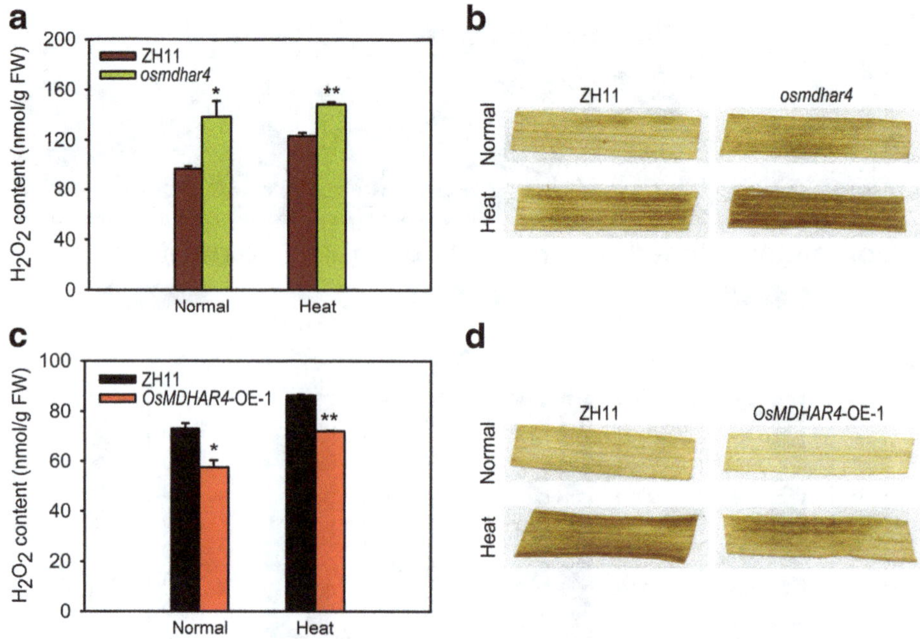

Fig. 6 Accumulation of H_2O_2 in the *osmdhar4* mutant and *OsMDHAR4*-overexpressing plants. **a** Detection of the H_2O_2 content of WT and the *OsMDHAR4* mutant under normal and heat stress conditions. **b** DAB staining in the seedling leaves of WT and the *osmdhar4* mutant under normal and heat stress conditions. **c** Detection of the H_2O_2 content of WT and *OsMDHAR4*-OE-1 plants under normal and heat stress conditions. **d** DAB staining in the seedling leaves of WT and *OsMDHAR4*-OE-1 plants under normal and heat stress conditions. Error bars indicate the SE based on three biological replicates. *, $P < 0.05$, and **, $P < 0.01$, by Student's *t*-test. FW, Fresh Weight

two most homologous MDARs from *Zea mays* and *Brachypodium distachyon*, were located in the cytoplasm. However, intracellular localization studies using YFP fusion confirmed chloroplast localization of OsMDHAR4 (Fig. 1). Plant chloroplasts are the most significant generators of ROS and proposed to be heat sensors. A large number of chloroplasts are distributed in mesophyll cells (Ishikawa and Shigeoka, 2008; Maruta et al., 2012). *OsMDHAR4* transcripts peaked in the blade of flag leaf, in which a lot of mesophyll cells were distributed, while it showed the lowest transcript level in the root (almost with no chloroplasts) (Fig. 2a), the results further confirmed the finding that OsMDHAR4 protein was located in chloroplast. Since 1995, several studies have reported that MDAR genes are regulated by abiotic stresses (Leterrier et al., 2005; Grantz et al., 1995; Eltelib et al., 2011). We found that *OsMDHAR4* transcripts were induced by heat and PEG (2.8-fold and 2.2-fold). Under both ABA and H_2O_2 treatment, the *OsMDHAR4* expression levels were increased in first and decreased at last, but salt and cold treatments had almost no effect on these (Fig. 2b). So, we assumed that *OsMDHAR4* may function in rice heat resistance.

Many scholars have focused on conferring abiotic stress tolerance through the overexpression of *MDAR* genes. Overexpression of *LeMDAR* gene alleviated photoinhibition of PSI and PSII and enhanced the tolerance to various abiotic stresses by elevating AsA level (Li et al., 2010). Overexpressing *AeMDHAR* gene increased MDHAR enzyme activity compared to untransformed plants under both NaCl and control conditions (Sultana et al., 2012). *OsMDHAR*-expressing yeast cells displayed enhanced tolerance to H_2O_2 by maintaining redox homoeostasis, proteostasis, and the ascorbate (AsA)-like pool (Kim et al., 2016). In the above studies, we can find that all the *MDAR* genes play a positive role in response to different abiotic stresses. In our study, we found that disruption of *OsMDHAR4* reduced the water loss rates and increased heat tolerance (Fig. 3), and overexpression of *OsMDHAR4* accelerated the water loss rates and decreased heat tolerance (Fig. 4). Unlike previous researches, these results showed that *OsMDHAR4* plays a negative role in heat tolerance in rice.

The importance of maintaining higher levels of AsA over DHA has been reported in previous studies. In transgenic tobacco plants overexpressing cucumber ascorbate oxidase, decreased AsA/DHA ratio increased ozone sensitivity (Sanmartin et al., 2003), whereas a higher AsA/DHA brought about by overexpression of DHAR and MDAR in cytosol resulted in enhanced tolerance to salt stress (Eltayeb et al., 2006). Higher AsA/DHA ratio, higher photosynthetic activity and lower H_2O_2 contents were reported in transgenic tobacco plants expressing ascorbate oxidase gene (Yamamoto et al., 2005). According to this, we measured the AsA and DHA contents in leaves. Consistently with previous study (Chen et al., 2003), the *osmdhar4* mutant showed lower AsA contents and AsA/DHA ratio than the WT before or after heat treatment,and the *OsMDHAR4*-OE-1 plants showed higher AsA contents and AsA/DHA ratio than the WT before or after treatment (Additional file 5: Figure S5). However, the changes of ascorbate redox state were not be able to explain the increased heat tolerance in the *osmdhar4* mutant and incremental heat sensitivity in *OsMDHAR4* overexpression lines.

Stomata control the exchange of gases-most importantly water vapor and CO_2-between the interior of the leaf and the atmosphere, and are involved in responses to abiotic stresses (Hetherington and Woodward, 2003). Stomatal closure degree makes major contribution to the ability of the plant to hold water in rice leaves. H_2O_2 is an essential signaling molecule involved in the regulation of stomatal movement (Wang and Song, 2008; Yao et al., 2013). Mutation of *DST* results in the down regulation of *peroxidase 24 precursor* (*Prx-24*, a scavenger of H_2O_2), might lead to the accumulation of H_2O_2 in guard cells and trigger stomatal closure, and enhances drought and salt tolerance (Huang et al., 2009). *Arabidopsis* mutants lacking either or both a cytosolic and chloroplastic ascorbate peroxidase (APX), which were responsible for H_2O_2 removal, were found to be more tolerant to salinity stress (Miller et al. 2007). Recently,one study showed that overexpression of *Prx-24* (the target gene of DST) enhanced the sensitivity to drought and salt stress in rice (Cui et al., 2015). In our case, *OsMDHAR4* might also be a target regulated by some transcript factors in response to heat stress, more research will be required to address this hypothesis. Liu et al. (2016) found that the *oshtas* mutant (gain-of-function) showed a strongly enhanced tolerance to heat stress with lower water loss rates, more closed stomata and significantly higher H_2O_2 contents after the treatment as compared with the WT. Our data also showed that the *osmdhar4* mutant had more closed stomata and significantly reduced stomatal conductance after heat treatment. Meanwhile, the *osmdhar4* mutant accumulated more H_2O_2 as compared with the WT before or after heat treatment. All of the above was found to be opposite in the *OsMDHAR4* overexpression line (*OsMDHAR4*-OE-1, Figs. 5, 6). These findings suggested that *OsMDHAR4* might play a negative regulation role in H_2O_2-induced stomatal closure.

Conclusions

Our results demonstrated that *OsMDHAR4* negatively regulates tolerance to heat stress in rice by promoting H_2O_2-induced stomatal closure. This study increases our insights into the molecular mechanisms of rice responses to heat stress and may ultimately be helpful for enhancing heat-tolerance via genetic engineering in crop breeding programs.

Methods

Plant materials and stress treatments

Overexpression transgenic plants and the *osmdhar4* mutant were all based on the *Oryza sativa* L. *ssp.* Japonica Zhonghua 11 rice background.

To check the expression levels of the *OsMDHAR4* gene under various abiotic stresses or phytohormone treatment, Zhonghua 11 rice plants were grown in Yoshida solution for approximately 3 weeks under normal conditions. The seedlings at the 3.5- to 4.5-leaf stage were treated with abiotic stresses, including heat stress (exposing plants to 45 °C), simulated drought stress (treated with 20% [*w*/*v*] PEG 6000), ABA treatment (100 μM ABA), oxidative stress (treated with 100 mM H_2O_2), salt stress (treated with 250 mM NaCl), and cold stress (seedlings were transferred to a growth chamber at 4 °C), followed by sampling at the designated time points (0 h, 0.5 h, 3 h, 6 h, 12 h and 24 h). We prepared each RNA sample with shoots from at least four seedlings and ensured that every sampled seedling was of the same growth stage.

Before treatment, *OsMDHAR4*-overexpressing transgenic plants were firstly selected by germinated seeds on 1/2 MS medium containing 50 mg L^{-1} hygromycin, WT and the *osmdhar4* mutant were grown on normal 1/2 MS medium 1 d later. For heat stress testing, the uniformly germinated seeds were sown in the same barrels filled with a well-mixed soil, which was then cultured in a growth chamber with a 13-h-light (28 °C)/11-h-dark (25 °C) photoperiod, 70% humidity. The 5.5- to 6.5-leaf stage seedlings were transferred to a growth chamber with a 13-h-light (45 °C)/11-h-dark (45 °C) photoperiod, 70% humidity for 48 h or 72 h. The treated seedlings were transferred back into the previous growth conditions for recovery, and seedlings with newly growing leaf blades were then counted as surviving plants.

Plasmid construction and Rice transformation

To generate the overexpression construct, the full-length CDS of *OsMDHAR4* was amplified from the cDNA of Zhonghua 11 by PCR. The PCR fragment was inserted into vector pEXT06/g (target gene driven by 35S promoter). The resulting construct was then confirmed by sequencing and transformed into Zhonghua 11, a japonica rice that can be easily transformed, by the *Agrobacterium tumefaciens*-mediated co-cultivation method. Transformed calli were selected on hygromycin medium. The primers used in this work are listed in Additional file 6: Table S1.

Subcellular localization of OsMDHAR4

Coding sequence of the *OsMDHAR4* gene was amplified by PCR and directionally inserted into pCAMBIA1300-35S:YFP-NOS vector. The cultures of the *Agrobacterium tumefaciens* strain EHA105 harboring empty and recombinant vector were used to infect the healthy leaves of *Nicotiana benthamiana* (4 weeks old). The fluorescence signals were observed using a confocal microscope (Leica TCS SP5 or Zeiss LSM710). The primers used in this work are listed in Additional file 6: Table S1.

RNA extraction and qRT-PCR analysis

The TRIZol reagent (Invitrogen, USA) was used according to the manufacturer's instructions to extract total RNA. Before reverse transcription, total RNA was treated with gDNA Eraser (TaKaRa, Japan) for 5 min at 42 °C to degrade possibly contaminated residual genomic DNA. The cDNA templates were synthesized using PrimeScript™ reagent Kit (TaKaRa, Japan) according to the manufacturer's instructions. Quantitative real-time PCR was performed on an optical 96-well plate with a CFX96 Real-time PCR Detection System (Bio-Rad, USA) using SYBR Premix Ex Taq (TaKaRa, Japan). The PCR thermal cycling protocol was as follows: 95 °C for 10 s, followed by 40 cycles at 95 ° C for 5 s and 60 °C for 30 s. Gene-specific primers for *OsMDHAR4* were designed using the Roche Web site (Roche Applied Science). The rice *Actin1* was used as the internal reference (Zhang et al., 2012), and data analyses with the 2^{-ddCt} method were performed as described (Livak and Schmittgen, 2001). The primers used in this work are listed in Additional file 6: Table S1.

Measurement of H_2O_2 contents

Leaves harvested from 3.5- to 4.5-leaf stage seedlings, with or without heat treatment, were used to measure H_2O_2 contents. The contents were measured spectrophotometrically after reaction with potassium iodide (KI). The reaction mixture consisted of 0.5 ml of 0.1% trichloroacetic acid (TCA), leaf extract supernatant, 0.5 ml of 100 mM potassium phosphate buffer (pH 7.8) and 1 ml reagent (1 M KI, *w*/*v* in fresh double-distilled water). The blank control consisted of 1 ml 0.1% TCA and 1 ml KI in the absence of leaf extract. After 1 h of reaction in darkness, the absorbance was measured at 390 nm. The amount of H_2O_2 was calculated using a standard curve prepared with known concentrations of H_2O_2.

DAB staining for H_2O_2 in leaves and detection of H_2O_2 production in guard cells

3,3′-Diaminobenzidine (DAB) staining was performed following a published method with some modifications (Liu et al., 2014). Rice seedlings at 5.5- to 6.5-leaf stage were treated with a 45 °C high temperature for 24 h or not treated, then the leaves (3–5 mm in width) at the same position were detached and immersed in 1% 3,3′-diaminobenzidine (DAB) solution in HCl-acidifed (pH 3.8). After 30 min under vacuum the samples were incubated at room temperature for 24 h in the dark. The samples were then bleached by boiling in ethanol in order to remove the chlorophyll and reveal the brown spots, which are indicative of the reaction of DAB with H_2O_2. The samples were observed and imaged under a scanner.

H_2O_2 production in guard cells was detected using 2′,7′-dichlorodihydrofluorescein diacetate (H2DCFDA; Molecular Probes) as described previously (Huang et al., 2009) with small modifications. Leaves of 5.5- to 6.5-leaf stage seedlings were immerged in 0.01% Tween-20 and vacuum-infiltrated for 10 min. After rinsing twice with distilled water, leaves were incubated in 3% (*w*/*v*) cellulase RS (Yakult Honsha, Japan) for 5 h at 40 °C without shaking to facilitate peeling off the epidermal strips. The epidermal strips were peeled off from leaves using tissue forceps. After washing with loading buffer (10 mM Tris-HCl, 50 mM KCl at pH 7.2), the epidermal strips were incubated in staining buffer (loading buffer containing 50 mM H2DCFDA) for 10 min at room temperature in the dark. The epidermal strips were washed with distilled water to remove the excess dye. The fluorescence was examined using a confocal laser-scanning microscope (Leica TCS SP5, Germany). All confocal images were taken under identical conditions and the guard cell region was selected to quantify the mean grey value of guard cells.

Observation of guard cells and detection of water loss rates

Leaves of 3.5- to 4.5-leaf stage plants with or without heat treatment for 24 h were fixed with 2.5% (*v*/*v*) glutaraldehyde, and stomatal images (20 kV, 2000×) were obtained using scanning electron microscopy (KYKY-EM3200).

Plants germinated under normal conditions for 4 weeks. The leaves were detached from various lines with same age and position, and weighed immediately as the initial fresh weight. They were then placed in clean filter papers, and incubated at 25 °C. The decreases in fresh weight were recorded at every 30 min for 6.5 h or 7 h. Water loss was presented as percentage of fresh weight loss versus the initial fresh weight.

Measurement of the AsA and DHA contents

Leaves harvested from 3.5- to 4.5-leaf stage seedlings, with or without heat treatment, were used to measure the total AsA (tAsA), AsA and DHA contents determined by the spectrophotometric method described previously (Gillespie and Ainsworth, 2007), with minor modifications. The frozen leaf samples were ground with inert sand and 10% trichloroacetic acid (TCA) solution using a mortar and pestle. The homogenate was centrifuged at 12,000 rpm for 20 min. The tAsA contents were determined in a reaction mixture consisting of crude extract and 150 mM KH_2PO_4 buffer (pH 7.4) containing 5 mM EDTA and 10 mM dithiothreitol (DTT) for the reduction of DHA to AsA. The reaction mixtures were incubated at room temperature for 10 min and 0.5% N-ethylmaleimide (NEM) was added. AsA was assayed in a similar manner, except that deionized H_2O was substituted for DTT and NEM. The color was developed in both the reaction mixtures by the addition of 10% TCA, 44% *o*-phosphoric acid, α,ά-dipyridyl in 70% ethanol, and 30% $FeCl_3$. The reaction mixtures were incubated at 37 °C for 1 h and quantified spectrophotometrically at 525 nm.

Additional files

Additional file 1: Figure S1. Alignment of OsMDHAR4 protein with monodehydroascorbate reductase proteins from *Zea mays* (ZmMDAR), *Brachypodium distachyon* (BdMDAR), *Triticum aestivum* (TaMDHAR4, TaMDAR6), *Arabidopsis thaliana* (AtMDAR4, AtMDAR6), *Glycine max* (GmMDAR), and *Solanum lycopersicum* (SlMDAR). Red line: FAD/NAD-binding domain, blue line: Pyr_redox_2 (pyridine nucleotide-disulfide oxidoreductase 2) domain, and green line: FAD/NAD-linked reductase, domain. Dark blue shadow: amino acids are conserved among all 9 sequences, pink shadow: amino acids are conserved among 7 or 8 sequences, light blue shadow: amino acids are conserved among 5 or 6 sequences Sequence alignment was performed using the DNA-MAN6 software.

Additional file 2: Figure S2. Genotyping of *osmdhar4* T-DNA insertion mutant. F1, R1 and L4 stand for the primers. W, wild type. M, homozygous mutant. H, heterozygous mutant.

Additional file 3: Figure S3. a *OsMDHAR4*-overexpressing construct. **b** PCR identification of transgenic plants from two overexpression lines (*OsMDHAR4*-OE-1 and OE-5). Positive transgenic plants have a PCR band about 1471 bp or 675 bp when using different primers combinations (*OsMDHAR4*-OE-F + R or Hyg-F + R). P, positive control. N, negative control.

Additional file 4: Figure S4. a Detection of H_2O_2 production in guard cells of WT and *osmdhar4* mutant plants with H2DCFDA. Scale bars = 10 μm. **b** Quantitative analysis of H_2O_2 production in guard cells of WT and *osmdhar4* mutant (3 repeats, 12 stomata in each repeat). Error bars indicate the SE based on three biological replicates. *, $P < 0.05$, by Student's *t*-test.

Additional file 5: Figure S5. a Detection of the AsA content of the WT, the *osmdhar4* mutant and *OsMDHAR4*-OE-1 plants under normal or heat stress conditions. **b** Detection of the DHA content of the WT, the *osmdhar4* mutant and *OsMDHAR4*-OE-1 plants under normal or heat stress conditions. **c** AsA/DHA ratio of WT, the *osmdhar4* mutant and *OsMDHAR4*-OE-1 plants under normal or heat stress conditions. Error bars indicate the SE based on three biological replicates. *, $P < 0.05$, **, $P < 0.01$, by Student's *t*-test. ns, no significant. FW, Fresh Weight.

Additional file 6: Table S1 List of primers used in this study (F, forward primer; R, reverse primer; q, quantitative RT-PCR).

Abbreviations

ABA: Abscisic acid; APX: Ascorbate peroxidase; AsA: Ascorbic acid; CAT: Catalase; CDS: Coding sequence; DAB: 3,3′-Diaminobenzidine; DHA: Dehydroascorbate; FAD: Flavin adeninedinucleotide; GPX: Glutathione peroxidase; GSH: Glutathione; H2DCFDA: 2′,7′-dichlorodihydrofluorescein diacetate; H_2O_2: Hydrogen peroxide; Hyg: Hygromycin; MDA: Monodehydroascorbate; MDAR or MDHAR: Monodehydroascorbate reductase; OE: Overexpression; PEG: Polyethylene glycol; qRT-PCR: Quantitative real time polymerase chain reaction; ROS: Reactive oxygen species; SOD: Superoxide dismutase; TOCs: Tocopherols; YFP: Yellow fluorescent protein; ZH11: Zhonghua 11

Acknowledgements

This work was supported by the National Natural Science Foundation of China (Grant number: 31601232, 31761130073), the Natural Science Foundation of Fujian Province (Grant number: 2017 J05046), China Postdoctoral Science Foundation (Grant number: 2017 M612108) and Research Grant of Fujian Agriculture and Forestry University (Grant number: KXGH17005). The authors would like to thank Prof. Jumin Tu (Zhejiang University) for providing the plasmid for subcellular localization.

Funding

This study was supported by the National Natural Science Foundation of China (Grant number: 31601232, 31761130073), the Natural Science Foundation of Fujian Province (Grant number: 2017 J05046), China Postdoctoral Science Foundation (Grant number: 2017 M612108) and Research Grant of Fujian Agriculture and Forestry University (Grant number: KXGH17005).

Authors' contributions

LJP, XWF and ZJH contributed to the experimental design. LJP, SXJ, XFY, ZYJ, ZQ and MR contributed to experiment performance and data analysis. LJP, XWF and ZJH drafted the manuscript. LJS contributed to good advice on designing the experiment. All authors read and approved the final manuscript.

Competing interests

The authors declare that they have no competing interests.

Author details

[1]Center for Plant Water-use and Nutrition Regulation and College of Life Sciences, Joint International Research Laboratory of Water and Nutrient in Crop, Fujian Agriculture and Forestry University, Jinshan Fuzhou 350002, China. [2]Department of Biology, Hong Kong Baptist University, Hong Kong, China. [3]Department of Biology, Southern University of Science and Technology, Shenzhen 518055, China.

References

Abou-Attia MA, Wang X, Nashaat Al-Attala M, Xu Q, Zhan G, Kang Z (2016) *TaMDAR6* acts as a negative regulator of plant cell death and participates indirectly in stomatal regulation during the wheat stripe rust-fungus interaction. Physiol Plant 156:262–277

Apel K, Hirt H (2004) Reactive oxygen species: metabolism, oxidative stress, and signal transduction. Annu Rev Plant Biol 55:373–399

Bright J, Desikan R, Hancock JT, Weir IS, Neill SJ (2006) ABA-induced NO generation and stomatal closure in *Arabidopsis* are dependent on H_2O_2 synthesis. Plant J 45:113–122

Chen Z, Young TE, Ling J, Chang SC, Gallie DR (2003) Increasing vitamin C content of plants through enhanced ascorbate recycling. Proc Natl Acad Sci 100:3525–3530

Cui LG, Shan JX, Shi M, Gao JP, Lin HX (2015) DCA1 acts as a transcriptional co-activator of DST and contributes to drought and salt tolerance in Rice. PLoS Genet 11:e1005617

Eltayeb AE, Kawano N, Badawi GH, Kaminaka H, Sanekata T, Morishima I, Shibahara T, Inanaga S, Tanaka K (2006) Enhanced tolerance to ozone and drought stresses in transgenic tobacco overexpressing dehydroascorbate reductase in cytosol. Physiol Plant 127:57–65

Eltayeb AE, Kawano N, Badawi GH, Kaminaka H, Sanekata T, Shibahara T, Inanaga S, Tanaka K (2007) Overexpression of monodehydroascorbate reductase in transgenic tobacco confers enhanced tolerance to ozone. salt and polyethylene glycol stresses Planta 225:1255–1264

Eltelib HA, Badejo AA, Fujikawa Y, Esaka M (2011) Gene expression of monodehydroascorbate reductase and dehydroascorbate reductase during fruit ripening and in response to environmental stresses in acerola (*Malpighia glabra*). J Plant Physiol 168:619–627

Feng H, Liu W, Zhang Q, Wang X, Wang X, Duan X, Li F, Huang L, Kang Z (2014) *TaMDHAR4*, a monodehydroascorbate reductase gene participates in the interactions between wheat and *Puccinia striiformis* f. Sp. *tritici*. Plant Physiol Biochem 76:7–16

Foreman J, Demidchik V, Bothwell JH, Mylona P, Miedema H, Torres MA, Linstead P, Costa S, Brownlee C, Jones JD, Davies JM, Dolan L (2003) Reactive oxygen species produced by NADPH oxidase regulate plant cell growth. Nature 422:442–446

Gill SS, Tuteja N (2010) Reactive oxygen species and antioxidant machinery in abiotic stress tolerance in crop plants. Plant Physiol Biochem 48:909–930

Gillespie KM, Ainsworth EA (2007) Measurement of reduced, oxidized and total ascorbate content in plants. Nat Protoc 2:871–874

Grantz AA, Brummell DA, Bennett AB (1995) Ascorbate free radical reductase mRNA levels are induced by wounding. Plant Physiol 108: 411–418

Hetherington AM, Woodward FI (2003) The role of stomata in sensing and driving environmental change. Nature 424:901–908

Huang XY, Chao DY, Gao JP, Zhu MZ, Shi M, Lin HX (2009) A previously unknown zinc finger protein, DST, regulates drought and salt tolerance in rice via stomatal aperture control. Genes Dev 23:1805–1817

Ishikawa T, Shigeoka S (2008) Recent advances in ascorbate biosynthesis and the physiological significance of ascorbate peroxidase in photosynthesizing organisms. Biosci Biotechnol Biochem 72:1143–1154

Kim IS, Kim YS, Kim YH, Park AK, Kim HW, Lee JH, Yoon HS (2016) Potential application of the Oryza sativa Monodehydroascorbate reductase gene (*OsMDHAR*) to improve the stress tolerance and fermentative capacity of Saccharomyces cerevisiae. PLoS One 11:e0158841

Kim JJ, Kim YS, Park SI, Mok JE, Kim YH, Park HM, Kim IS, Yoon HS (2017) Cytosolic monodehydroascorbate reductase gene affects stress adaptation and grain yield under paddy field conditions in *Oryza sativa* L. *japonica*. Mol Breed 37

Leterrier M, Corpas FJ, Barroso JB, Sandalio LM, del Rio LA (2005) Peroxisomal monodehydroascorbate reductase. Genomic clone characterization and functional analysis under environmental stress conditions. Plant Physiol 138:2111–2123

Levine A, Tenhaken R, Dixon R, Lamb C (1994) H_2O_2 from the oxidative burst orchestrates the plant hypersensitive disease resistance response. Cell 79:583–593

Li F, Wu QY, Sun YL, Wang LY, Yang XH, Meng QW (2010) Overexpression of chloroplastic monodehydroascorbate reductase enhanced tolerance to temperature and methyl viologen-mediated oxidative stresses. Physiol Plant 139:421–434

Liu H, Ma Y, Chen N, Guo S, Liu H, Guo X, Chong K, Xu Y (2014) Overexpression of stress-inducible OsBURP16, the beta subunit of polygalacturonase 1, decreases pectin content and cell adhesion and increases abiotic stress sensitivity in rice. Plant Cell Environ 37:1144–1158

Liu J, Zhang C, Wei C, Liu X, Wang M, Yu F, Xie Q, Tu J (2016) The RING finger ubiquitin E3 ligase OsHTAS enhances heat tolerance by promoting H_2O_2-induced stomatal closure in Rice. Plant Physiol 170:429–443

Liu X, Huang B (2000) Heat stress injury in relation to membrane lipid peroxidation in creeping Bentgrass. Crop Sci 40:503–510

Livak KJ, Schmittgen TD (2001) Analysis of relative gene expression data using real-time quantitative PCR and the $2^{-\Delta\Delta CT}$ method. Methods 25:402–408

Maruta T, Noshi M, Tanouchi A, Tamoi M, Yabuta Y, Yoshimura K, Ishikawa T, Shigeoka S (2012) H_2O_2-triggered retrograde signaling from chloroplasts to nucleus plays specific role in response to stress. J Biol Chem 287:11717–11729

McAinsh MR, Clayton H, Mansfield TA, Hetherington AM (1996) Changes in stomatal behavior and guard cell cytosolic free calcium in response to oxidative stress. Plant Physiol 111:1031–1042

Miller G, Suzuki N, Rizhsky L, Hegie A, Koussevitzky S, Mittler R (2007) Double mutants deficient in cytosolic and thylakoid ascorbate peroxidase reveal a complex mode of interaction between reactive oxygen species, plant development. and response to abiotic stresses Plant Physiol 144:1777–1785

Mittler R (2002) Oxidative stress. antioxidants and stress tolerance Trends Plant Sci 7:405–410

Mittler R, Vanderauwera S, Gollery M, Van Breusegem F (2004) Reactive oxygen gene network of plants. Trends Plant Sci 9:490–498

Noctor G, Foyer CH (1998) ASCORBATE AND GLUTATHIONE: keeping active oxygen under control. Annu Rev Plant Physiol Plant Mol Biol 49:249–279

Omoto E, Nagao H, Taniguchi M, Miyake H (2013) Localization of reactive oxygen species and change of antioxidant capacities in mesophyll and bundle sheath chloroplasts of maize under salinity. Physiol Plant 149:1–12

Pei ZM, Murata Y, Benning G, Thomine S, Klusener B, Allen GJ, Grill E, Schroeder JI (2000) Calcium channels activated by hydrogen peroxide mediate abscisic acid signalling in guard cells. Nature 406:731–734

Peng C, Ou Z, Liu N, Lin G (2005) Response to high temperature in flag leaves of super high-yielding rice Pei'ai 64S/E32 and Liangyoupeijiu. Rice Sci 12:179–186

Quan LJ, Zhang B, Shi WW, Li HY (2008) Hydrogen peroxide in plants: a versatile molecule of the reactive oxygen species network. J Integr Plant Biol 50:2–18

Sakihama Y, Mano J, Sano S, Asada K, Yamasaki H (2000) Reduction of phenoxyl radicals mediated by monodehydroascorbate reductase. Biochem Biophys Res Commun 279:949–954

Sanmartin M, Drogoudi PD, Lyons T, Pateraki I, Barnes J, Kanellis AK (2003) Over-expression of ascorbate oxidase in the apoplast of transgenic tobacco results in altered ascorbate and glutathione redox states and increased sensitivity to ozone. Planta 216:918–928

Sano S, Tao S, Endo Y, Inaba T, Hossain MA, Miyake C, Matsuo M, Aoki H, Asada K, Saito K (2005) Purification and cDNA cloning of chloroplastic monodehydroascorbate reductase from spinach. Biosci Biotechnol Biochem 69:762–772

Schroeder JI, Allen GJ, Hugouvieux V, Kwak JM, Waner D (2001) Guard cell signal transduction. Annu Rev Plant Biol 52:627–658

Sharma P, Dubey RS (2005) Modulation of nitrate reductase activity in rice seedlings under aluminium toxicity and water stress: role of osmolytes as enzyme protectant. J Plant Physiol 162:854–864

Sharma YK, Davis KR (1997) The effects of ozone on antioxidant responses in plants. Free Radic Biol Med 23:480–488

Sirichandra C, Gu D, Hu H-C, Davanture M, Lee S, Djaoui M, Valot B, Zivy M, Leung J, Merlot S (2009) Phosphorylation of the *Arabidopsis* AtrbohF NADPH oxidase by OST1 protein kinase. FEBS Lett 583:2982–2986

Song Y, Miao Y, Song CP (2014) Behind the scenes: the roles of reactive oxygen species in guard cells. New Phytol 201:1121–1140

Sultana S, Khew CY, Morshed MM, Namasivayam P, Napis S, Ho CL (2012) Overexpression of monodehydroascorbate reductase from a mangrove plant (AeMDHAR) confers salt tolerance on rice. J Plant Physiol 169:311–318

Wahid A, Gelani S, Ashraf M, Foolad MR (2007) Heat tolerance in plants: an overview. Environ Exp Bot 61:199–223

Wang P, Song CP (2008) Guard-cell signalling for hydrogen peroxide and abscisic acid. New Phytol 178:703–718

Wu C, Li X, Yuan W, Chen G, Kilian A, Li J, Xu C, Li X, Zhou DX, Wang S, Zhang Q (2003) Development of enhancer trap lines for functional analysis of the rice genome. Plant J 35:418–427

Yamamoto A, Bhuiyan MN, Waditee R, Tanaka Y, Esaka M, Oba K, Jagendorf AT, Takabe T (2005) Suppressed expression of the apoplastic ascorbate oxidase gene increases salt tolerance in tobacco and Arabidopsis plants. J Exp Bot 56:1785–1796

Yao Y, Liu X, Li Z, Ma X, Rennenberg H, Wang X, Li H (2013) Drought-induced H_2O_2 accumulation in subsidiary cells is involved in regulatory signaling of stomatal closure in maize leaves. Planta 238:217–227

Yokotani N, Ichikawa T, Kondou Y, Matsui M, Hirochika H, Iwabuchi M, Oda K (2008) Expression of rice heat stress transcription factor OsHsfA2e enhances tolerance to environmental stresses in transgenic *Arabidopsis*. Planta 227:957–967

You J, Zong W, Li X, Ning J, Hu H, Li X, Xiao J, Xiong L (2013) The SNAC1-targeted gene *OsSRO1c* modulates stomatal closure and oxidative stress tolerance by regulating hydrogen peroxide in rice. J Exp Bot 64:569–583

Zhang CC, Yuan WY, Zhang QF (2012) *RPL1*, a gene involved in epigenetic processes regulates phenotypic plasticity in rice. Mol Plant 5:482–493

Zhang J, Li C, Wu C, Xiong L, Chen G, Zhang Q, Wang S (2006) RMD: a rice mutant database for functional analysis of the rice genome. Nucleic Acids Res 34:D745–D748

11

OsRACK1A, encodes a circadian clock-regulated WD40 protein, negatively affect salt tolerance in rice

Dongping Zhang[1,4†], Yuzhu Wang[1†], Jinyu Shen[1], Jianfeng Yin[1], Dahong Li[3], Yan Gao[1], Weifeng Xu[2*] and Jiansheng Liang[4*]

Abstract

The receptor for activated C kinase 1 (RACK1) is a WD40 type protein that is involved in multiple signaling pathways and is conserved from prokaryotes to eukaryotes. Here we report that rice *RACK1A* (*OsRACK1A*) is regulated by circadian clocks and plays an important role in the salt stress response. *OsRACK1A* was found to follow a rhythmic expression profile under circadian conditions at both the transcription and the translation levels, although the expression was arrhythmic under salt stress. Analysis of plant survival rates, fresh weight, proline content, malondialdehyde, and chlorophyll showed that suppression of *OsRACK1A* enhanced tolerance to salt stress. The ion concentration in both roots and leaves revealed that *OsRACK1A*-suppressed transgenic rice could maintain low Na^+ and high K^+ concentrations. Furthermore, *OsRACK1A*-suppressed transgenic rice accumulated significantly more abscisic acid (ABA) and more transcripts of ABA- and stress-inducible genes compared with the wild-type plants. Real-time quantitative polymerase chain reaction analysis revealed that many stress-related genes, including APETALA 2/Ethylene Responsive Factor (AP2/ERF) transcription factors, were upregulated in the *OsRACK1A*-suppressed transgenic rice line. We identified putative interactors of OsRACK1A, and found that OsRACK1A interacted with many salt stress-responsive proteins directly. These results suggest that *OsRACK1A* is regulated by circadian rhythm, and involved in the regulation of salt stress responses.

Keywords: OsRACK1A, Salt tolerance, Circadian, Rice

Background

The receptor for activated C kinase 1 (RACK1) is a member of the WD repeat-containing scaffold proteins and is conserved from prokaryotes to eukaryotes (Zhang et al., 2013). As a scaffolding protein, RACK1 protein interacts with many proteins and is involved in multiple signaling pathways (McCahill et al., 2002; Zhang et al., 2013). In plants, *RACK1* is involved in diverse biological processes, such as seed germination, organ development, hormones and stress responses (Nakashima et al., 2008; Guo et al., 2009, 2011; Zhang et al., 2014). Compared with the advances made from studies in metazoans and yeast, little is known about the molecular mechanisms of *RACK1* in plants.

The *Arabidopsis* genome contains three *RACK1* orthologues, *RACK1A*, *RACK1B* and *RACK1C*, which are ~ 78% similar to mammalian *RACK1* (Guo and Chen, 2008). Using loss-of-function mutants of *RACK1A* in *Arabidopsis*, Chen et al.(2006) found that *AtRACK1A* plays a role in several plant hormonal responses, including abscisic acid (ABA), gibberellin (GA), indole-3-acetic acid (IAA), and brassinosteroid (BR). There is direct and indirect evidence that *RACK1s* are involved in the regulation of plant tolerance to abiotic and biotic stresses (Kundu et al., 2013; Cheng et al., 2015). In *Arabidopsis*, the *rack1a* mutant strongly tolerates soil drying, compared with the wild-type (Zhang et al., 2013). Moreover,

* Correspondence: wfxu@fafu.edu.cn; liangjs@sustc.edu.cn
†Dongping Zhang and Yuzhu Wang contributed equally to this work.
[2]College of Life Sciences, Fujian Agriculture and Forestry University, Jinshan, Fuzhou 350002, China
[4]Department of Biology, Southern University of Science and Technology, Shenzhen 518055, China
Full list of author information is available at the end of the article

water loss in detached leaves and stomatal conductance of *rack1* mutants were significantly lower than in the wild-type, and the endogenous ABA content of *rack1a* mutants was higher than in the wild-type (Guo et al., 2009; Zhang et al., 2013). In addition, *rack1a* mutants were hypersensitive to ABA in serval developmental processes, such as seed germination, cotyledon greening, and root growth, and some ABA-responsive marker genes were upregulated in *rack1a* mutants, while the *RACK1* genes were downregulated by ABA (Guo et al., 2009). These results suggest that *RACK1* functions as a negative regulator of ABA signaling and consequently enhances drought stress tolerance via ABA-dependent signaling in response to water stress in plants. Comparative proteomic analysis showed that the *Arabidopsis* RACK1C protein might play roles in regulating plant resistance to salt stress (Shi et al., 2011).

The rice genome contains two *RACK1* homologous genes that are ~ 80% similar to *Arabidopsis* RACK1 proteins at the amino acid level: *OsRACK1A* and *OsRACK1B* (Nakashima et al., 2008). Li et al. (2009) found that *OsRACK1A*-suppressed transgenic rice lines were more tolerant of soil drying, but the molecular mechanism remains unknown. Comparative phosphoproteomics studies revealed that the OsRACK1A protein is phosphorylated in response to exogenous ABA and drought treatment (He et al., 2008; Ke et al., 2009). These findings suggested that *OsRACK1A* plays essential roles in ABA signaling and is involved in ABA-dependent stress responses. In addition to the involvement of *RACK1* in the regulation of plant responses to abiotic stresses, it has been reported to function in plant innate immunity. Overexpression of *OsRACK1A* enhanced the production of reactive oxygen species (ROS) and increased resistance to blast fungus in rice (Nakashima et al., 2008). *OsRACK1A* regulated ROS levels not only in abiotic stress responses but also in the seed germination process. Previously, we found that *OsRACK1A* positively regulated seed germination by promoting H_2O_2 production and enhancing ABA catabolism (Zhang et al., 2014). Although *RACK1* functions in ABA signaling in both rice and *Arabidopsis*, it is still unclear whether *RACK1* is involved directly in ABA-dependent stress responses.

Circadian clocks are 24-h biological oscillators, which generally enable organisms to coordinate their activities with the external light/dark cycles by anticipating the onset of dawn or dusk. In mammals, RACK1 protein plays a crucial role in circadian clocks by interacting with BMAL1, a component of the heterodimeric CLOCK:BMAL1 circadian complex. However, the expression of *RACK1* itself showed little or no circadian variation across the circadian cycle (Robles et al., 2010). In plants, no clock component has been reported to interact with RACK1 protein and whether plant *RACK1* is involved directly in circadian clock regulation has yet to be investigated. In this study, our results indicated that *OsRACK1A* is a circadian rhythm gene and is involved in the response to salt stress. *OsRACK1A*-suppressed transgenic plants were hyposensitive to salt stress, compared with wild-type Nipponbare. OsRACK1A plays an important role in the tolerance to high salinity by regulating many stress-related genes and interacting directly with many stress-response proteins.

Methods

Plant materials and stress treatment

Rice (*Oryza sativa* L. cv. Nipponbare) was used as the wild-type (non-transgenic line; NTL) and in the generation of all transgenic plants. All transgenic rice lines were generated and kept in our laboratory. An *OsRACK1A* over-expressing transgenic line, OeTL3–8, and an RNA-interfered transgenic line, RiTL4–2, were used as experimental materials (Zhang et al., 2014). For NaCl treatment, 4-week-old hydroponic cultured rice plants were placed in different concentrations of NaCl solution (100, 150, 200 mM) for 10–20 d and finally determined 150 mM NaCl treated with 18 d and recovered for 10 d was the best condition for identifying stress phenotypes. All the plants grew in a plant growth chamber (Conviron atc26, 16 h light/ 8 h dark, 30 °C day/ 22 °C night).

Measurements of physiological index

For the tolerance experiments, all rice plants were cultured in a plant growth chamber (Conviron atc26) (30 ° C day / 22 °C night). The survival rate and fresh weight were calculated after 18 d of treatment with 150 mM NaCl and recovery in normal conditions for another 10 d. Lipid peroxidation was determined by measuring the MDA content (Dhindsa and Matowe, 1981). The content of free proline in leaves was determined as described previously. (Bates et al., 1973) Chlorophyll was extracted from the leaves in 10 mL of 80% acetone for 16 h in the dark and was determined by measuring the absorbance at 652 nm (Arnon, 1949). To measure the Na^+ and K^+ concentrations, 2-week-old hydroponic cultured rice seedlings were supplemented with 150 mM NaCl for 24, 48, or 72 h. Shoots and roots were harvested at the indicated times and all physiological measurements were based on the procedure described by Yang et al. (2015). Measurement of water loss form detached leaves was performed as previously described by Zhang et al. (2015). The detached leaves of non-transgenic and transgenic rice lines were weighed at room temperature (~ 23 °C) with 35% relative humidity. The endogenous ABA levels of rice leaves were measured based on the procedures described by Zhang et al. (2014). All of the data were subjected to Student's t-test analysis using SPSS ver. 13.0 (SPSS Company, Chicago, IL).

Gene expression analysis

The RNAprep Pure Plant Kit (cat. no. DP441; Tiangen Biotech) was used to extract total RNA from rice. Single-strand cDNAs were synthesized by using the HiScript Q RT SuperMix for qPCR kit (cat. no. R123; Vazyme). Transcript-level expression of each gene were measured by quantitative RT-PCR using a 7300 Real-Time PCR system (ABI), with the iTaq universal SYBR Green SuperMix (Bio-Rad), and normalized against the values obtained for housekeeping gene *OsActin1* (LOC_Os03g50890). Three biological replicates were performed for each experiment. Additional file 1: Table S2 lists the qRT-PCR primer sequences. All of the data were subjected to Student's t-test analysis using SPSS ver. 13.0 (SPSS Company, Chicago, IL).

Protein blot analysis

Rice leaves of seedlings were ground in liquid nitrogen and homogenized in PBS buffer (cat. no. CW0040S; CoWin Bioscience) containing complete protease inhibitor cocktail (cat. no. 04693132001; Roche). To prepare total protein, the homogenate was centrifuged (6000×*g*, 30 min, 4 °C) to remove cellular debris. Then, proteins were separated by sodium dodecyl sulfate polyacrylamide gel electrophoresis (SDS-PAGE) on 10% gels and blotted onto polyvinylidene difluoride (PVDF) membranes. The antibodies used were anti-β-actin antibody (cat. no. CW0264M; CoWin Bioscience), anti-green fluorescence protein (GFP) antibody (cat. no. ab290; Abcam), and anti-OsRACK1A antibody (cat. no. AbP80112-A-SE; Beijing Protein Innovation).

Co-IP assay

To identify the OsRACK1A interaction proteins, *UBI::GFP* and *UBI::GFP-OsRACK1A* transgenic rice were used for co-IP assays. Leaves from 4-week-old plants were harvested and ground in liquid nitrogen. Proteins were extracted with the buffer containing 50 mM Tris (pH 7.5), 150 mM NaCl, 0.1% IGEPAL CA-630, Proteinase Inhibitor Cocktail (cat. no. 04693159001; Roche) and Phophatase Inhibitor Cocktail (cat. no. 04906845001). The samples were centrifuged at 12,000 g for 15 min at 4 °C and the supernatant was incubated with anti-GFP magnetic beads (catalog no. D153–11; MBL) to overnight at 4 °C with gentle rotation. The beads were then washed four times with PBS. The immunoprecipitated proteins were eluted with 1 M glycine (pH 3.0). The presence of the corresponding proteins was detected by tandem liquid chromatograph-mass spectrometry (LC-MS/MS).

Statistical analyses

Statistical analyses were performed using IBM SPSS Statistics 21 software (Chicago, IL, USA), and analyzed with one-way ANOVA. $P < 0.05$ was considered significant.

Results

Expression of the *OsRACK1A* gene is controlled in a circadian-clock like manner

Information retrieved from the public microarray database (ArrayExpress, Accession: E-MTAB-275) showed that a circadian rhythm in *OsRACK1A* mRNA abundance occurred under photocycling (12 h light/12 h dark; 12 h hot/12 h hot; LDHH), thermocycling (12 h light/12 h light; 12 h hot/ 12 h cold; LLHC) or photocycling and thermocycling (12 h light/12 h dark; 12 h hot/12 h cold; LDHC) conditions (Additional file 1: Figure S1A). Another rice *RACK1* homolog, *OsRACK1B*, exhibited similar expression patterns (Additional file 1: Figure S1B). To confirm whether the expression of *OsRACK1A* was controlled by a circadian clock, the expression of *OsRACK1A* in a 24-h period was measured. Quantitative RT-PCR analysis showed that the transcript level of *OsRACK1A* started accumulating with the onset of light and reached a maximum level 10 h after the lights were switched on (ZT0 and ZT24) and then the transcript level declined gradually and reached a minimum 6 h after the lights were switched off (ZT16 and ZT40, Fig. 1a). We also examined levels of OsRACK1A protein during the light/dark cycle using western blot analysis and the results revealed that OsRACK1A protein accumulated in the light (ZT0 to ZT14) and declined in the dark (ZT16 to ZT22, Fig. 1b). Moreover, we tested the expression of *OsRACK1A* under the constant light conditions and found that *OsRACK1A* also displayed rhythmic expression (Additional file 1: Figure S1C).

Overexpression of *OsRACK1A* delays the time of heading

Some circadian clock-controlled genes have been reported to be involved in photoperiodic flowering regulation (Xue et al., 2008; Ishikawa et al., 2011; Matsubara et al., 2011). To investigate whether *OsRACK1A* plays a role in photoperiod-controlled heading, we generated several *OsRACK1A* RNA-interference (RNAi) and overexpressing lines (Li et al., 2009). From these transgenic rice lines, we chose the stable downregulated RNA-interfered transgenic line RiTL4–2, and the upregulated overexpressed transgenic line OeTL3–8. Compared with the non-transgenic line (NTL), the OsRACK1A protein level was higher in OeTL3–8 and lower in RiTL4–2, measured by Western blot analysis using an OsRACK1A-specific antibody (Fig. 1c). This

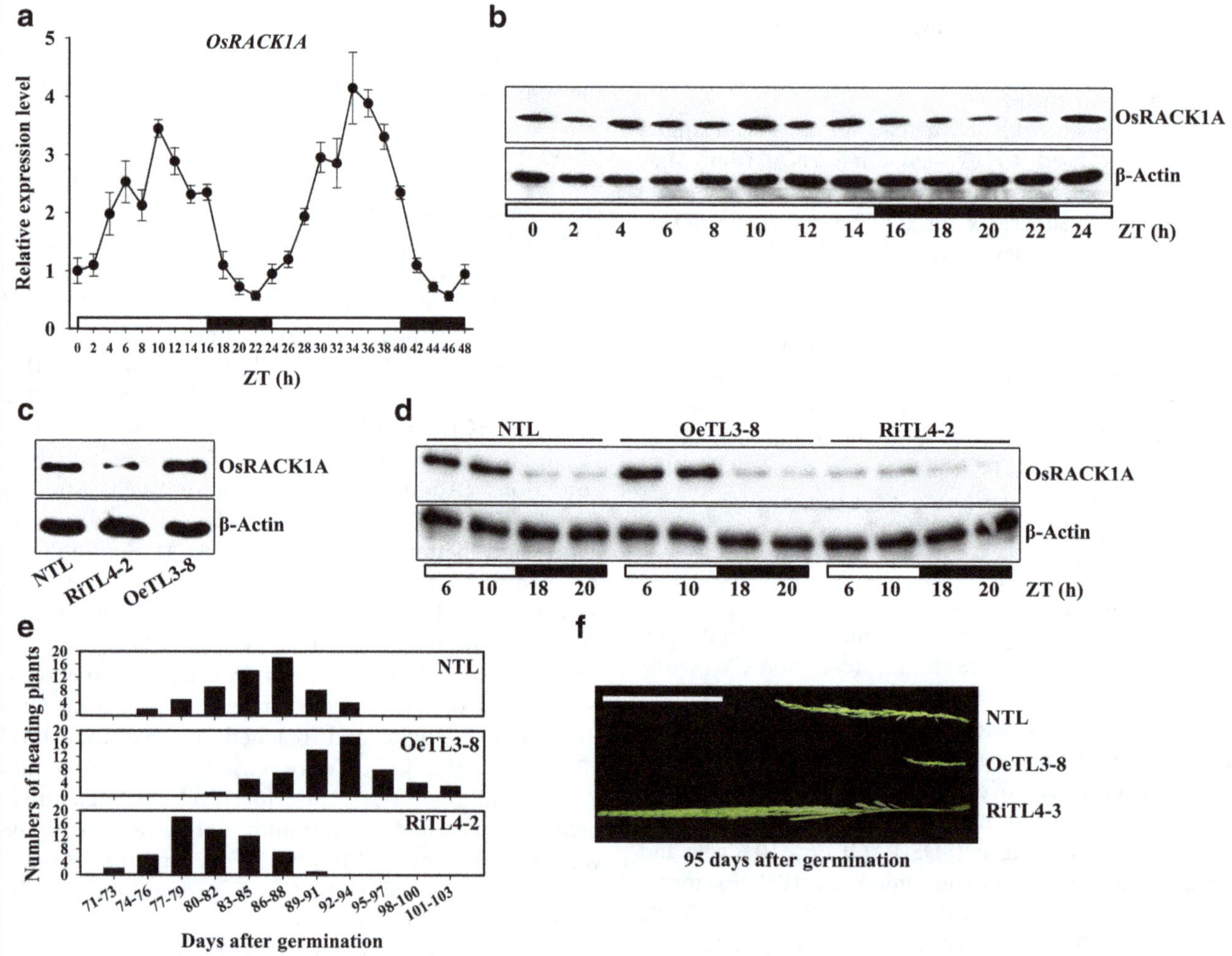

Fig. 1 Circadian clock-controlled *OsRACK1A* expression in rice. **a** Transcript analysis of the *OsRACK1A* gene by qRT-PCR over a 48-h period. Rice plants were grown under 16 h light and 8 h dark in a growth chamber at a constant temperature of 28 °C and sampled at regular intervals. **b** Immunoblot analysis of OsRACK1A protein over a 24-h period. **c** *OsRACK1A* expression was monitored by immunoblot analysis in seedlings of a non-transgenic line (NTL), an *OsRACK1A*-overexpressing transgenic line (OeTL3–8), and an RNA-interfered transgenic line (RiTLs4–2). **d** Protein levels of OsRACK1A in NTL, OeTL3–8 and RiTLs4–2 in light and dark. **e** Heading time distributions in transgenic rice plants grown in the field. During the heading period, the number of heading plants for each line was recorded every 3 days and compared. **f** Phenotypic comparison panicles during the heading period of transgenic plants. The rice plants had been grown in the field for 89 d. ZT, Zeitgeber Time. White and black rectangles indicate lights on and lights off, respectively

differential protein expression patterns of the three genotypes was occurred only in the light, whereas OsRACK1A protein was nearly undetectable in the dark (Fig. 1d). The heading date of field-grown plants was recorded and OeTL3–8 had a heading date nearly 1 week later than NTL, whereas RiTL4–2 showed a heading time approximately 1 week earlier than NTL (Fig. 1e). The panicles phenotypes of field-grown plants at 95 d after germination are shown in Fig. 1f.

NaCl treatment affects expression of the *OsRACK1A* gene

To investigate the expression profile of the *OsRACK1A* gene under salt stress, 2-week-old hydroponic cultured rice seedlings were exposed to 150 mM NaCl for different times and the transcript-level expression of this gene was monitored using quantitative RT-PCR. The *OsRACK1A* expression pattern changed significantly in response to NaCl treatment. The transcript level of *OsRACK1A* accumulated from ZT0 to ZT12 under both salt stress and control conditions, but expression of *OsRACK1A* was slightly higher under control than under salt stress conditions (Fig. 2a). Under normal conditions, the expression of *OsRACK1A* behaved like a circadian clock; when treated with NaCl, however, the transcription level of *OsRACK1A* first increased to a relatively high level and then maintained this level throughout the experiment (Fig. 2a), which means that the circadian clock of *OsRACK1A* expression disappeared when exposed to salt stress. The

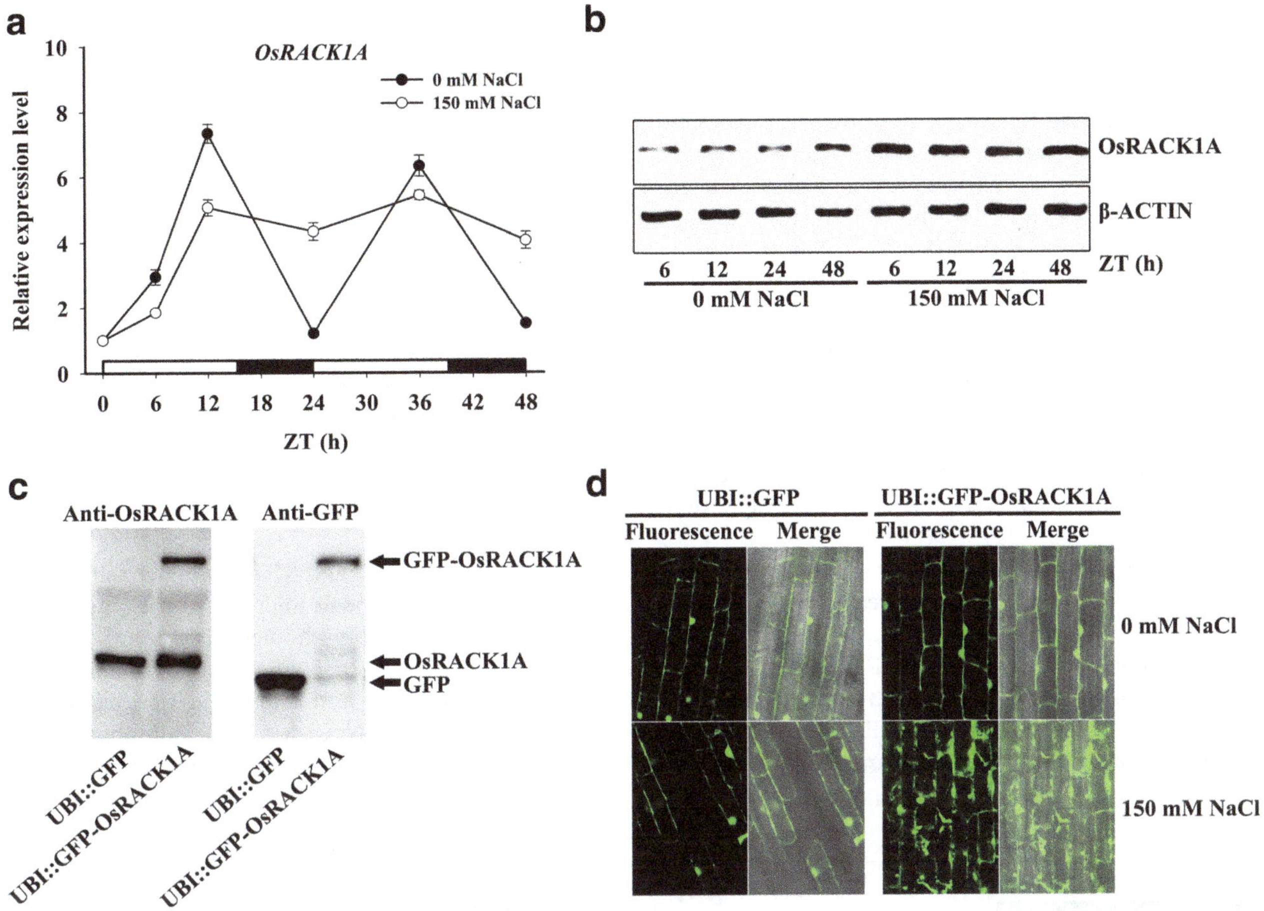

Fig. 2 Salt stress-controlled *OsRACK1A* expression in rice. **a** Transcript analysis of the *OsRACK1A* gene by qRT-PCR for a 48-h period under 0 and 150 mM NaCl treatment. **b** Immunoblot analysis of OsRACK1A protein for a 24-h period under 0 and 150 mM NaCl treatment. **c** GFP, OsRACK1A and a GFP-OsRACK1A fusion protein were monitored by immunoblot analysis in seedlings of *UBI*-promoted *GFP* and *GFP-OsRACK1A* transgenic rice. **d** GFP fluorescence was examined in roots of *UBI*-promoted *GFP* and *GFP-OsRACK1A* transgenic rice under 0 and 150 mM NaCl treatment for 6 h. ZT, Zeitgeber Time. White and black rectangles indicate lights on and lights off, respectively

similar expression pattern of *OsRACK1A* was shown under both light/dark cycle and constant light conditions with NaCl treatment (Additional file 1: Figure S1B). Interestingly, the protein level of OsRACK1A increased and was significantly higher than in the untreated control at 6, 12, 24, and 48 h after the onset of salt stress (Fig. 2b), whereas the transcript levels of *OsRACK1A* were lower at ZT6 and ZT12 in the NaCl treatment than under control conditions (Fig. 2a).

Therefore, we suspected that OsRACK1A protein levels were under post transcriptional and/or translational control. To test this, transgenic rice plants were generated that constitutively expressed GFP-OsRACK1A. As shown in Fig. 2c, the GFP-OsRACK1A fusion protein was detected in ubiquitin-promoted *GFP-OsRACK1A* transgenic plants, whereas the GFP protein was detected in *UBI::GFP* transgenic plants. Figure 2d shows fluorescence images of *GFP-OsRACK1A* transgenic plants in the presence or absence of 150 mM NaCl for 6 h. Control plants containing GFP alone showed no change in subcellular localization in response to salt stress (Fig. 2d). Furthermore, before salt treatment, plants containing GFP-OsRACK1A exhibited fluorescence that was detectable in the cytosolic fraction, as well as in the plasma membrane and nuclei. After treatment with 150 mM NaCl, GFP fluorescence from the GFP-OsRACK1A fusion was enhanced and appeared diffusely in the cytosol (Fig. 2d). These results supported the premise that OsRACK1A protein was controlled by post-transcriptional and/or translational regulation and accumulated under salt stress.

OsRACK1A negatively regulates salt tolerance

Because both mRNA and protein levels of *OsRACK1A* were induced by the high-salinity treatment, we used the *OsRACK1A*-overexpressing (OeTL3–8) and RNAi (RiTL4–2) lines to determine whether these different transgenic lines showed differences in performance under salt stress versus the NTL. Under normal conditions, transgenic plants showed no significant difference in growth versus the NTL. When 4-week-old plants were stressed with 150 mM NaCl for 18 d, the RiTL4–2 plants had more green leaves than the OeTL3–8 or NTL plants. After 18 d of high-salt treatment, all plants were subjected to normal irrigation (without salt stress) to allow recovery. Only RiTL4–2 plants survived and resumed growth, forming new tillers, while OeTL3–8 and NTL plants died during the 10-d recovery period (Fig. 3a). Twenty pots of plants were counted and the data showed that the survival rate of RiTL4–2 plants was ~ 50%, whereas only ~ 20% of NTL plants survived. The lowest survival rate (< 10%) was observed in OeTL3–8 plants (Fig. 3b). After 10 d of high-salinity treatment, the fresh weight of NTL was significantly higher than that of OeTL3–8 and lower than that of RiTL4–2 (Fig. 3c). These results supported the notion that *OsRACK1A* increases the salt stress response in rice.

To evaluate the effects of salt stress on cell membranes, 4-week-old seedlings were treated with 150 mM NaCl for 24, 48, or 72 h and the malondialdehyde (MDA) content was measured. The RiTL4–2 plants had lower MDA contents, whereas OeTL3–8 contained more MDA than NTL under salt stress (Fig. 3d). The MDA contents indicated that cell membrane stability was reduced in the *OsRACK1A*-overexpressing line and increased in the *OsRACK1A*-RNAi line, versus the NTL under high-salinity stress. Most plants showed increased proline contents under salt-stress conditions, which was considered to be correlated with their stress resistance. In this study, the content of proline was increased after salt stress in plants. Compared with NTL plants, RiTL4–2 plants accumulated higher levels of proline and OeTL3–8 accumulated lower levels under 150 mM NaCl treatment (Fig. 3e). It is known that salt stress causes chlorophyll degradation. We examined the chlorophyll content of rice plants exposed to 150 mM NaCl. As shown in Fig. 3f, the chlorophyll content declined after salt stress. Compared with NTL plants, chlorophyll contents in RiTL4–2 plants were higher, whereas those in OeTL3–8 were lower under 150 mM NaCl treatment. These results indicated that suppression of *OsRACK1A* enhanced salt-stress tolerance.

OsRACK1A regulates Na^+ and K^+ levels under salt stress

An important aspect of salt tolerance is the avoidance of Na^+ accumulation, and K^+ homeostasis is important for this process (Zhu, 2003). Four-week-old hydroponically grown transgenic and non-transgenic rice plants were subjected to 150 mM NaCl for 72 h. Subsequently, the leaves and roots were harvested at 0 h (before stress) and after 24, 48, and 72 h of salt stress, to measure Na^+

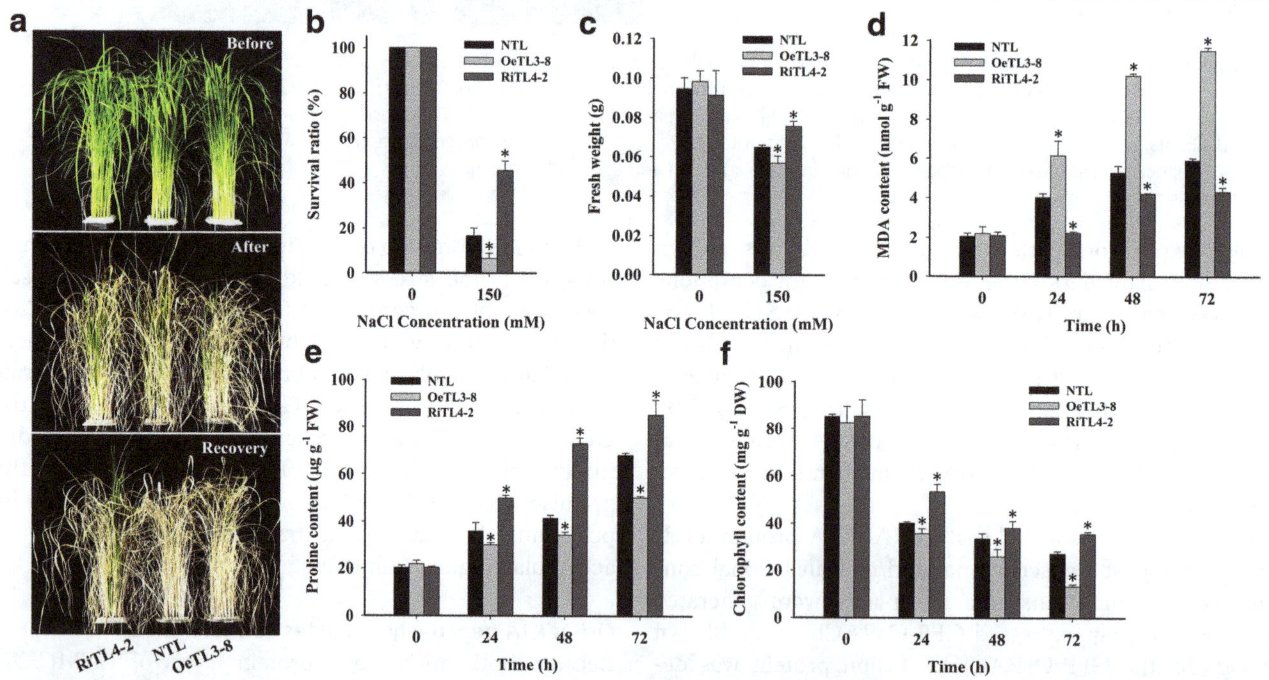

Fig. 3 *OsRACK1A* negatively regulates salt tolerance. **a** Comparison of the non-transgenic line (NTL), the *OsRACK1A*-overexpressing transgenic line (OeTL3–8), and the RNA-interfered transgenic line (RiTLs4–2) under 150 mM NaCl for 18 d (salt-stress recovery for 10 d). **b** Survival rate of plants after 150 mM NaCl treatment and recovery. **c** Fresh weight of plants after 150 mM NaCl treatment. **d** to **f** Contents of proline, MDA, and chlorophyll under 150 mM NaCl for 0, 24, 48 and 72 h. Sixty seedlings per genotype (twenty seedlings for each biological replicate) were used. Data shown are the means ± SE of three biological replicates. An asterisk indicates a significant difference ($P < 0.05$) versus stressed NTL

and K^+ contents. Before the NaCl treatment, Na^+ and K^+ levels in both shoots and roots of the plants with the three different genotypes were similar. The level of Na^+ increased continuously under salt stress in all plant lines tested. However, RiTL4–2 plants accumulated significantly less Na^+ than the NTL, whereas OeTL3–8 contained more Na^+ in both leaves and roots (Fig. 4a). In contrast to the Na^+ levels, the K^+ levels declined in the leaves and roots of all plant lines tested during salt stress. At 72 h of salt stress, the content of K^+ in RiTL4–2 plants was higher than in NTL, whereas OeTL3–8 contained less K^+ than NTL in leaves and roots (Fig. 4b). These results suggested that RiTL4–2 plants might have the ability to avoid Na^+ accumulation

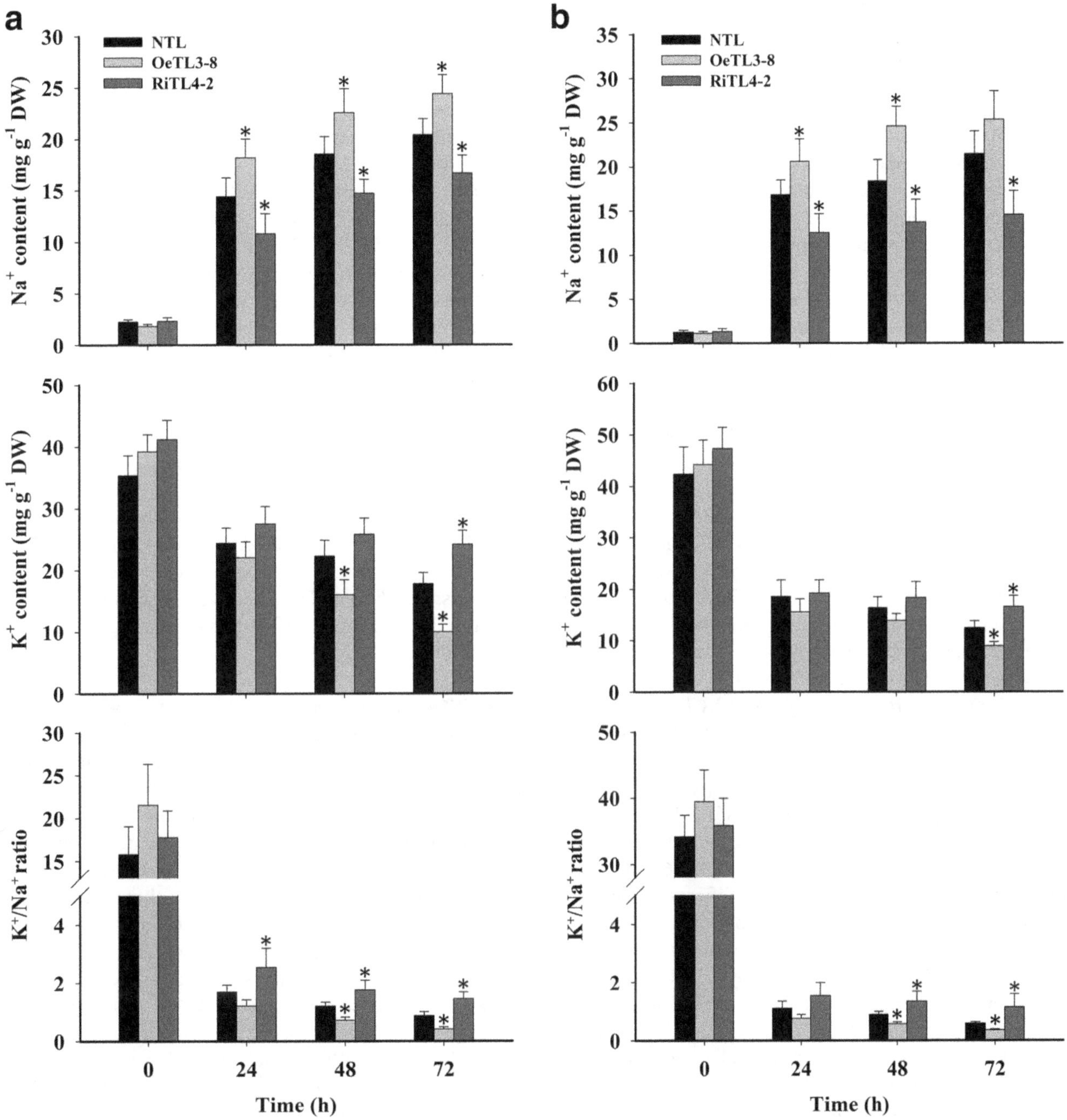

Fig. 4 Na^+ and K^+ contents in shoots and roots of various rice plants under salt stress. Ion content in leaf (**a**) and root tissue (**b**) of four-week-old NTL, OeTL3–8 and RiTL4–2 plants exposed to salt stress (150 mM NaCl) for 72 h. Values shown are for Na^+ and K^+ together with changes in the K^+/Na^+ ratio over time. Sixty seedlings per genotype (twenty seedlings for each biological replicate) were used. Data shown are the means ± SE of three biological replicates. An asterisk indicates a significant difference ($P < 0.05$) versus stressed NTL

and maintain K^+ homeostasis under high-salinity stress. It is generally accepted that the ability to maintain a high K^+/Na^+ ratio contributes to salt tolerance in plants (Zhu, 2003). A decreasing K^+/Na^+ ratio was detected in the roots and leaves of all plants under salt stress. However, a markedly higher K^+/Na^+ ratio was observed in both shoots and roots of the RiTL4–2 plants than those of the NTL and OeTL3–8 plants, whereas no significant difference was observed under normal conditions (Fig. 4a, b). These results imply that *OsRACK1A* negatively regulates rice tolerance of NaCl largely by controlling the Na^+ and K^+ accumulation in cells.

OsRACK1A regulates endogenous ABA content and ABA-responsive genes under salt stress

The phytohormone ABA is a crucial regulator of plant growth and development, and plays a critical role in controlling adaptive plant responses to environmental stresses, such as drought, high salt stress, cold stress, and pathogen infection (Cutler et al., 2010; Umezawa et al., 2010). ABA accumulation and some ABA biosynthesis genes are upregulated by NaCl, drought, and cold stress (Cutler et al., 2010). We determined the endogenous ABA content of leaves under salt stress and found that the endogenous ABA content was significantly lower in OeTL3–8, and significantly higher in RiTL4–2 compared with that of NTL (Fig. 5a). The ABA content induced under stress conditions is regulated by the ABA biosynthesis 9-cis-epoxycarotenoid dioxygenase (*NCED*) genes (Xiong et al., 2002). Our preliminary analysis showed that the transcript level of the *OsNCED4* and *OsNCED5* genes was dramatically induced under salt stress (Additional file 1: Figure S2). This study showed the transcript level of *OsNCED4* in RiTL4–3 was ~ 1.5-fold those in NTL and OeTL3–8, and the transcript level of *OsNCED5* was much higher in RiTL4–3 than in NTL and OeTL3–8 under stress conditions (Fig. 5b).

Next, we determined the transcript expression of three ABA response genes– *OsRAB16A*, *OsLEA3* and *OsLIP9* – under salt stress. As shown in Fig. 5c, without NaCl treatment, the transcript levels of *OsRAB16A*, *OsLEA3* and *OsLIP9* showed no significant difference between wild-type and transgenic plants. Upon 150 mM NaCl treatment, transcripts of these genes accumulated significantly in all three genotypes, while RiTL4–3 accumulated more transcripts than NTL and OeTL3–8 in response to salt stress (Fig. 5c). Because ABA is a key regulator of stomatal opening and closure, water loss from the detached leaves of NTL, OeTL3–8 and RiTL4–2 was compared. As shown in Fig. 5d, water loss in RiTL4–2 was much slower than in NTL and OeTL3–8. These results suggested that *OsRACK1A* negatively regulated the expression of ABA-dependent stress-inducible genes under salt treatment conditions.

OsRACK1A significantly changes expression of salt stress-related genes in rice plants

Next, we evaluated the expression of stress-related genes in NTL, OeTL3–8, and RiTL4–2 plants grown under both control and salt-stress conditions by real-time qPCR. As shown in Fig. 6a, under control conditions, the dehydration-responsive element-binding protein 1 (DREB1) genes, *OsDREB-A, -1B, -1C, –1E, -1G*and *-1H,* and the stress-related APETALA2/Ethylene Responsive Factor (AP2/ERF) gene AP59 were upregulated in RiTL4–2 and downregulated in OeTL3–8, in comparison with NTL. When treated with 150 mM NaCl for 24 h, the transcript levels of *OsDREB1A, 1B, -1C, –1E, –1G*, and *OsAP59* in RiTL4–2 were higher than those in NTL, and the expression of OsDREB1A, -1C and OsAP59 was lower in OeTL3–8 (Fig. 6a).

We selected another seven salt stress-responsive genes (*OsMYB2, SNAC1, OsTCP19, OsTPS1, OsMAPK5, OsSIK1*, and *OsCPK4*) that have been reported to improve salt-stress tolerance (Xiong and Yang, 2003; Hu et al., 2006; Ouyang et al., 2010; Li et al., 2011; Yang et al., 2012; Campo et al., 2014; Mukhopadhyay and Tyagi, 2015). The expression levels of these genes were all upregulated significantly in RiTL4–2, while the expression levels of *OsMAPK5, OsMYB2* and *SNAC1* were downregulated significantly in OeTL3–8. *OsRMC* has been reported to be a negative regulator of the salt-stress response in rice (Zhang et al., 2009; Serra et al., 2013) and the expression of *OsRMC* was downregulated in RiTL4–2 (Fig. 6b). Similar expression profile of these genes occurred under salt-stress conditions (Fig. 6b). These results may partially explain the phenotype of RiTL4–2 plants under stress conditions.

OsRACK1A interacts with salt-stress response proteins

As a scaffold protein, RACK1 interacts with numerous proteins and plays a critical role in many fundamental physiological processes, including stress responses (Zhang et al., 2013). In this study, we used co-immunoprecipitation (co-IP) to identify novel proteins that interact with OsRACK1A under both normal and salt-stress conditions. As shown in Fig. 7a and Additional file 2: Table S1, 12 and 20 proteins were detected to interact with OsRACK1A directly or indirectly in normal and stress conditions, respectively. Of these 32 identified proteins, two (Os07g37760 and Os01g25610) interacted with OsRACK1A in both normal and salt-stress conditions (Fig. 7a). Nine of these genes responded to NaCl treatment (Additional file 1: Figure S3). We also used the yeast two-hybrid assay to confirm these interactions and ultimately found that six of the identified proteins interacted directly with OsRACK1A (Fig. 7b). Interestingly, these six proteins were all identified in

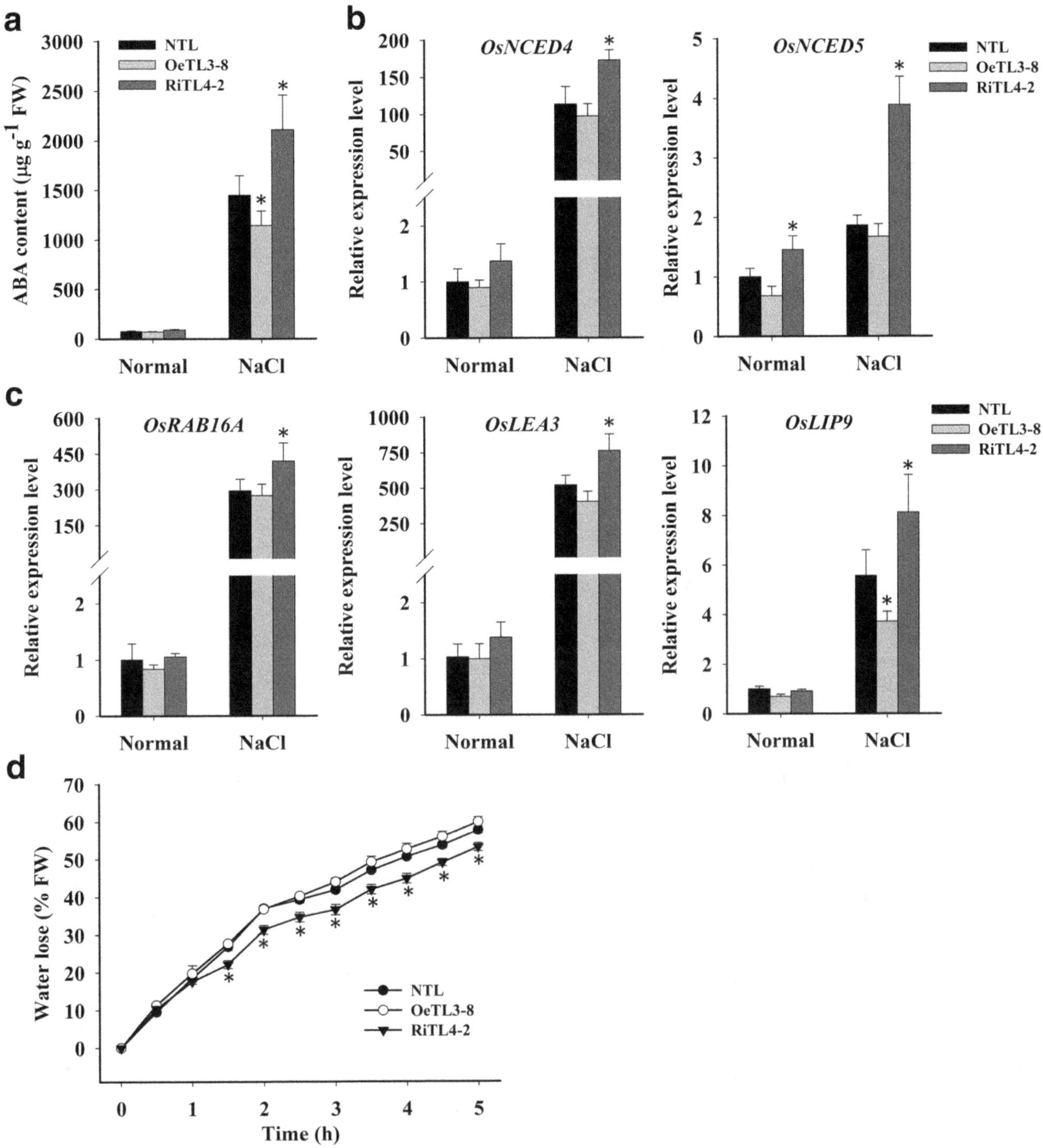

Fig. 5 *OsRACK1A* regulates ABA content and ABA-responsive genes under salt stress. **a** ABA content of various rice plants under normal and salt stress (150 mM NaCl). **b** Transcript level of ABA biosynthesis genes, *OsNCED4* and *OsNCED5* in various rice plants under normal and salt stress (150 mM NaCl for 24 h). **c** Transcript level of ABA- and stress-inducible genes in various rice plants under normal conditions and salt stress (150 mM NaCl for 24 h). **d** Water loss in various rice plants. Whole leaves of 14-d-old plants were cut off and used for the water loss assays. Sixty seedlings per genotype (twenty seedlings for each biological replicate) were used. Data shown are the means ± SE of three biological replicates. An asterisk indicates a significant difference ($P < 0.05$) versus stressed NTL

salt-stress conditions by co-IP assay and their mRNA levels were all downregulated with salt treatment (Fig. 7a, Additional file 1: Figure S3). The mRNA expression of these six OsRACK1A interacting proteins was also detected in NTL, OeTL3–8, and RiTL4–2 under both normal and salt-stress conditions. The transcript levels of

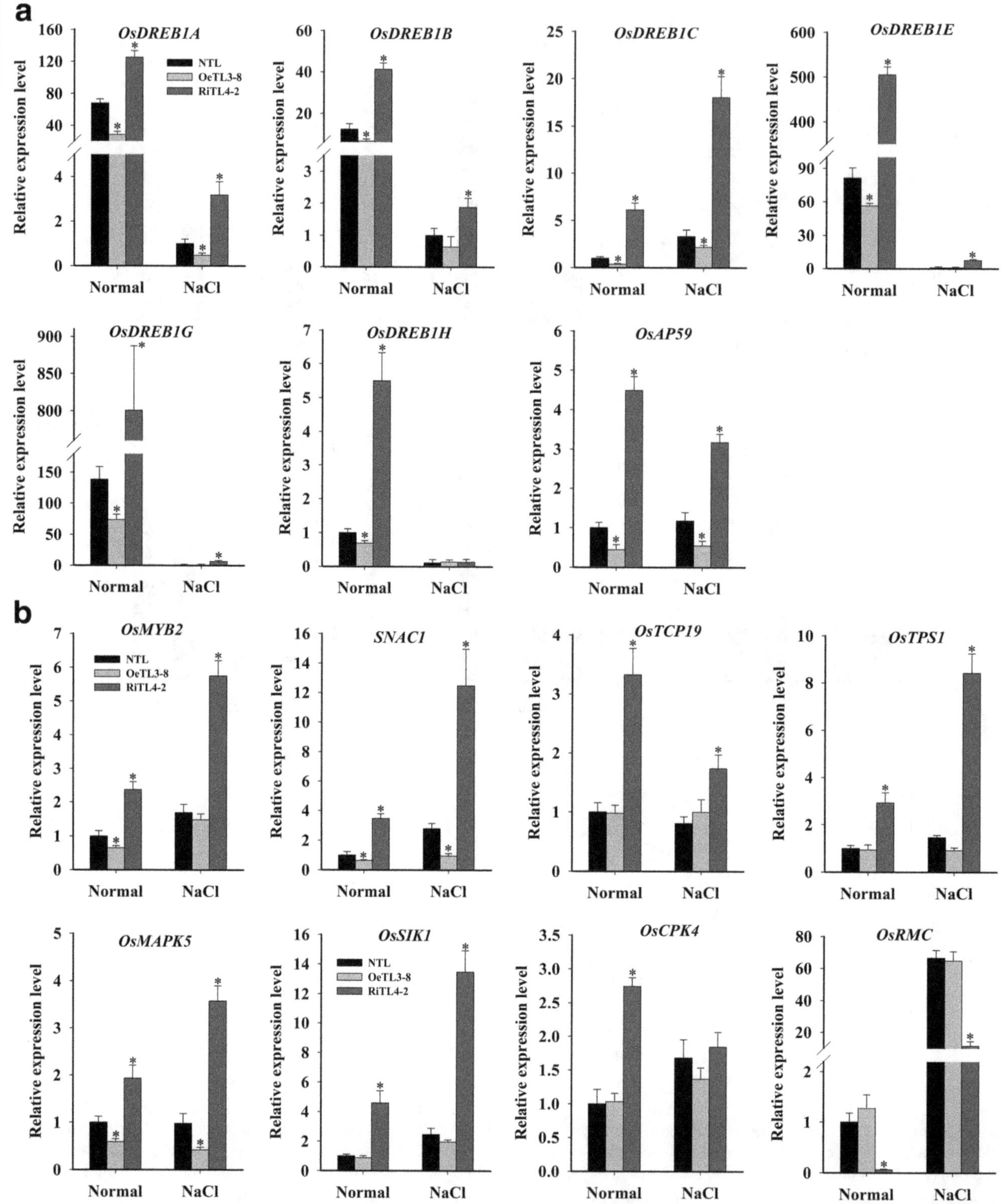

Fig. 6 *OsRACK1A* regulates the expression of stress-related genes. Expression of stress-related AP2/ERF genes (**a**) and other stress-related genes (**b**) in OeTL3–8 and RiTL4–2 compared with the NTL in both normal and salt-stress (150 mM NaCl for 24 h) conditions by qRT-PCR. Data shown are the means ± SE of four biological replicates. An asterisk indicates a significant difference ($P < 0.05$) versus stressed NTL

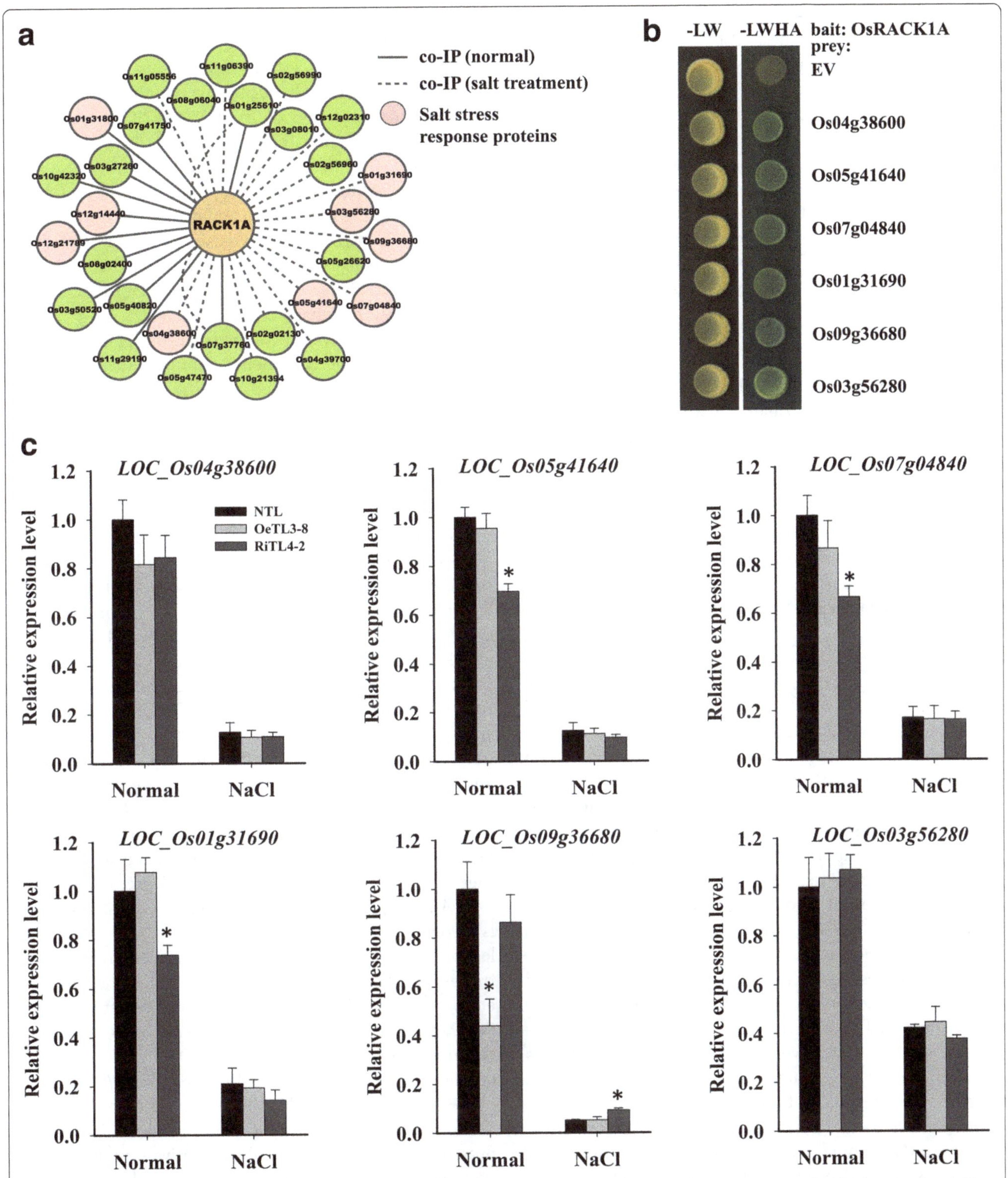

Fig. 7 OsRACK1A directly interacts with stress-response proteins. **a** OsRACK1A interaction proteins screening using co-IP in both normal and salt stress (150 mM NaCl for 24 h) conditions. **b** Verifying the protein interactions by yeast two hybrid assay. **c** Quantitative RT-PCR analysis of the expression of OsRACK1A interactors in response to salt stress (150 mM NaCl for 24 h). Data shown are the means ± SE of four biological replicates. An asterisk indicates a significant difference ($P < 0.05$) versus stressed NTL

Os05g41640, Os07g04840, and Os01g31690, which encode phosphoglycerate kinase, PsbP, and oxygen-evolving enhancer protein 1, respectively, were lower in RiTL4–2 compared with NTL and OeTL3–8 in normal conditions, whereas no significant change in the expression of these genes was observed in NTL and transgenic plants after NaCl treatment (Fig. 7c). The Os09g36680, which encodes a ribonuclease T2 family domain containing protein, was downregulated in OeTL3–8 with no treatment and up-regulated in RiTL4–2 under salt treatment (Fig. 7c). These results indicated that OsRACK1A regulates stress responses by interacting directly with many stress related proteins.

Discussion

RACK1 is a highly conserved scaffold protein that is expressed ubiquitously (Zhang et al., 2013). *RACK1* is involved in multiple signaling pathways, including growth and development and responses to external environmental stresses (McCahill et al., 2002; Zhang et al., 2013). However, the molecular mechanisms of RACK1 in plants is still in its infancy. In plant, *RACK1* is involved in the regulation of cell proliferation and elongation, and the responses to plant hormones and environmental factors (Chen et al., 2006; Nakashima et al., 2008; Guo et al., 2009; Li et al., 2009; Zhang et al., 2013; Zhang et al., 2014). The rice genome contains two *RACK1* ortholog genes, *OsRACK1A* and *OsRACK1B* (Nakashima et al., 2008). Although *OsRACK1A* and *OsRACK1B* are similar, *OsRACK1A* transcript levels are always significantly higher than those of *OsRACK1B* in leaves, roots, and mature seeds (Zhang et al., 2014). Previously, we reported that *OsRACK1A* negatively regulated the response of seed germination to exogenous ABA and the suppression of *OsRACK1A* improved drought tolerant in rice (Li et al., 2009; Zhang et al., 2014). In the present study, we showed that *OsRACK1A* negatively regulated salt stress tolerance and sought to explore the molecular mechanism(s) involved.

A circadian oscillator controls the timing of several physiological functions in living organisms. In plants, processes controlled by a circadian clock include the photoperiodic induction of flowering, rhythmic leaf movements and stomatal opening (Thines and Harmon, 2011). Recent research also suggests that a circadian clock may contribute to plant fitness, enhancing their ability to tolerate abiotic stress (Grundy et al., 2015). In maize, transcripts of many stress related genes exhibits a diurnal cycling pattern (Hayes et al., 2010; Khan et al., 2010). Some salt stress-responsive genes, such as *SOS1*, *RD29A* and *DREB2A*, exhibit a 24-h period of expression in *Arabidopsis*, suggesting that salt tolerance may also be affected by the circadian clock (Park et al., 2016a). In some cases, other stress, such as cold and drought, modifies the transcription pattern of a major portion of genes showing diurnal oscillation (Wilkins et al., 2010; Jończyk et al., 2017). Resent evidence indicates that plants respond to salt stress more strongly during the day than at night and salt-induced expression of RD29A and SOS1 was much higher in the daytime than at night (Park et al., 2016b). We found that both mRNA and protein levels of *OsRACK1A* exhibits a diurnal cycling pattern, and much higher during the day than at night (Fig. 1a, b). However, expression of OsRACK1A increased under salt stress and remained high in both the light and dark (Fig. 2a, b). It might be that higher expression levels of OsRACK1A in day caused more damage under salt stress.

In rice, OsRACK1A protein is phosphorylated under ABA and drought treatment, although the kinase responsible was not identified (He and Li, 2008; Ke et al., 2009). Recently, Urano et al. (2015) showed that *Arabidopsis* RACK1A (AtRACK1A) is also phosphorylated by an atypical serine/threonine protein kinase, WITH NO LYSINE 8 (WNK8), and phosphorylation of AtRACK1A rendered it unstable. Interestingly, in this present study, we found that OsRACK1A protein was controlled by post-transcriptional or translational regulation and consequently accumulated under salt stress. These results led us to the hypothesize that phosphorylation of OsRACK1A does not reduce the protein stability in rice and RACK1 protein may play distinct roles in different plant species. Guo and Sun (2017) found that sumoylation of *Arabidopsis* RACK1B (AtRACK1B) increased AtRACK1B stability and its tolerance to ubiquitin-mediated degradation in the ABA response, and consequently enhanced the interaction between RACK1B and RAP2.6. Combined, these findings illustrate that protein stability controlled by post-transcriptional modification is a critical regulatory mechanism for RACK1 in both *Arabidopsis* and rice.

In *Arabidopsis*, the clock component GIGANTEA (GI) is involved in salt-stress responses (Kim et al., 2013). Similar to *OsRACK1A*, *GI* transcription is under circadian control and peaked at 8–10 h after the start of the day (Park et al., 1999). Under normal conditions, GI interacts with SOS2, a key component of the SOS pathway, preventing the interaction between SOS2 and SOS3. Under salt stress conditions, GI is degraded and the free SOS2/SOS3 complex activates SOS1, a Na^+/H^+ antiporter, to export sodium (Na^+) ions from cells (Kim et al., 2013). In the present study, *OsRACK1A* also negatively regulated Na^+ accumulation and subsequently maintained a low K^+/Na^+ ratio in rice seedlings under NaCl stress (Fig. 4a, b). We investigated the proteins that interact with OsRACK1A and identified six salt-stress suppressed proteins that interacted with OsRACK1A directly (Fig. 7a, b). Unfortunately, none of these proteins were reported to be directly involved in salt-stress responses and the relationship between *OsRACK1A* and the Na^+/H^+ antiporter is still unclear. In eukaryotes, RACK1 regulates various signaling pathways and cellular processes through its interaction with numerous signaling proteins (Zhang et al., 2013). For example, OsRACK1A binds the active form of Racl and interacts with the N terminus of Rboh, RAR1, and SGT1, to form a complex in rice innate immunity (Nakashima et al., 2008). Similarly, OsRACK1A may form a complex with these salt-stress responses proteins, and active downstream molecules, such as salt-stress relative transcription factor. Future studies will reveal whether these OsRACK1A-interaction proteins are involved in salt-stress response.

Under high-salt-stress conditions, a key plant stress-signaling hormone, ABA, and numerous ABA-induced stress-responsive genes products accumulate (Yoshida et al., 2014). The *NCED* genes are known to encode key enzymes in ABA biosynthesis in plants (Nambara and Marion-Poll, 2005). In *Arabidopsis*, *AtNCED3* is induced by drought and high salinity, and the overexpression of *AtNCED3* in transgenic plants enhanced dehydration stress tolerance (Iuchi et al., 2001). Five *NCED* genes (*OsNCED1–5*) have been identified in the rice genome (Zhu et al., 2009). The qPCR analysis showed that the *OsNCED4* and *OsNCED5* were induced strongly under salt stress (Additional file 1: Figure S2), suggesting that transcriptional regulation of the *OsNCED4* and *OsNCED5* genes may be involved in salt-induced ABA accumulation in rice. Previous study has showed that *OsRACK1A* negatively regulated the response of seed germination to exogenous ABA (Zhang et al., 2014). Here we showed that the expression levels of *OsNCED4* and *OsNCED5* were higher in the *OsRACK1A*-suppressed line (RiTL4–2) than the non-transgenic line and the *OsRACK1A*-expressing line (OeTL3–8; Fig. 5b). Additionally, the ABA content was higher in RiTL4–2 than in NTL and OeTL3–8 (Fig. 5a), suggesting that the OsRACK1A protein suppressed ABA accumulation under salt stress by regulating the expression of ABA biosynthesis genes. Some typical ABA-dependent stress-inducible genes, such as *OsRAB16A*, *OsLEA3* and *OsLIP9*, show higher mRNA levels in RiTL4–2, indicating that *OsRACK1A* is involved in ABA-dependent stress pathways.

The AP2/ERF transcription factor superfamily is involved in responses to biotic and abiotic stresses, the regulation of metabolism, and developmental processes in various plant species (Dossa et al., 2016). We selected some *AP2/ERF* genes (*OsDREB-1A, 1B, -1C, –1E, –1G, -1H* and *OsAP59*) and found that these *AP2/ERF* genes were all upregulated in RiTL4–2 and some of them were downregulated in OeTL3–8 (Fig. 6a). Many of the upregulated *AP2/ERF* genes have been reported to play roles in salt-stress tolerance. Transgenic plants overexpressing *OsDREB1B* showed higher tolerances to drought, high salt, and freezing stresses (Dubouzet et al., 2003; Qin et al., 2006; Mao and Chen, 2012). The *OsAP59* gene was found to be induced after exposure to drought and high-salt conditions, and constitutive expression of *OsAP59* in rice increased the tolerance to drought and high salinity during vegetative development (Oh et al., 2009). Some of these *AP2/ERF* genes, such as *OsAP59*, were not induced by ABA (Oh et al., 2009). These results suggested that *OsRACK1A* is also involved in ABA-independent signaling in response to stress in rice. We found some other stress-related transcription factors, such as *OsMYB2*, *SNAC1* and *OsTCP19* were upregulated in *OsRACK1A* suppressed-expression plants in both normal and stress condition (Fig. 6b). OsMYB2-overexpressing plants were reported showing more tolerant to salt, cold, and dehydration stresses and more sensitive to abscisic acid than wild-type plants (Yang et al., 2012). Interestingly, two core circadian clock components, CIRCADIAN CLOCK ASSOCIATED 1 (CCA1) and LATE ELONGATED HYPOCOTYL 1 (LHY1), are also belone to MYB transcription factor family and involved in cold stress responses (Dong et al., 2011), suggesting that MYB transcription factors might be molecular link between circadian clock and stress responses. We also showed that suppression of *OsRACK1A* activated several known stress-related kinases, such as *OsSIK1*, *OsMAPK5*, and *OsCPK4* (Fig. 6b). These genes have been reported to be induced by cold, drought, salinity, ABA, and other abiotic stresses. Transgenic plants overexpressing these genes exhibited enhanced tolerance to various stresses (Xiong and Yang, 2003; Ouyang et al., 2010; Campo et al., 2014). In addition, we showed that expression of *OsRMC*, which negatively regulates salt-stress tolerance in rice (Serra et al., 2013), was suppressed in the RiTL4–2 line (Fig. 6b). Although the signal transduction pathway involving these gene products is unclear, we suggest that *OsRACK1A* participates in abiotic stress pathways, directly or indirectly, by altering the expression of these stress-related genes.

Conclusions

In summary, results presented in this study demonstrate that OsRACK1A functions as a stress-responsive gene and *OsRACK1A* RNAi transgenic rice can significantly improve salt stress tolerance through ABA-dependent and -independent pathway. As a negative regulator of salt stress response, *OsRACK1A* expresses rhythmically under normal conditions and shows the loss of cycling under salt stress. Although OsRACK1A interacts with many salt-responsive proteins, no directly evidence links OsRACK1A protein to salt stress related transcription factors, such as DREB and AP2/ERF. Further investigations on the identification of the functions of OsRACK1A interaction proteins will be helpful to elucidate the mechanism of *OsRACK1A* in regulating salt stress tolerance.

Additional files

Additional file 1: Table S2. Primers used for the qRT-PCR analysis of various genes. **Figure S1.** Public microarray data showing *OsRACK1A* (A) and *OsRACK1B* (B) expression is controlled by a circadian clock (http://www.ebi.ac.uk/arrayexpress/experiments/E-MTAB-275/). *OsRACK1A* expression in rice leaves under 16 h light/ 8 h dark (LD) or constant light (LL) conditions and under NaCl treatment (C). **Figure S2.** *OsNCED* gene expression in rice leaves under 150 mM NaCl treatment for 12 h. **Figure S3.** Quantitative RT-PCR analysis of the expression of OsRACK1 interactors in response to salt stress.

Additional file 2: Table S1. Identification of OsRACK1A interacting proteins.

Abbreviations

ABA: Abscisic acid; AP2/ERF: Apetala2/ethylene responsive factor; BMAL1: Brain and muscle arnt-like protein-1; BR: Brassinosteroid; DREB: Dehydration responsive element binding protein; ERFs: Ethylene response factors; GA: Gibberellin; H_2O_2: Hydrogen peroxide; IAA: Indole-3-acetic acid; MDA: Malondialdehyde; NCED: 9-cis-epoxycarotenoid dioxygenase; NTL: Non-transgenic line; OeTL: Overexpressing-transgenic line; qPCR: Quantitative PCR; RACK1: Receptor for activated C kinase 1; RiTL: RNA interference- transgenic line; ROS: Reactive oxygen species; SOS: salt overly sensitive; WD40: Tryptophan-aspartic acid 40; ZT: Zeitgeber time

Acknowledgments

We acknowledge Miss Min Li and Dr. Yujie Fang (Yangzhou University) for their suggestions during this study.

Funding

This work was supported by the National Science Foundation of China (31271622, 31422047, 31601230, and 31761130073), Natural Science Research Project for Colleges and Universities in Jiangsu Province (16KJB210013), the Jiangsu Province College Students Innovation Training Program in 2016 (X20160741).

Authors' contributions

JL, WX and DZ conceived and designed the experiments. DZ and YW performed plasmid construction and phenotypic analyses, quantitative real-time RT-PCR and western-blot assay. JS, JY and YG performed yeast two hybrid and co-immunoprecipitation (co-IP) experiment. DZ, YW, DL, and WX analyzed data. DZ, YW and JL wrote article. All authors read and approved the final manuscript.

Competing interests

The authors declare that they have no competing interests.

Author details

[1]Jiangsu Key Laboratory of Crop Genetics and Physiology/Co-Innovation Center for Modern Production Technology of Grain Crop, Yangzhou University, Yangzhou 225009, Jiangsu, China. [2]College of Life Sciences, Fujian Agriculture and Forestry University, Jinshan, Fuzhou 350002, China. [3]Department of Biological Engineering, Huanghuai University, Zhumadian 463000, Henan, China. [4]Department of Biology, Southern University of Science and Technology, Shenzhen 518055, China.

References

Arnon DI (1949) Copper enzymes in isolated chloroplasts. Polyphenoloxidase in Beta vulgaris Plant Physiol 24:1–15

Bates LS, Waldren RP, Teare ID (1973) Rapid determination of free proline for water-stress studies. Plant Soil 39:205–207

Campo S, Baldrich P, Messeguer J, Lalanne E, Coca M, San Segundo B (2014) Overexpression of a calcium-dependent protein kinase confers salt and drought tolerance in Rice by preventing membrane lipid peroxidation. Plant Physiol 165:688–704

Chen JG, Ullah H, Temple B, Liang J, Guo J, Alonso JM et al (2006) RACK1 mediates multiple hormone responsiveness and developmental processes in *Arabidopsis*. J Exp Bot 57:2697–2708

Cheng Z, Li JF, Niu Y, Zhang XC, Woody OZ, Xiong Y et al (2015) Pathogen-secreted proteases activate a novel plant immune pathway. Nature 521:213–216

Cutler SR, Rodriguez PL, Finkelstein RR, Abrams SR (2010) Abscisic acid: emergence of a core signaling network. Annu Rev Plant Biol 61:651–679

Dhindsa RS, Matowe W (1981) Drought tolerance in two mosses: correlated with enzymatic defence against lipid peroxidation. J Exp Bot 32:79–91

Dong MA, Farre EM, Thomashow MF (2011) Circadian Clock-Associated 1 and Late Elongated Hypocotyl regulate expression of the C-Repeat Binding Factor (CBF) pathway in Arabidopsis. Proc Natl Acad Sci U S A 108:7241–7246

Dossa K, Wei X, Li D, Fonceka D, Zhang Y, Wang L et al (2016) Insight into the AP2/ERF transcription factor superfamily in sesame and expression profiling of DREB subfamily under drought stress. BMC Plant Biol 16:171

Dubouzet JG, Sakuma Y, Ito Y, Kasuga M, Dubouzet EG, Miura S et al (2003) OsDREB genes in rice, Oryza sativa L., encode transcription activators that function in drought-, high-salt- and cold-responsive gene expression. Plant J 33:751–763

Grundy J, Stoker C, Carré IA (2015) Circadian regulation of abiotic stress tolerance in plants. Front Plant Sci 6:648

Guo J, Chen JG (2008) RACK1 genes regulate plant development with unequal genetic redundancy in *Arabidopsis*. BMC Plant Biol 8:108

Guo J, Wang J, Xi L, Huang WD, Liang J, Chen JG (2009) RACK1 is a negative regulator of ABA responses in *Arabidopsis*. J Exp Bot 60:3819–3833

Guo J, Wang S, Valerius O, Hall H, Zeng Q, Li JF et al (2011) Involvement of *Arabidopsis* RACK1 in protein translation and its regulation by abscisic acid. Plant Physiol 155:370–383

Guo R, Sun W (2017) Sumoylation stabilizes RACK1B and enhance its interaction with RAP2.6 in the abscisic acid response. Sci Rep 7:44090

Hayes KR, Beatty M, Meng X, Simmons CR, Habben JE, Danilevskaya ON (2010) Maize global transcriptomics reveals pervasive leaf diurnal rhythms but rhythms in developing ears are largely limited to the core oscillator. PLoS One 5:e12887

He H, Li J (2008) Proteomic analysis of phosphoproteins regulated by abscisic acid in rice leaves. Biochem Biophys Res Commun 371:883–888

Hu H, Dai M, Yao J, Xiao B, Li X, Zhang Q et al (2006) Overexpressing a NAM, ATAF, and CUC (NAC) transcription factor enhances drought resistance and salt tolerance in rice. Proc Natl Acad Sci U S A 103:12987–12992

Ishikawa R, Aoki M, Kurotani KI, Yokoi S, Shinomura T, Takano M et al (2011) Phytochrome B regulates heading date 1 (Hd1)-mediated expression of rice florigen Hd3a and critical day length in rice. Mol Gen Genomics 285:461–470

Iuchi S, Kobayashi M, Taji T, Naramoto M, Seki M, Kato T et al (2001) Regulation of drought tolerance by gene manipulation of 9-cis-epoxycarotenoid dioxygenase, a key enzyme in abscisic acid biosynthesis in *Arabidopsis*. Plant J 27:325–333

Jończyk M, Sobkowiak A, Trzcinska-Danielewicz J, Skoneczny M, Solecka D, Fronk J, Sowiński P (2017) Global analysis of gene expression in maize leaves treated with low temperature. II. Combined effect of severe cold (8 °C) and circadian rhythm. Plant Mol Biol 95:279–302

Ke Y, Han G, He H, Li J (2009) Differential regulation of proteins and phosphoproteins in rice under drought stress. Biochem Biophys Res Commun 379:133–138

Khan S, Rowe SC, Harmon FG (2010) Coordination of the maize transcriptome by a conserved circadian clock. BMC Plant Biol 10:126

Kim WY, Ali Z, Park HJ, Park SJ, Cha JY, Perez-Hormaeche J et al (2013) Release of SOS2 kinase from sequestration with GIGANTEA determines salt tolerance in *Arabidopsis*. Nat Commun 4:1352

Kundu N, Dozier U, Deslandes L, Somssich IE, Ullah H (2013) *Arabidopsis* scaffold protein RACK1A interacts with diverse environmental stress and photosynthesis related proteins. Plant Signal Behav 8:e24012

Li DH, Liu H, Yang YL, Zhen PP, Liang JS (2009) Down-regulated expression of RACK1 gene by RNA interference enhances drought tolerance in rice. Rice Sci 16:14–20

Li HW, Zang BS, Deng XW, Wang XP (2011) Overexpression of the trehalose-6-phosphate synthase gene OsTPS1 enhances abiotic stress tolerance in rice. Planta 234:1007–1018

Mao D, Chen C (2012) Colinearity and similar expression pattern of rice DREB1s reveal their functional conservation in the cold-responsive pathway. PLoS One 7:e47275

Matsubara K, Yamanouchi U, Nonoue Y, Sugimoto K, Wang ZX, Minobe Y et al (2011) Ehd3, encoding a plant homeodomain finger-containing protein, is a critical promoter of rice flowering. Plant J 66:603–612

McCahill A, Warwicker J, Bolger GB, Houslay MD, Yarwood SJ (2002) The RACK1 scaffold protein: a dynamic cog in cell response mechanisms. Mol Pharmacol 62:1261–1273

Mukhopadhyay P, Tyagi AK (2015) OsTCP19 influences developmental and abiotic stress signaling by modulating ABI4-mediated pathways. Sci Rep 5: 9998

Nakashima A, Chen L, Thao NP, Fujiwara M, Wong HL, Kuwano M et al (2008) RACK1 functions in rice innate immunity by interacting with the Rac1 immune complex. Plant Cell 20:2265–2279

Nambara E, Marion-Poll A (2005) Abscisic acid biosynthesis and catabolism. Annu Rev Plant Biol 56:165–185

Oh SJ, Kim YS, Kwon CW, Park HK, Jeong JS, Kim JK (2009) Overexpression of the transcription factor AP37 in rice improves grain yield under drought conditions. Plant Physiol 150:1368–1379

Ouyang SQ, Liu YF, Liu P, Lei G, He SJ, Ma B et al (2010) Receptor-like kinase OsSIK1 improves drought and salt stress tolerance in rice (Oryza sativa) plants. Plant J 62:316–329

Park DH, Somers DE, Kim YS, Choy YH, Lim HK, Soh MS et al (1999) Control of circadian rhythms and photoperiodic flowering by the *Arabidopsis* GIGANTEA gene. Science 285:1579–1582

Park HJ, Kim WY, Yun DJ (2016a) A New Insight of Salt Stress Signaling in Plant. *Mol Cells* 39:447–459

Park HJ, Qiang Z, Kim WY, Yun DJ (2016b) Diurnal and circadian regulation of salt tolerance in Arabidopsis. J Plant Biol 59:569–578

Qin QL, Liu JG, Zhang Z, Peng RH, Xiong AS, Yao QH et al (2006) Isolation, optimization, and functional analysis of the cDNA encoding transcription factor OsDREB1B in Oryza Sativa L. Mol Breeding 19:329–340

Robles MS, Boyault C, Knutti D, Padmanabhan K, Weitz CJ (2010) Identification of RACK1 and protein kinase Calpha as integral components of the mammalian circadian clock. Science 327:463–466

Serra TS, Figueiredo DD, Cordeiro AM, Almeida DM, Lourenço T, Abreu IA et al (2013) OsRMC, a negative regulator of salt stress response in rice, is regulated by two AP2/ERF transcription factors. Plant Mol Biol 82:439–455

Shi S, Chen W, Sun W (2011) Comparative proteomic analysis of the *Arabidopsis* cbl1 mutant in response to salt stress. Proteomics 11:4712–4725

Thines B, Harmon FG (2011) Four easy pieces: mechanisms underlying circadian regulation of growth and development. Curr Opin Plant Biol 14:31–37

Umezawa T, Nakashima K, Miyakawa T, Kuromori T, Tanokura M, Shinozaki K et al (2010) Molecular basis of the core regulatory network in ABA responses: sensing, signaling and transport. Plant Cell Physiol 51:1821–1839

Urano D, Czarnecki O, Wang X, Jones AM, Chen JG (2015) *Arabidopsis* receptor of activated c kinase1 phosphorylation by with no lysine8 kinase. Plant Physiol 167:507–516

Wilkins O, Bräutigam K, Campbell MM (2010) Time of day shapes Arabidopsis drought transcriptomes. Plant J 63:715–727

Xiong L, Schumaker KS, Zhu JK (2002) Cell signaling during cold, drought, and salt stress. Plant Cell 14(Suppl):S165–S183

Xiong L, Yang Y (2003) Disease resistance and abiotic stress tolerance in rice are inversely modulated by an abscisic acid-inducible mitogen-activated protein kinase. Plant Cell 15:745–759

Xue W, Xing Y, Weng X, Zhao Y, Tang W, Wang L et al (2008) Natural variation in Ghd7 is an important regulator of heading date and yield potential in rice. Nat Genet 40:761–767

Yang A, Dai X, Zhang WH (2012) A R2R3-type MYB gene, OsMYB2, is involved in salt, cold, and dehydration tolerance in rice. J Exp Bot 63:2541–2556

Yang C, Ma B, He SJ, Xiong Q, Duan KX, Yin CC et al (2015) Maohuzi6/Ethylene Insensitive3-Like1 and Ethylene Insensitive3-Like2 regulate ethylene response of roots and coleoptiles and negatively affect salt tolerance in Rice. Plant Physiol 169:148–165

Yoshida T, Mogami J, Yamaguchi-Shinozaki K (2014) ABA-dependent and ABA-independent signaling in response to osmotic stress in plants. Curr Opin Plant Biol 21C:133–139

Zhang D, Chen L, Li D, Lv B, Chen Y, Chen J et al (2014) OsRACK1 is involved in abscisic acid- and H2O2-mediated signaling to regulate seed germination in rice (Oryza sativa, L.). PLoS One 9:e97120

Zhang D, Zhou Y, Yin J, Yan X, Lin S, Xu W et al (2015) Rice G-protein subunits qPE9-1 and RGB1 play distinct roles in abscisic acid responses and drought adaptation. J Exp Bot 66:6371–6384

Zhang DP, Chen L, Lv B, Liang JS (2013) The scaffolding protein RACK1: a platform for divers functions in the plant kingdom. Plant Biology Soil Health 1:7–13

Zhang L, Tian LH, Zhao JF, Song Y, Zhang CJ, Guo Y (2009) Identification of an Apoplastic protein involved in the initial phase of salt stress response in Rice root by two-dimensional electrophoresis. Plant Physiol 149:916–928

Zhu G, Ye N, Zhang J (2009) Glucose-induced delay of seed germination in rice is mediated by the suppression of ABA catabolism rather than an enhancement of ABA biosynthesis. Plant Cell Physiol. 50:644–651

Zhu JK (2003) Regulation of ion homeostasis under salt stress. Curr Opin Plant Biol 6:441–445

12

Overexpression of the Tibetan Plateau annual wild barley (*Hordeum spontaneum*) *HsCIPKs* enhances rice tolerance to heavy metal toxicities and other abiotic stresses

Weihuai Pan[1,2†], Jinqiu Shen[3†], Zhongzhong Zheng[3†], Xu Yan[1], Jianxin Shou[2], Wenxiang Wang[3], Lixi Jiang[4] and Jianwei Pan[1*]

Abstract

Background: The calcineurin B-like protein (CBL) and CBL-interacting protein kinase (CIPK) signaling system plays a key regulatory role in plant stress signaling. The roles of plant-specific CIPKs, essential for CBL-CIPK functions, in the response to various abiotic stresses have been extensively studied so far. However, until now, the possible roles of the *CIPKs* in the plant response to heavy metal toxicities are largely unknown.

Results: In this study, we used bioinformatic and molecular strategies to isolate 12 *HsCIPK* genes in Tibetan Plateau annual wild barley (*Hordeum spontaneum* C. Koch) and subsequently identified their functional roles in the response to heavy metal toxicities. The results showed that multiple *HsCIPKs* were transcriptionally regulated by heavy metal toxicities (e.g., Hg, Cd, Cr, Pb, and Cu) and other abiotic stresses (e.g., salt, drought, aluminum, low and high temperature, and abscisic acid). Furthermore, the ectopic overexpression of each *HsCIPK* in rice (*Oryza sativa* L. cv Nipponbare) showed that transgenic plants of multiple *HsCIPKs* displayed enhanced tolerance of root growth to heavy metal toxicities (Hg, Cd, Cr, and Cu), salt and drought stresses. These results suggest that *HsCIPKs* are involved in the response to heavy metal toxicities and other abiotic stresses.

Conclusions: Tibetan Plateau annual wild barley *HsCIPKs* possess broad applications in genetically engineering of rice with tolerance to heavy metal toxicities and other abiotic stresses.

Keywords: Abiotic stresses, Cloning, Heavy metal toxicity, *HsCIPKs*, Overexpression, Tibetan plateau annual wild barley, Rice, Transformation

Background

In plant cells, the calcium ion (Ca^{2+}) is involved as a second messenger in the regulation of a variety of abiotic and biotic stress responses and the formation and development of plant organs (Dodd et al. 2010). The core components of Ca^{2+} signaling are calcium sensors, including calmodulins (CaMs), calmodulin-like proteins (CMLs), calcium-dependent protein kinases (CDPKs), and calcineurin B-like proteins (CBLs), which bind Ca^{2+} and activate downstream signaling components (Rudd and Franklin-Tong 1999; Singh and Parniske 2012; Zheng et al. 2013). Among these Ca^{2+}-bound calcium sensors, CBLs selectively interact with plant-specific CBL-interacting protein kinases (CIPKs) and thereby form a CBL-CIPK signaling system that has been demonstrated to serve as a key regulation node during stress signaling in plants (Luan 2009; Weinl and Kudla 2009; Shen et al. 2014). Thus, dissecting the mechanisms of the CBL-CIPK signaling system is one of the research priorities in the plant stress physiology field. Due to the lack of kinase activity in CBLs, different combinations with CIPKs largely determine the specificity, diversity, and complexity of the CBL-CIPK signaling system

* Correspondence: jwpan@lzu.edu.cn
†Weihuai Pan, Jinqiu Shen and Zhongzhong Zheng contributed equally to this work.
[1]MOE Key Laboratory of Cell Activities and Stress Adaptations, School of Life Sciences, Lanzhou University, Lanzhou 730000, China
Full list of author information is available at the end of the article

(Batistic et al. 2010). Therefore, functional identification of the CIPKs in distinct plant species will enhance the better understanding of the functional roles and modes of action of the CBL-CIPK signaling system.

Bioinformatic analysis has shown that there are 26 and 31 *CIPK* homologous genes in the model plant genomes of *Arabidopsis thaliana* and rice (*Oryza sativa*), respectively (Kolukisaoglu et al. 2004). Recently, multiple *CIPK* families have been bioinformatically identified in other plant species, including poplar (*Populus*) (Yu et al. 2007), cotton (*Gossypium spp*) (Wang et al. 2016a), soybean (*Glycine max*) (Zhu et al. 2016), canola (*Brassica napus*) (Zhang et al. 2014), eggplant (*Solanum melongena*) (Li et al. 2016), cassava (*Manihot esculenta*) (Hu et al. 2015), maize (*Zea mays*) (Chen et al. 2011), and wheat (*Triticum aestivum*) (Sun et al. 2015). However, genomic analysis of the cultivated barley (*Hordeum vulgare*) *HvCIPK* family remains lacking. Tibetan plateau wild barley, in particular, the annual wild barley (*Hordeum spontaneum* C. Koch), has suffered the extreme climate and environmental conditions for a long term and therefore has evolutionally generated abundant natural variations and/or unique gene networks for stress tolerance. Due to the close genetic homology of Tibetan plateau wild barley to cultivated barley, the Tibetan plateau was recently considered to be one of the centers of domestication of cultivated barley (Dai et al. 2012). Thus, Tibetan plateau wild barley is one of the few germplasm resources to utilize wild barley *CIPKs* to genetically engineer rice or other crops with higher stress tolerance.

Although CIPKs have been demonstrated to function in various responses to abiotic stresses, including salt, drought, flood, wounding, abscisic acid (ABA), low and high temperature (Guo et al. 2001; Kim et al. 2003; Lee et al. 2009; Li et al. 2012; Yan et al. 2014; Zhang et al. 2014), biotic stresses, such as pathogen infection (Kurusu et al. 2010; de la Torre et al. 2013; Meteignier et al. 2017), and nutrient deficiency (Xu et al. 2006; Pandey et al. 2007; Wang et al. 2016b; Straub et al. 2017), so far, no evidence has shown that CIPKs are involved in the plant response to heavy metal toxicities, which is one of the most dangerous types of toxic species for plants and therefore for animals and humans via the food chain. Heavy metals are defined as elements having a specific gravity above five and include mercury (Hg), cadmium (Cd), chromium (Cr), copper (Cu), and lead (Pb). Heavy metals-polluted soils cause irreversible harm to plant growth and development, and crop yield and quality (Mustafa and Komatsu 2016) due to their extremely stable and nonbiodegradable biochemical characteristics. Heavy metal ions enter the cell and tightly bind to intracellular protein enzymes by replacing specific cations from their binding sites, leading to the inactivation of enzymes and the induction of reactive oxygen species (ROS), (Sharma and Dietz 2009), which causes oxidative damage to plant cells. Recent studies have shown that Ca^{2+} or Ca^{2+}-dependent signaling is involved in plant tolerance to heavy metal stresses, including Cd and Cr (Fang et al. 2014; Huang et al. 2014; Ahmad et al. 2015), and aluminum (Al) toxicity (Zhang and Rengel 1999; Lan et al. 2016). However, whether CIPKs function in Ca^{2+}-dependent plant tolerance to heavy metal and Al toxicities is largely unknown.

In this study, we functionally identified the roles of 12 members of the Tibetan plateau annual wild barley *HsCIPK* family in the response to heavy metal toxicities, including Hg, Cd, Cr, Cu, and Pb, and other abiotic stresses such as salt, drought, Al, low and high temperature, and ABA. Our results demonstrate that multiple *HsCIPKs* are involved in plant tolerance to multiple heavy metal toxicities and salt and drought stresses.

Methods

Plant materials and growth conditions

Grains of wild barley, Tibetan Plateau annual wild barley X74 (*Hordeum spontaneum* C. Koch), and Nipponbare rice (*Oryza sativa* L. ssp. *japonica*) were surface sterilized by 70% ethanol for 10 min followed by 10% NaClO for 30 min and finally 8 rinses with water. The endosperms in barley and rice grains contain a large amount of nutrients and therefore provide enough nutrients to sustain the grains for one week following germination. Thus, a simple $CaCl_2$ solution (0.1 mM $CaCl_2$; pH 5.8; Pan et al. 2004) was used for barley and rice grain germination and seedling growth. To germinate the barley grains (Pan et al. 2004), sterilized grains were germinated between two layers of wet filter papers with the $CaCl_2$ solution for one day under darkness (25 °C), and the germinated grains were incubated for another four days in the $CaCl_2$ solution under darkness (25 °C). To germinate the rice grain (Pan et al. 2011), sterilized grains were treated for three days under darkness (4 °C) and subsequently incubated for 3 days at 37 °C for germination. Finally, the germinated rice grains were transferred to the $CaCl_2$ solution and incubated for four days under light conditions (14-h light/10-h dark, 28 °C light/25 °C dark). Four-day-old barley and rice seedlings with similar root lengths were used in this study.

In silico cloning, molecular cloning, and sequence analysis of *HsCIPKs*

Two approaches were used in *HsCIPK* in silico cloning. We first used available full-length cDNAs of rice *OsCIPK1* to *OsCIPK31* (http://www.ncbi.nlm.nih.gov/) (accession numbers of the *OsCIPK* cDNAs are shown in Additional file 1: Table S1) as probes to search for their homologous genes in a full-length cDNA library of a barley cultivar (http://earth.lab.nig.ac.jp/~dclust/cgi-bin/

barley_pub/). Second, full-length cDNAs, conserved motif in kinase domain, and the NAF/FISL domain motif of rice *OsCIPK1* to *OsCIPK31* as probes were subjected to a BLAST comparison with the barley nucleotide collection (nr/nt) database (http://blast.ncbi.nlm.nih.gov/) for their homologous fragment sequences, and the subsequent resulting homologous fragment sequences were spliced in silico and extended for multiple rounds via corresponding overlapping contigs. Finally, the resulting full-length cDNAs were analyzed by DNA STAR SeqMan and Megalign software.

Reverse transcription-polymerase chain reaction-(RT-PCR-) and sequencing-based approaches were used to clone the *HsCIPK* coding sequences (CDSs) from a total cDNA pool of the Tibetan Plateau annual wild barley X74. *HsCIPK*-specific primers for the RT-PCR assay (Additional file 1: Table S2) were designed based on the recovered in silico cDNA sequences of the *HvCIPKs*. Furthermore, 5'-RACE (rapid amplification of cDNA ends) and 3'-RACE were used to confirm individual 5'- and 3'-end sequences. Finally, the resulting full-length cDNAs of individual *HsCIPKs* were analyzed by sequence alignment and open reading frame (ORF) comparison with *OsCIPKs* and *AtCIPKs* (Accession numbers of *AtCIPK* cDNAs shown in Additional file 1: Table S1). Using the Clustal W Method in DNA STAR Megalign software, cDNA sequence-based predicted amino acid sequences were used to generate a phylogenetic tree of HsCIPKs, OsCIPKs, and AtCIPKs.

Chemicals and treatments

Unless specified, all reagents were from Sigma-Aldrich. All chemical stock solutions were prepared as follows: water was used to dissolve $CaCl_2$ (1 M), NaCl (5 M), $AlCl_3$ (40 mM), $HgCl_2$ (0.5 M), $CdCl_2$ (10 mM), $PbCl_2$ (10 mM), $K_2Cr_2O_7$ (0.5 M), and $CuSO_4$ (1 M). Abscisic acid (ABA; 10 mM) was first dissolved in a few drops of 1 M KOH and then diluted with water, whereas polyethylene glycol 6000 (PEG 6000) powders were directly prepared before use (see below). Unless otherwise indicated, in the qRT-PCR analysis of the transcriptionally induced expression levels of the endogenous *HsCIPKs* under heavy metal toxicities and other abiotic stresses, the final working concentrations were 400 mM for NaCl, 1 mM for $CuSO_4$, 0.5 mM for $K_2Cr_2O_7$, 20 μM for $HgCl_2$, $PbCl_2$, $CdCl_2$ $AlCl_3$, and ABA, and 20% (*w*/*v*) for PEG 6000. In the root growth assay of rice transgenic lines, based on the effects of different concentrations on root elongation in the wild-type Nipponbare rice (Additional file 1: Figure S1), the final working concentrations were 50 mM for NaCl, 50 μM for $AlCl_3$, 5 μM for $K_2Cr_2O_7$, $PbCl_2$, and $CdCl_2$, 1 μM for ABA, 0.5 μM for $HgCl_2$, 0.25 μM for CuSO4, and 10% (*w*/*v*) for PEG 6000. All stock solutions of these chemicals were diluted to the working solutions in a simple $CaCl_2$ solution (0.1 mM $CaCl_2$; pH 5.8), whereas PEG 6000 powders were directly dissolved in a simple $CaCl_2$ solution (0.1 mM $CaCl_2$; pH 5.8) to 10% and 20% (w/v; working concentration) before use. In addition, the time lengths of all the treatments, including temperature (4 °C and 35 °C), are indicated in the text.

Quantitative real-time RT-PCR (qRT-PCR) assay

To examine the effects of heavy metal toxicities and other abiotic stresses on the transcriptional expression levels of endogenous *HsCIPKs* in Tibetan Plateau annual wild barley, whole seedlings or roots after treatment were used to isolate the total RNAs using an RNeasy Plant Mini Kit (Qiagen). The first-strand synthesis of cDNA was synthesized with a SuperScript III First-Strand Synthesis System (Invitrogen). The qRT-PCR assay was performed with Thunderbird SYBR qPCR mix (Toyobo) and a StepOnePlus Real-Time PCR System (Applied Biosystems). The reactions were performed in a 20-μL volume containing 10 μL 2 × SYBR qPCR mix (Toyobo), 10 ng cDNA, and 1 μM of each gene-specific primer (Additional file 1: Table S3). The PCR cycles were performed as follows: one cycle of 95 °C for 3 min, 40 cycles of 95 °C for 5 s and 60 °C for 50 s. The resulting data were collected and analyzed using the StepOne Software v2.1. The transcriptional levels were normalized to the housekeeping gene *HvActin* (Additional file 1: Table S3). For each *HsCIPK*, the transcription levels upon stress treatment for different time lengths were presented as relative values of the 0-h time point (mock control; set as 1.0). For statistical analysis (Student's t-test, two tails; type 2), the transcription levels from three independent experiments at different time-points were compared with those of the mock control.

Constructs, transformation and molecular identification in rice

To overexpress each *HsCIPK* gene in rice, constructs of *35S::HsCIPKs* were generated individually using PCR, restriction digestion, and ligation with the plant transformation vector pCAMBIA2300S containing a 2× CAMV 35S promoter and a kanamycin-resistant marker (Xiong and Yang 2003). Finally, the resulting constructs were confirmed by sequencing. All the primer sequences for the *HsCIPK* constructs are indicated in Additional file 1: Table S2.

For rice transformation, Nipponbare rice mature embryos were used as the initial materials for callus induction. Briefly, surface-sterilized mature embryos were incubated on agar plates containing N6D media (Ozawa 2009; Toki et al. 2006) for callus induction. Light yellow, compact and hard calluses were used in the Agrobacterium-mediated transformation with

Agrobacterium tumefaciens EHA105 (Hiei and Komari 2008). Kanamycin-resistant vigorous calluses were recovered on selection solid media containing G418 (150 mg/L; Amresco) due to high-level kanamycin-resistance in the wild-type rice background (Dekeyser et al. 1989) and subsequently transferred to differentiation solid media with G418 (100 mg/L) for green shoot induction. Finally, the roots were induced on solid rooting media with G418 (70 mg/L). After acclimatization, regeneration plantlets (T1 generation) were cultivated in hydroponic conditions.

To determine whether these transgenic plants harboring *35S::HsCIPK* constructs are true overexpression lines, a RT-PCR assay was performed in T1 generation plants. RNA isolation and cDNA synthesis were conducted as described. RT-PCR primers were designed in the *OsCIPK* nonhomologous region of *HsCIPKs* (Additional file 1: Table S4). Rice *OsActin2*, a housekeeping gene, was used as an internal control (Additional file 1: Table S4), whereas the wild-type Nipponbare cDNA sample served as a negative control (NC). Twenty-seven cycles of PCR were used to amplify all the exogenous *HsCIPKs* and the endogenous *OsActin2* in the rice transgenic lines. Homozygous lines for all the transgenic lines overexpressing each *HsCIPK* were recovered in the T3 generation via G418-resistent selection.

Root length measurement

To determine the roles of the HsCIPKs in the plant response to heavy metal toxicities and other abiotic stresses, primary root growth was used to examine the effect of *HsCIPK* overexpression on the plant growth response to stress treatments. To quantify root elongation, the primary root lengths were individually measured before (0 h) and after 24 h of treatment. To reduce differential physiological effects on root growth before treatment (0 h), root relative elongation rates (RERs; %) were used to evaluate the stress effects on root growth. The RERs were estimated according to the following formula as previously described (Pan et al. 2004; Pan et al. 2011): $RER = (RL_{T24h}\text{-}RL_{T0h})/(RL_{M24h}\text{-}RL_{M0h}) \times 100\%$. RL_{T0h} and RL_{T24h} indicate root lengths (RL; mm) before (0 h) and after the 24-h stress treatment, respectively, whereas RL_{M0h} and RL_{M24h} represent the root lengths before (0 h) and after 24 h of mock treatment, respectively. The quantitative data of 45 seedlings for each treatment from three independent experiments were statistically evaluated using a Student's *t*-test (two tails; type 2) compared with the nontransgenic regeneration lines (NT; as the wild-type control). Multiple transgenic lines for each construct were examined in the stress treatment and the representative lines are presented.

Results

In silico assay of cultivated barley *HvCIPKs*

To obtain the *HvCIPK* cDNA sequences of cultivated barley, we performed an in silico assay to search for the corresponding homologous *CIPKs* in a full-length cDNA library of cultivated barley using the full-length cDNAs of rice *OsCIPK1* to *OsCIPK31* (Kolukisaoglu et al. 2004) as probes and subsequently recovered five rice homologous sequences of *HvCIPKs*, including *CIPK2, - 9, - 11, - 23,* and *- 28*.

Next, we used conserved motifs in the kinase and NAF/FISL domains of the OsCIPKs combined with full-length cDNAs of *OsCIPK1* to *OsCIPK31* to BLAST homologous sequences in a nonredundant nucleotide database of cultivated barley and finally obtained eight rice homologous sequences of *HvCIPKs*, including *CIPK5, - 14, - 15, - 17, - 24, - 29, - 30,* and *- 31*. Thus, we obtained 13 potential *HvCIPK* cDNA sequences.

Molecular cloning of the *HsCIPKs* in Tibetan plateau annual wild barley

To clone the *HsCIPK* CDS from the Tibetan Plateau annual wild barley, we used these 13 *HvCIPK* CDS sequences, combined with RT-PCR- and sequencing-based approaches, to isolate their homologous sequences from a total cDNA pool of Tibetan Plateau annual wild barley. We finally recovered 13 corresponding full-length CDS sequences of *HsCIPKs*. Previous findings indicated that plant-specific CIPKs possess typical conserved domains, including a kinase domain (activation loop) at the N terminal, a CBL-interacting domain (NAF/FISL motif), and a protein–phosphatase interaction (PPI) domain at the C terminal (Ohta et al. 2003). Analysis of the functional domain prediction confirmed that all of the 13 HsCIPKs indeed contain a conserved activation loop, an NAF/FISL motif, and a PPI domain (Fig. 1a) similar to the rice and Arabidopsis CIPKs (Additional file 1: Figure S2; HsCIPK5 as an example). Phylogenetic analysis of the deduced amino acid sequences showed that these 13 HsCIPKs were evolutionally divided into three branches, including (I) HsCIPK2, - 5, - 11, - 14, - 15, - 28, and - 30, (II) HsCIPK9, - 17, - 23, - 24, and - 31, and (III) HsCIPK29 (Fig. 1b). It is noteworthy that HsCIPK14 has an almost identical sequence to HsCIPK15 except for an additional four amino acids at its 3′-end, similar to the rice and Arabidopsis CIPK14/CIPK15 (Kolukisaoglu et al. 2004), and therefore, only the *HsCIPK14* gene was used in the following studies. In addition, except for *HsCIPK15*, 12 other *HsCIPKs* CDS and their deduced amino acid sequences have been deposited into the GenBank (Additional file 1: Figure S3a).

Next, we analyzed amino acid sequence homologies between the HsCIPKs versus the OsCIPKs and AtCIPKs. The results showed that the HsCIPKs share higher

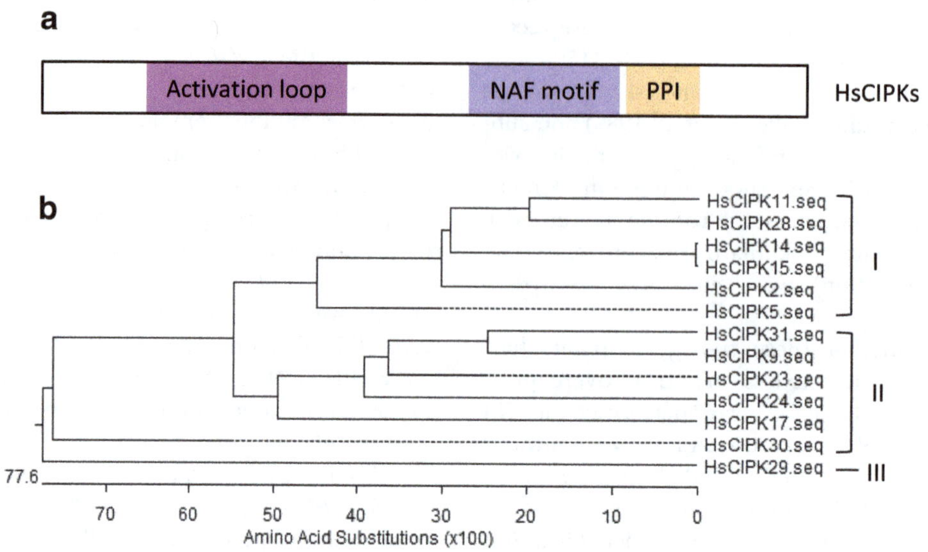

Fig. 1 Schematic diagram and sequence analysis of HsCIPKs. **a** Prediction of functional domains in HsCIPKs. **b** Phylogenetic analysis of 13 HsCIPKs

homology with the OsCIPKs (~ 70% to ~ 95%) than the AtCIPKs (~ 46% to ~ 77%) at the amino acid level (Additional file 1: Figure S3b and c). Consistently, phylogenetic tree analysis among the HsCIPKs, OsCIPKs, and AtCIPKs showed that the individual HsCIPKs and corresponding homologous OsCIPKs were located at the identical branches, whereas the AtCIPKs were distributed at the side branches (Additional file 1: Figure S4). These results suggest that the HsCIPKs have a closer genetic relationship to the OsCIPKs than to the AtCIPKs. Thus, we designated the *HsCIPK* numbers with the corresponding homologous *OsCIPK* number.

Responses of *HsCIPKs* to heavy metal toxicities

It is well known that plant CIPKs function in the regulation of various abiotic and biotic responses (Guo et al. 2001; de la Torre et al. 2013). However, whether CIPKs are involved in the plant response to heavy metal toxicities is largely unknown. To this end, we performed a time-course qRT-PCR analysis to examine whether heavy metal toxicities influence the transcriptional expression levels of the endogenous *HsCIPKs* in the Tibetan Plateau annual wild barley seedlings treated with $HgCl_2$ (20 μM), $CdCl_2$ (20 μM), $K_2Cr_2O_7$ (0.5 mM), $PbCl_2$ (20 μM), and $CuSO_4$ (1 mM), respectively. As shown in Fig. 2a, upon seedling exposure to $HgCl_2$ treatment, the transcriptional levels of 10 genes (*HsCIPK2, 11, 14, 17, 23, 24, 28, 29, 30,* and *31*) were dramatically elevated relative to the mock control (0 h time point). Consistently, the expression levels of 10 genes (*HsCIPK2, 9, 11, 14, 17, 23, 24, 29, 30,* and *31*) in the $CdCl_2$ treatment (Fig. 2b), 10 genes (*HsCIPK2, 5, 9, 11, 14, 17, 23, 24, 29,* and *30*) in the $K_2Cr_2O_7$ treatment (Fig. 2c), five genes (*HsCIPK9, 14, 17, 24,* and *29*) in the $PbCl_2$ treatment (Fig. 3a), and eight genes (*HsCIPK2, 5, 11, 17, 23, 29, 30,* and *31*) in the $CuSO_4$ treatment (Fig. 3b) were significantly increased compared to those in the corresponding mock control (0 h time point). In addition, the maximum induction levels of all 12 endogenous *HsCIPKs* during each heavy metal treatment are briefly summarized in Additional file 1: Table S5, whereas these *HsCIPKs* that were transcriptionally three times higher than the mock control are presented in Figs. 2 and 3. These results suggest that the *HsCIPKs* respond transcriptionally to heavy metal toxicities.

Responses of *HsCIPKs* to other abiotic stresses

In additional to heavy metal toxicities, we also examined the effects of other abiotic stresses on the transcriptional expression levels of these 12 *HsCIPKs* in the Tibetan Plateau annual wild barley seedlings, including treatments with NaCl (400 mM), polyethylene glycol 6000 (PEG 6000; 20%; artificial drought), $AlCl_3$ (20 μM), low and high temperature (4 °C and 35 °C), and ABA (20 μM), respectively. Time-course analysis showed that the transcriptional levels of seven genes (*HsCIPK2, 5, 17, 24, 29, 30,* and *31*) in the NaCl treatment (Fig. 4a), three genes (*HsCIPK9, 29,* and *31*) in the PEG6000 treatment (Fig. 4b), five genes (*HsCIPK24, 28, 29, 30,* and *31*) in the $AlCl_3$ treatment (Fig.3c), three genes (*HsCIPK9, 30* and *31*) in the 4 °C treatment (Fig. 4c), two genes (*HsCIPK5* and *9*) in the 35 °C treatment (Fig. 4d), and three genes (*HsCIPK2, 17,* and *31*) in the ABA treatment (Fig. 4e) were significantly elevated relative to the corresponding mock control (0 h time point). Furthermore,

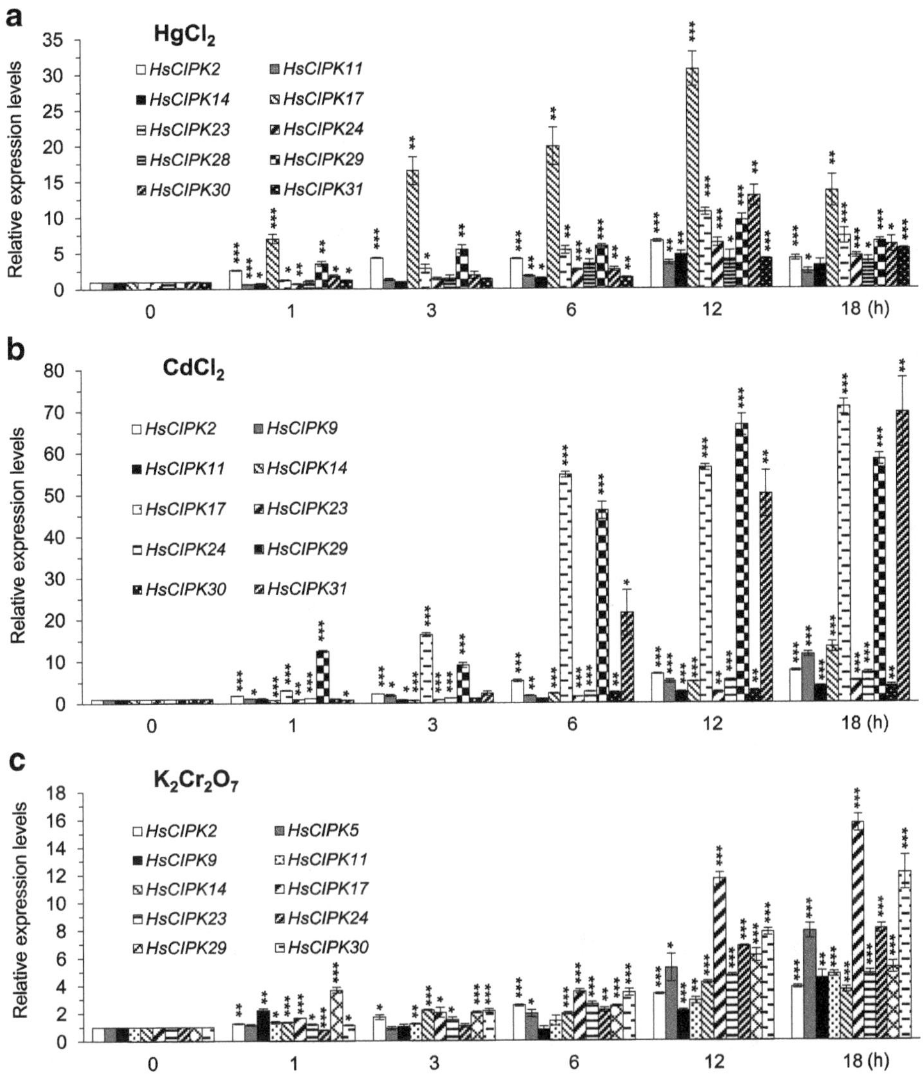

Fig. 2 Hg-, Cd-, and Cr-induced expression patterns of *HsCIPKs*. **a-c** Relative expression levels of *HsCIPKs* in four-day-old wild barley seedlings treated for different time lengths (0, 1, 3, 6, 12, and 18 h) with 20 μM $HgCl_2$ (**a**), 20 μM $CdCl_2$ (**b**), and 0.5 mM $K_2Cr_2O_7$ (**c**), respectively. Values shown are means ± SD. Single, double, and triple asterisks indicate $P < 0.01$, 0.001, and 0.0001, respectively (*t* test; compared to the corresponding 0-time point mock control). The expression level of the 0-time point mock control for each *HsCIPK* was set as 1.0

the maximum expression levels of all 12 endogenous *HsCIPKs* during salt, PEG, $AlCl_3$, temperature (4 °C and 35 °C), and ABA treatments are briefly summarized in Additional file 1: Table S6, whereas their induction levels, which were two times higher than the mock control are presented in Figs. 3c and 4. These results further confirmed the involvement of *HsCIPKs* in multiple abiotic stress responses.

Generation of rice transgenic lines overexpressing *HsCIPKs*

To provide further evidence for the roles of HsCIPKs in the plant response to heavy metal toxicities and other abiotic stresses, rice transgenic lines overexpressing individual *HsCIPKs* under the control of the CAMV 35S promoter were generated in the wild-type rice cultivar Nipponbare (*Oryza sativa* L. ssp. japonica) (Fig. 5a-c), which is a good transformation system and additionally possesses a closer genetic relationship to the wild barley than Arabidopsis at the CIPK levels (Additional file 1: Figsure S3 and S4). A RT-PCR assay with *HsCIPK*-specific PCR primers was used to examine the overexpression levels of the exogenous *HsCIPK* genes in the T1 generation plants. As shown in Additional file 1: Figure S5, multiple overexpression transgenic lines for each *HsCIPK* construct were recovered. In the T3 generation, homozygous transgenic lines with no further segregation were recovered via G418-resistant selection.

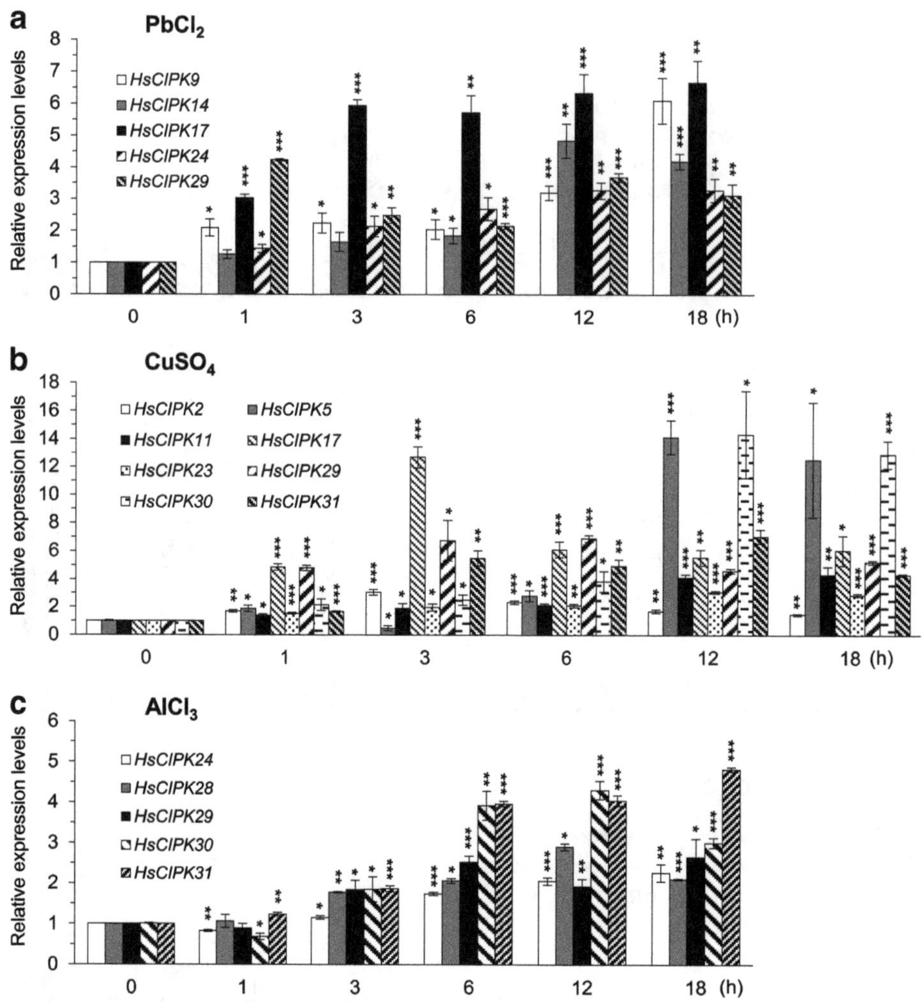

Fig. 3 Pb-, Cu-, and Al-induced expression patterns of *HsCIPKs*. **a-c** Relative expression levels of *HsCIPKs* in four-day-old wild barley seedlings treated for different time lengths (0, 1, 3, 6, 12, and 18 h) with 20 μM $PbCl_2$ (**a**), 1 mM $CuSO_4$ (**b**), and 20 μM $AlCl_3$ (**c**), respectively. Values shown are means ± SD. Single, double, and triple asterisks indicate P < 0.01, 0.001, and 0.0001, respectively (*t* test; compared to the corresponding 0-time point mock control). The expression level of the 0-time point mock control for each *HsCIPK* was set as 1.0

Functional analysis of *HsCIPK* overexpression in response to heavy metal toxicities

In plants, heavy metal toxicity-induced symptoms include the inhibition of seed germination and root elongation, wilting and stunted plant growth, chlorosis, leaf rolling and necrosis, and senescence, as well as low biomass (DalCorso et al. 2013), among which root elongation inhibition is an early symptom of heavy metal-induced responses. Thus, root elongation has been widely used as a sensitive indicator for plant responses to heavy metal toxicities (Wong and Bradshaw 1982) and other abiotic stresses (Llugany et al. 1995; Ishikawa et al. 1998). In this study, the relative elongation rate (RER; %) of the primary root was used to evaluate the root growth response to heavy metal toxicities. Heavy metal treatment experiments with different concentrations in four-day-old wild-type Nipponbare seedlings revealed that 0.5 μM $HgCl_2$, 5 μM $CdCl_2$, 5 μM $K_2Cr_2O_7$, 0.25 μM $CuSO_4$, and 5 μM $PbCl_2$ were suitable to examine the effects of *HsCIPK* overexpression on root growth response to heavy metal toxicities (Additional file 1: Figure S1). Upon transgenic seedling exposure to $HgCl_2$ treatment for 24 h, the overexpression of *HsCIPK2*, *HsCIPK14*, and *HsCIPK17* significantly enhanced root growth relative to the corresponding wild-type controls (nontransgenic regeneration lines; NT) (Fig. 5d). Similarly, transgenic lines individually overexpressing seven *HsCIPKs* (*HsCIPK9, 11, 14, 17, 23, 24,* and *29*) in the $CdCl_2$ treatment (Fig. 5e), five *HsCIPKs* (*HsCIPK2, 5, 9, 17,* and *30*) in the $K_2Cr_2O_7$ treatment (Fig. 5f), and four *HsCIPKs* (*HsCIPK5, 24, 29,* and *31*) in the $CuSO_4$

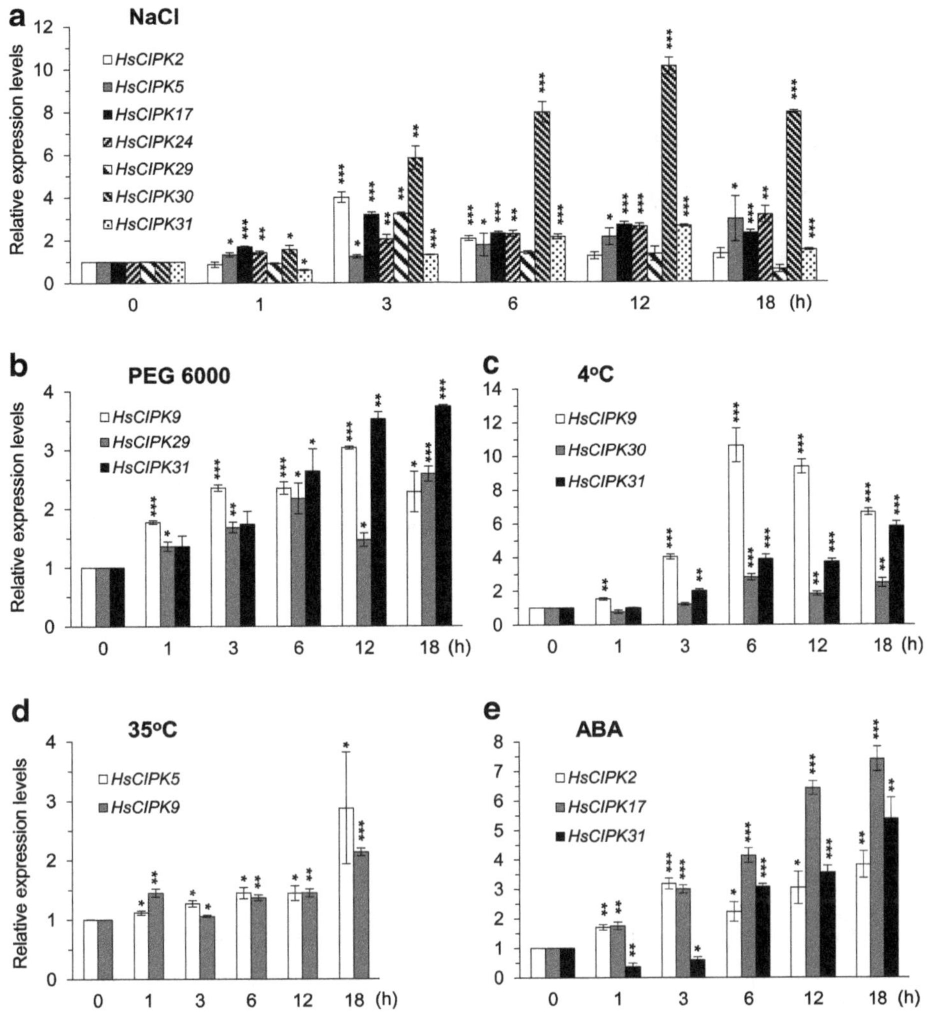

Fig. 4 Salt-, drought-, temperature-, and ABA-induced expression patterns of *HsCIPKs*. **a-e** Relative expression levels of *HsCIPKs* in four-day-old wild barley seedlings treated for different time lengths (0, 1, 3, 6, 12, and 18 h) with 400 mM NaCl (**a**), 20% PEG 6000 (**b**), 4 °C (**c**), 35 °C (**d**), and 20 μM ABA (**e**), respectively. Values shown are means ± SD. Single, double, and triple asterisks indicate $P < 0.01$, 0.001, and 0.0001, respectively (*t* test; compared to the corresponding 0-time point mock control). The expression level of the 0-time point mock control for each *HsCIPK* was set as 1.0

treatment (Fig. 5g) displayed a significant elevation in root growth relative to the corresponding NT. In the $PbCl_2$ treatment, surprisingly, no transgenic lines displayed enhanced effects in root growth relative to the NT (Fig. 5h). In contrast, the transgenic lines overexpressing *HsCIPK31* in the $K_2Cr_2O_7$ treatment (Fig. 5f) and *HsCIPK17, 28, 29,* and *30* in the $PbCl_2$ treatment (Fig. 5h), respectively, exhibited a significant reduction in root growth compared to the corresponding NT. The responses of root growth to these heavy metal toxicities in the transgenic lines of all 12 *HsCIPKs* are summarized in Additional file 1: Table S7. Taken together, these results suggest that multiple *HsCIPKs* are involved in the plant response to heavy metal toxicities.

Functional analysis of *HsCIPK* overexpression in response to other abiotic stresses

Next, to test whether the overexpression of *HsCIPKs* enhances plant tolerance to other abiotic stresses, we examined root growth of four-day-old transgenic rice seedlings treated for 24 h with salt (50 mM NaCl), drought (10% PEG6000), ABA (1 μM), and $AlCl_3$ (50 μM) (Additional file 1: Figure S1), respectively. As shown in Fig. 6a-c, upon transgenic seedling exposure to NaCl, PEG6000, and ABA treatments, respectively, the overexpression lines for six *HsCIPKs* (*HsCIPK2, 5, 17, 28, 29,* and *30*) in the NaCl treatment, four *HsCIPKs* (*HsCIPK17, 23, 29,* and *31*) in PEG 6000 treatment, and two *HsCIPKs* (*HsCIPK2* and *17*) in the ABA treatment displayed a significantly enhanced effect on root growth

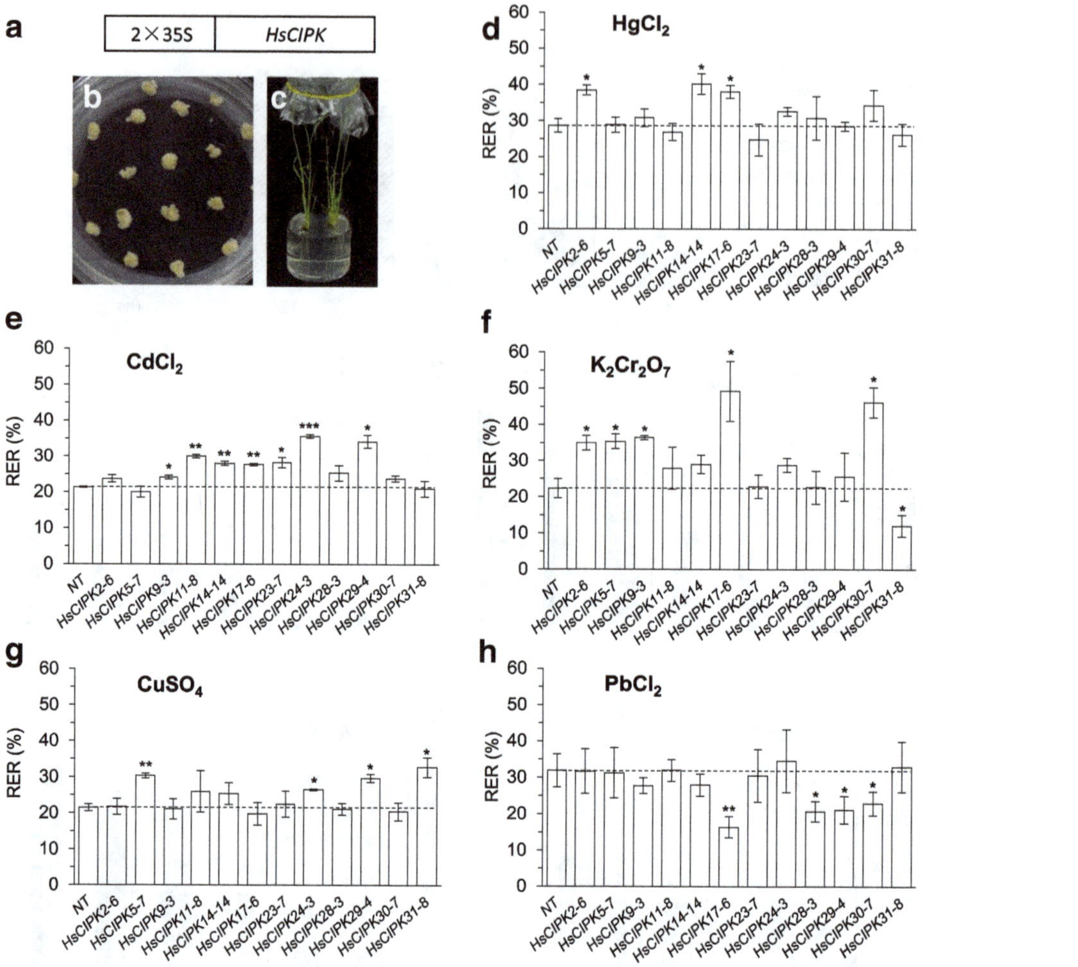

Fig. 5 Effects of heavy metal toxicities on root growth in rice seedlings overexpressing *HsCIPKs*. **a-c** Construct for *HsCIPK* overexpression (**a**), callus induction in rice mature embryos (**b**), and transgenic plants (**c**). d-h Root relative elongation rates (%; RER) of four-day-old rice seedlings overexpressing *HsCIPKs* treated for 24 h with 0.5 μM $HgCl_2$ (**d**), 5 μM $CdCl_2$ (**e**), 5 μM $K_2Cr_2O_7$ (**f**), 0.25 μM $CuSO_4$ (**g**), and 5 μM $PbCl_2$ (**h**), respectively. Values shown are means ± SD. Single, double, and triple asterisks indicate $P < 0.05$, 0.01, and 0.001, respectively (*t* test; compared to the wild-type control NT seedlings). NT, nontransgenic regeneration lines

relative to the corresponding NT, indicating a positive role of these *HsCIPKs* in plant tolerance to salt and drought stresses and ABA treatment. However, in the $AlCl_3$ treatment, no transgenic lines showed an enhanced effect on root growth relative to the NT (Fig. 6d). In contrast, the transgenic lines overexpressing two *HsCIPKs* (*HsCIPK23* and *29*) in the ABA treatment (Fig. 6c) and four *HsCIPKs (HsCIPK9, 14, 17,* and *30)* in the $AlCl_3$ treatment (Fig. 6d) exhibited an inhibitory effect on root growth relative to the corresponding NT, indicating a negative role of these *HsCIPKs* in the plant response to ABA and $AlCl_3$ stresses. The responses of root growth to these abiotic stresses in the transgenic lines of all 12 *HsCIPKs* are summarized in Additional file 1: Table S7. These results further confirmed the roles of *HsCIPKs* in the plant response to salt and drought stresses.

Discussion

Previous studies have demonstrated that plant-specific CIPKs function in plant responses to various abiotic and biotic stresses, including salt, drought, low and high temperature, wounding, low oxygen, and pathogen infection (Shen et al. 2014; Yu et al. 2014). However, the evidence for whether *CIPKs* are involved in the plant esponse to heavy metal toxicities currently remains lacking. In this study, we used an in silico assay and a molecular cloning strategy to isolate 12 *HsCIPKs* from Tibetan Plateau annual wild barley and subsequently examined the heavy metal toxicity-induced expression patterns of 12 endogenous *HsCIPK* genes and determined the role of their overexpression in the rice response to multiple heavy metal toxicities and other abiotic stresses.

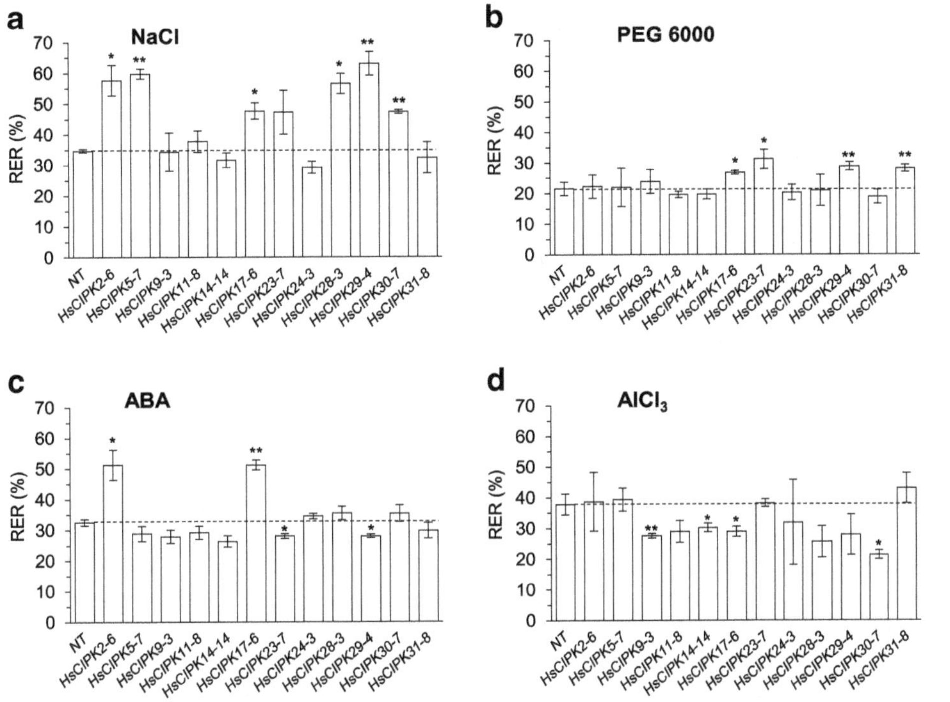

Fig. 6 Effects of salt, drought, ABA, and Al treatments on root growth in rice seedlings overexpressing *HsCIPKs*. **a-d** Root relative elongation rates (%; RER) of four-day-old rice seedlings overexpressing *HsCIPKs* treated for 24 h with 50 mM NaCl (**a**), 10% PEG6000 (**b**), 1 μM ABA (**c**), and 50 μM $AlCl_3$ (**d**), respectively. Values shown are means ± SD. Single and double asterisks indicate $P < 0.05$ and 0.01, respectively (*t* test; compared to the wild-type control NT seedlings). NT, nontransgenic regeneration lines

Heavy metal toxicities induce transcriptional expression of multiple *HsCIPKs*

Our time-course qRT-PCR analysis (Figs. 2, 3 and 4; Additional file 1: Tables S5 and S6) revealed that all of the 12 endogenous *HsCIPKs* examined can be induced by multiple heavy metal toxicities, including Hg, Cd, Cr, Pb, and Cu, and in addition, nine *HsCIPKs* (except for *HsCIPK11*, *14*, and *23*) are also induced by multiple other abiotic stresses, including salt, drought, cold, heat, and Al stresses, and ABA treatment. Heavy metal treatment experiments (Figs. 2 and 3; Additional file 1: Table S5) showed that *HsCIPK17* and *29* can be highly induced by all five heavy metal toxicities examined, whereas *HsCIPK2*, *11*, *14*, *23*, *24*, and *30* are highly induced by four heavy metal toxicities. In addition, *HsCIPK9* and *31* are also highly induced by three heavy metal toxicities. In addition to the response to heavy metal toxicities, *HsCIPKs* also respond to other abiotic stresses, including salt, drought, temperature, and Al stresses. Of note, the transcriptional induction levels corresponding to these stresses (Fig. 4; Additional file 1: Table S6) are relatively low compared to those corresponding to heavy metals (Figs. 2 and 3; Additional file 1: Table S5). These data suggest that multiple *HsCIPKs* are involved in the plant response to heavy metal toxicities and other abiotic stresses.

Previous studies have shown that wheat *TaCIPK14* and Arabidopsis *AtCIPK14* are upregulated by treatments with salt, PEG, and ABA (Deng et al. 2013; Yan et al. 2014). In contrast, our treatment experiments (Additional file 1: Tables S5 and S6; Figs. 2 and 3) revealed that *HsCIPK14* is downregulated by salt, PEG, heat, and ABA treatments but upregulated by multiple heavy metal toxicities, including Hg, Cd, Cr, and Pb. Similarly, *Brachypodium distachyon BdCIPK31* was found to be downregulated by salt, PEG, H_2O_2, and ABA treatments (Luo et al. 2017), whereas *HsCIPK31* is upregulated by not only salt, PEG, ABA, Al, and cold treatments but also multiple heavy metal toxicities, including Hg, Cd, and Cu (Figs. 2, 3 and 4; Additional file 1: Tables S5 and S6). These different responses clearly reflect the functional divergence of the *CIPK* paralog genes in distinct plant species in response to various abiotic stresses during long-term evolution.

Overexpression of the *HsCIPKs* enhances plant tolerance to heavy metal toxicities

The overexpression experiment in rice (Figs. 5 and 6; Additional file 1: Table S7) showed that the individual overexpression of 11 *HsCIPKs* (except for *HsCIPK28*) promotes root growth tolerance to multiple heavy metal toxicities (Hg, Cd, Cr, and Cu), and furthermore, plants individually overexpressing 8 *HsCIPKs* (except for *HsCIPK9, 11, 14,* and *24*) display enhanced tolerance of root growth to salt, drought, and ABA treatments, further confirming the involvement of the *HsCIPKs* in the plant response to heavy metal toxicities and other abiotic stresses, including salt and drought. However, we cannot rule out the possibility that the transgenic lines of some *HsCIPKs* with no positive role in root growth tolerance to most of stress treatments are due to their inadequate expression levels, for example, *HsCIPK11* and *HsCIPK23* (Additional file 1: Figure S5). In addition, $PbCl_2$ and $AlCl_3$ treatments in wild barley significantly induce the transcriptional expression of multiple *HsCIPKs* (*9, 14, 17, 24, 28, 29, 30,* and *31*; Fig. 3; Additional file 1: Tables S5 and S6), but the overexpression of *HsCIPKs* (*9, 14, 17, 28, 29,* and *30*) in rice inhibits the root growth tolerance to Pb and Al toxicities (Figs. 5 and 6; Additional file 1: Table S7). Similarly, the overexpression of *HsCIPK31* in rice inhibits the root growth tolerance to Cr toxicity, which does not visibly induce *HsCIPK31* transcriptional expression (Fig. 5; Additional file 1: Tables S6 and S7). Although Al and heavy metal toxicities affect plant root growth differentially, our results clearly indicate that these HsCIPKs act as a negative regulator in the plant tolerance to Pb, Al, and Cr toxicities. This finding is similar to the previous studies showing that ABA treatment induces endogenous *BnCIPK6* and *AtCIPK6* expression transcriptionally, although their overexpression in *Arabidopsis* enhances the sensitivity to the ABA treatment (Chen et al. 2012; Chen et al. 2013), and that the loss of *AtCIPK5* function mutants display an increased resistance response to pathogens (Meteignier et al. 2017), and therefore Bn/AtCIPK6 and AtCIPK5 are a negative regulator for plant tolerance to ABA treatment and pathogen attack, respectively. It is likely that during the plant response to Pb, Al, and Cr toxicities, these HsCIPKs with a negative role may exert feedback regulation on the positive functions of other HsCIPKs/CBLs and/or other calcium sensors (CaMs, CMLs, and CDPKs). This hypothesis will require additional study. Fortunately, a high-quality reference genome assembly for cultivated barley has been recently published (Mascher et al. 2017) and thereby will accelerate the functional identification of all *HsCIPK* family members.

It is widely acknowledged that multiple CIPKs in distinct plant species redundantly function in salt, drought, and ABA stress responses (Luan 2009; Weinl and Kudla 2009; Shen et al. 2014; Yu et al. 2014). Consistently, our findings (Figs. 5 and 6; Additional file 1: Table S7) reveal that the same single *HsCIPK* is involved in the response to multiple distinct heavy metal toxicities, whereas multiple *HsCIPKs* function in the response to the same single heavy metal toxicity, indicating that plant-specific *CIPKs* are functionally redundant in the regulation of the plant response to heavy metal toxicities. The identification of knockout or knockdown mutants for *HsCIPKs* will provide genetic evidence for HsCIPK function in heavy metal toxicities.

The traditional remediation strategies for heavy metals-contaminated soils primarily depend on physical and chemical methods and are expensive and relatively ineffective due to large-scale contamination farmlands and high costs. Thus, the best approaches are to genetically engineer crops with high resistance to heavy metal toxicities (Cao et al. 2014). Rice, as a primary food cereal in the world, is one of the major sources of heavy metal intakes for humans in inland China (Zhang et al. 2010), and therefore, developing rice cultivars with a higher tolerance to multiple heavy metal toxicities is critical for heavy metal-polluted soils that usually contain multiple heavy metal toxicities. Intriguingly, our results (Figs. 5 and 6; Additional file 1: Table S7) showed that the overexpression of the same single *HsCIPK* enhances rice growth tolerance to multiple heavy metal toxicities, indicating that these identified *HsCIPKs* (*2, 5, 9, 14, 17, 24,* and *29*) are important for genetically engineering crops with multiple heavy metal tolerances, which will have a broad application in heavy metal-contaminated soils. However, the mode of action by which HsCIPKs enhance plant tolerance to heavy metal toxicities via modulating heavy metal excretion and/or chelation mechanisms or scavenging heavy metal-induced ROS remain to be experimentally elucidated in future studies.

Conclusions

Using an in silico assay and a molecular cloning strategy to isolate 12 *HsCIPKs* in Tibetan Plateau annual wild barley enabled us to the effects of heavy metal toxicities on the transcriptional expression patterns of 12 endogenous *HsCIPK* genes and evaluate their functional roles in the response to multiple heavy metal toxicities and other abiotic stresses in rice. Multiple *HsCIPKs* were found to be involved in the response of plants to heavy metal toxicities, including Hg, Cd, Cr, Pb, and Cu and other abiotic stresses, including salt, drought, Al, low and high temperature, and ABA. The ectopic overexpression of *HsCIPK2, 5, 9, 14, 17, 24,* and *29* in Nipponbare rice (*Oryza sativa*) enhanced the tolerance of rice root growth to multiple heavy metal toxicities, whereas the overexpression of *HsCIPK2, 5, 17,* 23, *24,* 28, *29,* 30, and 31 promoted root growth tolerance to salt and/or drought stresses. These results suggest that plant-specific CIPKs function in heavy metal toxicities and these *HsCIPKs* examined will have a broad application in genetically engineered rice and other crops with tolerance to heavy metal toxicities and other abiotic stresses.

Additional file

Additional file 1: Figure S1. Effects of abiotic stresses on root elongation in rice. **Figure S2.** Alignment analysis of amino acid sequences among HsCIPK5, OsCIPK5, and AtCIPK5. **Figure S3.** GenBank accession numbers and homology analysis of HsCIPKs. **Figure S4.** Phylogenetic analysis of HsCIPKs, OsCIPKs, and AtCIPKs. **Figure S5.** Molecular identification of *HsCIPK* overexpression in rice. **Table S1.** Accession numbers of rice and *Arabidopsis CIPK* cDNA in NCBI database. **Table S2.** Primer sequences for *HsCIPKs* cloning. **Table S3.** Primer sequences for qRT-PCR assay in Tibetan Plateau annual wild barley. **Table S4.** Primer sequences for RT-PCR Assay in rice transgenic lines. **Table S5.** Summary of endogenous *HsCIPK* responses to heavy metal toxicities. **Table S6.** Summary of endogenous *HsCIPK* responses to other abiotic stresses. **Table S7.** Summary of root growth tolerance to abiotic stresses in rice transgenic lines overexpressing *HsCIPKs*.

Abbreviations

ABA: Abscisic acid; Al: Aluminum; Ca^{2+}: Calcium; CBL: Calcineurin B-like protein; Cd: Cadmium; CDS: Coding sequence; CIPK: CBL-interacting protein kinase; Cr: Chromium; Cu: Copper; Hg: Mercury; NAF/FISL: Conserved amino acid motif NAF/FISL; NC: Negative control; NT: Nontransgenic regeneration lines; Pb: Lead; PEG 6000: Polyethylene glycol 6000; PPI: Protein–phosphatase interaction domain; qRT-PCR: Quantitative Real-Time PCR; RER: Relative elongation rates; RT-PCR: Reverse transcription-polymerase chain reaction

Acknowledgements

We thank Guoping Zhang (Zhejiang University), Yinong Yang (Pennsylvania State University), and Lin Yang (Zhejiang Normal University) for sharing materials.

Funding

This study was supported by the National Science and Technology Major Program for New Varieties of Genetically Modified Crops (2009ZX08009076–003), the National Natural Science Foundation of China (No. 31670283 and 91754104), the Natural Science Foundation of Zhejiang Province (No. LY16C130004 and LY17C060001), and the Fundamental Research Funds for the Central Universities (lzujbky-2017-ot16).

Authors' contributions

WP, JS, ZZ, LJ, and JP conceived the study and designed the experiments. WP, JS, ZZ, XY, JS, and WW carried out the bioinformatic assays and the experiments. WP, JS, ZZ, LJ, and JP analyzed the data. WP and JP wrote the manuscript. WP, LJ, and JP revised the manuscript. All authors read and approved the final manuscript.

Competing interests

The authors declare that they have no competing interests.

Author details

[1]MOE Key Laboratory of Cell Activities and Stress Adaptations, School of Life Sciences, Lanzhou University, Lanzhou 730000, China. [2]College of Life Sciences, Shaoxing University, Shaoxing 312000, Zhejiang, China. [3]College of Chemistry and Life Sciences, Zhejiang Normal University, Jinhua 321004, Zhejiang, China. [4]Department of Agronomy, College of Agriculture and Biotechnology, Zhejiang University, Hangzhou 310058, China.

References

Ahmad A, Hadi F, Ali N (2015) Effective phytoextraction of cadmium (cd) with increasing concentration of total phenolics and free proline in *Cannabis sativa* (L) plant under various treatments of fertilizers, plant growth regulators and sodium salt. Int J Phytoremed 17:56–65

Batistic O, Waadt R, Steinhorst L, Held K, Kudla J (2010) CBL-mediated targeting of CIPKs facilitates the decoding of calcium signals emanating from distinct cellular stores. Plant J 61:211–222

Cao FB, Wang RF, Cheng WD, Zeng FR, Ahmed IM, Hu XN, Zhang GP, Wu FB (2014) Genotypic and environmental variation in cadmium, chromium, lead and copper in rice and approaches for reducing the accumulation. Sci Total Environ 496:275–281

Chen L, Ren F, Zhou L, Wang QQ, Zhong H, Li XB (2012) The *Brassica napus* Calcineurin B-like 1/CBL-interacting protein kinase 6 (CBL1/CIPK6) component is involved in the plant response to abiotic stress and ABA signaling. J Exp Bot 63:6211–6222

Chen L, Wang QQ, Zhou L, Ren F, Li DD, Li XB (2013) Arabidopsis CBL interacting protein kinase (CIPK6) is involved in plant response to salt/osmotic stress and ABA. Mol Biol Rep 40:4759–4767

Chen XF, Gu ZM, Xin DD, Hao L, Liu CJ, Huang J, Ma BJ, Zhang HS (2011) Identification and characterization of putative CIPK genes in maize. J Genet Genomics 38:77–87

Dai F, Nevo E, Wu DZ, Comadran J, Zhou MX, Qiu L, Cheng ZH, Belles A, Chen GX, Zhang GP (2012) Tibet is one of the centers of domestication of cultivated barley. Proc Natl Acad Sci U S A 109:16969–16973

DalCorso G, Manara A, Furini A (2013) An overview of heavy metal challenge in plants: from roots to shoots. Metallomics 5:1117–1132

Dekeyser R, Claes B, Marichal M, Van Montagu M, Caplan A (1989) Evaluation of selectable markers for rice transformation. Plant Physiol 90:217–223

Deng X, Zhou S, Hu W, Feng J, Zhang F, Chen L, Huang C, Luo Q, He Y, Yang G, He G (2013) Ectopic expression of wheat *TaCIPK14*, encoding a calcineurin B-like protein-interacting protein kinase, confers salinity and cold tolerance in tobacco. Physiol Plant 149:367–377

Dodd AN, Kudla J, Sanders D (2010) The language of calcium signaling. Annu Rev Plant Biol 61:593–620

Fang H, Jing T, Liu Z, Zhang L, Jin Z, Pei Y (2014) Hydrogen sulfide interacts with calcium signaling to enhance the chromium tolerance in *Setaria italica*. Cell Calcium 56:472–481

Guo Y, Halfter U, Ishitani M, Zhu JK (2001) Molecular characterization of functional domains in the protein kinase SOS2 that is required for plant salt tolerance. Plant Cell 13:1383–1400

Hiei Y, Komari T (2008) Agrobacterium-mediated transformation of rice using immature embryos or calli induced from mature seed. Nat Protoc 3:824–834

Hu W, Xia Z, Yan Y, Ding Z, Tie W, Wang L, Zou M, Wei Y, Lu C, Hou X, Wang W, Peng M (2015) Genome-wide gene phylogeny of CIPK family in cassava and expression analysis of partial drought-induced genes. Front Plant Sci 6:914

Huang TL, Huang LY, Fu SF, Trinh NN, Huang HJ (2014) Genomic profiling of rice roots with short-and long-term chromium stress. Plant Mol Biol 86:157–170

Ishikawa S, Wagatsuma T (1998) Plasma membrane permeability of root-tip cells following temporary exposure to Al ions is a rapid measure of Al tolerance among plant species. Plant Cell Physiol 39:516–525

Kim KN, Cheong YH, Grant JJ, Pandey GK, Luan S (2003) *CIPK3*, a calcium sensor-associated protein kinase that regulates abscisic acid and cold signal transduction in Arabidopsis. Plant Cell 15:411–423

Kolukisaoglu U, Weinl S, Blazevic D, Batistic O, Kudla J (2004) Calcium sensors and their interacting protein kinases: genomics of the Arabidopsis and rice CBL-CIPK signaling networks. Plant Physiol 134:43–58

Kurusu T, Hamada J, Nokajima H, Kitagawa Y, Kiyoduka M, Takahashi A, Hanamata S, Ohno R, Hayashi T, Okada K, Koga J, Hirochika H, Yamane H, Kuchitsu K (2010) Regulation of microbe-associated molecular pattern induced hypersensitive cell death, phytoalexin production, and defense gene expression by calcineurin B-like protein-interacting protein kinases, OsCIPK14/15, in rice cultured cells. Plant Physiol 153:678–692

Lan T, You J, Kong L, Yu M, Liu M, Yang Z (2016) The interaction of salicylic acid and Ca^{2+} alleviates aluminum toxicity in soybean (Glycine max L.). Plant Physiol Biochem 98:146–154

Lee KW, Chen PW, Lu CA, Chen S, Ho TH, Yu SM (2009) Coordinated responses to oxygen and sugar deficiency allow rice seedlings to tolerate flooding. Sci Signal 2:ra61. https://doi.org/10.1126/scisignal.2000333

Li J, Jiang MM, Ren L, Liu Y, Chen HY (2016) Identification and characterization of CBL and CIPK gene families in eggplant (*Solanum melongena* L.). Mol Gen Genomics 291:1769–1781

Li R, Zhang J, Wu G, Wang H, Chen Y, Wei J (2012) HbCIPK2, a novel CBL-interacting protein kinase from halophyte *Hordeum brevisubulatum*, confers salt and osmotic stress tolerance. Plant Cell Environ 35:1582–1600

Llugany M, Poschenrieder C, Barceló J (1995) Monitoring of aluminum-induced inhibition of root elongation in four maize cultivars differing in tolerance to aluminum and proton toxicity. Physiol Plant 93:265–271

Luan S (2009) The CBL-CIPK network in plant calcium signaling. Trends Plant Sci 14:37–42

Luo Q, Wei Q, Wang R, Zhang Y, Zhang F, He Y, Zhou S, Feng J, Yang G, He G (2017) BdCIPK31, a calcineurin B-like protein interacting protein kinase, regulates plant response to drought and salt stress. Front Plant Sci 8:1184

Mascher M et al (2017) A chromosome conformation capture ordered sequence of the barley genome. Nature 544:427–433

Meteignier LV, El Oirdi M, Cohen M, Barff T, Matteau D, Lucier JF, Rodrigue S, Jacques PE, Yoshioka K, Moffett P (2017) Translatome analysis of an NB-LRR immune response identifies important contributors to plant immunity in *Arabidopsis*. J Exp Bot 68:2333–2344

Mustafa G, Komatsu S (2016) Toxicity of heavy metals and metal-containing nanoparticles on plants. Biochim Biophys Acta 1864:932–494

Ohta M, Guo Y, Halfter U, Zhu JK (2003) A novel domain in the protein kinase SOS2 mediates interaction with the protein phosphatase2C ABI2. PNAS 100: 11771–11776

Ozawa K (2009) Establishment of a high efficiency agrobacterium-mediated transformation system of rice (*Oryza sativa* L.). Plant Sci 176:522–527

Pan JW, Zheng K, Ye D, Yi HL, Jiang ZM, Jing CT, Pan WH, Zhu MY (2004) Aluminum-induced ultraweak luminescence changes and sister-chromatid exchanges in root tip cells of barley. Plant Sci 167:1391–1399

Pan W, Shou J, Zhou X, Zha X, Guo T, Zhu M, Pan J (2011) Al-induced cell wall hydroxyproline-rich glycoprotein accumulation is involved in alleviating Al toxicity in rice. Acta Physiol Plant 33:601–608

Pandey GK, Cheong YH, Kim BG, Grant JJ, Li L, Luan S (2007) CIPK9: a calcium sensor-interacting protein kinase required for low-potassium tolerance in Arabidopsis. Cell Res 17:411–421

Rudd JJ, Franklin-Tong VE (1999) Calcium signaling in plants. Cell Mol Life Sci 55: 214–232

Sharma SS, Dietz KJ (2009) The relationship between metal toxicity and cellular redox imbalance. Trends Plant Sci 14:43–50

Shen JQ, Zheng ZZ, Pan WH, Pan JW (2014) Functions and action mechanisms of CBL-CIPK signaling system in plants. Plant Physiol J 50:641–650 (Chinese manuscript with English abstract)

Singh S, Parniske M (2012) Activation of calcium- and calmodulin-dependent protein kinase (CCaMK), the central regulator of plant root endosymbiosis. Curr Opin Plant Biol 15:444–453

Straub T, Ludewig U, Neuhauser B (2017) The kinase CIPK23 inhibits ammonium transport in *Arabidopsis thaliana*. Plant Cell 29:409–422

Sun T, Wang Y, Wang M, Li T, Zhou Y, Wang X, Wei S, He G, Yang G (2015) Identification and comprehensive analyses of the CBL and CIPK gene families in wheat (*Triticum aestivum* L.). BMC Plant Biol 15:269

Toki S, Hara N, Ono K, Onodera H, Tagiri A, Oka S, Tanaka H (2006) Early infection of scutellum tissue with agrobacterium allows high-speed transformation of rice. Plant J 47:969–976

de la Torre F, Gutierrez-Beltran E, Pareja-Jaime Y, Chakravarthy S, Martin GB, del Pozo O (2013) The tomato calcium sensor Cbl10 and its interacting protein kinase Cipk6 defense a signaling pathway in plant immunity. Plant Cell 25: 2748–2764

Wang JJ, Lu XK, Yin ZJ, Mu M, Zhao XJ, Wang DL, Wang S, Fan WL, Guo LX, Ye WW, Yu SX (2016a) Genome-wide identification and expression analysis of CIPK genes in diploid cottons. Genet Mol Res 15(4). https://doi.org/10.4238/gmr15048852

Wang XP, Chen LM, Liu WX, Shen LK, Wang FL, Zhou Y, Zhang Z, Wu WH, Wang Y (2016b) AtKC1 and CIPK23 synergistically modulate AKT1-mediated low-potassium stress responses in Arabidopsis. Plant Physiol 170:2264–2277

Weinl S, Kudla J (2009) The CBL-CIPK Ca^{2+}-decoding signaling network: function and perspectives. New Phytol 184:517–528

Wong MH, Bradshaw AD (1982) A comparison of the toxicity of heavy metals using root elongation of rye grass *lolium perenne*. New Phytol 91:255–262

Xiong L, Yang Y (2003) Disease resistance and abiotic stress tolerance in rice are inversely modulated by an abscisic acid–inducible mitogen-activated protein kinase. Plant Cell 15:745–759

Xu J, Li HD, Chen LQ, Wang Y, Liu LL, He L, Wu WH (2006) A protein kinase, interacting with two calcineurin B-like proteins, regulates K+ transporter AKT1 in Arabidopsis. Cell 125:1347–1360

Yan J, Niu F, Liu WZ, Zhang H, Wang B, Lan W, Che Y, Yang B, Luan S, Jiang YQ (2014) Arabidopsis CIPK14 positively regulates glucose response. Biochem Biophys Res Commun 450:1679–1683

Yu Q, An L, Li W (2014) The CBL-CIPK network mediates different signaling pathways in plants. Plant Cell Rep 33:203–214

Yu YH, Xia XL, Yin WL, Zhang HC (2007) Comparative genomic analysis of CIPK gene family in *Arabidopsis* and *Populus*. Plant Growth Regul 52:101–110

Zhang H, Feng XB, Larssen T, Qiu GL, Vogt RD (2010) In inland China, rice, rather than fish, is the major pathway for methylmercury exposure. Environ Health Perspect 118:1183–1188

Zhang H, Yang B, Liu WZ, Li H, Wang L, Wang B, Deng M, Liang W, Deyholos MK, Jiang YQ (2014) Identification and characterization of CBL and CIPK gene families in canola (*Brassica napus* L.). BMC Plant Biol 14:8

Zhang WH, Rengel Z (1999) Aluminum induces an increase in cytoplasmic calcium in intact wheat root tip cells. Aust J Plant Physiol 26:401–409

Zheng ZZ, Shen JQ, Pan WH, Pan JW (2013) Calcium sensors and their stress signaling pathways in plants. Yi Chuan 35:875–884 (Chinese manuscript with English abstract)

Zhu K, Chen F, Liu J, Chen X, Hewezi T, Cheng ZM (2016) Evolution of an intron-poor cluster of the CIPK gene family and expression in response to drought stress in soybean. Sci Rep 6:28225

13

The transcription factor OsbHLH035 mediates seed germination and enables seedling recovery from salt stress through ABA-dependent and ABA-independent pathways, respectively

Hung-Chi Chen[1], Wan-Hsing Cheng[2], Chwan-Yang Hong[3], Yu-Sen Chang[4] and Men-Chi Chang[1*]

Abstract

Background: Many transcription factors (TFs), such as those in the basic helix-loop-helix (bHLH) family, are important for regulating plant growth and plant responses to abiotic stress. The expression of *OsbHLH035* is induced by drought and salinity. However, its functional role in rice growth, development, and the salt response is still unknown.

Results: The bHLH TF *OsbHLH035* is a salt-induced gene that is primarily expressed in germinating seeds and seedlings. Stable expression of GFP-fused OsbHLH035 in rice transgenic plants revealed that this protein is predominantly localized to the nucleus. *Osbhlh035* mutants show delayed seed germination, particularly under salt-stress conditions. In parallel, abscisic acid (ABA) contents are over-accumulated, and the expression of the ABA biosynthetic genes *OsABA2* and *OsAAO3* is upregulated; furthermore, compared with that in wild-type (WT) seedlings, the salt-induced expression of *OsABA8ox1*, an ABA catabolic gene, in germinating *Osbhlh035* mutant seeds is downregulated. Moreover, *Osbhlh035* mutant seedlings are unable to recover from salt-stress treatment. Consistently, sodium is over-accumulated in aerial tissues but slightly reduced in terrestrial tissues from *Osbhlh035* seedlings after salt treatment. Additionally, the expression of the sodium transporters *OsHKT1;3* and *1;5* is reduced in *Osbhlh035* aerial and terrestrial tissues, respectively. Furthermore, genetic complementation can restore both the delayed seed germination and the impaired recovery of salt-treated *Osbhlh035* seedlings to normal growth.

Conclusion: OsbHLH035 mediates seed germination and seedling recovery after salt stress relief through the ABA-dependent and ABA-independent activation of *OsHKT* pathways, respectively.

Keywords: ABA, bHLH, OsHKT, Salt stress, Transcription factor

Background

Salt stress severely affects plant growth and development. In fact, more than 20% of the land used for agriculture is of poor quality due to soil salinity (FAO, http://www.fao.org/home/en/). The adverse effects of salt stress on plant growth and development can be attributed to initial osmotic dehydration and subsequent ion toxicity. Osmotic stress signaling is known to be transduced via both ABA-dependent (e.g., mitogen-activated protein kinases [MAPKs]) and ABA-independent (e.g., via the OsDREB1 and OsDREB2 TFs) pathways (Kumar et al. 2013). Additionally, the maintenance of cellular ion homeostasis is important for plant survival under salt stress, especially in response to high cytosolic K^+/Na^+ ratios. In fact, many transporters, including HKTs (high-affinity potassium transporters), NHXs (Na^+/H^+ exchangers), and SOS (salt overly sensitive), act collectively to regulate intracellular sodium levels

* Correspondence: menchi@ntu.edu.tw
[1]Department of Agronomy, National Taiwan University, No. 1, Section 4, Roosevelt Road, Taipei, Taiwan, Republic of China
Full list of author information is available at the end of the article

(Deinlein et al. 2014). The regulation of *HKTs* expression is critical for mediating salt stress tolerance in crops (Hamamoto et al. 2015). HKT transporters can be divided into two subfamilies. Subfamily I *HKT* genes (*HKT1s*) encode Na^+-selective transporters, which are mostly located at the plasma membrane in xylem parenchyma cells. The functional role of HKT1s is responsible for retrieving Na^+ from the xylem sap, which prevent aerial tissues from Na^+ over-accumulation and toxicity. Subfamily II HKT transporter (HKT2) is involved in uptake of Na^+ from the external medium (soil) when K^+ is limiting. Therefore, when plants are exposed to salt, the phytohormone ABA, TFs, and transporters act as key components that integrate cellular signal transduction and stress-responsive gene expression to confer salt tolerance.

ABA plays a critical role in regulating seed dormancy, germination, and stress tolerance (Karssen et al. 1983; Koornneef et al. 1989; Groot and Karssen 1992; Qin and Zeevaart 2002; Finkelstein et al. 2008). To date, most genes related to ABA metabolism have been cloned and characterized (for review, see Taylor et al. 2000; Seo and Koshiba 2002; Nambara and Marion-Poll 2005; Ye et al. 2012). The first committed step of ABA biosynthesis involves the oxidative cleavage of a C_{40} 9-*cis*-epoxycarotenoid, such as 9'-*cis*-neoxanthin or 9-*cis*-violaxanthin, to a C_{15} xanthoxin (the primary precursor) and a C_{25} by-product, which is catalyzed in plastids by 9-*cis*-epoxycarotenoid dioxygenase (NCED) (Schwartz et al. 2003; Xiong and Zhu 2003). At least 5 *NCED* genes have been documented in *Arabidopsis*, and 3 have been reported in rice (Tan et al. 2003; Welsch et al. 2008). Of the five *AtNCEDs* (*AtNCED2, 3, 5, 6,* and *9*), only *AtNCED3* is significantly induced by drought stress, and it controls the levels of endogenous ABA under drought-stress conditions (Iuchi et al. 2001). In rice, the expression of *OsNCED3, OsNCED4,* and *OsNCED5* was induced after 1 h of salt stress and closely correlated with the endogenous ABA contents in roots (Welsch et al. 2008). ABA biosynthesis subsequently involves the conversion of xanthoxin to abscisic aldehyde (an intermediate), and the final steps of ABA production are catalyzed in the cytosol by AtABA2 and AtAAO3 in *Arabidopsis* (Seo et al. 2000; Rook et al. 2001; Cheng et al. 2002; Gonzalez-Guzman et al. 2002). For full catalytic activity, AtAAO3 requires a molybdenum cofactor, which is derived from AtABA3 function (Bittner et al. 2001; Xiong et al. 2001). Indeed, loss-of-function mutations in *AtABA2, AtAAO3,* and *AtABA3* lead to ABA deficiency and early germination. Cellular ABA homeostasis is regulated by both the biosynthesis and degradation of ABA. ABA catabolism occurs via two main types of chemical reactions, hydroxylation and conjugation, with the hydroxylation of ABA at position C-8' being the key regulatory step (Kushiro et al. 2004; Saito et al. 2004; Nambara and Marion-Poll 2005). *Arabidopsis* and rice possess 4 and 3 known *ABA8ox* genes, respectively (Kushiro et al. 2004; Saito et al. 2004; Saika et al. 2007). The four *AtABA8ox* genes, also known as *CYP707As* (*CYP707A1, 2, 3,* and *4*), display complex spatiotemporal expression patterns and are important for many physiological processes. For example, *CYP707A1* and *CYP707A2* but not *CYP707A3* or *CYP707A4* are responsible for relieving the ABA-mediated inhibition of seed germination (Okamoto et al. 2006). These *CYP707As* are induced by osmotic stress presumably because the maintenance of endogenous ABA homeostasis under stress is necessary for plant survival (Saito et al. 2004). Although the biochemical and physiological roles of these ABA metabolic genes have been well characterized and documented, the mechanisms of their transcriptional regulation are still poorly understood.

Deciphering the roles of TFs is essential for revealing abiotic stress tolerance mechanisms in plants (Singh et al. 2002). In fact, several different types of TFs have been demonstrated to be involved in regulating abiotic stress responses (for review, see Nakashima et al. 2009; Todaka et al. 2012). For example, AtbHLH116/ICE1-*CBF3/DREB1A* [*AtERF#031* (At4g25480)] acts in the cold-responsive signaling pathway via an ABA-independent pathway (Chinnusamy et al. 2003); ABI3, 4, and 5 (B3, AP2/ERF, and bZIP, respectively) act in the osmotic stress response pathway via an ABA-dependent pathway (Giraudat et al. 1992; Finkelstein et al. 1998; Finkelstein and Lynch 2000); and SUB1A (AP2/ERF) acts in submergence- and drought-responsive pathways (Fukao et al. 2011). Additionally, many TF types, including bHLH (Zhou et al. 2009; Li et al. 2010), WRKY (Tao et al. 2011), NAC (Hu et al. 2006), and AP2/ERF (Schmidt et al. 2013), are involved in the regulation of salt stress tolerance in rice. Approximately 7% of plant genes encode TFs. The *Arabidopsis* and rice (*Oryza sativa subsp. japonica*) genomes contain 2016 and 2424 TFs, respectively, that can be categorized into 58 and 56 families based on their DNA binding domains (Zhang et al. 2011). Basic helix-loop-helix (bHLH) TFs comprise the second largest TF family in plants and can affect many developmental and physiological processes (Feller et al. 2011). To date, at least 167 *Arabidopsis* and 177 rice *bHLHs* have been identified (Li et al. 2006; Carretero-Paulet et al. 2010). These genes are classified based on the proteins they encode, which usually contain a conserved bHLH domain (approximately 60 amino acids long), with the exception of a few atypical *bHLH* genes known as *HLHs* that lack the basic region (Li et al. 2006). A typical bHLH domain comprises two functionally distinct regions: a basic region for DNA binding and an HLH region for protein homodimerization or heterodimerization (Massari and Murre 2000; Toledo-Ortiz et al. 2003; Feller et al. 2011). Thus, HLH proteins can heterodimerize with bHLH proteins, thereby disrupting bHLH-bHLH interactions and preventing DNA binding (Sun et al. 1991; Toledo-Ortiz et al. 2003). bHLH proteins can recognize two types of target *cis*-acting elements: the E-box (5'-CANNTG-3') and the G-box (5'-CACGTG-3'). Previous studies have demonstrated that several AtbHLHs are involved in regulating abiotic stress responses. AtbHLH116/ICE1 regulates the expression of *CBF3/DREB1A*, an AP2/ERF TF, that provides freezing tolerance through an ABA-independent pathway that is also known as the ICE-CBF regulon (Chinnusamy et al. 2003). OrbHLH1 and OrbHLH2, two ICE-like proteins in wild rice (*Oryza rufipogon*), positively regulate salt-stress responses in transgenic *Arabidopsis* through an ICE/CBF-independent and ABA-independent pathway, respectively (Zhou et al. 2009; Li et al. 2010). More recently, AtbHLH122 has been shown to confer tolerance to drought, salt, and osmotic stresses through an ABA-dependent pathway via the transcriptional repression of *CYP707A3*, thereby blocking ABA degradation (Liu et al. 2014). Notably, only approximately 10% of *OsbHLHs* have been characterized, in contrast to 38% of *AtbHLHs* (Heang and Sassa 2012).

To investigate which uncharacterized *OsbHLHs* may also be involved in the regulation of abiotic stress responses, we analyzed the gene expression profiles from publicly available microarray data (GSE6901) in the NCBI-GEO database, and several abiotic stress-responsive *OsbHLHs* were found, including *OsbHLH035* (Additional file 1: Figure S1). In this study, we examined the roles of *OsbHLH035* in rice during the germination and seedling stages. Using a reverse genetic approach with a *Tos17*-tagged mutant line, we found that *OsbHLH035* mediates germination and confers recovery after salt stress relief through ABA-dependent and ABA-independent pathways, respectively.

Results

Characterization of the OsbHLH035 gene and its Tos17-tagged mutant line, NG7221

In silico analyses of gene expression profiles revealed that *OsbHLH035* gene expression is upregulated by drought and salt treatments (Additional file 1: Figure S1). The *OsbHLH035* gene contains two exons, and the encoded protein, which putatively binds G-boxes, contains a typical bHLH domain (residues 64 to 113) (Fig. 1a, Additional file 1: Figure S2A). To verify the microarray data, we compared the expression pattern of *OsbHLH035* in aerial tissues from WT rice (*Oryza sativa* L. cv. Nipponbare, Nip) under normal and salt-treated conditions using RT-PCR. As shown in Fig. 1b, *OsbHLH035* transcripts were barely detectable in aerial tissues under normal conditions, whereas they were present under salt-treated conditions. These data indicate that *OsbHLH035* is reliably induced by salt stress in rice seedlings.

To investigate the roles of *OsbHLH035* in plant growth and salt-stress responses, we searched the rice *Tos17* insertion mutant database and identified a retrotransposon insertional *Osbhlh035* mutant line, NG7221. Based on its annotation, the retrotransposon is located

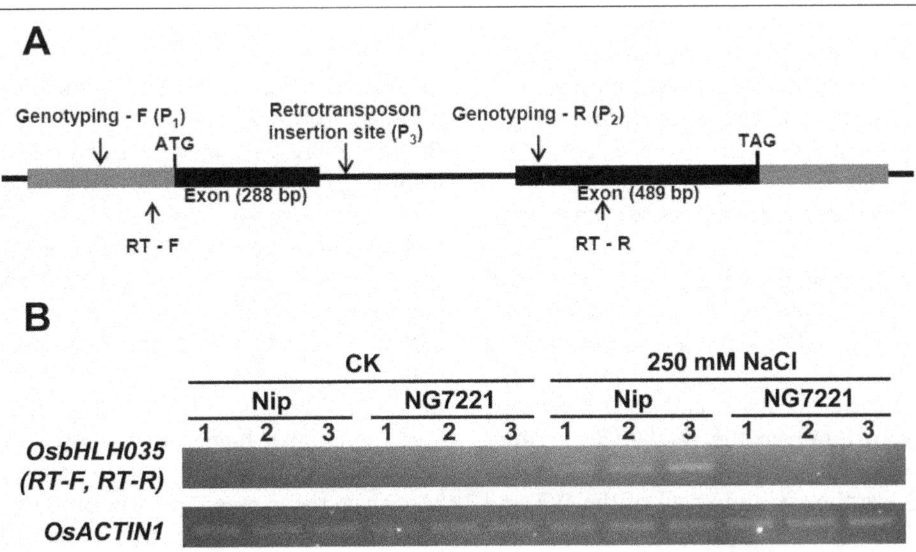

Fig. 1 *OsbHLH035* expression is induced by salt and is abolished in the NG7221 line. **a** The structure of the *OsbHLH035* gene and the retrotransposon insertion site. **b** *OsbHLH035* expression patterns in aerial tissues from third-leaf-stage WT and NG7221 seedlings. The seedlings were grown on basal medium for 13 days and then transferred to basal medium containing 0 (CK) or 250 mM NaCl for an additional day. The primer positions and sequences are shown in (**a**) and Additional file 2: Table S1. Arabic numerals in (**b**) represent three independent biological replicates within each genotype

in an *OsbHLH035* gene intron (Fig. 1a, Additional file 1: Figure S2B). Homozygous mutants were identified from the segregating population by genomic DNA genotyping (Additional file 1: Figure S2C). RT-PCR analysis showed that *OsbHLH035* transcripts were absent in aerial tissues from third-leaf-stage NG7221 seedlings under both normal and salt-treated conditions (Fig. 1b). Therefore, the NG7221 line is used as an *Osbhlh035* mutant in this study.

Spatiotemporal OsbHLH035 expression and OsbHLH035 localization

Analyzing spatiotemporal gene expression patterns and protein localization can provide valuable insights into gene function. Thus, we monitored *OsbHLH035* gene expression by detecting β-glucuronidase (GUS) signals in *OsbHLH035::GUS* transgenic plants from the germination stage until the third-leaf stage. In the germinating seeds, time-dependent expression of *GUS* was observed in the embryo, scutellum, aleurone layer, and endosperm after imbibition (Fig. 2a). In the aerial tissues, *GUS* was expressed in the hypocotyl and coleoptile at the postgermination stage but was difficult to detect in the third-leaf-stage seedlings (Fig. 2a, b). In terrestrial tissues, GUS signals were present in the root tip, vascular tissue, and lateral root initiation sites of both postgermination and third-leaf-stage seedlings (Fig. 2a, b).

To validate the spatial expression of *OsbHLH035* and the subcellular localization of OsbHLH035 protein in rice, an *OsbHLH035::GFP-OsbHLH035* construct was transformed into the *Osbhlh035* mutant via stable *Agrobacterium*-mediated transformation. Similar to its transcriptional expression pattern, the GFP-OsbHLH035 fusion protein was present in the root tip and vascular tissue (Fig. 2b vs. c) and was predominantly localized to the nucleus in rice cells and calli (Fig. 2d, Additional file 1: Figure S3).

Excess ABA leads to reduced germination rates in Osbhlh035 mutants

To investigate whether OsbHLH035 plays a role in regulating germination, we compared the germination rates of both WT and *Osbhlh035* plants under normal and salt-treated conditions. Under normal conditions, the WT germination rates were 63.46% and 90.38% on days 2 and 3, respectively; however, the *Osbhlh035* germination rates were reduced to 55.93% and 81.64%, respectively (Fig. 3a, b). Nevertheless, almost all of the WT and the *Osbhlh035* seeds had germinated by day 4. Under salt-treated conditions, the WT germination rates were 17.31%, 44.23%, 51.92%, and 63.46% on days 4, 5, 6, and 7, respectively; however, they were 1.92%, 21.96%, 30.58%, and 36.75%, respectively, in the *Osbhlh035* mutant. Notably, when normalized to the corresponding WT germination rates, the *Osbhlh035* germination rates under normal conditions were reduced by 11.87% and 9.67% on days 2 and 3, respectively. However, the *Osbhlh035* germination rates in the salt-treated conditions were reduced by 88.91%, 50.35%, 41.1%, and 42.09% on days 4, 5, 6, and 7, respectively. Taken together, these data indicate that OsbHLH035 may promote rice seed germination, especially under salt stress.

Because seed germination is inhibited by ABA, which antagonizes gibberellin signaling, we then measured the endogenous ABA contents in germinating WT and *Osbhlh035* seeds. After 2 days under normal conditions, the ABA levels in germinating *Osbhlh035* seeds were approximately 1.84-fold higher than those in WT seeds, and the corresponding germination rate in the *Osbhlh035* mutant was lower than in WT (Fig. 3b, Additional file 1: Figure S4). Additionally, salt-induced ABA accumulation was observed in germinating WT and *Osbhlh035* seeds after 5 days under salt-treated conditions (Fig. 3c). However, the ABA levels were approximately 1.45-fold higher in germinating *Osbhlh035* seeds than in WT seeds, and the corresponding germination rate of the *Osbhlh035* mutant was also lower than that of WT (Fig. 3b, c). To validate that the delayed-germination phenotype of *Osbhlh035* mutants is due to excess ABA accumulation, the biosynthetic inhibitor fluridone was used in further tests. Indeed, the germination rates of WT and *Osbhlh035* seeds were not statistically significantly different after 2 days under normal conditions or 5 days under salt-treated conditions in the presence of fluridone (Fig. 3d). Taken together, these data suggest that OsbHLH035 is involved in the repression of ABA-inhibited germination.

Expression of ABA metabolic genes in germinating WT and Osbhlh035 seeds

Because ABA metabolic genes have been well characterized in rice, we used q-PCR to investigate the expression profiles of these genes in germinating WT and *Osbhlh035* seeds. In germinating WT seeds, the expression levels of the ABA biosynthetic genes *OsABA2* and *OsAAO3* under salt-treated conditions were approximately 2.01- and 2.06-fold, respectively, higher than those under normal conditions (Fig. 4a, b). However, the *OsABA2* and *OsAAO3* expression levels in germinating *Osbhlh035* seeds were also higher than those in germinating WT seeds under either normal or salt-treated conditions. Moreover, the expression levels of the ABA catabolic gene *OsABA8ox1* in germinating WT and *Osbhlh035* seeds were comparable under normal conditions; however, the salt-induced expression of *OsABA8ox1* was abolished in germinating *Osbhlh035* seeds (Fig. 4c). These results agreed with the corresponding ABA levels detected under these conditions (Fig. 3c vs. Fig. 4). Taken together, these

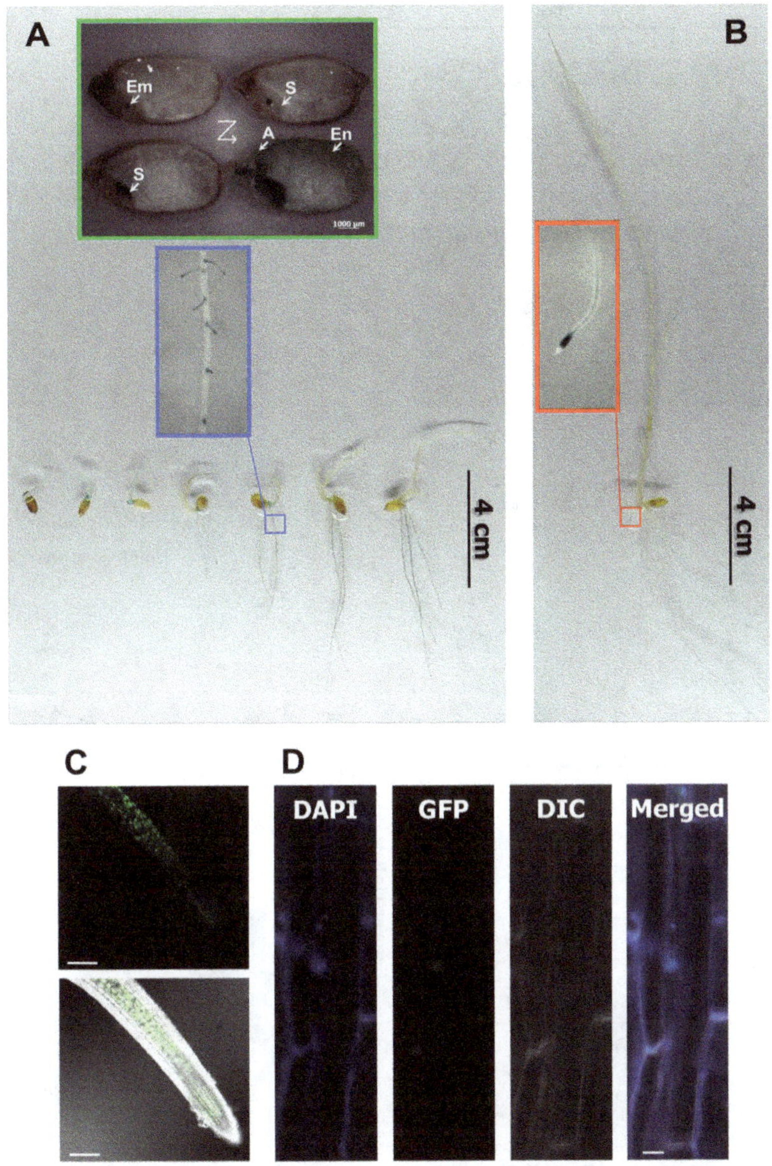

Fig. 2 The spatiotemporal expression of *OsbHLH035* and the subcellular localization of GFP-OsbHLH035 in stable rice transformants. **a** The germinating seeds and post-germination-stage seedlings. The Z-scheme represents the staining order of the water-imbibed transgenic seeds, which were collected within 2 days. Em, Embryo; S, Scutellum; A, Aleurone layer; and En, Endosperm. **b** A third-leaf-stage seedling. **c** The presence of GFP-OsbHLH035 in the root tip. The upper panel is fluorescence image; the lower panel is overlaid with transmitted light image. Scale bars, 200 μm. **d** GFP-fused OsbHLH035 protein is localized with the DAPI, a nuclear affinity dye, in rice cells. Scale bars, 5 μm

data reveal that the accumulation of excess ABA in germinating *Osbhlh035* seeds may be caused not only by elevated ABA biosynthesis but also by reduced ABA catabolism under salt-treated conditions.

OsbHLH035 is not involved in regulating plant growth during the seedling stage

To understand the roles of salt-inducible *OsbHLH035* during the seedling stage, we compared the plant growth and salt-stress responses in both WT and *Osbhlh035* seedlings under normal and salt-treated conditions. Under normal conditions, there were no significant differences in the growth patterns of WT and *Osbhlh035* aerial tissues on days 14 (Fig. 5a). Interestingly, there were also no obvious differences in the growth patterns and salt responses of the WT and *Osbhlh035* aerial tissues under salt-treated conditions (Fig. 5a). Moreover, the water-loss rate and ABA contents in both WT and *Osbhlh035* aerial tissues were comparable under either normal or salt-treated conditions (Fig. 5b and e). Furthermore, the primary root length and ABA contents of the *Osbhlh035* terrestrial tissues were similar to those of

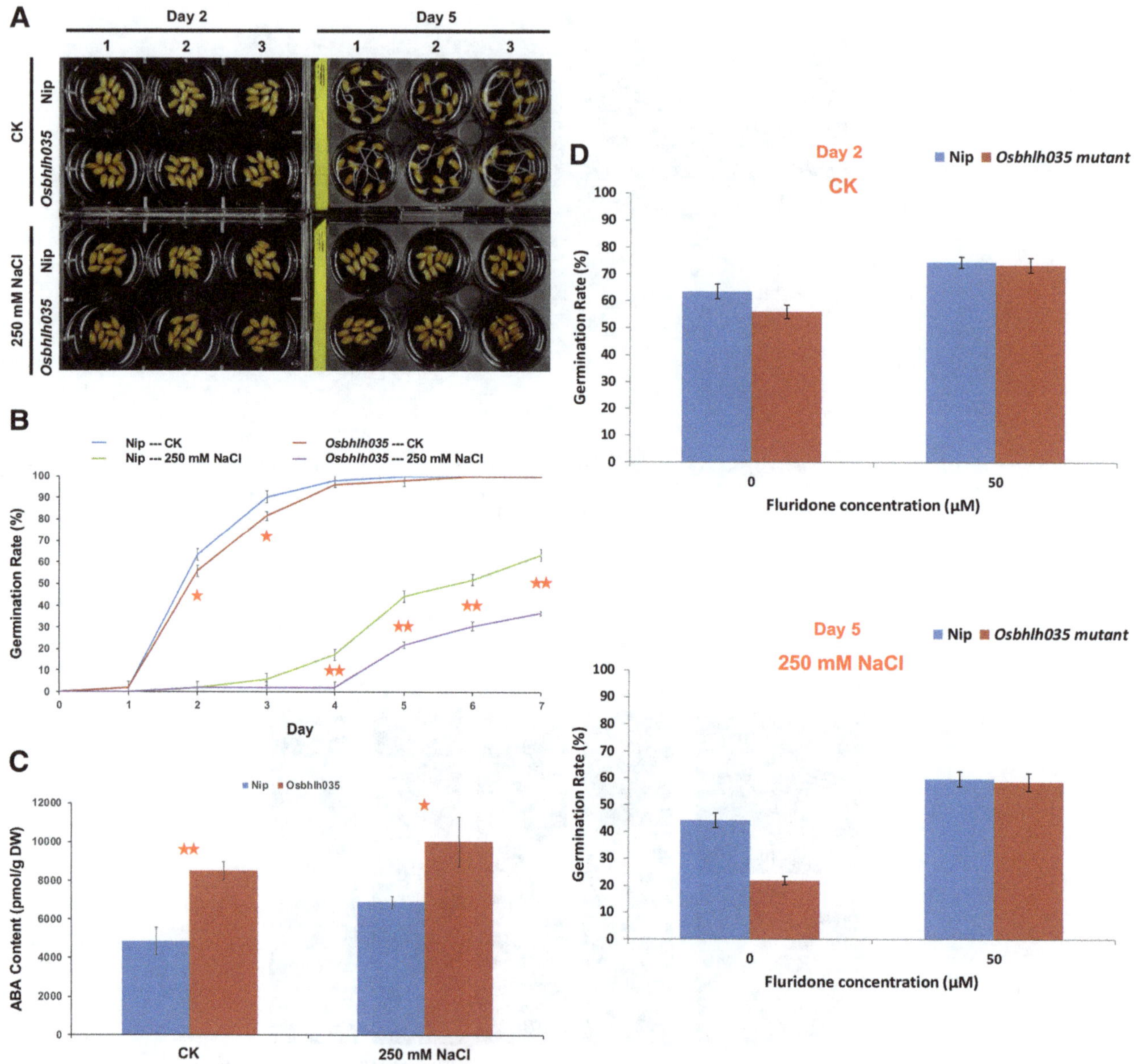

Fig. 3 The delayed-germination phenotype of *Osbhlh035* mutants is accompanied by the excess accumulation of endogenous ABA. **a** The germination of WT and *Osbhlh035* seeds under normal and salt-treated conditions at 2 and 5 days after imbibition. Arabic numerals represent three independent biological replicates, and the seeds in each independent biological replicate were harvested from an individual rice plant. **b** The germination rates of WT and *Osbhlh035* seeds under normal and salt-treated conditions. All germination rates represent the mean ± SD of two independent experiments performed with three independent biological replicates; 100 seeds were used in each independent biological replicate. **c** The endogenous ABA contents in both germinating WT and *Osbhlh035* seeds grown under normal and salt-treated conditions on day 5. **d** The delayed-germination phenotype of *Osbhlh035* mutants can be rescued by fluridone. The germination rates of WT and *Osbhlh035* seeds under normal conditions at day 2 (upper panel) and salt-treated conditions at day 5 (lower panel). The seeds were grown on basal medium containing 0 (normal conditions, CK) or 250 mM NaCl (salt-treated conditions) supplemented with or without 50 μM fluridone. Samples for the ABA ELISA were harvested on day 5. Asterisks indicate significant differences in comparison with WT ($^*P < 0.05$ and $^{**}P < 0.01$) based on Student's *t*-test

the WT terrestrial tissues under either normal or salt-treated conditions (Fig. 5c-e). These data indicate that OsbHLH035 does not appear to be involved in regulating plant growth during the seedling stage.

Excess sodium accumulation in Osbhlh035 aerial tissues

As mentioned above, *OsbHLH035* is expressed in aerial and terrestrial tissues from the postgermination stage until the third-leaf stage (Fig. 2a), and its expression is

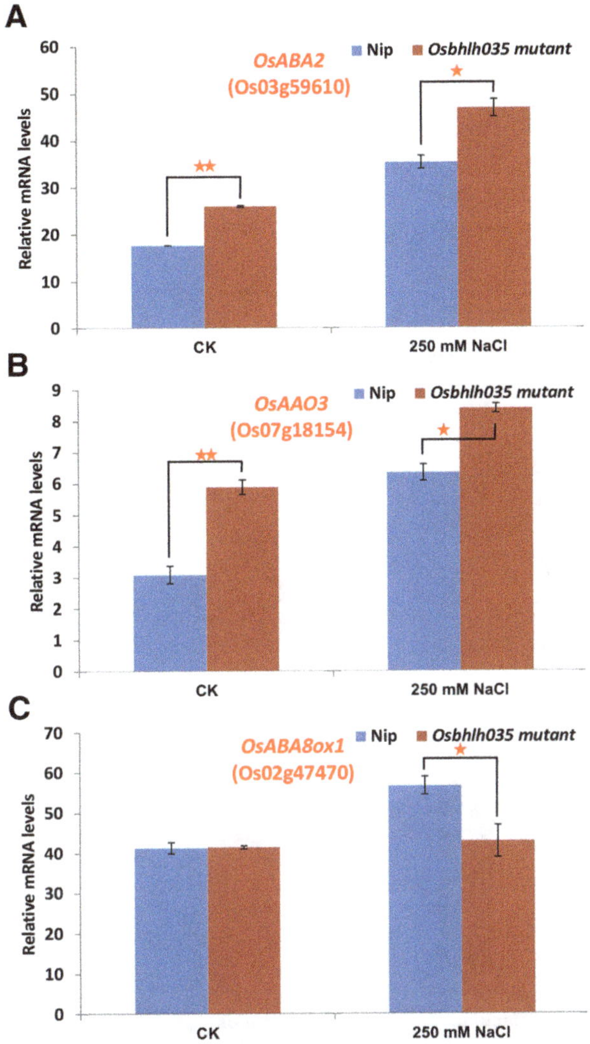

Fig. 4 The expression of ABA metabolic genes in both germinating WT and *Osbhlh035* seeds. ABA over-accumulation in germinating *Osbhlh035* seeds is due to the upregulation of the biosynthetic genes *OsABA2* (**a**) and *OsAAO3* (**b**) and abolition of salt-induced *OsABAox1* (**c**) expression. The experimental materials and methods are the same as those described in Fig. 3. Samples for the q-PCR assay were harvested on day 5. Asterisks indicate significant differences in comparison with WT (*$P < 0.05$ and **$P < 0.01$) based on Student's *t*-test

enhanced by salt treatment (Fig. 1b). However, the roles of *OsbHLH035* do not appear to include regulating seedling growth or salt tolerance at these stages and under these conditions (Fig. 5). To investigate the roles of *OsbHLH035* in detail, we detected the sodium and potassium contents in both WT and *Osbhlh035* seedlings by using inductively coupled plasma-optical emission spectrometry (ICP-OES). As shown in Fig. 6a, the sodium/potassium (Na^+/K^+) ratios in the aerial or terrestrial tissues from the WT and *Osbhlh035* seedlings were comparable under either normal or salt-treated conditions; however, sodium was excessively accumulated in the *Osbhlh035* aerial tissues, while it was slightly reduced in its terrestrial tissues (Fig. 6b). These data suggest that Na^+ exclusion from the transpiration stream is impaired in the *Osbhlh035* seedlings. Furthermore, q-PCR analysis revealed that *OsHKT1;3* and *OsHKT1;5* mRNA levels in *Osbhlh035* aerial and terrestrial tissues, respectively, were lower than those in the corresponding WT tissues under either normal or salt-treated conditions (Fig. 7a).

OsbHLH035 contributes to seedling recovery following the removal of salt stress

As mentioned previously, the ability to tolerate and recover from stress are both important for rice to manage abiotic stresses (Lenka et al. 2011; Zhang et al. 2012). Therefore, we further investigated the growth responses of WT and *Osbhlh035* seedlings after the removal of salt stress. After a 7-day recovery period, new leaves sprouted in the WT seedlings but not in the *Osbhlh035* seedlings (Fig. 8a). The primary roots of WT seedlings that had recovered from salt stress were longer than those of the corresponding *Osbhlh035* seedlings (Fig. 8b, c). Moreover, the survival rate of WT seedlings after the removal of salt stress was approximately 61.67%, while the rate of *Osbhlh035* seedlings was only approximately 30% (Fig. 8d). Although the endogenous ABA content was still high in both the WT and *Osbhlh035* seedlings that had recovered from salt stress compared with controls, there were no significant difference between either the salt-stressed WT or *Osbhlh035* seedlings postrecovery (Fig. 8e). Notably, sodium was continuously over-accumulated in salt-relieved *Osbhlh035* aerial tissues (Fig. 6d). Moreover, the ratio of excessive Na^+ in salt-relieved *Osbhlh035* aerial tissues is higher than that in salt-treated *Osbhlh035* aerial tissues (Fig. 6b vs. d). Besides, the expression of *OsHKT1;3* and *1;5* was also reduced in salt-relieved *Osbhlh035* aerial and terrestrial tissues, respectively (Fig. 7b). These data show that OsbHLH035 may play a positive, ABA-independent role in seedling recovery after salt stress.

Delayed germination and impaired salt-stress recovery of Osbhlh035 mutants can be rescued via genetic complementation

Genetic complementation was performed to validate the roles of *OsbHLH035* in regulating seed germination and seedling recovery, and two representative complemented transformants (C#13 and C#14) were investigated. Compared with WT seeds germinated for 1 day under normal conditions or 4 days under salt-treated conditions, the corresponding *Osbhlh035* seeds displayed a delayed-germination phenotype with excess ABA accumulation on day 4 after the beginning of imbibition. However, the

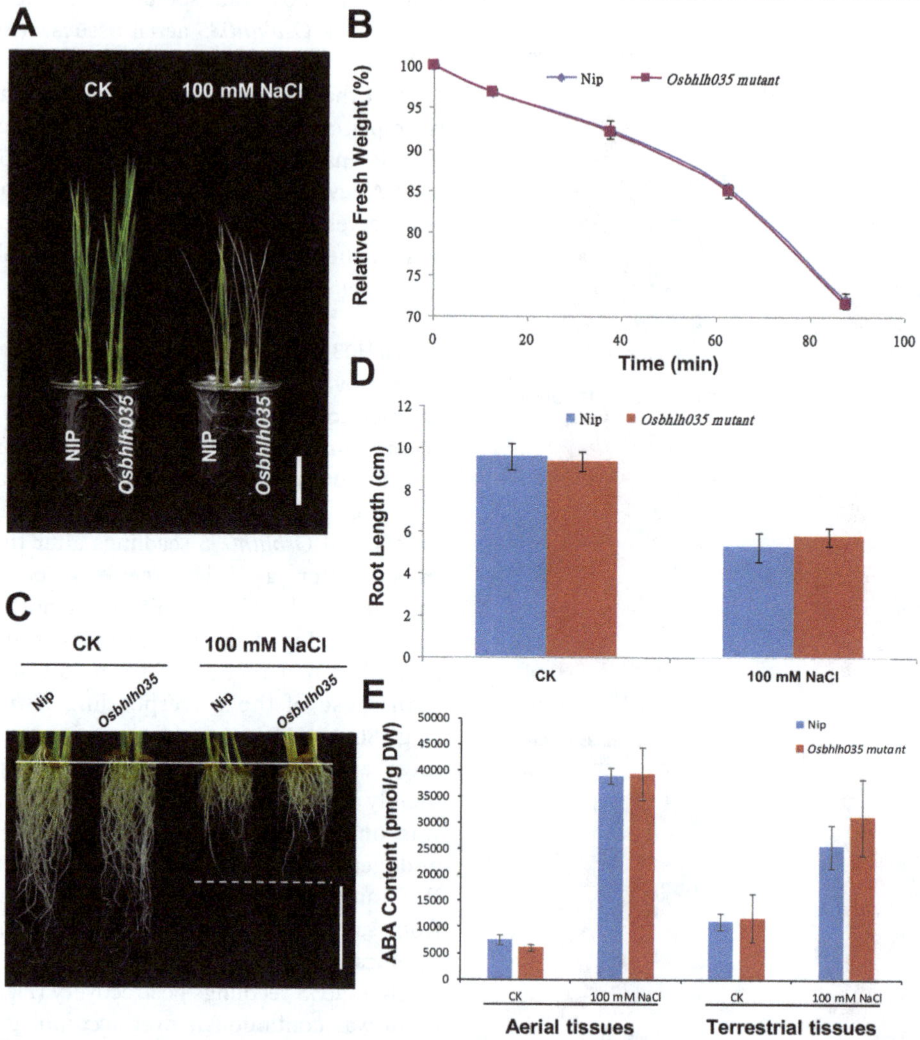

Fig. 5 A comparison of the growth patterns, salt responses, and ABA contents in both WT and *Osbhlh035* seedlings under normal (CK) and salt-treated (100 mM NaCl) conditions. The seedlings were grown on basal medium for 7 days and then transferred to basal medium supplemented with 0 (CK) or 100 mM NaCl for an additional 7 days. **a** The growth patterns and salt responses of 14-day-old WT and *Osbhlh035* aerial tissues. **b** The rates of water loss in 14-day-old WT and *Osbhlh035* aerial tissues. **c-d** Characterization and quantification of root growth in both WT and *Osbhlh035* seedlings. **e** The ABA contents in 14-day-old WT and *Osbhlh035* seedlings. Scale bars in (**a**) and (**c**) represent 4 cm

germination rate and ABA contents of the germinating C#13 and C#14 seeds were comparable to those of WT seeds (Fig. 9a-c). Moreover, *OsABA2*, *OsAAO3*, and *OsABA8ox1* expression in germinating C#13 and C#14 seeds was restored to levels similar to those of germinating WT seeds after 4 days under normal conditions or salt-treated conditions (Fig. 9d). These data demonstrate that OsbHLH035 plays a role in promoting germination by negatively regulating ABA-mediated seed germination.

Additionally, the excessive accumulation of sodium in *Obhlh035* aerial tissues and the impaired ability of *Osbhlh035* seedlings to recover from salt stress could also be rescued by genetic complementation. Under salt-treated conditions, the sodium levels in both the C#13 and C#14 aerial tissues were similar to those in corresponding WT tissues, with comparable levels of *OsHKT1;3* and *OsHKT1;5* expression (Fig. 10). After a 7-day recovery period, the primary root length and survival rates of both the C#13 and C#14 transformants were also comparable to those of WT seedlings (Fig. 11a-c). As expected, no obvious changes in endogenous ABA contents were observed between the WT, *Osbhlh035*, and two genetically complemented seedlings after 21 days under normal or post-salt-stressed conditions (Fig. 11d). These data suggest that OsbHLH035 might contribute to the ability of rice seedlings to recover from salt stress in an ABA-independent manner.

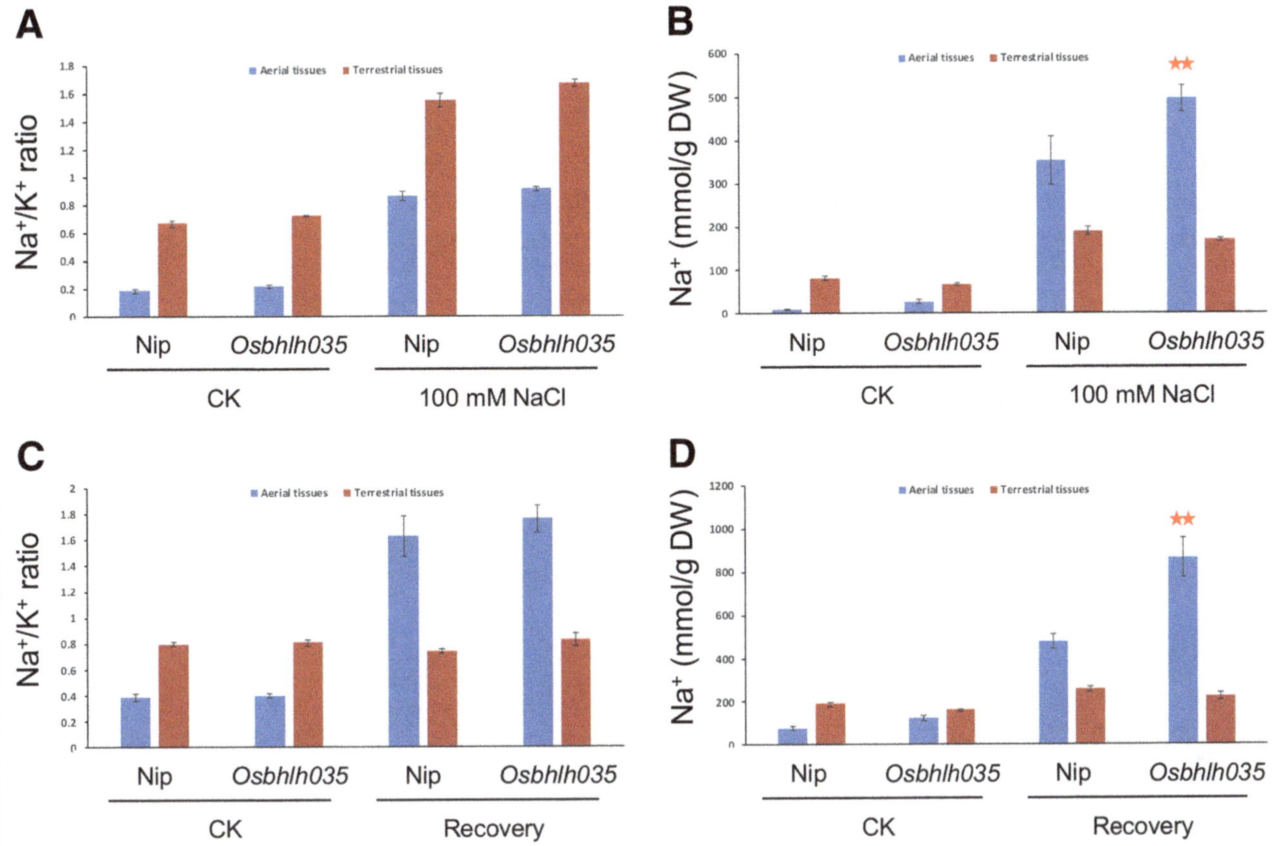

Fig. 6 Effects of salinity on the Na^+/K^+ ratio and Na^+ content in both WT and *Osbhlh035* seedlings. **a-b** Na^+/K^+ ratios and Na^+ contents in aerial and terrestrial tissues from both salt-treated WT and *Osbhlh035* seedlings. **c-d** Na^+/K^+ ratios and Na^+ contents in aerial and terrestrial tissues from both salt-relieved WT and *Osbhlh035* seedlings. The seedlings were grown on basal medium for 11 days and then transferred to basal medium containing 0 (CK) or 100 mM NaCl (salt-treated conditions) for an additional 3 days. After day 14, the seedlings were returned to basal medium for 7 days (Recovery). Asterisks indicate significant differences in comparison to WT ($*P < 0.05$ and $**P < 0.01$) based on Student's *t*-test

Discussion

Plants are sessile and therefore must effectively cope with environmental changes to grow or survive. Thus, plants have evolved complicated gene regulatory networks, particularly diverse TFs, to respond to the surrounding environment. However, the functions of most TFs, as well as their involvement in plant growth, development and stress responses, are unknown. In this article, we conducted a functional study to address the roles of the TF *OsbHLH035* on seed germination, salt tolerance, and the ability to recover from salt stress in rice.

OsbHLH035 mediates seed germination through an ABA-dependent pathway

ABA is important for coordinating the regulation of plant growth, development, and abiotic stress responses. However, the mechanisms that regulate the transcription of ABA metabolic genes in *Arabidopsis* and rice are poorly understood. In *Arabidopsis*, the ABA biosynthetic genes *AtNCED6* and *AtNCED9* contribute to ABA-mediated seed dormancy (Lefebvre et al. 2006). Notably, phylogenetic analysis reveals that plant NCEDs have undergone divergent evolution (Chen et al. 2011). In the rice genome, there are three putative NCEDs (see below for annotations) that are highly homologous to maize ZmVP14, the first plant NCED to be identified (Schwartz et al. 1997; Tan et al. 1997). To date, the rice orthologs of *AtNCED6* and *AtNCED9* have proven difficult to identify. In fact, both *OsNCED1* (Os03g44380) and *OsNCED3* (Os07g05940) expressions show no significant difference in germinating WT and *Osbhlh035* seeds, and *OsNCED2* (Os12g42280) transcripts are barely detectable under both normal and salt-treated conditions (data not shown). Apart from *AtNCEDs*, other biosynthetic genes such as *AtABA2* and *AtAAO3* as well as catabolic genes such as *CYP707A1/AtABA8ox1* and *CYP707A2/AtABAox2* are also involved in regulating seed germination. In germinating WT seeds, the expression levels of *OsABA2*, *OsAAO3*, and *OsABA8ox1*, which is a homolog of *CYP707A1*, *2*, and *3*

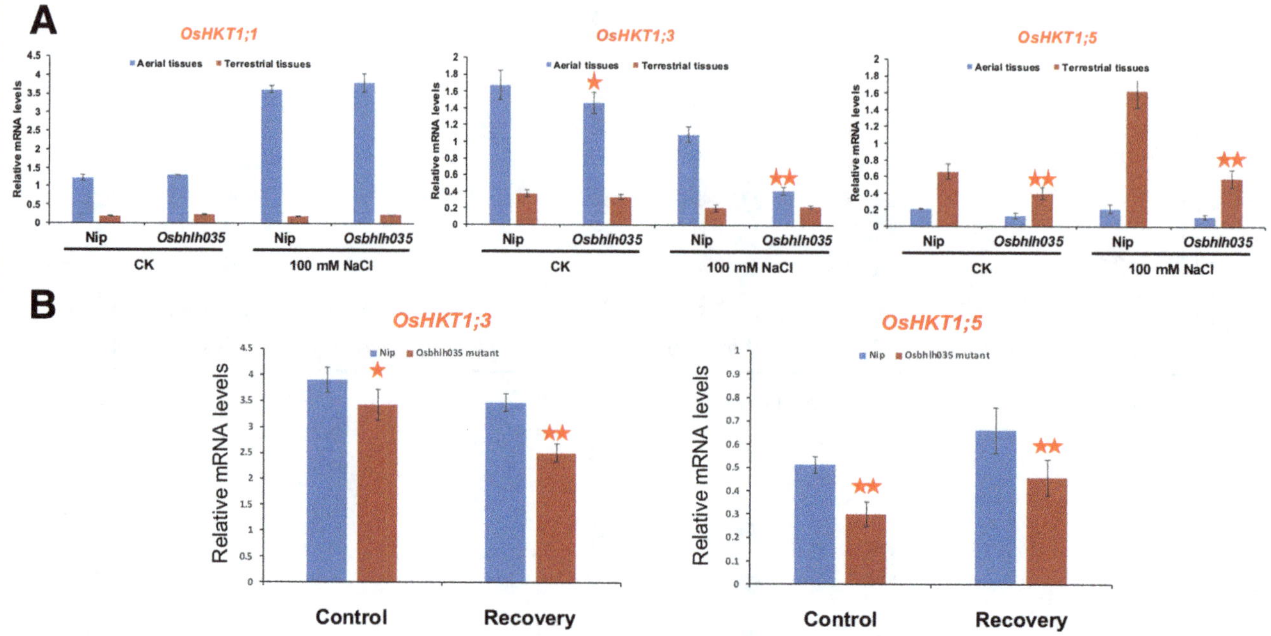

Fig. 7 Effects of salinity on the *OsHKT1s* expression profile in both WT and *Osbhlh035* seedlings. **a** Quantification of *OsHKT1;1*, *OsHKT1;3*, and *OsHKT1;5* mRNAs in aerial and terrestrial tissues from both salt-treated WT and *Osbhlh035* seedlings. **b** Quantification of *OsHKT1;3* and *OsHKT1;5* mRNAs in aerial (left panel) and terrestrial (right panel) tissues, respectively, from both salt-relieved WT and *Osbhlh035* seedlings. The seedlings were grown on basal medium for 11 days and then transferred to basal medium containing 0 (CK) or 100 mM NaCl (salt-treated conditions) for an additional 3 days. After day 14, the seedlings were returned to basal medium for 7 days (Recovery). Asterisks indicate significant differences in comparison to WT (*$P < 0.05$ and **$P < 0.01$) based on Student's *t*-test

(Additional file 1: Figure S5), are upregulated under salt-treated conditions (Fig. 4). However, under either normal or salt-treated conditions, both *OsABA2* and *OsAAO3* are more highly expressed in germinating *Osbhlh035* seeds than in corresponding WT seeds. No obvious differences in *OsABA8ox1* transcription were detected between germinating WT and *Osbhlh035* seeds under normal conditions, but the salt-induced expression of *OsABA8ox1* is abolished in germinating *Osbhlh035* seeds. Notably, the expression of *OsbHLH035-GFP* driven by a 2.0-kb *OsbHLH035* native promoter can restore *OsABA2*, *OsAAO3*, and *OsABA8ox1* expression and compensate for the germination deficiency in *Osbhlh035* mutants (Fig. 9). These data show that the excess ABA accumulation in *Osbhlh035* seeds germinating under salt stress is regulated at the transcriptional level due to not only the increased expression of ABA biosynthetic genes but also the inhibition of ABA catabolic gene expression. Given that OsbHLH035 is a G-box-binding TF, we searched for putative *cis*-acting elements in the upstream promoter regions of ABA metabolic genes. The 3807-bp and 3-kb regions upstream of the *OsAAO3* and *OsABA8ox1* initiation codons, respectively, were each found to contain one G-box element with a CACGTG core sequence. Because no G-box element was found in the promoter of *OsABA2* despite its increased transcription in the germinating *Osbhlh035* seeds, OsbHLH035 seems to suppress *OsABA2* expression in an indirect manner. Interestingly, a functional G-box element was identified to be necessary for drought-induced *AtNCED3* expression (Behnam et al. 2013). If this is the case, it cannot be excluded that OsbHLH035 suppresses *OsAAO3* and *OsABA8ox1* expression by interacting with G-box elements in their promoters. However, whether the transcriptional activity of *OsAAO3* and *OsABAox1* is directly regulated by OsbHLH035 needs to be confirmed.

OsbHLH035 provides the ability to recover from salt stress in an ABA-independent pathway

In this article, our data showed that no obvious differences were observed in plant growth, water-loss rate, or endogenous ABA contents in both WT and *Osbhlh035* seedlings grown under normal or salt-treated conditions (Fig. 5). However, the *Osbhlh035* mutant accumulates an excess of Na^+ ions in the aerial tissues and is unable to recover after salt stress relief (Figs. 6 and 8). These data indicate that OsbHLH035 seems to play a fine-tuning role in regulating salt stress response. Thus, the salt-susceptibility phenotype of *Osbhlh035* mutant becomes visible after salt stress relief. Besides, compared with that in aerial and terrestrial WT tissues, the expression of sodium transporter genes *OsHKT1;3* and *OsHKT1;5* is repressed in *Osbhlh035* tissues, respectively (Fig. 7). Notably, genetic complementation can restore

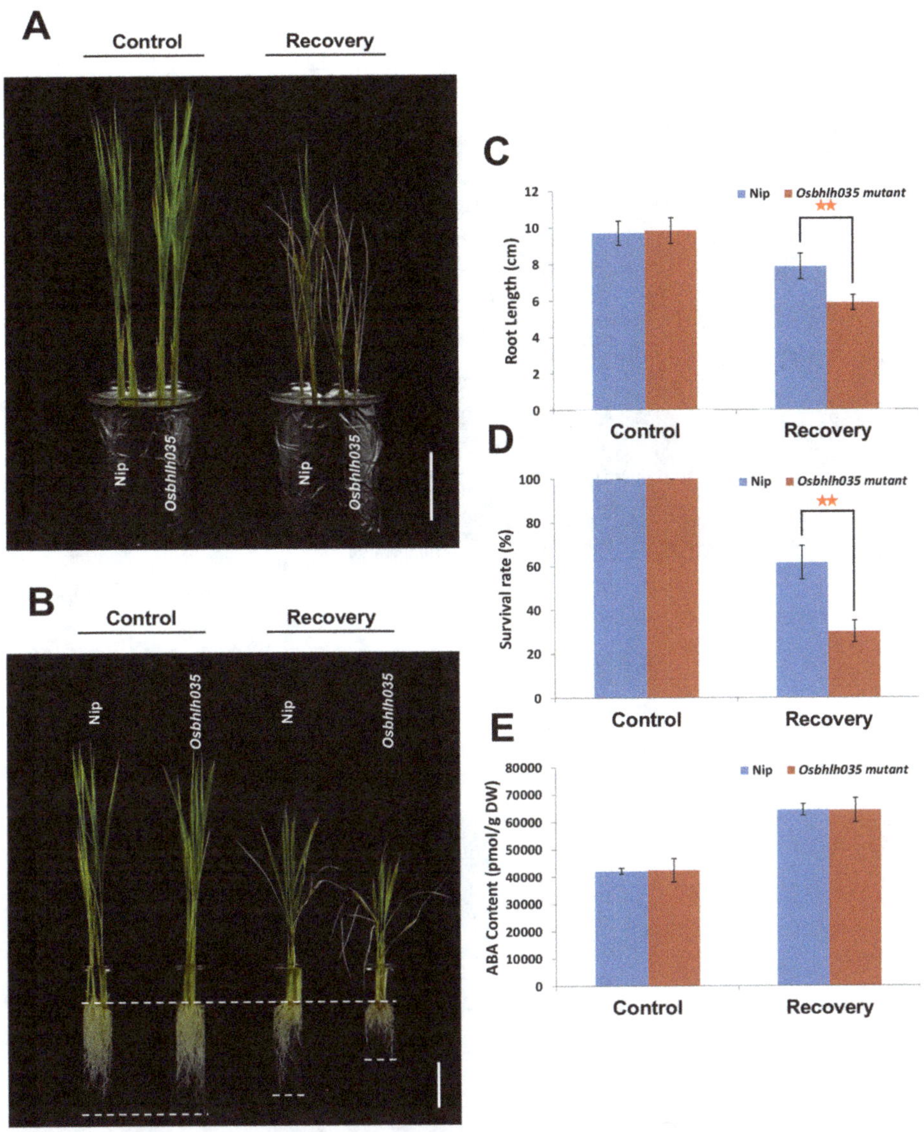

Fig. 8 Investigating growth responses, survival rate, and endogenous ABA contents in both WT and *Osbhlh035* seedlings after recovery from salt stress. **a-b** The growth responses of aerial and terrestrial tissues in both WT and *Osbhlh035* seedlings after relief from salt stress. **c** The primary root lengths of WT and *Osbhlh035* seedlings after relief from salt stress. **d** The survival rate of WT and *Osbhlh035* seedlings after relief from salt stress. All survival rates are shown as the mean ± SD of two independent experiments ($n = 20 \times 3$ independent biological replicates within each genotype and/or experimental condition). **e** The endogenous ABA contents of WT and *Osbhlh035* seedlings after recovery from salt stress. Scale bars in (**a**) and (**b**) represent 4 cm. The seedlings were grown on basal medium for 7 days and then transferred to basal medium containing 0 (control) or 100 mM NaCl (recovery) for an additional 7 days. After day 14, the seedlings were returned to basal medium for 7 days. Samples for the ABA ELISA were harvested on day 21. Asterisks indicate significant differences in comparison to WT (*$P < 0.05$ and **$P < 0.01$) based on Student's *t*-test

OsHKT1;3 and *OsHKT1;5* expression and rescue the impaired salt-stress recovery of the *Obhlh035* mutant (Figs. 10 and 11). These data suggest that OsbHLH035 confers the seedling recovery from salt stress through the ABA-independent activation of *OsHKT1*s. In *Arabidopsis* and rice, the HKT1s play a vital role in transferring Na^+ from the xylem into xylem parenchyma cells (for review, see Su et al. 2015). This mechanism is used to prevent aerial tissues from Na^+ over-accumulation and toxicity. Indeed, loss-of-function mutations in *OsHKT1*s, such as *OsHKT1;1* and *1;5*, lead to Na^+ overaccumulation in the aerial tissues and display hypersensitivity to salt stress (Wang et al. 2015; Kobayashi et al. 2017). In fact, the NaCl-induced expression of *OsHKT1;1* is directly regulated by a TF, OsMYBc (Wang et al. 2015). *OsHKT1;3* and *OsHKT1;5* are mainly expressed in the aerial and terrestrial tissues of rice seedlings, respectively (Garciadeblas et al. 2003; Kobayashi et al. 2017). Although

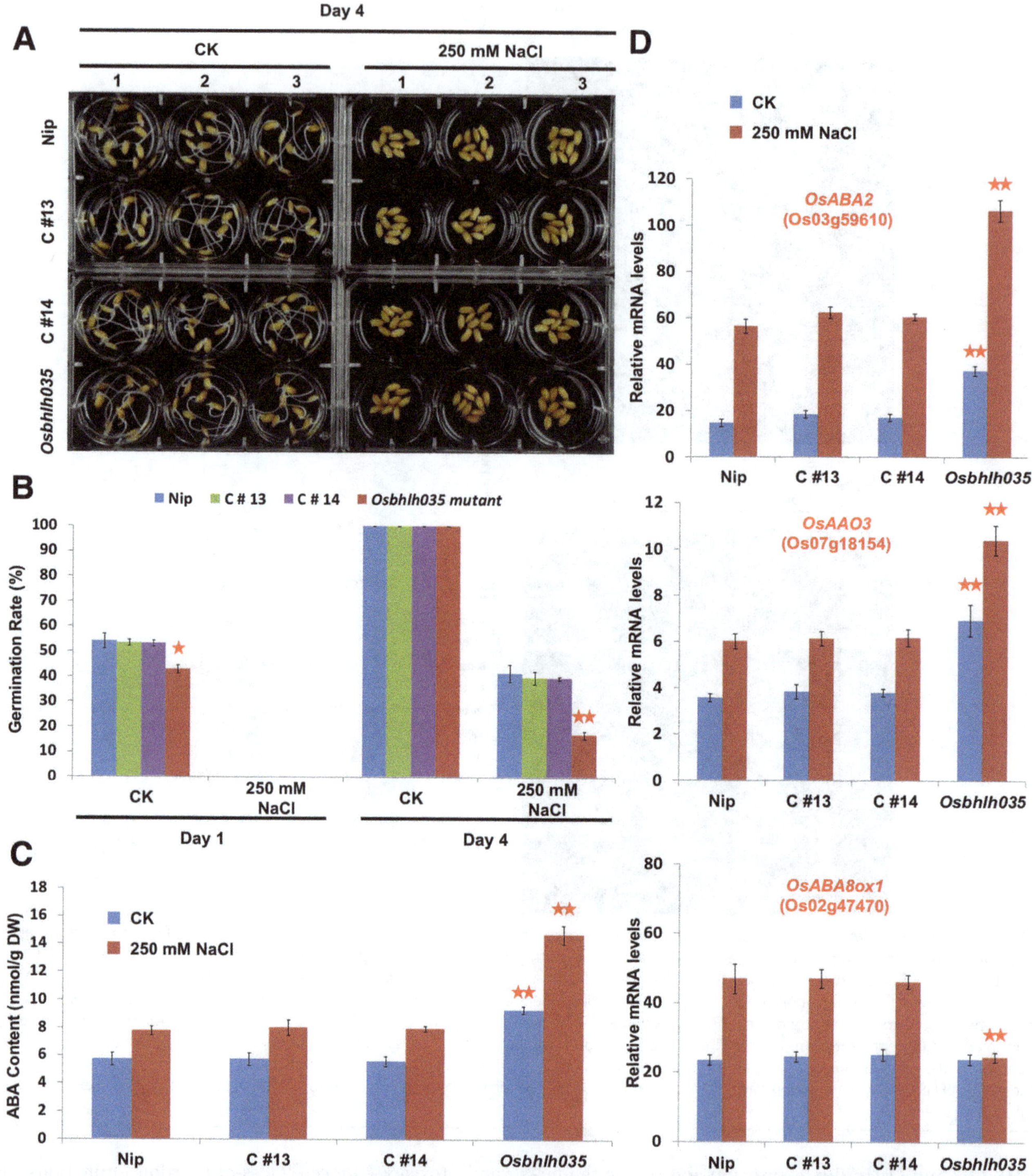

Fig. 9 Examination of seed germination, endogenous ABA contents, and ABA metabolic gene expression in WT, *Osbhlh035*, and genetically complemented transformant rice. **a-b** The germination characteristics and rates of seeds from WT, *Osbhlh035*, and genetically complemented lines under normal and salt-treated conditions. Arabic numerals represent three independent biological replicates, which were harvested from three different rice plants per genotype. All germination rates are shown as the mean ± SD of two independent experiments; $n = 50 \times 3$ biological replicates within each genotype and/or experimental condition. **c** The endogenous ABA contents of WT, *Osbhlh035*, and the genetically complemented transformant lines under normal and salt-treated conditions on day 4. **d** *OsABA2*, *OsAAO3*, and *OsABA8ox1* expression in WT, *Osbhlh035*, and the genetically complemented transformant lines under normal and salt-treated conditions on day 4. The seeds were grown on basal medium supplemented with 0 (CK) or 250 mM NaCl. Samples for the ABA ELISA and q-PCR assays were harvested on day 4. Asterisks indicate significant differences in comparison to WT (*$P < 0.05$ and **$P < 0.01$) based on Student's *t*-test

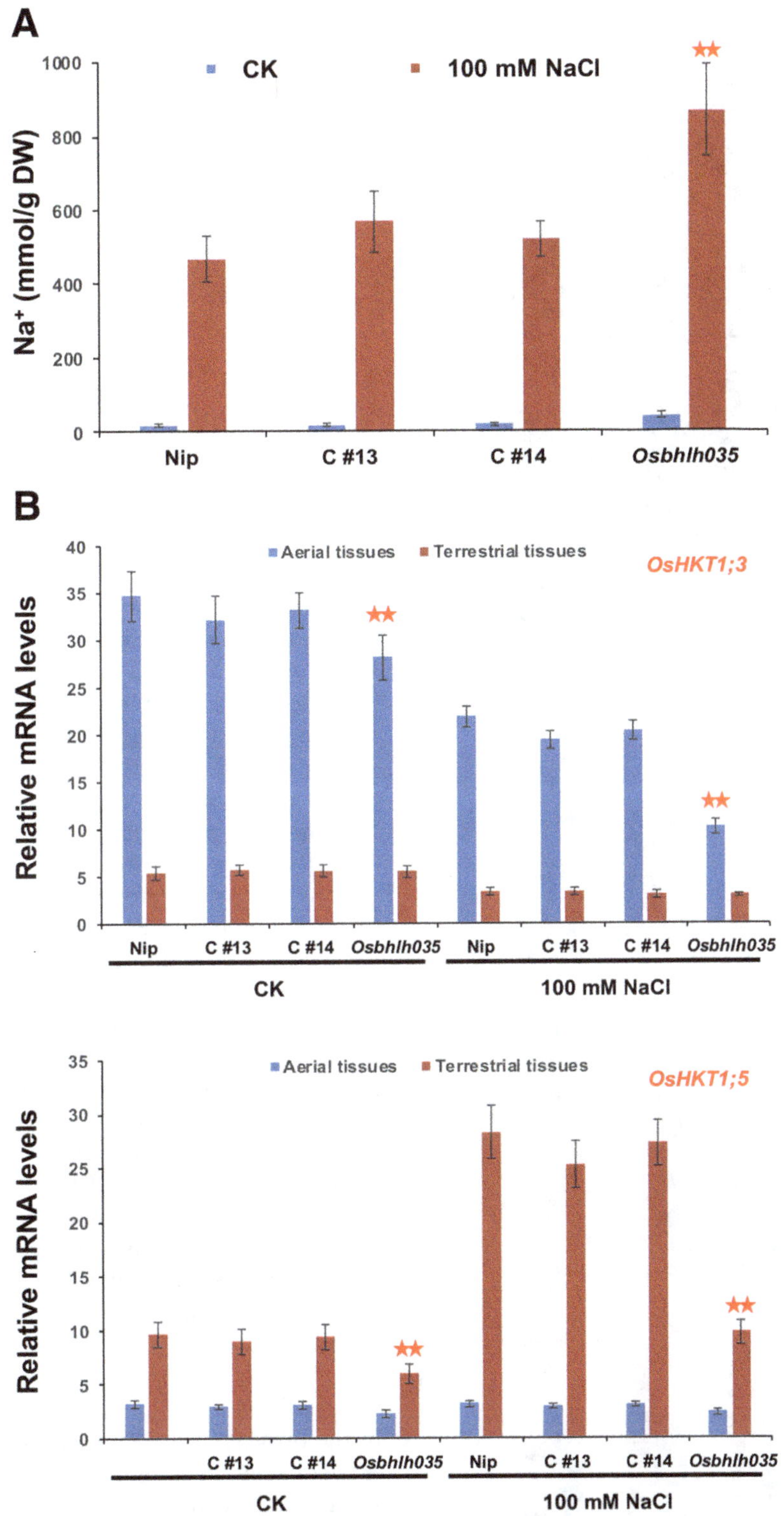

Fig. 10 Na^+ content **a** of the aerial tissues and the expression of *OsHKT1;3* **b** and *OsHKT1;5* **c** in WT, *Osbhlh035*, and genetically complemented transformant lines. The seedlings were grown on basal medium for 11 days and then transferred to basal medium supplemented with 0 (CK) or 100 mM NaCl for an additional 3 days. Asterisks indicate significant differences in comparison to WT (*$P < 0.05$ and **$P < 0.01$) based on Student's *t*-test

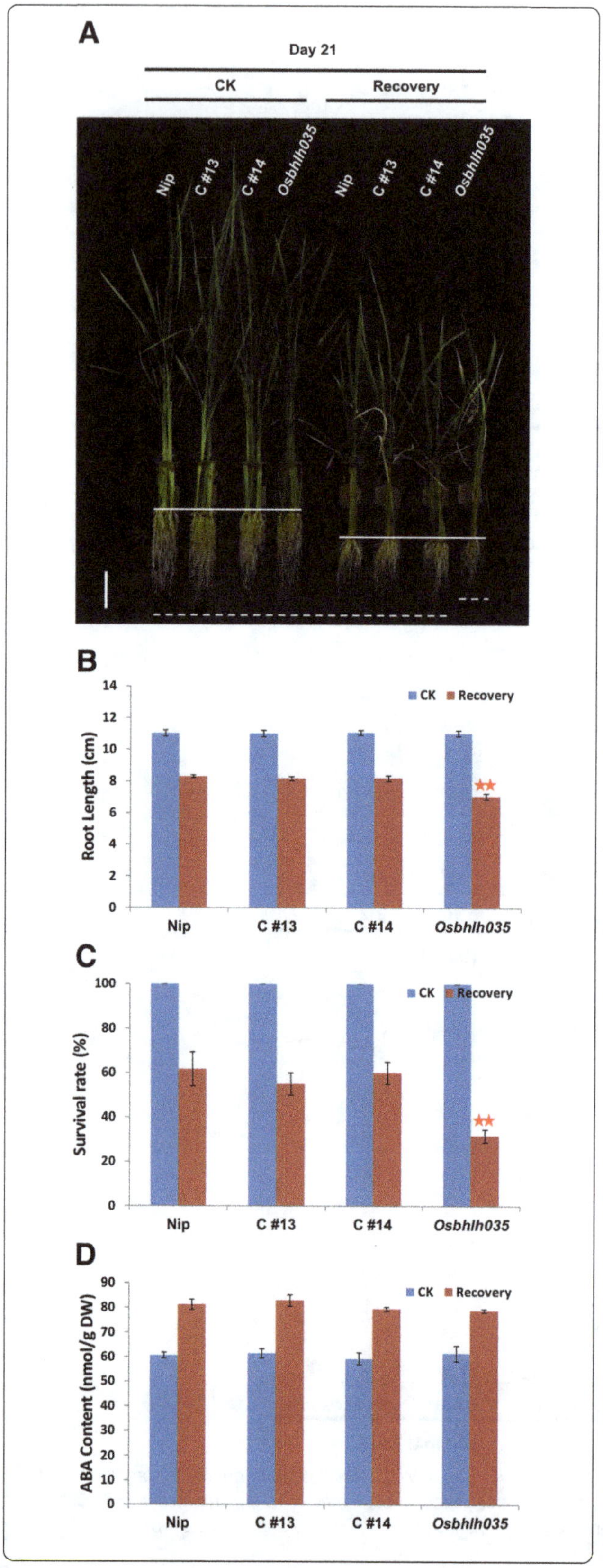

Fig. 11 Phenotypic characterization and ABA contents after recovery from salt treatment. **a-d** Root elongation, survival rates, and endogenous ABA contents in WT, *Osbhlh035*, and genetically complemented transformant lines after salt stress removal. The seedlings were grown on basal medium for 7 days and then transferred to basal medium supplemented with 0 (CK) or 100 mM NaCl (recovery) for an additional 7 days, after which the seedlings were returned to basal medium for 7 days. Samples for the ABA ELISA and q-PCR assays were harvested on day 21. Asterisks indicate significant differences in comparison to WT (* $P < 0.05$ and ** $P < 0.01$) based on Student's *t*-test

the expression of *OsHKT1;3* and *1;5* is downregulated in *Osbhlh035* aerial and terrestrial tissues, respectively, no G-box element was found in their 3-kb promoter. These data indicate that OsbHLH035 mediates the expression of *OsHKT1;3* and *1;5*, most likely in an indirect manner.

Pleiotropic role of OsbHLH035 in regulating seed germination and salt-treated seedling recovery

In the plant *bHLH* TF gene family, the formation of homodimers or heterodimers is important for determining target gene expression. Dimeric molecules can specifically bind to certain *cis*-acting elements within the promoter regions of target genes, whose products can then mediate divergent physiological responses in plants. Therefore, plant bHLHs can play a dual or pleiotropic role in regulating growth and development based on their dimeric forms. For example, the POSITIVE REGULATOR OF GRAIN LENGTH 1-ANTAGONIST OF PGL1 (PGL1-APG) heterodimer, an OsHLH-OsbHLH complex, increases grain length and weight. However, the APG homodimer decreases grain length and weight (Heang and Sassa 2012). The function of OsbHLH068 not only is involved in regulating salt-stress responses but also may mediate flowering time in *Arabidopsis* (Chen et al. 2017). Thus, we assume that OsbHLH035 plays a pleiotropic role in regulating seed germination and seed recovery, presumably due to interactions with different TFs during the germination and seedling stages.

Conclusions

In the present study, we propose that OsbHLH035 plays a pleiotropic role in regulating seed germination and salt-treated seedling recovery in rice. During the germination stage, OsbHLH035 alters the expression of ABA metabolic genes, which reduces ABA levels and relieves the inhibitory effect of ABA on germination. However, OsbHLH035 also mediates the expression of *OsHKT1* genes without altering ABA levels, which promotes seedling recovery from salt stress.

Methods

Plant materials, growth conditions, and experimental methods

Rice (*Oryza sativa* L.) cv. Nipponbare, an *Osbhlh035* mutant (*Tos17* line NG7221, obtained from the National Institute of Agrobiological Sciences [NIAS], Japan), and two independent transgenic complementation lines (C#13 and C#14, generated by *Agrobacterium*-mediated transformation) were analyzed in this study. Except in the germination test, the seeds used in all experiments were sterilized and imbibed at 37 °C for 2 days in the dark and then grown on a wire stand in a beaker at 28 °C under long-day conditions (16-h light/8-h dark cycle) with a light intensity of approximately 270 μE/s m^2. For the germination test, seeds harvested from three independent plants were sterilized and grown directly in a Petri dish at 37 °C in the dark. The basal medium used in all of the experiments was Kimura B solution (Yoshida et al. 1976).

RNA extraction, RT-PCR, and q-PCR

Total RNA was extracted from various rice tissues using an RNeasy Plant Mini Kit (Qiagen) according to the manufacturer's instructions. To avoid genomic DNA contamination, up to 3 μg of total RNA was treated after extraction with Turbo DNA-*free*™ DNase (Ambion) following the manufacturer's instructions, and 1 μg of DNA-free total RNA was then subjected to first-strand cDNA synthesis (Invitrogen, Cat No: 18080–051). The q-PCRs were carried out in an ABI 7500 system using the SYBR® Green PCR Master Mix Kit (Applied Biosystems [ABI]). The initial amount of template cDNA in each RT-PCR and q-PCR was 50 and 10 μg, respectively. At least three independent biological replicates were performed for each experiment, and *OsACTIN1* was used as an internal control for q-PCR normalization. The primer sequences used are provided in Additional file 2: Table S1.

Water loss test

The detached aerial tissues from 14-day-old rice seedlings were placed in plastic weigh boats under ambient conditions. The fresh weight of the tissue was measured after 12.5 min and every 25 min thereafter for a total of 87.5 min. At least three independent biological replicates were performed for each experiment.

ABA ELISA

Prior to the assay, the samples were vacuum dried at − 20 °C at least overnight for normalization to the final dry weight. ABA extraction, purification, and quantification were carried out as described previously (Lin et al. 2007), except that the samples were first ground using an SH-100 tissue homogenizer (Kurabo) before the addition of extraction buffer. At least three independent biological replicates containing two technical replicates each were performed for each experiment, and the data were presented as the average of two independent experiments.

Measurement of ion contents

Samples were harvested, washed twice with ultrapure water, and subsequently vacuum dried at least overnight to measure the dry weight. The dried samples were digested using nitric and hydrochloric acids in an approximately 4:1 ratio and boiled at 200 °C for 2 h. Ion contents were determined using an inductively coupled plasma optical emission spectrometer.

Transgene constructs and isolation of transgenic rice

The full-length coding sequences of *OsbHLH035* and *GFP* were PCR-amplified either with or without stop codons and cloned into the pGEM-T Easy vector. As shown in parenthesis below, these fragments were subcloned into the binary vector pCAMBIA-1300 where their expression was driven either by a 2.0-kb native promoter (*OsbHLH035::OsbHLH035-GFP* and *OsbHLH035::GFP-OsbHLH035*) or a CaMV 35S promoter (*35S::OsbHLH035-GFP* and *35S::GFP-OsbHLH035*). After the constructs were confirmed by sequencing, the *OsbHLH035::GFP-OsbHLH035* construct was transformed into the *Osbhlh035* mutant for subcellular OsbHLH035 localization and genetic complementation assays. Because the stable T_0 rice transformants were chimeric and to avoid the undesirable side effects of multiple T-DNA insertions, T_1 seeds were harvested from individual panicles of independent transformant lines that showed hygromycin resistance and sensitivity at a 3:1 ratio, normalized to seed viability, and used to further screen for homozygous transgenic rice. Additionally, the 2.0-kb *OsbHLH035* promoter was cloned into pCAMBIA-1305.1 (OsbHLH035::-GUS) and transformed into the rice cv. Tainung 67 background for further investigation of the spatiotemporal expression of the *OsbHLH035* gene.

Additional files

Additional file 1: Figure S1. Abiotic stress-responsive *OsbHLHs*. The fold change among each *OsbHLH* gene under different abiotic stresses is calculated as a ratio normalizing the data to it corresponding nontreated control. Red and green colors represent up- and downregulation, respectively. C, cold stress; D, drought stress; S, salt stress; and H, heat stress. **Figure S2.** Prediction of the conserved domains in OsbHLH035 and characterization of the NG7221 mutant line. (A) The ScanProsite tool in ExPASy (http://www.expasy.org/) predicts the presence of a typical bHLH domain (residues 64 to 113) in OsbHLH035. (B) Based on its annotation in the rice *Tos17* insertion mutant database (https://tos.nias.affrc.go.jp/), NG7221 is a single retrotransposon insertional line. (C) The identification of homozygous NG7221 mutants via PCR genotyping of gDNA. The primer positions and sequences are shown in Fig. 1a and Additional file 2: Table S1, respectively. Arabic numbers in (C) represent three independent biological replicates within each genotype. **Figure S3.** GFP-fused

OsbHLH035 protein is predominantly localized to the nucleus in rice calli. The husk-removed seeds harboring *OsbHLH035::GFP-OsbHLH035* were placed on callus induction medium (Tran and Sanan-Mishra 2015) for 7 days and then the calli were subjected to GFP visualization by a confocal microscopy. Scale bars, 10 µm. **Figure S4.** The endogenous ABA contents in both germinating WT and *Osbhlh035* seeds. The husk-removed seeds were grown on basal medium for 2 days and then subjected to ABA ELISAs. **Figure S5.** Phylogenetic analysis of AtCYP707As and OsABA8oxs using the neighbor-joining method. Numbers next to the descendant indicate confidence values based on the bootstrap method.

Additional file 2: Table S1. Primers used in this study.

Abbreviations
ABA: Abscisic acid; bHLH: Basic helix-loop-helix; GFP: Green fluorescent protein; GUS: β-glucuronidase; Nip: Nipponbare; q-PCR: Quantitative PCR; RT-PCR: Reverse transcription-PCR; TFs: Transcription factors

Acknowledgments
We thank the Live-Cell Imaging Core Lab at the Institute of Plant and Microbial Biology (IPMB), Academia Sinica, Taipei, Taiwan, for performing the confocal microscopy and Mr. Jiang-Tai Wu (IPMB, Academia Sinica) for assistance with the transient expression assays. We are also grateful to Miss Lin-Yun Kuang (Transgenic Plant Core Facility, Academia Sinica) for providing *Agrobacterium*-mediated stable transformation services.

Funding
This work was supported by the Ministry of Science and Technology, Taipei, Taiwan (Grant No. NSC-102-2811-B-002-146 to M.-C.C.).

Authors' contributions
H-CC was responsible for designing and conducting the experiments, data assembly, and writing the manuscript. W-HC, C-YH, and Y-SC were responsible for providing the experimental facilities and mentoring the experiments. M-CC conceived the study and supervised the experiments. All authors have read and approved the submission of the final manuscript.

Competing interests
The authors declare that they have no competing interests.

Author details
[1]Department of Agronomy, National Taiwan University, No. 1, Section 4, Roosevelt Road, Taipei, Taiwan, Republic of China. [2]Institute of Plant and Microbial Biology, Academia Sinica, Taipei, Taiwan, Republic of China. [3]Department of Agricultural Chemistry, National Taiwan University, Taipei, Taiwan, Republic of China. [4]Department of Horticulture and Landscape Architecture, National Taiwan University, Taipei, Taiwan, Republic of China.

References
Behnam B, Iuchi S, Fujita M, Fujita Y, Takasaki H, Osakabe Y, Yamaguchi-Shinozaki K, Kobayashi M, Shinozaki K (2013) Characterization of the promoter region of an gene for 9-*cis*-epoxycarotenoid dioxygenase involved in dehydration-inducible transcription. DNA Res 20:315–324

Bittner F, Oreb M, Mendel RR (2001) ABA3 is a molybdenum cofactor sulfurase required for activation of aldehyde oxidase and xanthine dehydrogenase in *Arabidopsis thaliana*. J Biol Chem 276:40381–40384

Carretero-Paulet L, Galstyan A, Roig-Villanova I, Martinez-Garcia JF, Bilbao-Castro JR, Robertson DL (2010) Genome-wide classification and evolutionary analysis of the bHLH family of transcription factors in *Arabidopsis*, poplar, rice, moss, and algae. Plant Physiol 153:1398–1412

Chen HC, Hsieh-Feng V, Liao PC, Cheng WH, Liu LY, Yang YW, Lai MH, Chang MC (2017) The function of *OsbHLH068* is partially redundant with its homolog, *AtbHLH112*, in the regulation of the salt stress response but has opposite functions to control flowering in *Arabidopsis*. Plant Mol Biol 94:531–548

Chen HC, Hwang SG, Chen SM, Shii CT, Cheng WH (2011) ABA-mediated heterophylly is regulated by differential expression of *9-cis-epoxycarotenoid dioxygenase 3* in lilies. Plant Cell Physiol 52:1806–1821

Cheng WH, Endo A, Zhou L, Penney J, Chen HC, Arroyo A, Leon P, Nambara E, Asami T, Seo M, Koshiba T, Sheen J (2002) A unique short-chain dehydrogenase/reductase in *Arabidopsis* glucose signaling and abscisic acid biosynthesis and functions. Plant Cell 14:2723–2743

Chinnusamy V, Ohta M, Kanrar S, Lee BH, Hong X, Agarwal M, Zhu JK (2003) ICE1: a regulator of cold-induced transcriptome and freezing tolerance in *Arabidopsis*. Genes Dev 17:1043–1054

Deinlein U, Stephan AB, Horie T, Luo W, Xu G, Schroeder JI (2014) Plant salt-tolerance mechanisms. Trends Plant Sci 19:371–379

Feller A, Machemer K, Braun EL, Grotewold E (2011) Evolutionary and comparative analysis of MYB and bHLH plant transcription factors. Plant J 66:94–116

Finkelstein R, Reeves W, Ariizumi T, Steber C (2008) Molecular aspects of seed dormancy. Annu Rev Plant Biol 59:387–415

Finkelstein RR, Lynch TJ (2000) The *Arabidopsis* abscisic acid response gene *ABI5* encodes a basic leucine zipper transcription factor. Plant Cell 12:599–609

Finkelstein RR, Wang ML, Lynch TJ, Rao S, Goodman HM (1998) The *Arabidopsis* abscisic acid response locus *ABI4* encodes an APETALA 2 domain protein. Plant Cell 10:1043–1054

Fukao T, Yeung E, Bailey-Serres J (2011) The submergence tolerance regulator SUB1A mediates crosstalk between submergence and drought tolerance in rice. Plant Cell 23:412–427

Garciadeblas B, Senn ME, Banuelos MA, Rodriguez-Navarro A (2003) Sodium transport and HKT transporters: the rice model. Plant J 34:788–801

Giraudat J, Hauge BM, Valon C, Smalle J, Parcy F, Goodman HM (1992) Isolation of the *Arabidopsis ABI3* gene by positional cloning. Plant Cell 4:1251–1261

Gonzalez-Guzman M, Apostolova N, Belles JM, Barrero JM, Piqueras P, Ponce MR, Micol JL, Serrano R, Rodriguez PL (2002) The short-chain alcohol dehydrogenase ABA2 catalyzes the conversion of xanthoxin to abscisic aldehyde. Plant Cell 14:1833–1846

Groot SP, Karssen CM (1992) Dormancy and germination of abscisic acid-deficient tomato seeds : studies with the *sitiens* mutant. Plant Physiol 99:952–958

Hamamoto S, Horie T, Hauser F, Deinlein U, Schroeder JI, Uozumi N (2015) HKT transporters mediate salt stress resistance in plants: from structure and function to the field. Curr Opin Biotechnol 32:113–120

Heang D, Sassa H (2012) Antagonistic actions of HLH/bHLH proteins are involved in grain length and weight in rice. PLoS One 7:e31325

Hu H, Dai M, Yao J, Xiao B, Li X, Zhang Q, Xiong L (2006) Overexpressing a NAM, ATAF, and CUC (NAC) transcription factor enhances drought resistance and salt tolerance in rice. Proc Natl Acad Sci U S A 103:12987–12992

Iuchi S, Kobayashi M, Taji T, Naramoto M, Seki M, Kato T, Tabata S, Kakubari Y, Yamaguchi-Shinozaki K, Shinozaki K (2001) Regulation of drought tolerance by gene manipulation of 9-cis-epoxycarotenoid dioxygenase, a key enzyme in abscisic acid biosynthesis in *Arabidopsis*. Plant J 27:325–333

Karssen CM, Brinkhorst-van der Swan DL, Breekland AE, Koornneef M (1983) Induction of dormancy during seed development by endogenous abscisic acid: studies on abscisic acid deficient genotypes of *Arabidopsis thaliana* (L.) Heynh. Planta 157:158–165

Kobayashi NI, Yamaji N, Yamamoto H, Okubo K, Ueno H, Costa A, Tanoi K, Matsumura H, Fujii-Kashino M, Horiuchi T, Nayef MA, Shabala S, An G, Ma JF, Horie T (2017) OsHKT1;5 mediates Na^+ exclusion in the vasculature to protect leaf blades and reproductive tissues from salt toxicity in rice. Plant J 91:657–670

Koornneef M, Hanhart CJ, Hilhorst HW, Karssen CM (1989) *In vivo* inhibition of seed development and reserve protein accumulation in recombinants of abscisic acid biosynthesis and responsiveness mutants in *Arabidopsis thaliana*. Plant Physiol 90:463–469

Kumar K, Kumar M, Kim SR, Ryu H, Cho YG (2013) Insights into genomics of salt stress response in rice. Rice 6:27

Kushiro T, Okamoto M, Nakabayashi K, Yamagishi K, Kitamura S, Asami T, Hirai N, Koshiba T, Kamiya Y, Nambara E (2004) The *Arabidopsis* cytochrome P450 CYP707A encodes ABA 8′-hydroxylases: key enzymes in ABA catabolism. EMBO J 23:1647–1656

Lefebvre V, North H, Frey A, Sotta B, Seo M, Okamoto M, Nambara E, Marion-Poll A (2006) Functional analysis of *Arabidopsis NCED6* and *NCED9* genes indicates that ABA synthesized in the endosperm is involved in the induction of seed dormancy. Plant J 45:309–319

Lenka SK, Katiyar A, Chinnusamy V, Bansal KC (2011) Comparative analysis of drought-responsive transcriptome in *Indica* rice genotypes with contrasting drought tolerance. Plant Biotechnol J 9:315–327

Li F, Guo S, Zhao Y, Chen D, Chong K, Xu Y (2010) Overexpression of a homopeptide repeat-containing bHLH protein gene (*OrbHLH001*) from Dongxiang wild rice confers freezing and salt tolerance in transgenic *Arabidopsis*. Plant Cell Rep 29:977–986

Li X, Duan X, Jiang H, Sun Y, Tang Y, Yuan Z, Guo J, Liang W, Chen L, Yin J, Ma H, Wang J, Zhang D (2006) Genome-wide analysis of basic/helix-loop-helix transcription factor family in rice and *Arabidopsis*. Plant Physiol 141: 1167–1184

Lin PC, Hwang SG, Endo A, Okamoto M, Koshiba T, Cheng WH (2007) Ectopic expression of *ABSCISIC ACID 2/GLUCOSE INSENSITIVE 1* in *Arabidopsis* promotes seed dormancy and stress tolerance. Plant Physiol 143:745–758

Liu W, Tai H, Li S, Gao W, Zhao M, Xie C, Li WX (2014) *bHLH122* is important for drought and osmotic stress resistance in *Arabidopsis* and in the repression of ABA catabolism. New Phytol 201:1192–1204

Massari ME, Murre C (2000) Helix-loop-helix proteins: regulators of transcription in eucaryotic organisms. Mol Cell Biol 20:429–440

Nakashima K, Ito Y, Yamaguchi-Shinozaki K (2009) Transcriptional regulatory networks in response to abiotic stresses in *Arabidopsis* and grasses. Plant Physiol 149:88–95

Nambara E, Marion-Poll A (2005) Abscisic acid biosynthesis and catabolism. Annu Rev Plant Biol 56:165–185

Okamoto M, Kuwahara A, Seo M, Kushiro T, Asami T, Hirai N, Kamiya Y, Koshiba T, Nambara E (2006) *CYP707A1* and *CYP707A2*, which encode abscisic acid 8'-hydroxylases, are indispensable for proper control of seed dormancy and germination in *Arabidopsis*. Plant Physiol 141:97–107

Qin X, Zeevaart JA (2002) Overexpression of a *9-cis-epoxycarotenoid dioxygenase* gene in *Nicotiana plumbaginifolia* increases abscisic acid and phaseic acid levels and enhances drought tolerance. Plant Physiol 128:544–551

Rook F, Corke F, Card R, Munz G, Smith C, Bevan MW (2001) Impaired sucrose-induction mutants reveal the modulation of sugar-induced starch biosynthetic gene expression by abscisic acid signalling. Plant J 26:421–433

Saika H, Okamoto M, Miyoshi K, Kushiro T, Shinoda S, Jikumaru Y, Fujimoto M, Arikawa T, Takahashi H, Ando M, Arimura S, Miyao A, Hirochika H, Kamiya Y, Tsutsumi N, Nambara E, Nakazono M (2007) Ethylene promotes submergence-induced expression of *OsABA8ox1*, a gene that encodes ABA 8'-hydroxylase in rice. Plant Cell Physiol 48:287–298

Saito S, Hirai N, Matsumoto C, Ohigashi H, Ohta D, Sakata K, Mizutani M (2004) *Arabidopsis* CYP707As encode (+)-abscisic acid 8'-hydroxylase, a key enzyme in the oxidative catabolism of abscisic acid. Plant Physiol 134:1439–1449

Schmidt R, Mieulet D, Hubberten HM, Obata T, Hoefgen R, Fernie AR, Fisahn J, Segundo BS, Guiderdoni E, Schippers JH, Mueller-Roeber B (2013) Salt-responsive ERF1 regulates reactive oxygen species-dependent signaling during the initial response to salt stress in rice. Plant Cell 25:2115–2131

Schwartz SH, Qin X, Zeevaart JA (2003) Elucidation of the indirect pathway of abscisic acid biosynthesis by mutants, genes, and enzymes. Plant Physiol 131: 1591–1601

Schwartz SH, Tan BC, Gage DA, Zeevaart JA, McCarty DR (1997) Specific oxidative cleavage of carotenoids by VP14 of maize. Science 276:1872–1874

Seo M, Koshiba T (2002) Complex regulation of ABA biosynthesis in plants. Trends Plant Sci 7:41–48

Seo M, Peeters AJ, Koiwai H, Oritani T, Marion-Poll A, Zeevaart JA, Koornneef M, Kamiya Y, Koshiba T (2000) The *Arabidopsis aldehyde oxidase 3* (*AAO3*) gene product catalyzes the final step in abscisic acid biosynthesis in leaves. Proc Natl Acad Sci U S A 97:12908–12913

Singh K, Foley RC, Onate-Sanchez L (2002) Transcription factors in plant defense and stress responses. Curr Opin Plant Biol 5:430–436

Su Y, Luo W, Lin W, Ma L, Kabir MH (2015) Model of cation transportation mediated by high-affinity potassium transporters (HKTs) in higher plants. Biol Proced Online 17:1

Sun XH, Copeland NG, Jenkins NA, Baltimore D (1991) Id proteins Id1 and Id2 selectively inhibit DNA binding by one class of helix-loop-helix proteins. Mol Cell Biol 11:5603–5611

Tan BC, Joseph LM, Deng WT, Liu L, Li QB, Cline K, McCarty DR (2003) Molecular characterization of the *Arabidopsis* *9-cis* epoxycarotenoid dioxygenase gene family. Plant J 35:44–56

Tan BC, Schwartz SH, Zeevaart JA, McCarty DR (1997) Genetic control of abscisic acid biosynthesis in maize. Proc Natl Acad Sci U S A 94:12235–12240

Tao Z, Kou Y, Liu H, Li X, Xiao J, Wang S (2011) *OsWRKY45* alleles play different roles in abscisic acid signalling and salt stress tolerance but similar roles in drought and cold tolerance in rice. J Exp Bot 62:4863–4874

Taylor IB, Burbidge A, Thompson AJ (2000) Control of abscisic acid synthesis. J Exp Bot 51:1563–1574

Todaka D, Nakashima K, Shinozaki K, Yamaguchi-Shinozaki K (2012) Toward understanding transcriptional regulatory networks in abiotic stress responses and tolerance in rice. Rice 5:6

Toledo-Ortiz G, Huq E, Quail PH (2003) The *Arabidopsis* basic/helix-loop-helix transcription factor family. Plant Cell 15:1749–1770

Tran TN, Sanan-Mishra N (2015) Effect of antibiotics on callus regeneration during transformation of IR 64 rice. Biotechnol Rep (Amst) 7:143–149

Wang R, Jing W, Xiao L, Jin Y, Shen L, Zhang W (2015) The rice high-affinity potassium transporter1;1 is involved in salt tolerance and regulated by an MYB-type transcription factor. Plant Physiol 168:1076–1090

Welsch R, Wust F, Bar C, Al-Babili S, Beyer P (2008) A third phytoene synthase is devoted to abiotic stress-induced abscisic acid formation in rice and defines functional diversification of phytoene synthase genes. Plant Physiol 147:367–380

Xiong L, Ishitani M, Lee H, Zhu JK (2001) The *Arabidopsis LOS5/ABA3* locus encodes a molybdenum cofactor sulfurase and modulates cold stress- and osmotic stress-responsive gene expression. Plant Cell 13:2063–2083

Xiong L, Zhu JK (2003) Regulation of abscisic acid biosynthesis. Plant Physiol 133:29–36

Ye N, Jia L, Zhang J (2012) ABA signal in rice under stress conditions. Rice 5:1

Yoshida S, Forno DA, Cock JH, Gomez KA (1976) Routine procedure for growing rice plants in culture solution. In: Laboratory manual for physiological studies of rice. International Rice Research Institute, Manila

Zhang H, Jin J, Tang L, Zhao Y, Gu X, Gao G, Luo J (2011) PlantTFDB 2.0: update and improvement of the comprehensive plant transcription factor database. Nucleic Acids Res 39:D1114–D1117

Zhang T, Zhao X, Wang W, Pan Y, Huang L, Liu X, Zong Y, Zhu L, Yang D, Fu B (2012) Comparative transcriptome profiling of chilling stress responsiveness in two contrasting rice genotypes. PLoS One 7:e43274

Zhou J, Li F, Wang JL, Ma Y, Chong K, Xu YY (2009) Basic helix-loop-helix transcription factor from wild rice (OrbHLH2) improves tolerance to salt- and osmotic stress in *Arabidopsis*. J Plant Physiol 166:1296–1306

14

Constitutive expression of *REL1* confers the rice response to drought stress and abscisic acid

Jiayan Liang, Shaoying Guo, Bo Sun, Qing Liu, Xionghui Chen, Haifeng Peng, Zemin Zhang* and Qingjun Xie*

Abstract

Leaf rolling is one of the most significant symptoms of drought stress in plant. Previously, we identified a dominant negative mutant, termed *rolled and erect 1* (hereafter referred to *rel1-D*), regulating leaf rolling and erectness in rice. However, the role of *REL1* in drought response is still poorly understood. Here, our results indicated that *rel1-D* displayed higher tolerance to drought relative to wild type, and the activity of superoxide dismutase (SOD) and drought responsive genes were significantly up-regulated in *rel1-D*. Moreover, our results revealed that *rel1-D* was hypersensitive to ABA and the expression of ABA associated genes was significantly increased in *rel1-D*, suggesting that *REL1* likely coordinates ABA to regulate drought response. Using the RNA-seq approach, we identified a large group of differentially expressed genes that regulate stimuli and stresses response. Consistently, we also found that constitutive expression of *REL1* alters the expression of biotic and abiotic stress responsive genes by the isobaric tags for relative and absolute quantification (iTRAQ) analysis. Integrative analysis demonstrated that 8 genes/proteins identified by both RNA-seq and iTRAQ would be the potential targets in term of the *REL1*-mediated leaf morphology. Together, we proposed that leaf rolling and drought tolerance of *rel1-D* under normal condition might be caused by the endogenously perturbed homeostasis derived from continuous stressful dynamics.

Keywords: Rice, *REL1*, Leaf morphology, Genomics, ABA, Stress response

Background

Crop yield is adversely challenged by drought stress, one of the most major environmental stresses, of which the occurrence and severity are both increased due to the recent climate change, inadequate water supply and world population growth worldwide. Therefore, improving drought tolerance of crop is an important objective to overcome such issue and provide enough world food (Yamaguchi-Shinozaki and Shinozaki, 2006). One of the most significant symptoms of drought stress in plant is the leaf rolling. Plant leaf generally is polarized along the adaxial-abaxial axis (Itoh et al. 2005), thereby generating two types of leaf forms under unfavorable conditions: abaxially leaf rolling and adaxially leaf rolling. Moderate leaf rolling promotes rice yield by increasing the photosynthetic efficiency and reducing the transpiration (Lang et al. 2004; Zhang et al. 2009; Zou et al. 2011). In addition, moderate leaf rolling also facilitates the survival and development of plant under stress conditions (Kadioglu et al. 2012). Therefore, manipulation of adaxial and abaxial leaf rolling would be one of the most important strategies to increase the rice productivity and tolerance to drought and other stresses in coming years (Zou et al. 2011).

Generally, leaf rolling is largely regulated by several factors: hormones (Fujino et al. 2008; Lee et al. 2006), metabolic changes (Talukdar and Talukdar, 2014) and formation of specific cells, such as bulliform cell (Fang et al. 2012; Xiang et al. 2012). To date, more than 30 genes responsible for the rice leaf rolling have been identified. For example, *Abaxially Curled Leaf 1* (*ACL1*), *Abaxially Curled Leaf 2* (*ACL2*) (Li et al. 2010) and *Rice outermost cell-specific gene 5* (*Roc5*) (Zou et al. 2011) are involved in abaxial leaf rolling, whereas *SEMI-ROLLED LEAF 1* (*SRL1*) (Xiang et al. 2012), *ROLLED LEAF 9* (*RL9*)/*SHALLOT-LIKE1* (*SLL1*) (Zhang et al. 2009; Yan

* Correspondence: zmzhang@scau.edu.cn; qjxie@scau.edu.cn
State Key Laboratory for Conservation and Utilization of Subtropical Agro-Bioresources, Guangdong Provincial Key Laboratory of Plant Molecular Breeding, South China Agricultural University, Guangzhou 510642, China

et al. 2008), *OsAGO7* (Shi et al. 2007), *Rolling-leaf 14* (*RL14*) (Fang et al. 2012), *R2R3-MYB transcription factor* (*OsMYB103L*) (Yang et al. 2014), *ADAXIALIZED LEAF 1* (*ADL1*) (Hibara et al. 2009) and *Narrow and Rolled Leaves 1* (*NRL1*) (Wu et al. 2010) participates in adaxial leaf rolling. In addition, other regulators, such as *CONSTITUTIVELY WILTED1* (*COW1*)/*NAL7* (Fujino et al. 2008), has also been well-characterized in term of their involvements in leaf rolling. Even thought, it is still poorly understood whether these leaf rolling genes participate in stress response, in particular drought.

To cope with drought stress, plants have been evolved a sophisticated adaptation mechanisms to increase their chance of survival through coordinating the expression of drought responsive genes, which retarded leaf rolling, wilting, and loss of chlorophyll via ABA-dependent or independent pathways (Yamaguchi-Shinozaki and Shinozaki, 2006; Weiner et al. 2010; Cutler et al. 2010). Under drought stress, ABA accumulates in plant cells and stimulates the core ABA signaling transduction through the ABA receptors *PYR/PYL/RCAR*, *Protein Phosphatases 2C* (*PP2Cs*), and subclass III *SNF 1-RELATED PROTEIN KINASE 2* (*SnRK2*) protein kinases (Weiner et al. 2010). For example, loss of the rice subclass-I and -II SnRK2s (*OsSAPK2*) confers the more sensitivity to drought stress and insensitive to ABA (Lou et al. 2017). Over-expression of *OsDT11* enhanced drought tolerance thought the ABA signaling pathway in rice (Li et al. 2017). In contrast, a group of regulators modulate drought tolerance by ABA-independent pathway. Fox example, *OsMADS50* was markedly induced by low water-deficit treatment but not ABA, and the *osmads50* knockout mutant exhibited significant delays in flowering compared with the WT, indicating *OsMADS50* had an ABA-independent and positive role in drought response (Du et al. 2018).

Here, we were particularly interested in elucidating the relationship between the *rel1-D* mediated leaf rolling and drought stress in rice. Our results suggested that up-regulation of *REL1* confers more drought tolerance by eliminating the ROS level and triggers the ABA response. Similarly, transcriptomic and proteomic profiling of *rel1-D* and wild type also revealed that most of the differentially expressed genes (DEGs) and differentially expressed proteins (DEPs) are mainly involved in metabolic changes and stress responses. Collectively, we proposed that leaf rolling of *rel1-D* is caused by the altered dynamics of stress response endogenously.

Methods

Plant materials and growth conditions

All plants (wild type and *rel1-D* mutants) used in this study were derived from our previous study (Chen et al. 2015). All rice seeds in this study were propagated in the paddy field in Guangzhou, China. For laboratory work, rice plants were grown in a greenhouse under a16-h-light/8-h-dark cycle at 30 °C. No significant differences were observed when plants were grown in the greenhouse compared to the paddy field.

Drought response assay

For drought tolerance test, wild type and *rel1-D* seedlings were grown in soil for 2 months, and then treated by withholding water for 14 days, followed by recovery irrigation for 7 days. Leaves from 1-month-old seedling were selected for dark-induced leaf senescence and ABA treatment. Seeds of wild type and *rel1-D* were germinated at 37 °C, and then transferred into 1/2 MS medium for 14 days at 30 °C under a16-h-light/8-h-dark cycle. Seedlings from 3rd leaf stage were selected for PEG treatment: 1) plants treated with 0%, 10%, 20%, 30% 40% PEG4000 for 24 h, and the rolling index was measured. The rolling index was measured as previous study (Shi et al. 2007). 2) Plants treated with 20% PEG4000 with 5 time courses, including 0 h, 3 h, 6 h, 12 h and 24 h. The SOD activity was measured as previous study (Zhang et al. 2005).

ABA treatment and chlorophyll content measurement

Leaves were treatment with distillation-distillation water with 0 μM or 20 μM ABA at 28 °C in darkness for 5 days. Chlorophyll content was measured as followed: leaves were milled in 95% ethanol; homogenate solution adjusted to 50 ml final volume; measure the absorbance at 652 nm. Calculate the chlorophyll content according to the following formula: chlorophyll content (mg/g) = $(A_{652} \times V)/\ (34.5 \times m)$; V was the final volume of homogenate solution, and m was the weight of leaves.

RNA extraction and quantitative real-time PCR

Total RNA was extracted using the RNeasy Plant Mini Kit (Qiagen) according to the manufacturer's instructions. The first strand of cDNA was synthesized using TransScript First-Strand cDNA Synthesis SuperMix (TransGen Biotech) and quantitative real-time polymerase chain reaction (qRT-PCR) was performed as previously described (Chen et al. 2015). The relative expression level of a target gene was normalized to that of rice *UBC*. All primers used in qRT-PCR are listed in Additional file 1: Table S12.

Sequence reads mapping

Raw data (raw reads) of fastq format were firstly processed through in-house perl scripts. In this step, the clean data (clean reads) were obtained by removing reads containing adapter, reads containing poly-N and low quality reads from raw data. At the same time, quality parameters of clean data including Q20, Q30, GC-content and sequence duplication level were used for data filtering. All the succeeding analyses were carried out using high quality clean data. Reference genome and gene model annotation files were downloaded from The MSU Rice Genome Annotation Project Database website at http://rice.plantbiology.msu.edu. An index of the reference genome was built using Bowtie2 v2.2.5 (Langmead et al. 2009) and paired-end clean reads were aligned to the reference genome using TopHat v2.0.14 (Trapnell et al. 2010). TopHat was chosen as the mapping tool, because it can generate a database of splice junctions based on the gene model annotation file, and thus give a better mapping result than other non-splice mapping tools.

Quantification and differential expression analysis of transcripts

HTSeq v0.6.1 (http://www-huber.embl.de/users/anders/HTSeq) was used to count the reads numbers mapped to each transcript. The parameter FPKM (expected number of Fragments per kilo-base of transcript sequence per Millions base pairs sequenced) was used to quantify transcripts expression. FPKM was calculated based on the mapped transcript fragments, transcript length and sequencing depth. Currently, this is the most commonly used method for estimating transcript expression (Trapnell et al. 2010). Differential expression analysis of two conditions/groups was performed using the DESeq R package (1.10.1) (Anders and Huber, 2010). DESeq provides statistical routines to determine differential expression in digital gene expression data using a model based on the negative binomial distribution. The resulting *q*-values (*p*-adjusted) were adjusted using the Benjamini and Hochberg's approach for controlling the false discovery rate (Anders and Huber, 2010). Genes with an adjusted $|\log_2 (FC)| \geq 1$ and $FDR < 0.05$ were assigned as differentially expressed.

iTRAQ assay and data analysis

The iTRAQ assay was performed by the BGI Company. Briefly, about 100 μg protein was subjected to the LC-ESI-MS/MS analysis based on the Triple TOF 5600. The proteins identification was performed by using Mascot search engine (Matrix Science, London, UK; version 2.3.02). For protein quantitation, it was required that a protein contains at least two unique spectra. The quantitative protein ratios were weighted and normalized by the median ratio in Mascot. We used ratios of *p*-values < 0.05, and fold changes > 1.5 or < 0.67 was considered as significant. Functional annotations of the proteins were conducted using Blast2GO program against the non-redundant protein database (https://www.ncbi.nlm.nih.gov/protein/). The keg database (http://www.genome.jp/kegg/) and the COG database (http://www.ncbi.nlm.nih.gov/COG/) were used to classify and group these identified proteins.

Results

Response of *rel1-D* to drought stress

Leaf rolling generally leads to reduction of water loss, and thereby enhances tolerance to multiple stresses (Lang et al. 2004; Zhang et al. 2009; Zou et al. 2011). Therefore, we were specifically interested in investigating the involvement of *REL1* in stress response. To address this issue, WT and *rel1-D* seedlings were undergone the drought assay. Phenotypic analysis showed that almost all of the WT plants displayed severe growth retardation and wilting while the *rel1-D* plants exhibited less abnormal phenotypes by withholding irrigation for 14 days, and then the *rel1-D* but not the WT was recovered upon re-watering for 7 days (Fig. 1a and b). To further explore the drought tolerance of *rel1-D*, leaves from WT and *rel1-D* were treated by polyethylene glycol (PEG) 4000 at different concentrations. Our results indicated that leaves of *rel1-D* were much more insensitive to the treatment than those of WT (Fig. 1c). The superoxide dismutase (SOD) activity is generally used as an important indicator for drought tolerance. Therefore, the SOD activity of WT and *rel1-D* that treated by 20% PEG4000 was measured at 5 time courses. Our results demonstrated that the change pattern of SOD activity was similar between WT and *rel1-D*, but the corresponding levels were higher in *rel1-D* than those in WT (Fig. 1d), suggesting that up-regulating *REL1* suppressed the boost of reactive oxygen species (ROS) under drought stress. To further gain insight into the *REL1*-mediated drought resistance, we then evaluated the expression of drought-responsive marker genes. Our results demonstrated that *OsDT11*, *OsSAPK2*, *OsMYB2*, *OsDREB1A* and *OsbHLH148*, functioning as positive regulators in drought response (Lou et al. 2017; Li et al. 2017; Chen et al. 2008; Dubouzet et al. 2003; Seo et al. 2011; Yang et al. 2012), were significantly up-regulated in the *rel1-D* (Fig. 1e-i). Consistently, the similar patterns of above marker genes were also found in the RNA-seq result (Additional file 2: Table S1). These results suggested that enhanced the expression of *REL1* leads to the substantial higher expression of drought-responsive genes endogenously, eventually resulting in the tolerance of *rel1-D* to drought.

Taking into account that *REL1* is an abaxial leaf rolling regulator, we wondered whether other abaxial leaf rolling associated genes is also involved in the response to drought stress. To address this issue, we examined their expression pattern in WT and *rel1-D* at different time courses by PEG treatment. Surprisingly, *REL1* exhibited normal expression during the treatment in WT, implying it is not related with drought stress. However, the

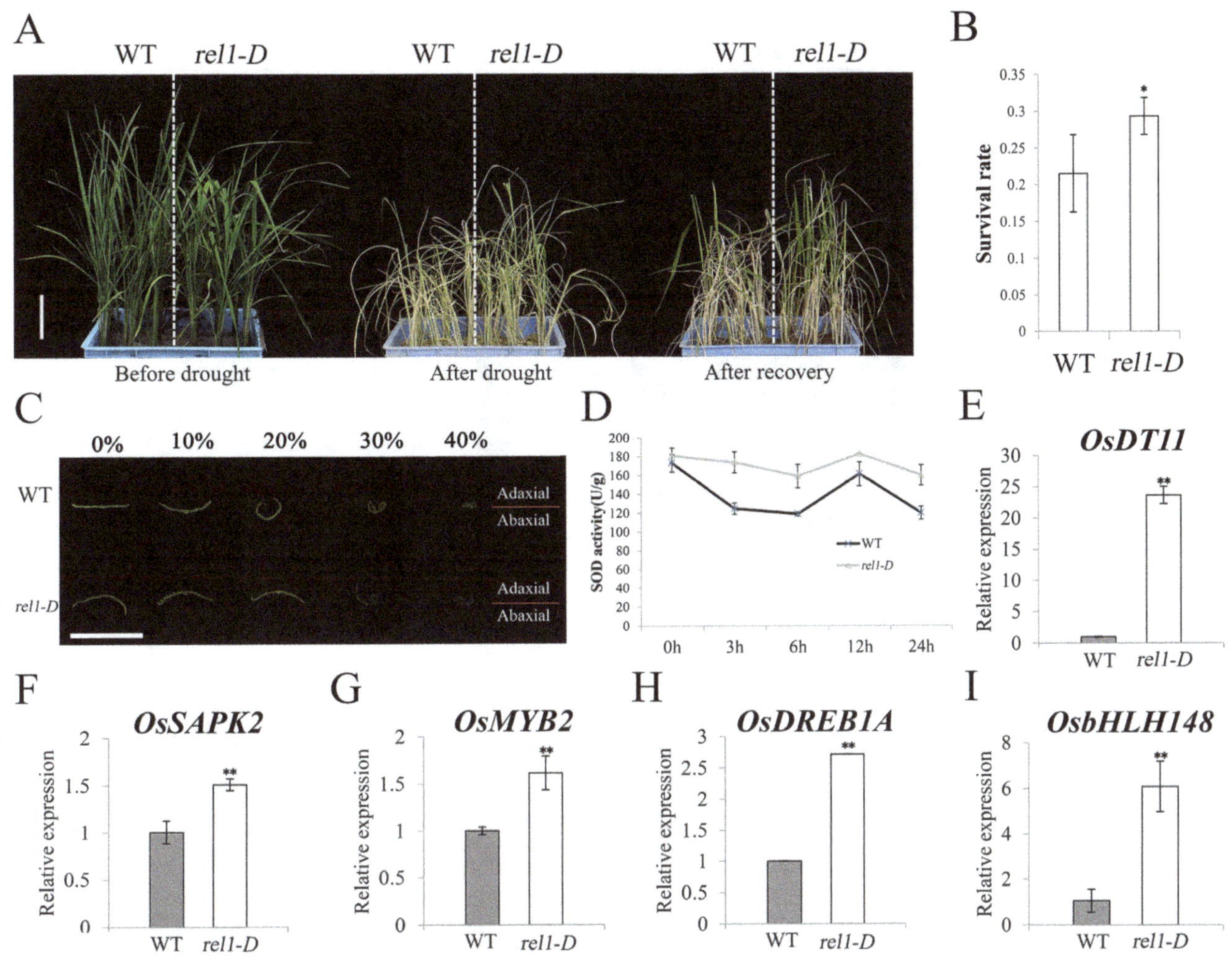

Fig. 1 Response of *rel1-D* to drought stress. **a**, Phenotype of WT and *rel1-D* with drought treatment. Two-month-old seedlings were used, bars = 10 cm. **b**, Survival rate of WT and *rel1-D* were derived from the (A). **c**, Third leaf stage seedling of WT and *rel1-D* were treated by different concentration of PEG, bars = 5 mm. **d**, Measurement of SOD activity in wild type and *rel1-D*. E-I, Expression of drought-responsive genes in wild type and *rel1-D*. **b** and **e-i**, Data were presented as mean ± S.E., * *p*-value < 0.05, ** *p*-value < 0.01, two-tailed, two sample Student's t test

expression level of *REL1* was substantial higher in *rel1-D* than that in WT even it was attenuated at 3 h but gradually increased afterward in *rel1-D* (Additional file 3: Figure S1A). The *ACL1* was not changed by the PEG treatment in both WT and *rel1-D* at 24 h (Additional file 3: Figure S1B), suggesting this leaf rolling gene is not involved in drought response. The *ACL2* was significantly enhanced at 24 h in WT, implying it might participate in drought stress. Notably, *ACL2* was retarded from 3 h to 12 h but increased to the normal level at 24 h in *rel1-D* with the similar level of that in WT (Additional file 3: Figure S1C), suggesting that constitutive expression of *REL1* would repress *ACL2* during drought stress. The *Roc5* was down-regulated at 12 h and increased to the normal level at 24 h in WT, but gradually decreased in *rel1-D* during treatment (Additional file 3: Figure S1D), demonstrating *Roc5* was suppressed by up-regulation of *REL1* under drought stress. Taken together, we proposed that *rel1-D* mediated drought tolerance independent on the three leaf rolling genes, but *ACL2* may be also involved in drought stress in a distinct pathway.

Response of *rel1-D* to ABA

Since ABA both regulates drought response and leaf senescence (Lou et al. 2017; Li et al. 2017; Chen et al. 2008; Dubouzet et al. 2003; Seo et al. 2011; Yang et al. 2012), we wondered whether *REL1* was also involved in the response to ABA. To address this issue, leaves from WT and *rel1-D* were treated by 0 and 20 μM ABA from 1 to 5 days by the dark-induced leaf senescence assay, respectively. Under the ABA treatment, *rel1-D* displayed early senescence phenotype and rapid degradation of chlorophyll (Chl) rather than that in WT (Fig. 2a and b), indicating it was hypersensitive to ABA. Notably, *REL1*

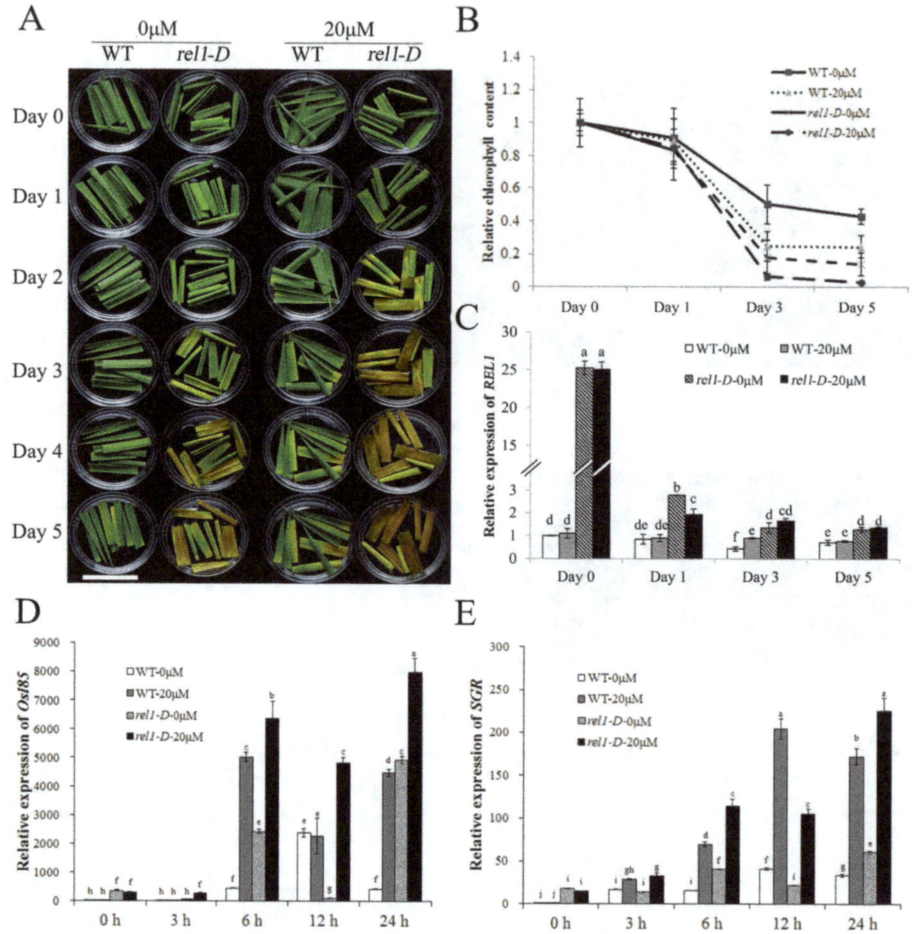

Fig. 2 Response of *rel1-D* to ABA. **a**, Response of *rel1-D* to ABA during dark-induced leaf senescence. Leaves from 1-month-old seedling of WT and *rel1-D* were incubated with 20 μM ABA for 1 to 5 days, bars = 5 cm. **b**, Chlorophyll content of WT and *rel1-D* by ABA treatments. **c**, Expression of *REL1* in response to ABA treatment. **d**, Expression of *OsI85* in response to ABA treatment. **e**, Expression of *SGR* in response to ABA treatment. B-E, Data were presented as mean ± S.E.. **c-e**, Multiple comparisons, Duncan, *p*-value < 0.01

is significantly repressed by the ABA treatments (Fig. 2c). Therefore, we concluded that boosting *REL1* accelerates ABA-induced leaf senescence. It was worthy to figure out that *rel1-D* leaves started to turn yellowish while the wild-type leaves still remained green at the 3 days. Five days after dark treatment, the Chl content in the *rel1-D* leaves was about 2-fold less than in the wild-type leaves (Fig. 2a and b). To further explore the relationship between *REL1* and ABA-induced senescence, we detected the expression of the senescence marker genes *OsI85* and *SGR*, and found that they were induced by dark-treatment assay as previously reported (Lee et al. 2001; Jiang et al. 2011). Notably, their levels were significantly higher in *rel1-D* as compared to WT, as well as with ABA rather than without ABA treatment (Fig. 2d and e). Therefore, we proposed that overexpression of *REL1* also triggers the natural leaf senescence and ABA would accelerate this response.

Transcriptomic profiling of the *rel1-D* mutant

To further investigate the regulatory mechanism of *REL1*-mediated leaf rolling and bending, we performed a RNA-seq analysis with the leaves of *rel1-D* mutants and wild type plants at the tillering stage, since the most obvious leaf rolling and bending phenotypes were occurred at this stage. In total, 487 differentially expressed genes (DEGs) were identified with the stringent criteria ($|\log_2(FC)| \geq 1$, and $FDR < 0.05$). To verify these DEGs, 10 randomly selected DEGs were detected by the qRT-PCR, and their trends were similar as RNA-seq (Additional file 4: Figure S2), indicating that the RNA-seq result was qualified for following study. Of these DEGs, 247 and 240 transcripts were up-regulated or down-regulated in *rel1-D* as compared to wild type, respectively (Additional file 5: Table S2). Even the statistical examination of most BR genes was not significant, there were still two BR signaling genes were induced in *rel-D* (Additional file 6: Figure S3), suggesting that BR was also involved in regulating the leaf morphology of *rel1-D*.Notably, the ABA pathway was significantly changed (Additional file 7: Figure S4), further supporting the note that ABA was involved in *rel1-D* mediated leaf morphology and response. To further characterize these DEGs, we performed the gene ontology (GO) enrichment analysis. In respect to the up-regulated DEGs, they were significantly assigned to certain cellular component GO terms, including cell wall, external encapsulating structure and cell periphery (Fig. 3a; Additional file 8: Table S3). In terms of the molecular function GO term, these DEGs were mainly associated with hydrolase activity, transcription factor activity and catalytic activity (Fig. 3b; Additional file 9: Table S4). In respect to the biological process GO term, these DEGs were significantly involved in multiple stimuli and stresses (Fig. 3c; Additional file 10: Table S5). Investigation of the down-regulated DEGs showed that they were also significantly associated with vacuole and endoplasmic reticulum (Fig. 3d; Additional file 11: Table S6), catalytic activity and transport activity (Fig. 3e; Additional file 12: Table S7), and response to various stimuli and metabolic processes (Fig. 3f; Additional file 13: Table S8). Taken together, our results suggested that boosting *REL1* apparently perturbed the homeostasis of stress dynamics in specific organelles, such as cell wall and vacuole, eventually leading to the abnormal leaf morphology.

Proteomics analysis of *rel1-D* mutant

To further analyze the function of *REL1* in leaf morphology at protein level, we then performed the isobaric tags for relative and absolute quantitation (iTRAQ) analysis on the above materials. In total, 3657 peptides were identified (Additional file 14: Table S9). Using the p-value < 0.05 and fold change > 1.5 or < 0.67 as significant cutoff, 20 and 29 differentially expressed proteins (DEPs) were up-regulated or down-regulated, respectively. To verify these DEPs, 10 randomly selected DEPs were detected by the qRT-PCR, and their changing pattern were similar as iTRAQ (Additional file 15: Figure S5), indicating that the iTRAQ result was qualified for following study. Subsequently, GO enrichment analysis of these DEPs revealed that they were enriched for cellular component GO terms related to inter- and intra- cellular organelles, ribosome and plastid envelope (Fig. 4a; Additional file 16: Table S10). In respect to the molecular function GO term, significant enrichments were found in amylase activity, transport activity and ion binding (Fig. 4b; Additional file 16: Table S10). Regarding the biological process, the DEPs were grouped into multiple catabolic processes, response to water transport and stimuli (Fig. 4c; Additional file 16: Table S10). It was worthy to mention that 24 out of 49 DEPs were annotated or predicted to be plastid-localized proteins (Additional file 17: Table S11), while REL1 was previously

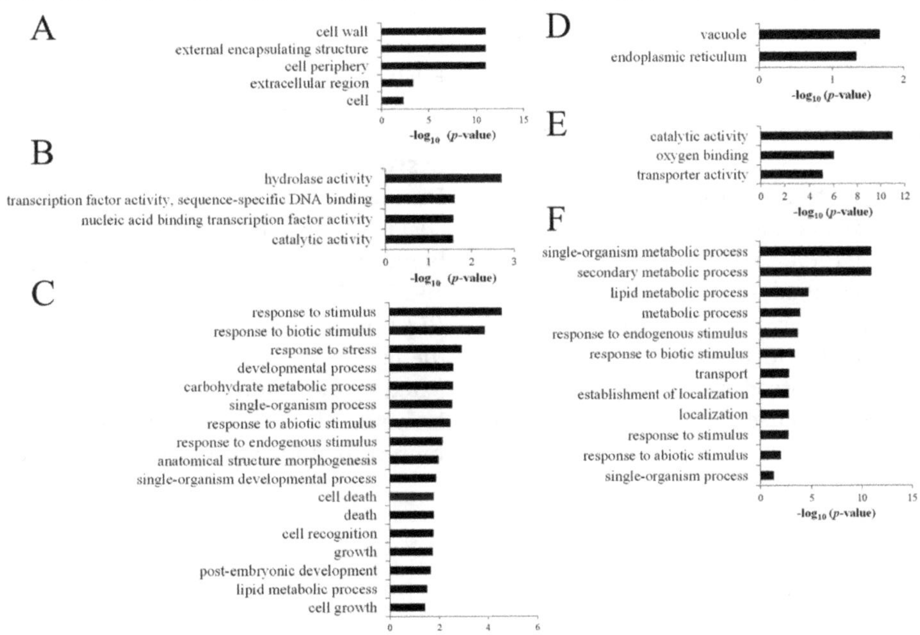

Fig. 3 Gene Ontology (GO) analysis for differentially expressed genes (DEGs). **a**, Significant cellular component GO terms of the up-regulated DEGs. **b**, Significant molecular function GO terms of the up-regulated DEGs. **c**, Significant biological process GO terms of the up-regulated DEGs. **d**, Significant cellular component GO terms of the down-regulated DEGs. **e**, Significant molecular function GO terms of the down-regulated DEGs. **f**, Significant biological process GO terms of the down-regulated DEGs

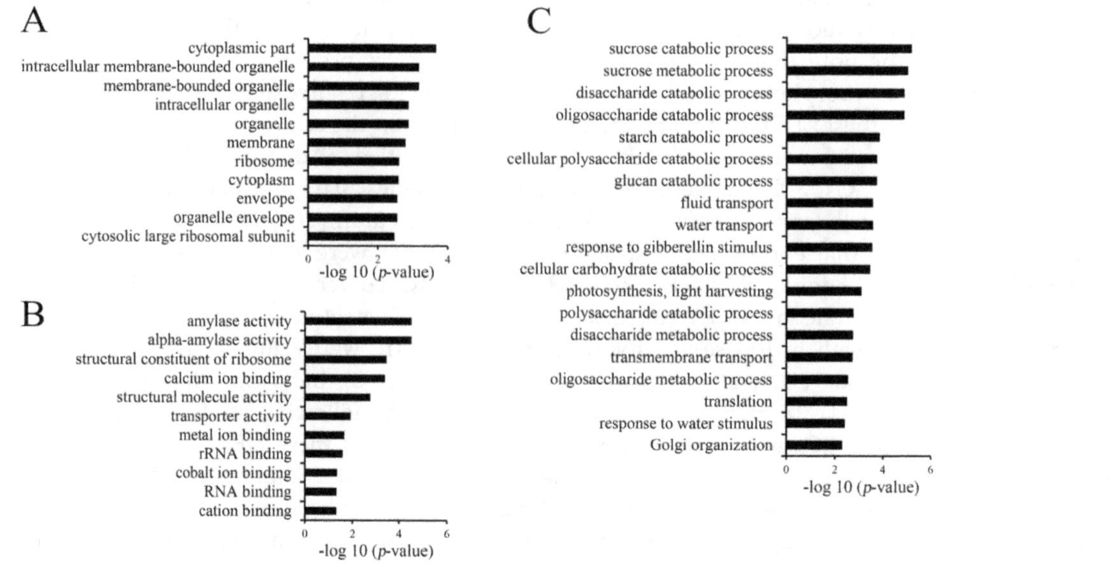

Fig. 4 Gene Ontology (GO) analysis for differentially expressed proteins (DEPs). **a**, The 11 significant cellular component GO terms of the DEPs. **b**, The 11 molecular function GO terms of the DEPs. **c**, The 19 significant biological process GO terms of the DEPs

implicated as plastid protein (Chen et al. 2015). Notably, 4 DEPs have been reported to regulate stress response (Table 1), including *OsPIP1;1*, *OsPIP1;2*, *SUS2* and *OsGLP8–7* (Liu et al. 2013; Mosa et al. 2012; Xiao et al. 2014; Breen and Bellgard 2010), further supporting the notion that *rel1-D* actives the endogenous stress responses. Collectively, we proposed that REL1 may coordinate the chloroplast DEPs to regulate leaf morphology by altering the metabolic process and stressful dynamics.

Integrative analysis of transcriptome and proteome for *rel1-D*

Integrative analysis of transcriptome and proteome may provide new insights into the identification of interest key genes. In total, 3575 co-expressed genes and proteins were identified (Fig. 5a). Then only 2 co-expressed genes/proteins, *LOC_Os05g09740* and *LOC_Os02g37654*, were found between the 487 DEGs and 49 DEPs (Fig. 5b). To broad view the genome-wide change, 246 DEPs were screened by a less stringent criteria (with fold change > 1.5 or < 0.67), and 234 out of these low criteria DEPs (DEPs-low) were identified in the RNA-seq data (Fig. 5c). Integrating the DEGs and the 246 DEPs-low, there were 8 genes/proteins found in each other (Fig. 5d). These 8 genes modulated stresses response and had distinct expression pattern, of which 5 genes showed similar trends in transcription and translation level but 3 genes exhibited opposite trends (Table 2). Taken together, we proposed that the molecular mechanism underlying *REL1*-mediated leaf phenotype was likely different between transcriptional and post-translation levels.

Discussion

During plant growth and development, leaf rolling is an adaptive rather than passive response to the abiotic and biotic stresses in plants (Kadioglu et al. 2012; Kadioglu and Terzi, 2007). Our previous study has implicated that *REL1* positively regulates leaf rolling through altering the profile of bulliform cells and leaf bending by coordinated expression of BR related genes (Chen et al. 2015). However, the biological function of *REL1* and the relevant regulatory mechanism still remains to be further

Table 1 The differentially expressed proteins (DEPs) related to the stress response

Gene locus	Symbol	Fold change[a]	Description (reference)
LOC_Os02g44630	*OsPIP1;1*	0.57 ± 0.13	Regulating the salt tolerance (Liu et al. 2013)
LOC_Os04g47220	*OsPIP1;2*	0.59 ± 0.02	Regulating the tolerance and transport of arsenate (Mosa et al. 2012)
LOC_Os06g09450	*SUS2*	0.58 ± 0.01	Induced by immerge during germination (Xiao et al. 2014)
LOC_Os08g09010	*OsGLP8–7*	1.62 ± 0.15	Induced by wounding and blast (Breen and Bellgard, 2010)

[a]The fold change was calculated by the *rel1-D* versus WT

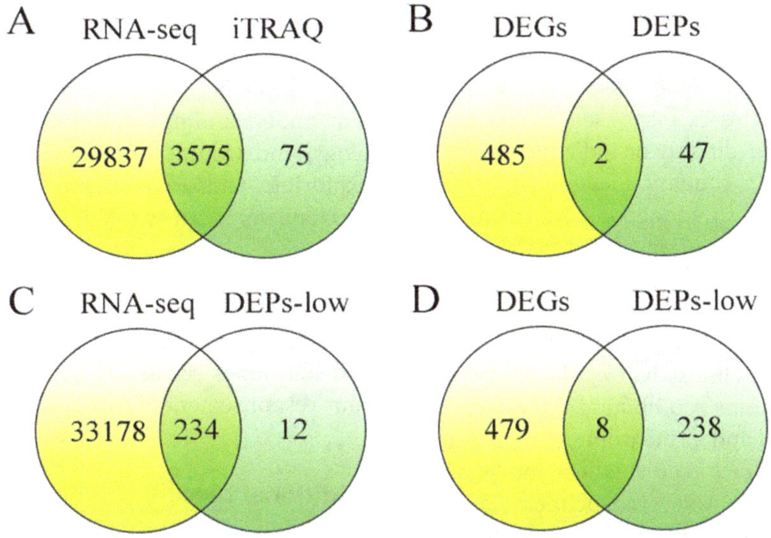

Fig. 5 Correlation between the transcript and protein levels. **a**, Venn analysis of all expressed mRNA and all expressed proteins. **b**, Venn analysis of differentially expressed genes (DEGs) and differentially expressed proteins (DEPs) with stringent criteria (p-value < 005). **c**, Venn analysis of all expressed transcripts and differentially expressed proteins with less stringent criteria. **d**, Venn analysis of differentially expressed transcripts and differentially expressed proteins with less stringent criteria

elucidated. Here, we further explored its regulatory role in leaf morphology by transcriptomic and proteomic analyses, as well as co-expression analysis. Our results may provide a new insight into the *REL1*-mediated leaf morphology in rice.

Advance studies have also been evident that water deficiency is one of the most major reasons for the formation of leaf rolling, which effectively reduces transpiration and thus is potentially useful in drought tolerance (Kadioglu et al. 2012). As expected, up-regulation of drought resistance marker genes and tolerance to drought treatment in *rel1-D* suggested that *REL1* positively participates in drought resistance. In addition, our result also indicated that *REL1*-mediated drought tolerance might be integrated the ABA pathway. Therefore, we proposed that *REL1* might coordinate ABA pathway to regulate drought tolerance. Comparative analysis of *REL1* and other abaxial leaf rolling genes demonstrated that they might play opposite roles in response to drought tolerance, suggesting that they might also function independently during the formation of leaf rolling under stresses. Surprisingly, we also found that *REL1* is involved in leaf senescence and ABA response, which would be another interesting issue to be addressed regarding the biological function of *REL1*.

Although *REL1* encodes an unknown but species-specific protein, genome-wide profiling may facilitate our

Table 2 Expression pattern of DEGs and DEPs identified in both transcriptomic and proteomic analyses

Gene locus	Transcript[a]	Protein[b]	Description
LOC_Os01g51570	2.45 ± 1.36	1.58 ± 0.21	Glycosyl hydrolases family 17, putative, expressed
LOC_Os08g39840	1.19 ± 0.10	0.61 ± 0.10	*OsHI-LOX; RLL; OsLOX9*
LOC_Os02g33820	−1.05 ± 0.28	1.52 ± 0.42	*Asr3; OsASR3; ASR1*
LOC_Os02g37654	−2.02 ± 0.47	0.58 ± 0.11	Lecithin: cholesterol acyltransferase, putative, expressed
LOC_Os03g40670	1.20 ± 0.34	0.63 ± 0.10	Glycerophosphoryl diester phosphodiesterase family protein, putative, expressed
LOC_Os05g09740	−4.16 ± 0.65	0.51 ± 0.06	HAD superfamily phosphatase, putative, expressed
LOC_Os05g10210	−5.67 ± 0.31	0.62 ± 0.00	HAD superfamily phosphatase, putative, expressed
LOC_Os10g36170	−1.81 ± 0.81	0.52 ± 0.02	LTPL160 - Protease inhibitor/seed storage/LTP family protein precursor, expressed

[a] Represent the fold change of the DEGs in RNA-seq analysis
[b] Represent the fold change of DEPs in iTRAQ analysis

understanding on the biological function of *REL1* in determining leaf morphology. Transcriptomic and proteomic profiling demonstrated that the change of leaf morphology in *rel1-D* was highly associated with the metabolic changes and stresses response. However, it is still unclear whether *REL1* directly or indirectly catalyzes specific primary and/or secondary metabolism so that generating stressful dynamics endogenously. Taking into account the chloroplast localization of REL1, the chloroplast DEGs may be functionally associated with *REL1*. Unexpectedly, only 4 DEGs were grouped into chloroplast GO term under the stringent statistical criteria. Differing from the DEGs, almost half of the DEPs were chloroplast localized proteins. Combining the integrative analyses of transcriptom and proteasome, two possibilities were proposed: 1) *REL1* regulates leaf morphology at the post-transcriptional level independent on the chloroplast genes; 2) *REL1* regulates leaf morphology at the post-translation levels through direct or indirect regulation of chloroplast proteins. These issues would be quite interesting for further study. Expectedly, a large part of the DEGs and DEPs were both enriched into the stress response GO term, further demonstrating that up-regulation of *REL1* generates endogenous stresses for the plant. Several genes/proteins identified in both DEGs and DEPs would be the candidates for future study in terms of leaf rolling and bending, particularly the response to multiple stresses.

Interestingly, a recent study reported that another gene *REL2*, encoding a DUF630 and DUF632 domains containing protein, regulates leaf rolling and bending as well (Yang et al. 2016). It was worthy to figure out that the gene *LOC_Os06g44610*, significantly down-regulated in *rel1-D*, also encodes a membrane associated DUF588 domain containing protein, suggesting a possible role of DUF family genes in the leaf development. Although *REL1* has a functional relationship with *REL2*, it is still challenged by: 1) the distinct localization pattern of these two proteins since REL2 is a plasma membrane localized protein while REL1 is a chloroplast protein; 2) *REL2* likely regulates bulliform cell through auxin pathway while *REL1*might be much more related to BR pathway. In addition, much more transcriptomics datasets would benefit the further understanding of the regulatory network between *REL1* and other leaf development genes by co-expression analysis. Meanwhile, genetic analysis by constructing double (or multiple) mutant would also benefit our knowledge of the correlation among these leaf morphology genes.

Conclusions

Leaf rolling is generally related to stress response. By the drought assay, we identified the involvement of *rel1-D* in such biological process. Furthermore, the RNA-seq and iTRAQ-based profiling of WT and *rel1-D* also indicated that enhanced expression of *REL1* resulted in the alteration of stress responsive genes and hormone genes. These genes might be the potential targets for extending our understanding of the *REL1*-mediated leaf morphology, and would also be valuable resources for further exploring and/or genetic engineering the molecular mechanism of stress tolerance in rice.

One sentence summary

REL1 is involved in drought tolerance, leaf senescence and hormones sensitivity by altering the profiles of relevant response genes in rice.

Additional files

Additional file 1: Table S12. Primers used in this study.

Additional file 2: Table S1. Expression of drought-responsive genes in WT and *rel1-D*.

Additional file 3: Figure S1. Expression of abaxial leaf rolling genes in response to PEG treatment. **a** Expression of *REL1* during different time courses by PEG treatment. **b** Expression of *ACL1* during different time courses by PEG treatment. **c** Expression of *ACL2* during different time courses by PEG treatment. **d** Expression of *Roc5* during different time courses by PEG treatment. **a-d** Multiple comparisons, Duncan, p-value < 0.01.

Additional file 4: Figure S2. Verification of the RNA-seq. Modes for qPCR were cDNA of WT and the *rel1-D* leaves in tillering stage, three biological repetitions, error bars were S.E.; the X axis was gene names, the Y axis was $\log_2$ (Fold change).

Additional file 5: Table S2. Differentially expressed genes in *rel1-D* mutant.

Additional file 6: Figure S3. Expressions of BR-related genes. A Expressions of 10 BR synthesis-related genes. B Expressions of 31 BR signaling genes. A-B The colors indicated the mean expression level (FPKM), the red color was the highest and the green color was the lowest; red panes highlighted p-value < 0.05.

Additional file 7: Figure S4. Expressions of ABA-related genes. A Expressions of 15 ABA synthesis-related genes (positive-related). B Expressions of 3 ABA synthesis-related genes (negative-related). C Expressions of 43 ABA signaling genes. A-C The colors indicated the mean expression level (FPKM), the red color was the highest and the green color was the lowest; red panes highlighted p-value < 0.05.

Additional file 8: Table S3. GO enrichment (Cellular Component) of the up-regulated DEGs in *rel1-D*.

Additional file 9: Table S4. GO enrichment (Molecular Function) of the up-regulated DEGs in *rel1-D*.

Additional file 10: Table S5. GO enrichment (Biological Process) of the up-regulated DEGs in *rel1-D*.

Additional file 11: Table S6. GO enrichment (Cellular Component) of the down-regulated DEGs in *rel1-D*.

Additional file 12: Table S7. GO enrichment (Molecular Function) of the down-regulated DEGs in *rel1-D*.

Additional file 13: Table S8. GO enrichment (Biological Process) of the down-regulated DEGs in *rel1-D*.

Additional file 14: Table S9. Peptides identified in iTRAQ.

Additional file 15 Figure S5. Verification of the iTRAQ. Modes for qPCR were cDNA of WT and the *rel1-D* leaves in tillering stage, three biological repetitions, error bars were S.E.; the X axis was gene names, the Y axis was $\log_2$ (Fold change of transcript).

Additional file 16: Table S10. Plastid associated differentially expressed proteins in *rel1-D*.

Additional file 17: Table S11. Primers used in this study.

Acknowledgements

We are thankful to Dr. Rui Xia (South China Agricultural University) and Dr. Jedrzej Szymanski (Weizmann Institute of Science) for helpful discussions and critical comments on the manuscript. We also thank the National Natural Science Foundation of China and the Science and Technology Program of Guangzhou for supporting this work.

Funding

This work was supported by grants from the National Natural Science Foundation of China (31671645) to Z. Z., and the Science and Technology Program of Guangzhou (201704030021) and the National Natural Science Foundation of China (31741095) to Q. X.

Authors' contributions

QX and ZZ together designed the experiments. JL performed most of the experiments assisted by SG, BS, QL, XC, and HP. JL and SG performed the drought and ABA response. JL and QL analyzed the RNA-seq and iTRAQ datasets. JL, XC and HP conducted the plant growth in greenhouse and paddy field. JL and QX wrote the manuscript. All authors have discussed the results and contributed to the drafting of the manuscript. All authors read and approved the final manuscript.

Competing interests

The authors declare that they have no competing interests.

References

Anders S, Huber W (2010) Differential expression analysis for sequence count data. Genome Biol 11:R106

Breen J, Bellgard M (2010) Germin-like proteins (GLPs) in cereal genomes: gene clustering and dynamic roles in plant defence. Funct Integr Genomics 10: 463–476

Chen JQ, Meng XP, Zhang Y, Xia M, Wang XP (2008) Over-expression of *OsDREB* genes lead to enhanced drought tolerance in rice. Biotechnol Lett 30:2191–2198

Chen Q, Xie Q, Gao J, Wang W, Sun B, Liu B, Zhu H, Peng H, Zhao H, Liu C, Wang J, Zhang J, Zhang G, Zhang Z (2015) Characterization of *Rolled and Erect Leaf 1* in regulating leave morphology in rice. J Exp Bot 66:6047–6058

Cutler SR, Rodriguez PL, Finkelstein RR, Abrams SR (2010) Abscisic acid: emergence of a core signaling network. Annu Rev Plant Biol 61:651–679

Du H, Huang F, Wu N, Li X, Hu H, Xiong L (2018) Integrative regulation of drought escape through ABA-dependent and -independent pathways in rice. Mol Plant 11:584–597

Dubouzet JG, Sakuma Y, Ito Y, Kasuga M, Dubouzet EG, Miura S, Seki M, Shinozaki K, Yamaguchi-Shinozaki K (2003) *OsDREB* genes in rice, *Oryza sativa* L., encode transcription activators that function in drought-, high-salt- and cold-responsive gene expression. Plant J 33:751–763

Fang L, Zhao F, Cong Y, Sang X, Du Q, Wang D, Li Y, Ling Y, Yang Z, He G (2012) Rolling-leaf14 is a 2OG-Fe (II) oxygenase family protein that modulates rice leaf rolling by affecting secondary cell wall formation in leaves. Plant Biotechnol J 10:524–532

Fujino K, Matsuda Y, Ozawa K, Nishimura T, Koshiba T, Fraaije MW, Sekiguchi H (2008) *NARROW LEAF 7* controls LEAF shape mediated by auxin in rice. Mol Gen Genomics 279:499–507

Hibara K, Obara M, Hayashida E, Abe M, Ishimaru T, Satoh H, Itoh J, Nagato Y (2009) The *ADAXIALIZED LEAF1* gene functions in LEAF and embryonic pattern formation in rice. Dev Biol 334:345–354

Itoh J, Nonomura K, Ikeda K, Yamaki S, Inukai Y, Yamagishi H, Kitano H, Nagato Y (2005) Rice plant development: from zygote to spikelet. Plant Cell Physiol 46:23–47

Jiang H, Chen Y, Li M, Xu X, Wu G (2011) Overexpression of *SGR* results in oxidative stress and lesion-mimic cell death in rice seedlings. J Integr Plant Biol 53:375–387

Kadioglu A, Terzi R (2007) A dehydration avoidance mechanism: leaf rolling. Bot Rev 73:290–302

Kadioglu A, Terzi R, Saruhan N, Saglam A (2012) Current advances in the investigation of leaf rolling caused by biotic and abiotic stress factors. Plant Sci 182:42–48

Lang Y, Zhang Z, Gu X, Yang J, Zhu Q (2004) Physiological and ecological effects of crimpy leaf character in rice (*Oryza sativa* L.) II. Photosynthetic character, dry mass production and yield forming. Acta Agron Sin 30:883–887

Langmead B, Trapnell C, Pop M, Salzberg SL (2009) Ultrafast and memory-efficient alignment of short DNA sequences to the human genome. Genome Biol 10:R25

Lee RH, Wang CH, Huang LT, Chen SC (2001) Leaf senescence in rice plants: cloning and characterization of senescence up-regulated genes. J Exp Bot 52: 1117–1121

Lee Y, Park J, Im K, Kim K, Lee J, Lee K, Park JA, Lee TK, Park DS, Yang JS, Kim D, Lee S (2006) *Arabidopsis* leaf necrosis caused by simulated acid rain is related to the salicylic acid signaling pathway. Plant Physiol Biochem 44:38–42

Li L, Shi ZY, Li L, Shen GZ, Wang XQ, An LS, Zhang JL (2010) Overexpression of *ACL1* (*Abaxially Curled Leaf 1*) increased bulliform cells and induced abaxial curling of leaf blades in rice. Mol Plant 3:807–817

Li X, Han H, Chen M, Yang W, Liu L, Li N, Ding X, Chu Z (2017) Overexpression of *OsDT11*, which encodes a novel cysteine-rich peptide, enhances drought tolerance and increases ABA concentration in rice. Plant Mol Biol 93:21–34

Liu C, Fukumoto T, Matsumoto T, Gena P, Frascaria D, Kaneko T, Katsuhara M, Zhong S, Sun X, Zhu Y, Iwasaki I, Ding X, Calamita G, Kitagawa Y (2013) Aquaporin OsPIP1;1 promotes rice salt resistance and seed germination. Plant Physiol Biochem 63:151–158

Lou D, Wang H, Liang G, Yu D (2017) *OsSAPK2* confers abscisic acid sensitivity and tolerance to drought stress in rice. Front Plant Sci 8:993

Mosa KA, Kumar K, Chhikara S, Mcdermott J, Liu Z, Musante C, White JC, Dhankher OP (2012) Members of rice plasma membrane intrinsic proteins subfamily are involved in arsenite permeability and tolerance in plants. Transgenic Res 21:1265–1277

Seo JS, Joo J, Kim MJ, Kim YK, Nahm BH, Song SI, Cheong JJ, Lee JS, Kim JK, Choi YD (2011) OsbHLH148, a basic helix-loop-helix protein, interacts with OsJAZ proteins in a jasmonate signaling pathway leading to drought tolerance in rice. Plant J 65:907–921

Shi Z, Wang J, Wan X, Shen G, Wang X, Zhang J (2007) Over-expression of rice *OsAGO7* gene induces upward curling of the leaf blade that enhanced erect-leaf habit. Planta 226:99–108

Talukdar D, Talukdar T (2014) Leaf rolling and stem fasciation in grass pea (*Lathyrus sativus* L.) mutant are mediated through glutathione-dependent cellular and metabolic changes and associated with a metabolic diversion through cysteine during phenotypic reversal. Biomed Res Int 2014:479180

Trapnell C, Williams BA, Pertea G, Mortazavi A, Kwan G, van Baren MJ, Salzberg SL, Wold BJ, Pachter L (2010) Transcript assembly and quantification by RNA-Seq reveals unannotated transcripts and isoform switching during cell differentiation. Nat Biotechnol 28:511–515

Weiner JJ, Peterson FC, Volkman BF, Cutler SR (2010) Structural and functional insights into core ABA signaling. Curr Opin Plant Biol 13:495–502

Wu C, Fu Y, Hu G, Si H, Cheng S, Liu W (2010) Isolation and characterization of a rice mutant with narrow and rolled leaves. Planta 232:313–324

Xiang JJ, Zhang GH, Qian Q, Xue HW (2012) *Semi-rolled leaf1* encodes a putative glycosylphosphatidylinositol-anchored protein and modulates rice leaf rolling by regulating the formation of bulliform cells. Plant Physiol 159:1488–1500

Xiao X, Tang C, Fang Y, Yang M, Zhou B, Qi J, Zhang Y (2014) Structure and expression profile of the sucrose synthase gene family in the rubber tree: indicative of roles in stress response and sucrose utilization in the laticifers. FEBS J 281:291–305

Yamaguchi-Shinozaki K, Shinozaki K (2006) Transcriptional regulatory networks in cellular responses and tolerance to dehydration and cold stresses. Annu Rev Plant Biol 57:781–803

Yan S, Yan CJ, Zeng XH, Yang YC, Fang YW, Tian CY, Sun YW, Cheng ZK, Gu MH (2008) *ROLLED LEAF 9*, encoding a GARP protein, regulates the leaf abaxial cell fate in rice. Plant Mol Biol 68:239–250

Yang A, Dai X, Zhang W (2012) A R2R3-type MYB gene, *OsMYB2*, is involved in salt, cold, and dehydration tolerance in rice. J Exp Bot 63:2541–2556

Yang C, Li D, Liu X, Ji C, Hao L, Zhao X, Li X, Chen C, Cheng Z, Zhu L (2014) *OsMYB103L*, an R2R3-MYB transcription factor, influences leaf rolling and mechanical strength in rice (*Oryza sativa* L.). BMC Plant Biol 14:158

Yang SQ, Li WQ, Miao H, Gan PF, Qiao L, Chang YL, Shi CH, Chen KM (2016) *REL2*, a gene encoding an unknown function protein which contains DUF630 and DUF632 domains controls leaf rolling in rice. Rice 9:37

Zhang GH, Xu Q, Zhu XD, Qian Q, Xue HW (2009) *SHALLOT-LIKE1* is a KANADI transcription factor that modulates rice leaf rolling by regulating leaf abaxial cell development. Plant Cell 21:719–735

Zhang Q, Li Y, Pang C, Lu C, Wang B (2005) NaCl enhances thylakoid-bound SOD activity in the leaves of C3 halophyte *Suaeda salsa* L. Plant Sci 168:423–430

Zou LP, Sun XH, Zhang ZG, Liu P, Wu JX, Tian CJ, Qiu JL, Lu TG (2011) Leaf rolling controlled by the homeodomain leucine zipper class IV gene *Roc5* in rice. Plant Physiol 156:1589–1602

15

Progressive drought alters architectural and anatomical traits of rice roots

Mohamed Hazman[1,2] and Kathleen M. Brown[1*]

Abstract

Background: Root architectural and anatomical phenotypes are important for adaptation to drought. Many rice-growing regions face increasing water scarcity. This study describes drought responses of 11 Egyptian rice cultivars with emphasis on plastic root responses that may enhance drought adaptation.

Results: Eleven Egyptian rice cultivars were phenotyped for root architectural and anatomical traits after 6 weeks growth in soil mesocosms under well-watered conditions. Four of these cultivars were more intensively phenotyped under progressive drought stress in mesocosms, using a system where more moisture was available at depth than near the surface. In response to drought stress, all cultivars significantly reduced nodal root number while increasing large lateral root branching density and total lateral root length in the deepest portions of the mesocosm, where moisture was available. Nodal root cross-sectional area, but not stele area, was reduced by drought stress, especially in the basal segments of the root, and the number of late metaxylem vessels was reduced in only one cultivar. Alterations in deposition of lignin were detected by UV illumination from laser ablation tomography, enhanced by digital staining, and confirmed with standard histochemical methods. In well-watered plants, the sclerenchyma and endodermis were heavily lignified, and lignin was also visible throughout the epidermis and cortex. Under drought stress, very little lignin was detected in the outer cell layers and none in the cortex of nodal roots, but lignin deposition was enhanced in the stele. Root anatomical phenes, including cross-section area and metaxylem vessel number and lignin deposition varied dramatically along large lateral root axes under drought stress, with increasing diameter and less lignification of the stele in successive samples taken from the base to the root apex.

Conclusions: Root architectural and anatomical traits varied significantly among a set of Egyptian cultivars. Most traits were plastic, i.e. changed significantly with drought treatment, and, in many cases, plasticity was cultivar-dependent. These phenotypic alterations may function to enhance water uptake efficiency. Increased large lateral root branching in the deep soil should maintain water acquisition, while water transport during drought should be secured with a more extensively lignified stele.

Keywords: Rice roots, Drought, Laser ablation tomography images, Architecture, Anatomy, Lignin

Background

Drought avoidance is one of the most important strategies for maintaining crop yields in water-limited environments. Drought avoidance is most often attributed to root phenes that support better water capture and transport to the shoot (Clark et al. 2002; Lynch et al., 2014). Investigating the architectural and anatomical phenes that contribute to rooting depth is essential for improving crop performance under drought stress (Lynch 2014). Compared to vegetative growth and yield, root traits have not been popular breeding objectives, partly due to the labor-intensive nature of root system phenotyping under agronomically relevant conditions. The recent development of high-throughput phenotyping platforms has increased the potential for associating root phenes with water acquisition from drying soil in cereals, including rice (Henry et al. 2012; Kadam et al. 2017) and maize (Lynch et al., 2014; Lynch 2014; Lynch 2018).

Production of rice consumes large amounts of water, two to three times more than dry-land cereals. Attempts to reduce water usage in rice, a semi-aquatic crop, have been limited by its sensitivity to drought stress

* Correspondence: kbe@psu.edu
[1]Department of Plant Science, The Pennsylvania State University, 102 Tyson Building, University Park, University Park, PA 16802-4200, USA
Full list of author information is available at the end of the article

(Wassmann et al. 2009). Since roots are responsible for water acquisition, root architectural and anatomical traits are key to breeding strategies aimed at drought avoidance. Root traits associated with improved drought avoidance include steeper nodal root angle to increase root depth, and larger diameter roots, which are associated with greater ability to penetrate hard soils to access deep water (Gowda et al. 2011).

Rice possesses a fibrous root system comprised of a mix of embryonic and post-embryonic roots with multiple branching orders (Rebouillat et al. 2009). Post-embryonic roots consist of nodal roots arising from each tiller and numerous lateral roots branching from these axes. There are two types of lateral roots in rice, small lateral roots and large lateral roots. Small lateral roots are shorter, more abundant, ageotropic and do not ramify, while large lateral roots are much longer, less abundant, geotropically positive and highly branched. Root length density in deeper soil layers is highly correlated with soil exploration and water uptake efficiency (Kamoshita et al. 2000; Siopongco et al. 2005). Since rice root architecture is characterized by relatively short nodal roots (shorter than maize or barley), the large lateral roots are likely to contribute substantially to deep soil exploration. Increased lateral root formation under drought stress was suggested as a potentially useful adaptation to drought in lowland rice (Morita et al. 2002; Henry et al. 2012). Several studies have associated improved shoot biomass, water uptake, and photosynthesis under drought with plasticity in lateral root development (Suralta et al. 2010; Kano et al. 2011; Kano-Nakata et al. 2011).

Root anatomical phenes influence radial and axial water transport in roots, which would be expected to influence the efficiency of water uptake and distribution (Lynch et al., 2014). Xylem vessel traits (number, diameter and area) affect axial water conductance while cortical traits and the presence of suberized cell layers may affect radial conductance. Larger xylem vessels and thicker roots are characteristic of upland rice and associated with improved drought tolerance (Gowda et al. 2011).

Like many rice-growing areas, Egypt is faced with increasingly limited irrigation water to support rice production. Egypt is the largest rice-producing country in the West Asia/ North Africa region, producing about 4.8 million tons of paddy rice in 2016 (U.S. Department of Agriculture FAS 2018). Rice production in Egypt has always been constrained by the availability of irrigation water, since the region receives minimal rainfall. Production has declined in recent years and is predicted to decline dramatically this year due to government-imposed 33% reduction in allotted rice growing area for 2018 (Wally and Beillard 2018). These restrictions are motivated by declining water resources to support this popular but water-intensive crop. Sustainable rice production will require cultivars adapted to water-saving management strategies that do not involve continuous flooding, which is the currently most common practice. We hypothesize that cultivars with the ability to adapt their roots to reduced moisture by maintaining root depth and deep branching will be better suited to progressive drought scenarios. Therefore, we investigated the architectural and anatomical traits of several rice cultivars in well-watered conditions and in response to progressive drought in greenhouse mesocosms.

Material and Methods

Plant Materials, Growth Conditions and Drought Treatment

Eleven publicly available lowland rice cultivars (*Oryza sativa*) developed by the scientific agricultural community in Egypt were obtained from the U.S. National Plant Germplasm System (https://npgsweb.ars-grin.gov/gringlobal/

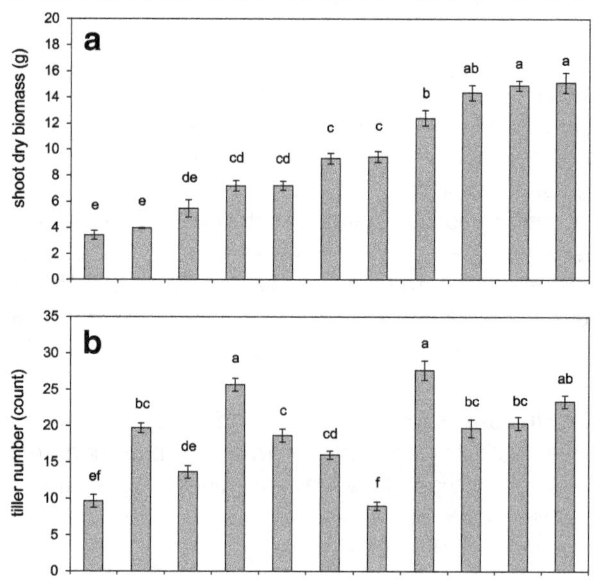

Fig. 1 **a** Shoot dry biomass, **b** tiller number, and **c** maximum root depth of 11 Egyptian rice cultivars grown in the greenhouse under well-watered conditions. Values shown are means of three replications ± SE. Means with the same letter are not significantly different according to Tukey's Honest Significant Differences (HSD) test ($P \leq 0.05$)

search.aspx). These cultivars varied for days to flowering and represent a range of varietal groups and subpopulations (Additional file 1: Table S1). In the first experiment, all 11 rice cultivars were grown under well-watered conditions to examine root architectural and anatomical traits. In the second experiment, four cultivars with high and comparable vigor in our growth system, Egypt 1, Egypt 5, Nabatat Asmar, and Nahda, were chosen for drought stress (DS) experiments. The experiments were performed in a greenhouse located on the campus of Pennsylvania State University, University Park, PA (40°48' N, 77°51' W). The caryopses were dehusked and surface sterilized according to (Hazman et al. 2015) and pre-germinated for 1 week prior to transplanting three healthy seedlings (reduced to one after 3 days) into 1.2 × 0.15 m plastic mesocosms. The mesocosms were lined with transparent high-density polyethylene film to facilitate root system excavation and sampling, and filled with a mixture of 40% (*v*/v) medium size commercial grade sand, 45% vermiculite, 5% perlite, and 10% clay topsoil (Hagerstown silt loam, mesic typic Hapludalf) collected from the Russell E. Larson Agricultural Research Center in Rock Spring, Pennsylvania. The mixture was fertilized with 50 g per mesocosm of Osmocote Plus Fertilizer, which includes micronutrients (Scotts Miracle-Gro Company, Marysville, OH USA). For well-watered treatments, plants were irrigated two times per day with water via drip irrigation (80–100% field capacity), and in addition manually received 50 ml per mesocosm per week of Yoshida nutrient solution (Yoshida et al. 1971). Plants were grown to the V8 stage (4 weeks) then harvested. In a separate experiment, four cultivars with similar vigor, Egypt 1, Egypt 5, Nabatat Asmar, and Nahda were grown under well-watered and drought stress conditions. Drought stress treatment was initiated after 2 weeks of growth by allowing the surface to dry gradually for 4 weeks. Soil moisture was monitored digitally with TDR (Time Domain Reflectometer) probes inserted 25 and 100 cm from the top of the medium of representative mesocosms (Campbell Scientific Inc., Utah, USA). There were no significant differences in volumetric moisture content among cultivars, as measured from the upper or lower sampling positions at the beginning of the drought stress or at the end of the experiment (data not shown). Both well-watered and drought stressed plants were harvested after 6 weeks growth.

Root and Shoot Growth Measurements

At harvest, tillers were counted and shoots removed for biomass determination. For root sampling, root systems were removed from the mesocosms, facilitated by the presence of a plastic liner. The maximum root depth was recorded in the intact root system, then the root

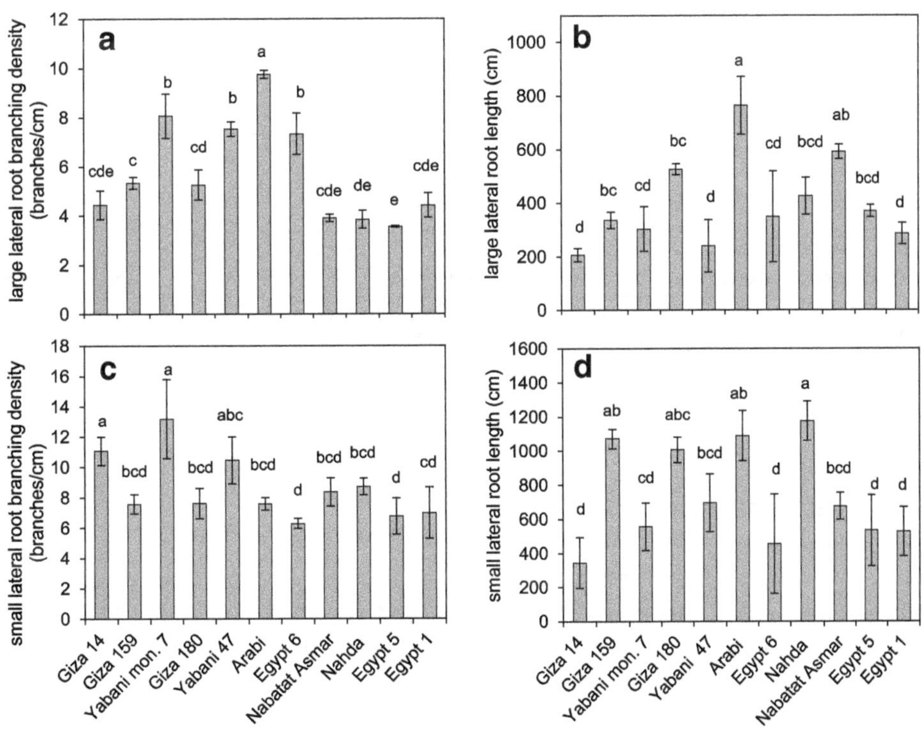

Fig. 2 Genotypic variation among 11 Egyptian rice cultivars in **a** large lateral root branching density, **b** total large lateral root length per nodal root, **c** small lateral root branching density, and **d** total small lateral root length per nodal root in well-watered conditions. Values shown are means of three replications ± SE. Means with the same letter are not significantly different according to Tukey's Honest Significant Differences (HSD) test ($P \leq 0.05$)

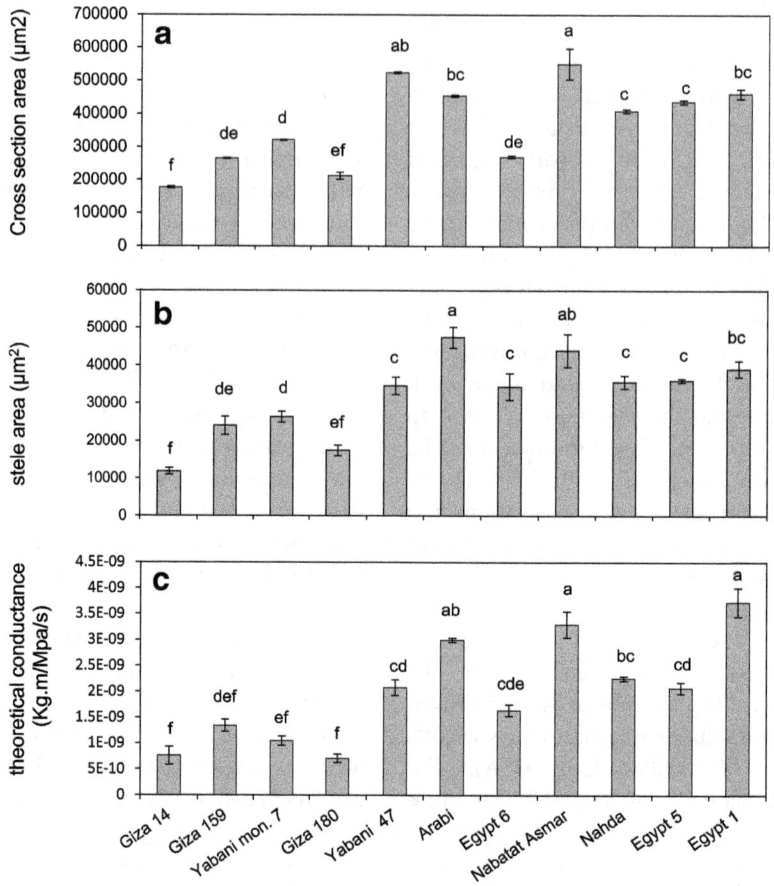

Fig. 3 Nodal root anatomical phenes of 11 Egyptian rice cultivars grown in the greenhouse under well-watered conditions. **a** cross section area, **b** stele area, and **c** theoretical axial conductance. Values shown are means of three replications ± SE. Means with the same letter are not significantly different according to Tukey's Honest Significant Differences (HSD) test (P ≤ 0.05)

Table 1 Pearson's correlations among root and shoot traits of 11 Egyptian rice cultivars under well-watered conditions

	Max. root depth	Large LRBD	Small LRBD	Large LRL	Small LRL	Cross section area	Stele area	Median MV area	Total MV area	MV No.	Theor. cond.	Shoot dry biomass	Tiller No.
Max. root depth	1												
Large LRBD	−0.135	1											
Small LRBD	−0.187	0.192	1										
Large LRL	0.245	0.208	−0.276	1									
Small LRL	−0.183	0.063	−0.111	0.733**	1								
Cross section area	0.484**	0.017	−0.081	0.173	0.068	1							
Stele area	0.414*	−0.271	−.538**	0.275	0.205	0.563**	1						
Median MV area	0.224	−0.008	0.303	0.145	0.259	0.342	0.267	1					
Total MV area	0.548**	0.020	−0.364*	0.346*	0.170	0.839**	0.630**	0.080	1				
MV No.	−0.398*	0.279	0.201	0.349*	0.504**	−0.199	−0.003	0.237	−0.280	1			
Theor. cond.	0.023	0.109	0.395*	0.207	0.337	0.137	0.191	0.871**	−0.120	0.631**	1		
Shoot dry biomass	0.624**	−0.399*	−0.475**	0.275	0.034	0.616**	0.568**	0.006	0.752**	−0.270	−0.161	1	
Tiller No.	0.259	−0.436*	−0.302	0.200	0.339	0.456**	0.664**	0.502**	0.378*	−0.016	0.372*	0.451**	1

LRBD Lateral root branching density, *LRL* Lateral root length, *MV* Metaxylem vessel, *Theor. cond* Theoretical conductance, *Max* Maximum, and *No.* Number significance indicated by * $p \leq 0.05$; ** $p \leq 0.01$

system was washed with a hose to remove the medium. Nodal root number was recorded, and a representative nodal root was removed and stored in 70% ethanol for analysis of additional architectural and anatomical traits. The remainder of the root system and the harvested shoots were dried in an 80 °C oven for 72 h for dry biomass determinations. Percentage of dry biomass reduction was calculated according to the formula: (WW-DS/WW) × 100, where WW = well-watered and DS = drought stress.

Root Architecture Measurements

Total length and branching density of large lateral roots (LLR) and small lateral roots (SLR) on the preserved nodal root sample were measured using WinRhizo (Regent Instruments, Quebec, Canada). The nodal root was divided into several lengths and each length was scanned individually using a flatbed scanner at a resolution of 600 dpi (HP ScanJet, Hewlett Packard, USA). WinRhizo was used to quantify lengths of small and large lateral roots according to the following diameter classes: less than 0.25 mm SLR, 0.25 to 0.60 mm for LLR, and larger than 0.60 mm for nodal roots. SLR were distinguished from LLR based on small diameter and lack of secondary branching. Branching density was calculated by number of tips relative to the length of the nodal root. These measurements were calculated separately

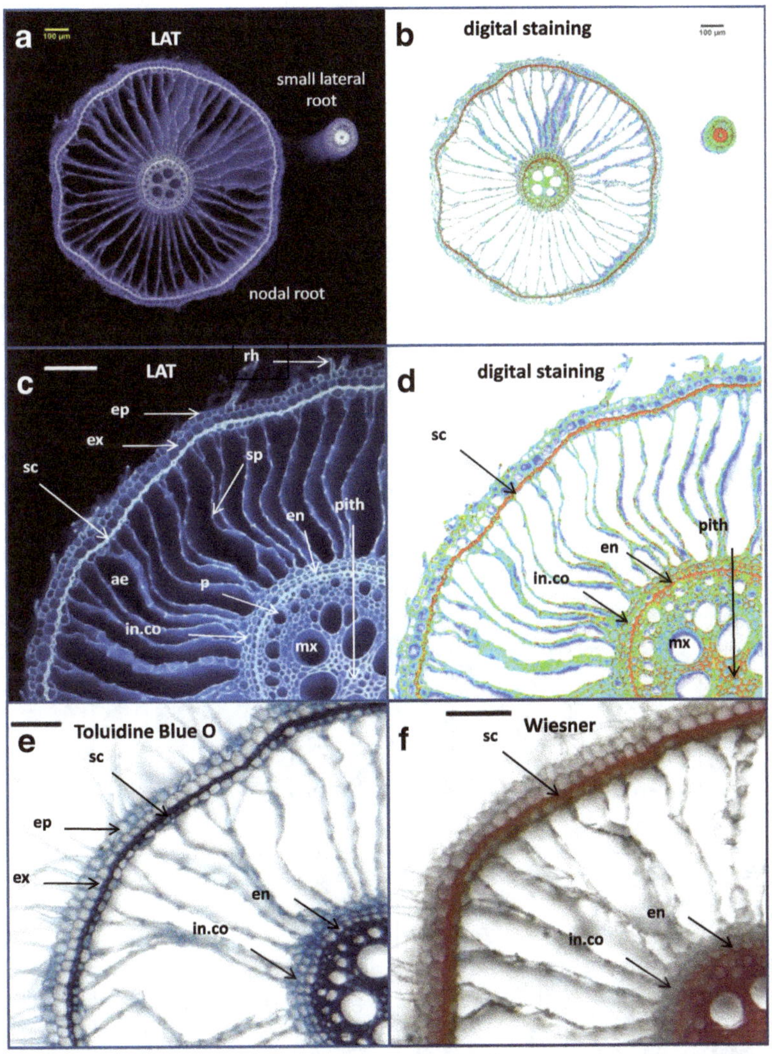

Fig. 4 Comparison of Laser Ablation Tomography (LAT) images plus digital processing with standard histological methods for lignin assessment. Nodal root samples were collected from 4-week-old well-watered Arabi. **a** LAT image showing a nodal root and a small lateral root; **b** the same image digitally "stained"; **c** enlarged view of a nodal root cross section showing detailed anatomical structures including root hairs (rh), epidermis (ep), exodermis (ex), sclerenchyma (sc), aerenchyma (ae), septa (sp), inner cortical cells (in.co), endodermis (en), phloem (p), metaxylem vessel (mx), and pith; **d** digital staining of image "**c**" showing lignin deposition in cell walls of sc, in.co, en and pith. **e** Toluidine Blue O staining of hand cross sections of fresh nodal roots showing lignin deposition (blue color) in cells of sc, in co, and en, and **f** Wiesner or phloroglucinol-HCl staining showing lignin deposition (reddish-brown color) in sc and en. Horizontal scale bars represent 100 μm

for the apical 20 cm and for the remaining basal portion of the nodal root.

Measurements of Root Anatomical Phenes

Dissected rice root segments were imaged using laser ablation tomography (LAT). Preserved root segments from nodal roots were collected at 20 cm from root base and 10 cm from root apex and dried with an automated critical point dryer (CPD, Leica EM CPD 300, Leica Microsystem, Vienna, Austria) according to the manufacturer protocol. For LLR anatomy, several LLR of one rice cultivar (Egypt 1) were gently dissected from the nodal root axis and divided with a razor blade into three parts for well-watered plants and five parts for drought-stressed plants and dried using CPD as previously described. Segments of dried nodal and LLR roots were ablated by a laser beam (Avia 7000, 355 nm pulsed laser) to vaporize the root at the camera focal plane ahead of an imaging stage, then cross-section images were taken using a Canon T3i camera with a 5x micro lens (MP-E 65 mm) on the laser-illuminated surface (Chimungu et al. 2015). The resulting images were analyzed by the software MIPAR (Sosa et al. 2014) to obtain areas of the cross-section, stele, and metaxylem vessels. The theoretical axial water conductance of nodal roots was calculated according to Tyree and Ewers (Tyree and Ewers 2018). Picasa software was used to convert the semi-monochromatic high-quality LAT images into complementary diadic colored images with a bright white background using the following recipe: a) Heat map the laser image to 0% hue and nearly 50% fade, b) invert the colors, c) cross process, then finally d) Orton-ish by 0% bloom, almost 50% brightness and 0% fade. This enabled us to qualitatively distinguish the secondary cell wall elements, and is hereafter referred to as *"digital staining"*.

Histochemical Staining of Secondary Cell Wall Elements

For histochemistry, ethanol-preserved nodal root segments of one representative cultivar (Egypt 5) were washed gently with deionized water and hand sectioned using a razor blade. Thin sections were stained with three different lignin dyes: 0.02% toluidine blue, 3% phloroglucinol-HCL

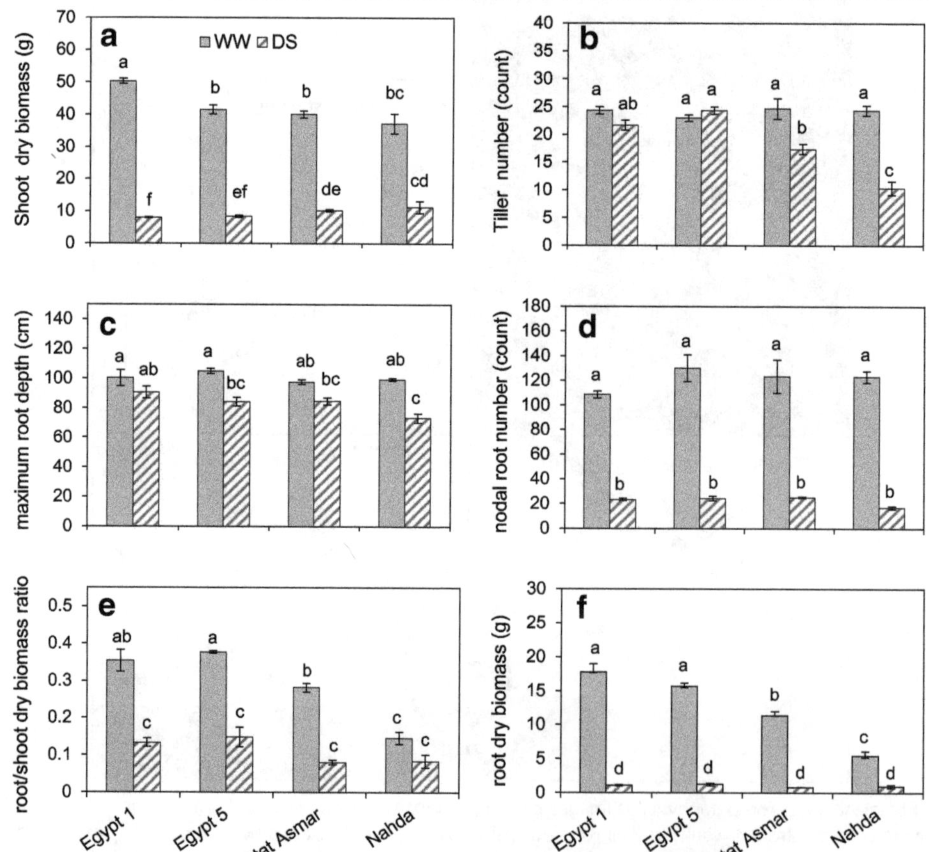

Fig. 5 Effects of drought stress on **a** shoot dry biomass, **b** tiller number, **c** maximum root depth, **d** nodal root number, **e** root to shoot ratio, and **f** root dry biomass of cultivars Egypt 1, Egypt 5, Nabatat Asmar and Nahda. Values shown are means of three replications ± SE. Means with the same letter are not significantly different according to Tukey's Honest Significant Differences (HSD) test ($P \leq 0.05$)

(Weisner staining) and 0.5% potassium permanganate (Mäule staining) (Sigma-Aldrich, USA). Images of the root sections were acquired with a Nikon SMZ 1500 stereoscope (Nikon, Japan) with 50× and 100× magnification (Pradhan Mitra and Loqué 2014).

Experimental Design and Statistical Analysis

A randomized complete block design was with at least three independent biological replications. SPSS (IBM Statistics, USA) software was used for statistical tests including mean separations by Tukey's Honestly Significant Difference (HSD) test, with a significance level of $P \leq 0.05$, Pearson correlation coefficient, and ANOVA analysis.

Results

Root Morphology and Anatomy of 11 Egyptian Cultivars Under Well-watered Conditions

We investigated root architectural and anatomical phenes of 11 cultivars of Egyptian paddy rice grown under well-watered, aerobic conditions (Additional file 1: Table S1). There was significant genetic variation in shoot dry biomass, tiller number and maximum root depth of 4-week-old plants (Fig. 1). Both root architectural (Fig. 2) and anatomical (Fig. 3) features showed significant genetic variation. Nahda, Egypt 1 and Egypt 5 produced the greatest shoot dry biomass. Tiller number was weakly correlated with shoot dry biomass and not correlated with maximum root depth (Fig. 1a, c and Table 1). Egypt 5 and Nabatat Asmar had the deepest roots while Yabani 47 had the shortest maximum root depth (Fig. 1c).

Root architecture phenes varied significantly among the rice cultivars (Fig. 2). Arabi had the greatest large lateral root branching density and length. Large and small lateral root branching densities were negatively correlated with shoot dry biomass and large lateral root branching density was negatively correlated with tiller number (Table 1). Small and large lateral root length were highly correlated with each other but were not correlated with lateral root branching density or with shoot biomass (Table 1).

Table 2 Analysis of variance for effects of cultivar and drought treatment on shoot and root traits

	Cultivar	Treatment	Cultivar x Treatment
Growth traits			
Shoot dry biomass	0.99	265.02**	23.35**
Tiller number	15.35**	60.84**	20.56**
Root dry biomass	44.80**	770.00**	41.40**
Root/shoot dry biomass ratio	29.51**	209.95**	10.11**
Root architectural traits			
Maximum root depth	1.16	474.88**	1.14
Nodal root number	3.69*	63.78**	2.89
Basal large lateral root branching density	0.20	12.68**	0.46
Basal small lateral root branching density	15**	10.11**	18.38**
Apical large lateral root branching density	2.66	201.42**	3.74*
Apical small lateral root branching density	6.59**	3.053	6.138**
Apical large lateral root length	1.78	40.28**	1.22
Apical small lateral root length	11.76**	74.06**	12.05**
Root anatomical traits			
Basal cross section area	10.02**	83.38**	5.84**
Basal stele area	2.31	0.16	6.28**
Ratio of basal cross-section to stele areas	12.46**	491.60**	10.85**
Basal median metaxylem vessel area	4.38*	0.29	1.33
Basal metaxylem vessel number	5.83**	12.50**	16.50**
Apical cross section area	10.02**	83.38**	5.84**
Apical stele area	3.09	8.08	0.97
Ratio of apical cross section to stele area	21.47**	322.40**	10.92**
Apical median metaxylem vessel area	1.88	5.67*	0.10
Apical metaxylem vessel number	5.86**	7.00*	3.57*

Values shown are *F* values, and *P* values are indicated by * $p˂0.05$ and ** $p˂0.01$

Root anatomy also varied among cultivars (Fig. 3, Additional file 2: Figure S1). Nodal root cross-sectional area and stele area were positively correlated with shoot dry biomass, tiller number, maximum root depth, and each other (Table 1). Nodal root stele area was negatively correlated with small lateral root branching density and positively correlated with maximum root depth. While median metaxylem vessel area was positively correlated with tiller number, it was not correlated with shoot biomass or metaxylem vessel number. Theoretical axial water conductance, calculated from vessel size and number, was positively correlated with tiller number but not shoot biomass (Table 1).

In order to take the advantage of the variation in auto-fluorescence created by UV-laser beam excitation for lignin visualization, we developed simple approach to convert high quality laser ablation tomography (LAT) images of rice roots (Fig. 4a) into multichromatic images reflecting pixel intensities. These *digitally stained* images permit high-resolution assessment of tissue-level distribution of lignin in the sclerenchyma, endodermis and stele tissues (Fig. 4a-d). Comparison of these digitally stained images with conventional histochemical staining of lignin using toluidine blue O and Wiesner (phloroglucinol-HCl) (Fig. 4e-f) demonstrates the coincident indications of lignification. To show that this method is useful for other species, lignin was also digitally visualized in sweet corn primary roots (Additional file 3: Figure S2), where lignin deposition around outer cortical cells and stele cells was clearly visualized using digital staining of a high-quality LAT image.

Drought Responses of Four Egyptian Cultivars: Shoot Growth and Root Architecture

Four cultivars with the greatest shoot dry biomass without drought (Egypt 1, Egypt 5, Nabatat Asmar and Nahda, Fig. 1) were selected for investigation of the effect of drought on rice root architecture and anatomy. Drought stress was imposed by stopping irrigation so that the upper part of the growth medium became gradually dry while the deep medium retained some moisture. Time domain reflectometry (TDR) probes showed that the volumetric water content (θ_v) of droughted mesocosms was less than that of well-watered mesocosms, and drier at the top than at the bottom (Additional file 4: Figure S3). Drought stress significantly reduced shoot and root dry biomass and tiller number (Fig. 5, Table 2). Shoot biomass reduction ranged from 70% in Nahda to 84% in Egypt 1, and root biomass reduction ranged from 82.65% in Nahda to 92.81% in Nabatat Asmar (Fig. 5). Tiller number was significantly reduced in Nahda (58%) and Nabatat Asmar (30%).

Maximum root depth did not vary among cultivars in well-watered plants but declined with drought by 20% to 26% (Fig. 5, Table 2). Drought stress reduced nodal root number to a much greater extent, averaging more than 80% reduction in all cultivars (Fig. 5). Drought stress significantly reduced root to shoot ratio in all cultivars except Nahda.

Lateral root branching varied along nodal root axes, with greater branching in the deeper segments of the roots under drought (Fig. 6). We therefore separately evaluated

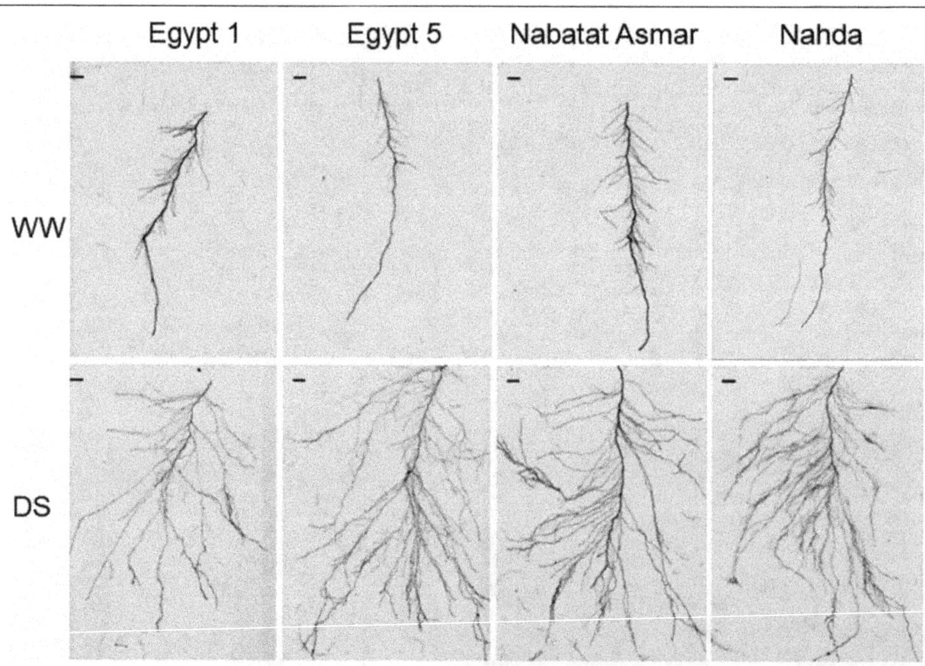

Fig. 6 Representative images of the apical 20 cm of nodal roots from well-watered (WW) and drought stressed (DS) plants. Scale bar represents 10 mm

lateral root branching in apical (deepest 20 cm) and basal (the rest of the root) nodal root segments (Fig. 7). Drought stress significantly increased large lateral root branching density especially in apical segments, and there was a significant cultivar x treatment interaction (Fig. 7, Table 2). The drought-induced increase in large lateral root branching density of the apical segment of nodal roots ranged from 3.0 (Nabatat Asmar) to 7.2-fold (Egypt 5). There was significant cultivar x treatment interaction for small lateral root branching density, but in this case the greater effect was on the basal segment (Table 2). Basal segments had greater small lateral root branching density than apical segments in all four cultivars (Fig. 7). Only Nahda had a significant increase in small lateral root branching density under drought stress, in both apical (3.2-fold) and basal segments (2.6-fold).

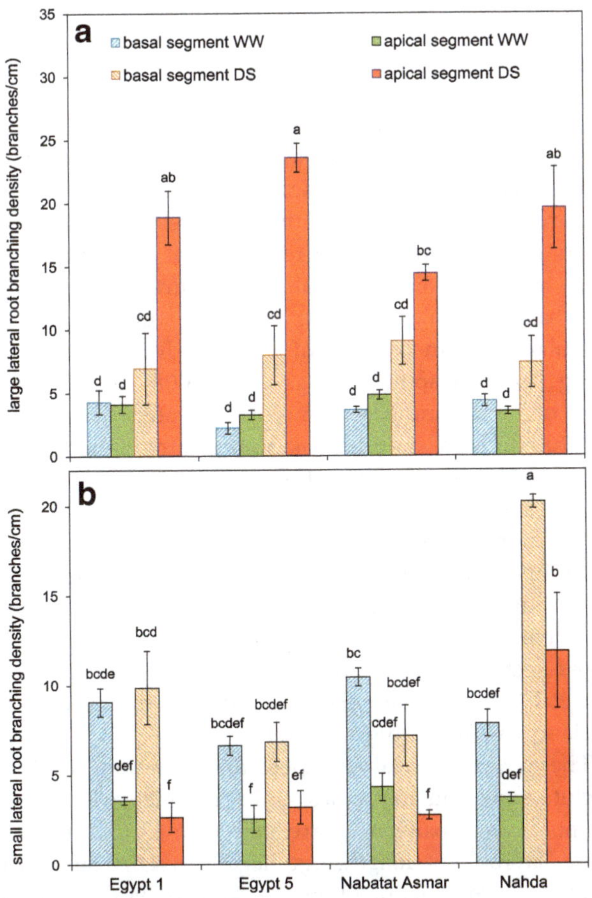

Fig. 7 Lateral root branching density basal and apical segments of well-watered (WW) and drought stressed (DS) nodal roots of four rice cultivars. The basal segment consisted of the entire nodal root except for the apical 20 cm segment, which was evaluated separately. **a** large lateral root branching density and **b** small lateral root branching density. Values shown are means of three replications ± SE. Means with the same letter are not significantly different according to Tukey's Honest Significant Differences (HSD) test ($P \leq 0.05$)

Large lateral root length was significantly increased by drought to a similar extent in all cultivars, while small lateral root length was strongly affected by drought, cultivar and their interaction (Table 2, Fig. 8). Nahda displayed the greatest increase in small lateral root length (14-fold) in the apical segment in response to drought compared to well-watered plants (Fig. 8).

Root Anatomy Responses to Drought

Nodal Roots are Thinner and Have a More Lignified Stele

Under drought treatment, nodal root cross sectional area was 32% (Nabatat Asmar) to 71% (Egypt 1) less than that of well-watered treatment in the basal root segments, and 31.9% (Egypt 1) to 61.6% (Egypt 5) less in the apical segments (Figs. 9 and 10a). The stele area near the root apex was not significantly affected by cultivar or treatment, but there was a significant treatment x cultivar interaction for basal segment stele area (Table 2), caused by a significantly larger stele under drought in Nabatat Asmar (Fig. 10b). Stele areas were preserved at the expense of cortical areas, so that they became a much greater proportion of the total cross-sectional area (Fig. 9). The mean proportion of cross-sectional area as stele was 5.8% and 13.07% for well-watered and drought stressed basal segments, and 6.54% and 15.31% for well-watered and drought stressed apical segments, respectively.

The number of late metaxylem vessels in basal and apical nodal root segments was significantly affected by cultivar, treatment, and their interaction (Fig. 10, Table 2). Egypt 1 displayed the greatest variation between treatments and sampling positions, with fewer vessels in apical than in basal segments, and fewer vessels in basal segments under drought. Theoretical axial water conductance was reduced by 66% under drought stress in basal segments of Egypt 1, driven by the number but not the median area (Additional file 5: Figure S4) of vessels. Egypt 5 under drought, in contrast, displayed reduced conductance of the apical segments, driven by both number and median area of vessels.

Drought stress led to a reduction in lignification (auto-fluorescence) of the epidermis, exodermis, and sclerenchyma in both basal and apical nodal root segments (Fig. 9). Likewise, the "spokes" of cortical cells separating aerenchyma lacunae lacked the lignification found in well-watered conditions and were no longer visible under UV illumination. We compared lignification patterns assessed by digital staining of LAT images with conventional histochemistry in nodal roots of one representative cultivar (Egypt 5) (Fig. 11, Additional file 6: Figure S5). The histochemical staining of nodal roots from dry soil by toluidine blue O, Wiesner and Mäule stains confirmed that the sclerenchyma

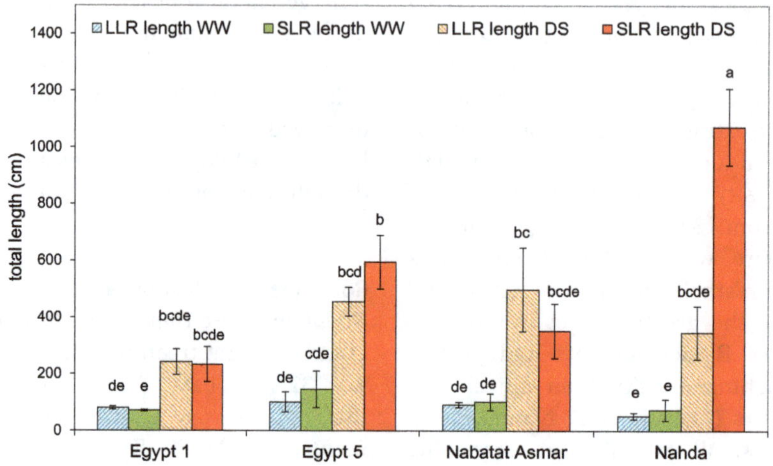

Fig. 8 Effect of drought stress on large and small lateral root lengths in four rice cultivars. Root lengths are for the apical (deepest) 20 cm nodal root segment. Values shown are means of three replications ± SE. Means with the same letter are not significantly different according to Tukey's Honest Significant Differences (HSD) test ($P \leq 0.05$)

layer and the cortical tissue between aerenchyma lacunae lacked lignin, while stele tissue accumulated more lignin than nodal roots from wet soil.

Anatomical Variation Along a Single Large Lateral Root

Observation of the anatomy of large lateral roots revealed increasing cross-sectional area from the root base to the apex, especially under drought (Fig. 12, Additional file 7: Figure S6, Additional file 8: Figure S7). Despite their smaller cross-sectional area (0.05–0.102 mm^2) compared with nodal roots (0.087–0.53 mm^2), the anatomy of large lateral roots and nodal roots was very similar (compare Figs. 9 and 12). Lignification was assessed by digital staining, which detected lignin with red color (Additional file 9: Figure S8). Like nodal roots, large lateral roots showed lignification of sclerenchyma, endodermis and inner cortical cells under well-watered conditions, while roots from drought stressed plants had less lignification outside the stele and more lignification in the stele in the older parts of the lateral root (Fig. 12, Additional file 9: Figure S8). Aerenchyma was observed in both treatments, but in drought stressed roots, it was less developed in medial segments (Fig. 12). Large lateral roots typically had a single late metaxylem vessel (Fig. 12, Additional file 8: Figure S7).

Discussion

The Egyptian cultivars displayed a wide variation in root phenotypes when grown in large mesocosms. The accessions represent a variety of backgrounds (Additional file 1: Table S1) which might explain some of the variation in root traits. For example, varietal group Japonica cultivars are typically considered to have deeper and thicker roots than those of Indica, facilitating deep soil penetration and water acquisition in upland conditions (Gowda et al. 2011). In this study, the thickest nodal roots in this study were indeed Japonica cultivars, i.e. Yabani 47 and Nabatat Asmar. However, the two cultivars with the greatest root depth were Japonica (Nabatat Asmar) and *indica* (Egypt 5), and the shortest was Japonica (Yabani 47, Fig. 1).

Correlation among root thickness-related anatomical traits has been previously reported for rice and maize (Uga et al. 2008; Uga et al. 2009; Burton et al. 2015; Vejchasarn et al. 2016). In previous studies of an F3 population from a cross of IR64 (Indica) and Kinandang Patong (Japonica) (Uga et al. 2008) and a collection of 59 diverse accessions of cultivated rice (Uga et al. 2009), high correlations were found among stele area, total area of late metaxylem vessels, and number of metaxylem vessels. Here we confirmed correlations between root cross sectional area and stele area, and between both those traits and total metaxylem vessel area, but not with median metaxylem area or number of metaxylem vessels (Table 2). Variation related to cross-sectional area could be partially explained by varietal group differences, but clearly there is wide variation within groups as well.

There were several unexpected correlations between architectural and anatomical traits, e.g. metaxylem vessel number was positively correlated with both small and large lateral root length, but negatively correlated with maximum root depth (Table 1). In a previous study of 15 rice cultivars, metaxylem vessel number was significantly correlated with small lateral root length under low phosphorus, but not high phosphorus conditions (Vejchasarn et al. 2016). Studies with a greater number of genotypes would be required to resolve whether these correlations are typical of rice germplasm and have functional significance.

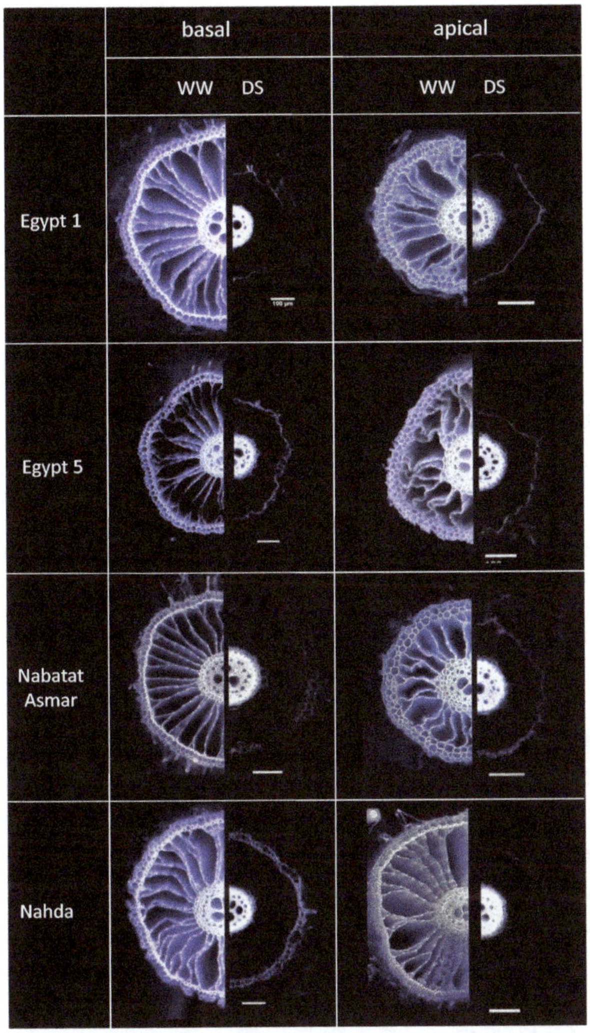

Fig. 9 Laser Ablation Tomography (LAT) images of rice nodal root segments sampled 20 cm from the root base (basal) and 10 cm from nodal root tip (apical), under well-watered (WW) and drought stressed (DS) conditions. Each scale bar represents 100 μm

In this study, most root traits were plastic, i.e. they responded to reduced moisture availability, and in some cases the extent of the change depended on the cultivar (Table 2). Under drought, nodal root number was reduced to a similar extent in all four cultivars (Fig. 5) but nodal roots branched extensively in the deeper part of the mesocosm where more moisture was available (Figs. 6, 7 and 8). Total length per nodal root of both large and small lateral roots increased in the apical segments of nodal roots, resulting primarily from increased branching density and elongation of large lateral roots (Figs. 6, 7 and 8). Just one cultivar, Nahda, displayed increased small lateral root branching density under drought, and this occurred mostly in the basal segment exposed to drier medium (Fig. 7).

Lateral root branching in response to spatially variable water availability has been termed "hydropatterning" (Bao et al. 2014). While the term was coined to describe radially variable lateral rooting patterns when different sides of axial roots were exposed to differential moisture, plants have long been known to respond to locally available limiting resources with greater branching (Gowda et al. 2011; Rich and Watt 2013). In rice fields subjected to periodic drawdown of water levels, water is generally more available at depth. While spatial variation in lateral branching responses of deep nodal roots under stratified moisture availability has not been described in detail, others have suggested that lateral root proliferation could be important for drought tolerance in rice (Henry et al. 2011; Kano et al. 2011; Kano-Nakata et al. 2013; Tran et al. 2015). In one study of four chromosome segment substitution lines, the line that performed best under drought was the one with greatest lateral root proliferation (Kano et al. 2011). In a more recent study, root branching plasticity at different depths was investigated in 20 genotypes in the field (Sandhu et al. 2016). The authors found a positive relationship between deep root length density (driven by all root classes) and yield stability across water regimes and management strategies. Our work demonstrates that this type of plasticity can be measured in a controlled environment where it is possible to recover the entire root system, and suggests that plasticity of lateral root branching specifically in deep, higher-moisture soil could be a potential breeding objective.

Nodal roots and large lateral roots that are exposed to drought experience variation in water availability along their axes, with potential for moisture loss from the shallower portion of the root to dry soil along a water potential gradient. In this study, both the basal and apical segments of nodal roots altered their anatomical structure and composition in response to drought. Basal segments of roots exposed to drought were much thinner than those in well-watered conditions (Fig. 10). The reduction in cross-sectional area resulted from a much smaller cortex, since the stele was not significantly affected by drought (Fig. 10). Similar results were reported in anatomical evaluations of basal segments of nodal roots of Indica rice varieties exposed to well-watered or drought conditions for 30 days in two greenhouse studies (Kadam et al. 2015; Kadam et al. 2017). We also observed anatomical variation along large lateral root axes, but in this case the apical segments, which were in a wetter part of the growth medium, were similar to large lateral root segments from well-watered treatments, while the older parts of the root displayed anatomical changes similar to those of the nodal roots (Fig. 12), supporting a role for these roots in nutrient and water uptake from the deeper soil strata.

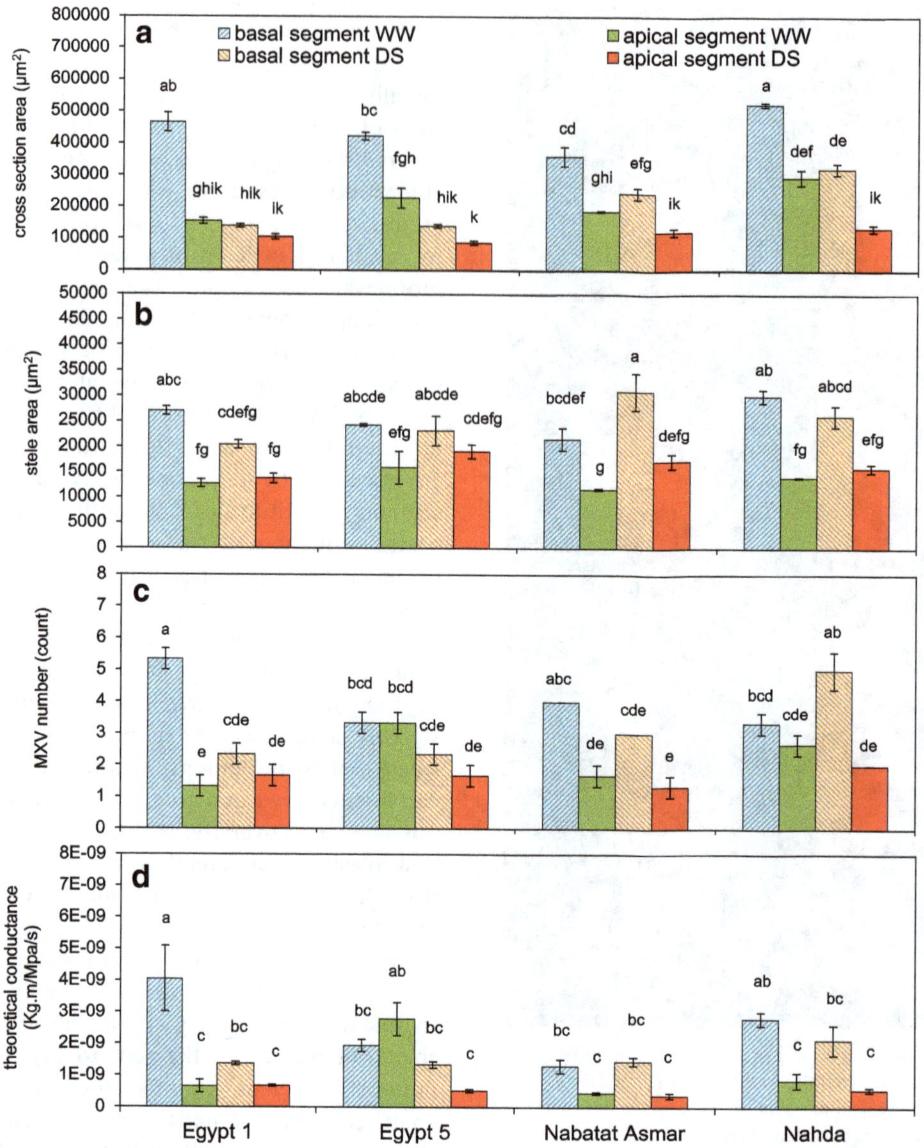

Fig. 10 Effects of drought stress on anatomy of nodal roots of four rice cultivars grown under well-watered (WW) and drought-stressed (DS) conditions. Basal segments were sampled from 20 cm from the root base and apical segments were sampled 10 cm from the root tips. **a** cross section area; **b** stele area; **c** number of late metaxylem vessels; **d** theoretical axial conductance. Values shown are means of three replications ± SE. Means with the same letter are not significantly different according to Tukey's Honest Significant Differences (HSD) test ($P \leq 0.05$)

Lignification patterns in rice roots were revealed by ultraviolet (UV) illumination provided by the laser used for ablation of root tissue. UV has previously been used to visualize root anatomy based on the ability of plant cells to autofluoresce (Rebouillat et al. 2009). We adapted this method for use with laser ablation tomography, demonstrated the coincidence of UV-induced autofluorescence with conventional Maüle and phloroglucinol staining, and presented a method for digitally processing the laser ablation tomography images for enhanced visualization of lignification patterns in two species (Figs. 9, 11 and 12, Additional file 3: Figure S2 and Additional file 6: Figure S5).

Basal and apical segments of nodal roots exposed to drought displayed more lignification of the stele and less lignification of the cortex and outer layers (epidermis, exodermis, and sclerenchyma) compared with corresponding segments from well-watered plants (Fig. 9). Similar responses were found in large lateral roots, but only in the basal segments (Fig. 12). Apical segments of large lateral roots lacked an obvious sclerenchyma layer and displayed fairly uniform lignin distribution across cell types with the exception of the early metaxylem vessels. To our knowledge, there have been no previous reports of reduced lignin in outer cell layers of rice roots and increased lignification of the stele under drought.

Fig. 11 Lignin distribution pattern in response to drought stress (DS) in basal nodal root segments of cultivar Egypt 5. Variation in lignin deposition in response to DS were visualized by, from top, LAT, LAT plus digital staining, and three lignin-specific histochemical stains. Horizontal bars represent 100 μm

However, suberization was reported to decline in the sclerenchyma and increase in the endodermis during drought for several rice cultivars grown in the greenhouse and field (Henry et al. 2012). Suberization and lignification of the outer cell layers of rice are typically correlated (Enstone et al. 2002; Kotula et al. 2009).

The distribution and environmental responsiveness of suberized and lignified endodermal and outer cell layers suggest a role protecting root function in waterlogged and drying soils. Their importance for development of a barrier to radial oxygen loss during waterlogging is well-established (Colmer 2003). However, greater lignification and suberization of the exodermis and endodermis in stagnant solutions relative to aerated solutions were not related to hydraulic conductivity, although permeability to NaCl was significantly reduced (Garthwaite et al. 2006; Ranathunge et al. 2011). Greater thickening of sclerenchyma layers was found in irrigated compared with submerged rice roots (Mostajeran and Rahimi-Eichi 2008), which is probably related to lignification. Kondo et al. (2000) suggested that thickening of the sclerenchyma layer could play a role in structural support when the root is exposed to drying, hardening soil. In other species, e.g. sorghum and maize, the exodermis was observed to become more lignified with drought stress (Enstone et al. 2002). The fact that rice behaves differently, fortifying the endodermis and stele rather than the outer cell layers during drought, may be related to its adaptation to waterlogged conditions, including the presence of extensive aerenchyma. Fortification of the stele would maintain root function as a conduit for water and nutrient movement from lateral roots, while preventing loss of ions and water to the soil.

Basal segments of nodal roots from all four cultivars were thinner when exposed to drought, an effect that has been observed in other studies (Kadam et al. 2015; Kadam et al. 2017). The basal segments of large lateral roots were also thinner under drought, but these roots became thicker as they elongated into medium that had increasing moisture with depth (Additional file 7: Figure S6). Thinner roots may be beneficial for soil exploration because they have less nutrient and carbon cost per unit length, increasing resources available for root elongation into new soil domains (Lynch 2013), and they can explore small pores and crevices in the soil (Bengough et al. 2011). On the other hand, thicker roots have been associated with better soil penetration ability, which is important for root growth into hardened soils or through hardpans under drought conditions (Clark et al. 2008; Gowda et al. 2011; Lynch et al., 2014). In rice, these functions may depend on root class, i.e. thicker nodal roots may penetrate harder soils, while thinner lateral roots may explore fissures and pores.

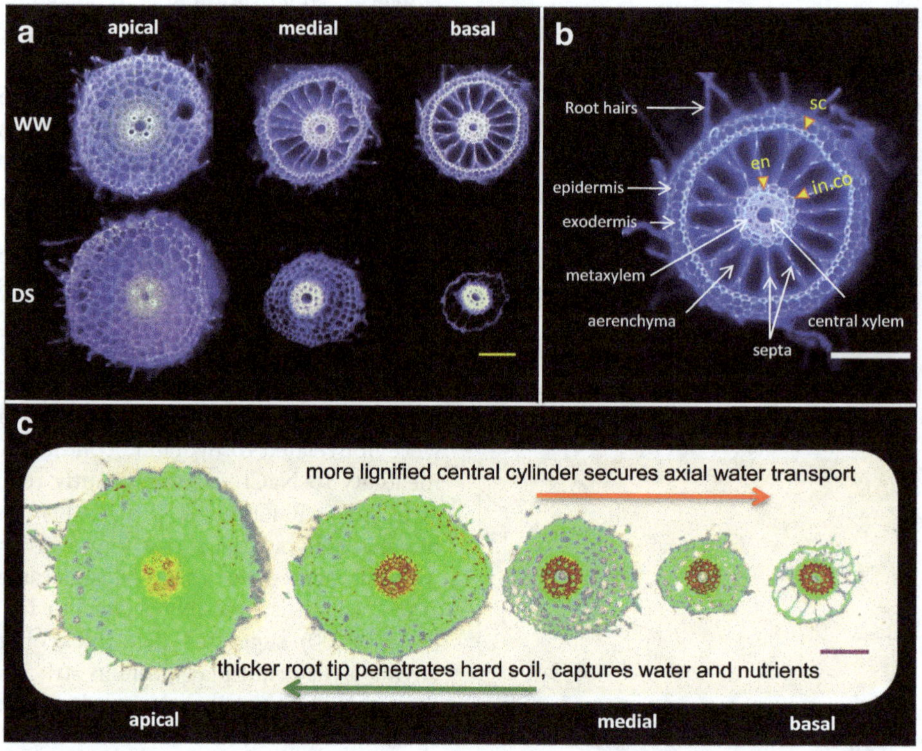

Fig. 12 Drought stress (DS) impact on large lateral root (LLR) anatomy of Egypt 1. **a** Laser Ablation Tomography images of LLR segments sampled at three different locations: apical (2–5 mm from root tip), medial (close to the midpoint of the root), and basal (2–5 mm from point of attachment to nodal root axis). **b** anatomy of basal segment of LLR after 6 weeks in well-watered (WW) conditions showing lignification of sclerenchyma (sc), inner cortical cells (in. co), and endodermis (en). **c** model suggesting functions of observed anatomical variation using digitally stained images of Egypt 1. Red color reveals heavy lignin deposition pattern in the stele of the basal segments. Scale bars represent 100 μm

Drought did not significantly affect apical stele area, and in one cultivar, Nabatat Asmar, it significantly increased basal stele area (Table 2, Fig. 10), so that the proportion of cross-sectional area as stele was significantly increased in all four cultivars. In another study, stele area was observed to increase during drought in both the greenhouse and the field (Henry et al. 2012). Conservation of stele area under drought could be beneficial for maintaining root penetration ability. In maize, stele area was positively associated with root tensile strength (Chimungu et al. 2015). In rice, maintenance of stele area and heavy fortification of the stele with lignin during drought could help roots to continue to grow as soils become harder.

The effect of drought on number of late metaxylem vessels varied among cultivars, and the median metaxylem vessel area varied in the basal segment under drought (Fig. 10, Additional file 5: Figure S4). Kadam et al. (2015) found no significant effect of drought on metaxylem diameter or number at several nodal root positions in three rice cultivars, but others have found that some rice cultivars had smaller vessels when exposed to drought (Yambao et al. 1992; Mostajeran and Rahimi-Eichi 2008; Henry et al. 2012). Smaller vessels would be expected protect the xylem from cavitation, moderate water movement to the shoot, and help maintain a moist rhizosphere for continued root growth and water and nutrient uptake. Plasticity of vessel size could be useful to provide these advantages under drought while permitting adequate water transport to the shoot to support rapid growth under well-watered conditions.

Cultivated rice displays wide variation for root architectural and anatomical traits, as well as variation for plasticity of these traits. We propose that under moderate progressive drought and under reduced water-usage management strategies, where water potential in the upper soil layers declines but more moisture is available at depth, rice crops will benefit from deep soil exploration via extension and branching of large lateral roots, while minimizing development of additional nodal roots. Patterns of lignification and suberization would additionally help to maintain water uptake from deep soil horizons while minimizing losses to dry soil in the shallower layers. These traits require further examination to confirm their utility in agricultural fields.

Additional files

Additional file 1: Table S1. Egyptian cultivars used in the study, with accession numbers and flowering times from the U.S. National Plant Germplasm System, and subpopulation assignments.

Additional file 2: Figure S1. Xylem anatomy of 11 Egyptian rice cultivars grown in the greenhouse under well-watered conditions: **a** median metaxylem vessel area and **b** metaxylem vessel number. Values shown are means of three replications ± SE. Means with the same letter are not significantly different according to Tukey's Honest Significant Differences (HSD) test ($P \leq 0.05$).

Additional file 3: Figure S2. a Laser Ablation Tomography (LAT) image of a sweet corn nodal root segment sampled 15 cm from the root base, and **b** the same image digitally stained. Lignin deposition is indicated as red colored areas. Scale bar represents 100 µm.

Additional file 4: Figure S3. Schematic representation of mesocosms used for rice growth and imposition of drought. Volumetric water content (θv) of the medium in the upper and lower parts of the mesocosms are indicated. These measurements were recorded using TDR probes after 6 weeks growth, including the final 4 weeks without additional water. Probes were inserted 25 cm from the surface under well-watered conditions, and at 25 cm and 100 cm depth for drought stress treatments. Values shown are ranges of three replications.

Additional file 5: Figure S4. Effects of drought stress on median metaxylem vessel area in basal segments of nodal roots of four rice cultivars grown under drought-stressed (DS) conditions. Values shown are means of three replications ± SE. Means with the same letter are not significantly different according to Tukey's Honest Significant Differences (HSD) test ($P \leq 0.05$). There were no significant differences in median metaxylem vessel areas in apical segments of DS plants or in any segments of well-watered plants.

Additional file 6: Figure S5. Nodal root cross sections taken from basal segments and stained with Weisner stain show variation in lignin deposition in the stele and endodermis between well-watered (WW) and drought stress (DS) conditions. Black arrowheads indicate the endodermis.

Additional file 7: Figure S6. Cross sectional areas of large lateral roots of cultivar Egypt 1 at distal, medial and basal positions. Values shown are means of three replications ± SE. Means with the same letter are not significantly different according to Tukey's Honest Significant Differences (HSD) test ($P \leq 0.05$).

Additional file 8: Figure S7. LAT (**a**) and digitally stained (**b**) images of a basal segment cross-section of a large lateral root under well-watered conditions showing lignification of sclerenchyma (sc), inner cortical area (in.co) and endodermis (en). Horizontal bars represent 100 µm.

Additional file 9: Figure S8. a Portion of scanned root system of Egypt 1 grown under drought stress with large lateral root (LLR) anatomical sampling positions indicated. **b** LAT images from a single LLR sampled from apical (1) to basal (5) positions. Scale bar in **b** represents 100 µm.

Abbreviations
CPD: Critical point dryer; DS: Drought-stressed; HSD: Honest significant difference; LAT: Laser ablation tomography; LLR: Large lateral root; LRBD: Lateral root branching density; SLR: Small lateral root; TDR: Time domain reflectometry; UV: Ultraviolet; WW: Well-watered

Acknowledgements
We thank Jenna Reeger and Amelia Henry for useful comments on the manuscript, and Robert Snyder and Jenna Reeger for technical assistance.

Funding
We thank the Binational Fulbright Commission in Egypt and Dr. Haneya Al Etrieby of the Agricultural Development Fund, Ministry of Agriculture and Land Reclamation, Egypt, for supporting Dr. Hazman's visit to Penn State. This material is based upon work that is supported by the National Institute of Food and Agriculture, U.S. Department of Agriculture, Hatch project PENO4582, accession # 1005492.

Authors' contributions
KB and MH designed the experiments, MH performed the experiments and analyzed the data, MH and KB interpreted the data and wrote the manuscript. Both authors read and approved the manuscript.

Competing interests
The authors declare that they have no competing interests.

Author details
[1]Department of Plant Science, The Pennsylvania State University, 102 Tyson Building, University Park, University Park, PA 16802-4200, USA. [2]Agricultural Genetic Engineering Research Institute (AGERI), Agricultural Research Centre (ARC), 9 Gamma St., Giza 12619, Egypt.

References
Bao Y, Aggarwal P, Robbins NE et al (2014) Plant roots use a patterning mechanism to position lateral root branches toward available water. Proc Natl Acad Sci. https://doi.org/10.1073/pnas.1400966111

Bengough AG, McKenzie BM, Hallett PD, Valentine TA (2011) Root elongation, water stress, and mechanical impedance: a review of limiting stresses and beneficial root tip traits. J Exp Bot 62:59–68. https://doi.org/10.1093/jxb/erq350

Burton AL, Johnson J, Foerster J et al (2015) QTL mapping and phenotypic variation of root anatomical traits in maize (Zea mays L.). Theor Appl Genet. https://doi.org/10.1007/s00122-014-2414-8

Chimungu JG, Loades KW, Lynch JP (2015) Root anatomical phenes predict root penetration ability and biomechanical properties in maize (Zea mays). J Exp Bot 66:3151–3162. https://doi.org/10.1093/jxb/erv121

Clark L, Cope R, Whalley W et al (2002) Root penetration of strong soil in rainfed lowland rice: comparison of laboratory screens with field performance. F Crop Res 76:189–198

Clark LJ, Price AH, Steele KA, Whalley WR (2008) Evidence from near-isogenic lines that root penetration increases with root diameter and bending stiffness in rice. Funct Plant Biol 35:1163–1171. https://doi.org/10.1071/fp08132

Colmer TD (2003) Long-distance transport of gases in plants: a perspective on internal aeration and radial oxygen loss from roots. Plant Cell Environ 26:17–36

Enstone DE, Peterson CA, Ma F (2002) Root endodermis and exodermis: structure, function, and responses to the environment. J Plant Growth Regul 21:335–351. https://doi.org/10.1007/s00344-003-0002-2

Garthwaite AJ, Steudle E, Colmer TD (2006) Water uptake by roots of Hordeum marinum: formation of a barrier to radial O-2 loss does not affect root hydraulic conductivity. J Exp Bot 57:655–664

Gowda VRP, Henry A, Yamauchi A et al (2011) Root biology and genetic improvement for drought avoidance in rice. F Crop Res 122:1–13

Hazman M, Hause B, Eiche E et al (2015) Increased tolerance to salt stress in OPDA-deficient rice ALLENE OXIDE CYCLASE mutants is linked to an increased ROS-scavenging activity. J Exp Bot 66:3339–3352

Henry A, Cal AJ, Batoto TC et al (2012) Root attributes affecting water uptake of rice (*Oryza sativa*) under drought. J Exp Bot 63:4751–4763. https://doi.org/10.1093/jxb/ers150

Henry A, Gowda VRP, Torres RO et al (2011) Variation in root system architecture and drought response in rice (Oryza sativa): phenotyping of the OryzaSNP panel in rainfed lowland fields. F Crop Res 120:205–214

Kadam NN, Tamilselvan A, Lawas LMF et al (2017) Genetic control of plasticity in root morphology and anatomy of Rice in response to water deficit. Plant Physiol 174:2302 LP–2302315

Kadam NN, Yin X, Bindraban PS et al (2015) Does morphological and anatomical plasticity during the vegetative stage make wheat more tolerant of water deficit stress than Rice? Plant Physiol 167:1389–1401. https://doi.org/10.1104/pp.114.253328

Kamoshita A, Wade LJ, Yamauchi A (2000) Genotypic variation in response of Rainfed lowland Rice to drought and Rewatering. III. Water extraction during the drought period. Plant Prod Sci 3:189–196. https://doi.org/10.1626/pps.3.189

Kano M, Inukai Y, Kitano H, Yamauchi A (2011) Root plasticity as the key root trait for adaptation to various intensities of drought stress in rice. Plant Soil 342: 117–128

Kano-Nakata M, Gowda VRP, Henry A et al (2013) Functional roles of the plasticity of root system development in biomass production and water uptake under rainfed lowland conditions. F Crop Res 144:288–296. https://doi.org/10.1016/j.fcr.2013.01.024

Kano-Nakata M, Inukai Y, Wade LJ et al (2011) Root development, water uptake, and shoot dry matter production under water deficit conditions in two CSSLs of Rice: functional roles of root plasticity. Plant Prod Sci 14:307–317. https://doi.org/10.1626/pps.14.307

Kondo M, Aguilar A, Abe J, Morita S (2000) Anatomy of nodal roots in tropical upland and lowland rice varieties. Plant Prod Sci 3:437–445

Kotula L, Ranathunge K, Schreiber L, Steudle E (2009) Functional and chemical comparison of apoplastic barriers to radial oxygen loss in roots of rice (Oryza sativa L.) grown in aerated or deoxygenated solution. J Exp Bot 60:2155–2167. https://doi.org/10.1093/jxb/erp089

Lynch J, Chimungu J, Brown K (2014) Root anatomical phenes associated with water acquisition from drying soil: targets for crop improvement. J Exp Bot 65:6155–6166

Lynch JP (2013) Steep, cheap and deep: an ideotype to optimize water and N acquisition by maize root systems. Ann Bot 112:347–357. https://doi.org/10.1093/aob/mcs293

Lynch JP (2014) Root phenes that reduce the metabolic costs of soil exploration: opportunities for 21st century agriculture. Plant Cell Environ 38:1775–1784. https://doi.org/10.1111/pce.12451

Lynch JP (2018) Rightsizing Root Phenotypes for Drought resistance. J Exp Bot. https://doi.org/10.1093/jxb/ery048

Morita MT, Kato T, Nagafusa K et al (2002) Involvement of the vacuoles of the endodermis in the early process of shoot gravitropism in Arabidopsis. Plant Cell 14:47–56

Mostajeran A, Rahimi-Eichi V (2008) Drought stress effects on root anatomical characteristics of rice cultivars (Oryza sativa L.). Pakistan J Biol Sci 11:2173–2183

Pradhan Mitra P, Loqué D (2014) Histochemical staining of Arabidopsis thaliana secondary Cell Wall elements. J Vis Exp:51381. https://doi.org/10.3791/51381

Ranathunge K, Lin J, Steudle E, Schreiber L (2011) Stagnant deoxygenated growth enhances root suberization and lignifications, but differentially affects water and NaCl permeabilities in rice (Oryza sativa L.) roots. Plant Cell Environ 34:1223–1240. https://doi.org/10.1111/j.1365-3040.2011.02318.x

Rebouillat J, Dievart A, Verdeil JL et al (2009) Molecular genetics of rice root development. Rice 2:15–34. https://doi.org/10.1007/s12284-008-9016-5

Rich SM, Watt M (2013) Soil conditions and cereal root system architecture: review and considerations for linking Darwin and weaver. J Exp Bot 64:1193–1208. https://doi.org/10.1093/jxb/ert043

Sandhu N, Raman KA, Torres RO et al (2016) Rice root architectural plasticity traits and genetic regions for adaptability to variable cultivation and stress conditions. Plant Physiol 171:2562 LP–2562576

Siopongco JD, Yamauchi A, Salekdeh H et al (2005) Root growth and water extraction response of doubled-haploid Rice lines to drought and Rewatering during the vegetative stage. Plant Prod Sci 8:497–508. https://doi.org/10.1626/pps.8.497

Sosa J, Huber D, Weld B, Fraser H (2014) Development and application of MIPARTM: a novel software package for two- and three-dimensional microstructural characterization. Integr Mater 3:10. doi: https://doi.org/10.1186/2193-9772-3-10

Suralta RR, Inukai Y, Yamauchi A (2010) Dry matter production in relation to root plastic development, oxygen transport, and water uptake of rice under transient soil moisture stresses. Plant Soil 332:87–104. https://doi.org/10.1007/s11104-009-0275-8

Tran T, Kano-Nakata M, Suralta R et al (2015) Root plasticity and its functional roles were triggered by water deficit but not by the resulting changes in the forms of soil N in rice. Plant Soil 386:65–76. https://doi.org/10.1007/s11104-014-2240-4

Tyree MT, Ewers FW (2018) The hydraulic architecture of trees and other woody plants. New Phytol 119:345–360. https://doi.org/10.1111/j.1469-8137.1991.tb00035.x

U.S. Department of Agriculture FAS (2018) World agricultural production, p 31

Uga Y, Ebana K, Abe J et al (2009) Variation in root morphology and anatomy among accessions of cultivated rice (*Oryza sativa* L.) with different genetic backgrounds. Breed Sci 59:87–93

Uga Y, Okuno K, Yano M (2008) QTLs underlying natural variation in stele and xylem structures of rice root. Breed Sci 58:7–14

Vejchasarn P, Lynch JP, Brown KM (2016) Genetic variability in phosphorus responses of Rice root phenotypes. Rice. https://doi.org/10.1186/s12284-016-0102-9

Wally A, Beillard MJ (2018) Egypt Grain and Feed Annual 2018. Wheat is up; Corn is steady, Egyptian Government Restricts Rice Planted Area. Foreign Agric Serv 2018:1–11

Wassmann R, Jagadish SVK, Heuer S, et al. (2009) Climate Change Affecting Rice Production: The Physiological and Agronomic Basis for Possible Adaptation Strategies. Adv Agron 101:59–122. doi: https://doi.org/10.1016/S0065-2113(08)00802-X

Yambao EB, Ingram KT, Real JG (1992) Root xylem influence on the water relations and drought resistance of rice. J Exp Bot 43:925–932

Yoshida S, Forno DA, Bhadrachalam A (1971) Zinc deficiency of the rice plant on calcareous and neutral soils in the philippines. Soil Science and Plant Nutrition 17 (2):83-87

16

Salinity tolerance in Australian wild *Oryza* species varies widely and matches that observed in *O. sativa*

Yoav Yichie[1*], Chris Brien[2,3], Bettina Berger[2,3], Thomas H. Roberts[1] and Brian J. Atwell[4]

Abstract

Background: Soil salinity is widespread in rice-producing areas globally, restricting both vegetative growth and grain yield. Attempts to improve the salt tolerance of Asian rice, *Oryza sativa*—the most salt sensitive of the major cereal crops—have met with limited success, due to the complexity of the trait and finite variation in salt responses among *O. sativa* lines. Naturally occurring variation among the more than 20 wild species of the *Oryza* genus has great potential to provide breeders with novel genes to improve resistance to salt. Here, through two distinct screening experiments, we investigated variation in salinity tolerance among accessions of two wild rice species endemic to Australia, *O. meridionalis* and *O. australiensis*, with *O. sativa* cultivars Pokkali and IR29 providing salt-tolerant and sensitive controls, respectively.

Results: Rice plants were grown on soil supplemented with field-relevant concentrations of NaCl (0, 40, 80, and 100 mM) for 30 d, a period sufficient to reveal differences in growth and physiological traits. Two complementary screening approaches were used: destructive phenotyping and high-throughput image-based phenotyping. All genotypes displayed clear responses to salt treatment. In the first experiment, both salt-tolerant Pokkali and an *O. australiensis* accession (*Oa*-VR) showed the least reduction in biomass accumulation, SES score and chlorophyll content in response to salinity. Average shoot Na^+/K^+ values of these plants were the lowest among the genotypes tested. In the second experiment, plant responses to different levels of salt stress were quantified over time based on projected shoot area calculated from visible red-green-blue (RGB) and fluorescence images. Pokkali grew significantly faster than the other genotypes. Pokkali and *Oa*-VR plants displayed the same absolute growth rate under 80 and 100 mM, while *Oa*-D grew significantly slower with the same treatments. *Oa*-VR showed substantially less inhibition of growth in response to salinity when compared with *Oa*-D. Senescence was seen in *Oa*-D after 30 d treatment with 40 mM NaCl, while the putatively salt-tolerant *Oa*-VR had only minor leaf damage, even at higher salt treatments, with less than a 40% increase in relative senescence at 100 mM NaCl compared to 120% for *Oa*-VR.

Conclusion: The combination of our two screening experiments uncovered striking levels of salt tolerance diversity among the Australian wild rice accessions tested and enabled analysis of their growth responses to a range of salt levels. Our results validate image-based phenotyping as a valuable tool for quantitative measurement of plant responses to abiotic stresses. They also highlight the potential of exotic germplasm to provide new genetic variation for salinity tolerance in rice.

Keywords: *Oryza sativa*, *Oryza australiensis*, *Oryza meridionalis*, Salt, Australian native rice

* Correspondence: yoav.yichie@sydney.edu.au
[1]Sydney Institute of Agriculture, University of Sydney, Sydney, Australia
Full list of author information is available at the end of the article

Introduction

Salinity, drought and heat are major abiotic stresses limiting the productivity of crop plants. Accumulation of toxic levels of salt as well as osmotic stress constitute a major threat to rice production worldwide, particularly in coastal rice-growing regions. Modern rice hybrids are some of the most salt-sensitive genotypes (Grattan et al. 2002; Munns et al. 2008; Qadir et al. 2014), with yield reductions evident above 30 mM NaCl (Ismail and Horie 2017) and survival of salt-sensitive genotypes compromised at 70 mM NaCl (Yeo et al. 1990). Rice is particularly vulnerable to salinity during the early seedling and reproductive stages (Zeng et al. 2002). The impact of salinity will be further exacerbated by factors such as marine inundation (Takagi et al. 2015). This has vast implications for food security because rice is the staple for much of Asia (Khush 2005) and throughout pantropical countries.

The basis of salt tolerance is polygenic, determined by a complex network of interactions involving signalling, stress-induced gene expression and membrane transporters (Wang et al. 2003). This complexity has complicated the search for physiological salt tolerance in rice because genotypes with tolerance in one trait are often intolerant in another (Yeo et al. 1990). Moreover, different developmental phases are characterised by distinct salt-tolerance mechanisms (Munns and Tester 2008), requiring breeding for genotypes with a suite of morphological, physiological and metabolic responses. Attempts to improve the salt tolerance of *O. sativa* have met with limited success due to these complexities as well as the interaction with nutritional factors, heterogeneity of field sites and other environmental factors such as heat and periodic drought (Flowers 2004; Yeo et al. 1990). Notwithstanding, the improvement of salt tolerance of rice at the seedling stage is a major breeding goal in many Asian countries, where seedlings must often establish in soils already contaminated by salt. While other crops might be better suited to salt-affected soils, few are suitable alternatives to rice because of its unique ability to grow when flooded.

Even though *O. sativa* represents less than 20% of the genetic diversity that exists in the 27 *Oryza* species (Zhu et al. 2007; Stein et al. 2018), there is still substantial variability in the tolerance to NaCl within this species (Gregorio et al. 1993; Lutts et al. 1995; Munns et al. 2016). In *O. sativa*, transport of Na^+ to the shoot is a major determinant of salt tolerance (Yeo et al. 1987; Yadav et al. 1996; Ochiai et al. 2002). The activity of a vacuolar antiporter was found to increase salt tolerance (Fukuda et al. 2004). More recently, a novel quantitative trait locus (QTL) named *Saltol* was found to encode a trans-membrane protein, OsHKT1;5, which regulates K^+/Na^+ homeostasis under salt stress, increasing tolerance to salt (Ren et al. 2005; Thomson et al. 2010). Additional studies have identified other QTL and mutations for salt tolerance within *O. sativa* (Lang et al. 2001; Yao et al. 2005; Sabouri et al. 2008; Islam et al. 2011; Takagi et al. 2015) but the mechanisms of the proteins encoded in these loci are yet to be revealed.

The diversity of wild rice relatives would suggest that a novel salt-tolerance mechanism for rice breeding programs should come from the examination of *Oryza* species from natural populations, of which four are indigenous to Australia: *O. meridionalis, O. officinalis, O. rufipogon* and *O. australiensis* (Henry et al. 2010; Atwell et al. 2014). While the best evidence thus far for the ability of *Oryza* species to contribute stress-tolerance genes is the case of resistance to brown leaf hopper (Khush 1997; Rahman et al. 2009), abiotic factors have been powerful selective forces on these species in northern Australia, encouraging our search for tolerance to physical constraints on growth. For example, *O. meridionalis* and *O. australiensis* have superior heat tolerance compared with *O. sativa* (Scafaro et al. 2010) with the wild allelic form of the Rubisco activase gene responsible for this trait in *O. australiensis* (Scafaro et al. 2016).

Although the Australian endemic rices are poorly characterised, trials demonstrate the potential of using wild rice species introgressions to enhance the growth of *O. sativa* (Ballini et al. 2007). A recent study showed that Australia may be a centre of origin and segregation of the AA genome of *Oryza* and underlined the wide genetic diversity within the *Oryza* species that share this genome (Brozynska et al. 2016). Further diversity could be expected in the phylogenetic outlier *O. australiensis*, which is the sole species with an EE genome (Jacquemin et al. 2013). The discovery of many domesticated alleles within the wild species reinforces the hypothesis that wild relatives are a key asset for crop improvement (Brozynska et al. 2016).

Over recent years, several studies in cereals and legumes have utilised high-throughput phenotyping technology under controlled environments to gain a better understanding of the genetic architecture and the physiological processes associated with salinity stress (Hairmansis et al. 2014; Campbell et al. 2015, 2017; Atieno et al. 2017). However, this approach had not been applied to crop wild relatives. In a large-scale, non-destructive phenotyping facility ('The Plant Accelerator') we assembled shoot images of *O. sativa, O. meridionalis* and *O. australiensis* exposed to a range of salt treatments for five weeks during the early vegetative stage. We sought to examine developmentally specific salinity responses, growth dynamics and the complex relationship between different traits under salt stress in Australian wild rices pre-selected for inherent tolerance to salinity. Comparisons were made between these genotypes and *O. sativa* genotypes Pokkali (salt-tolerant) and IR29 (salt-sensitive). The broader context of this work was to gain insights into abiotic stress tolerance of exotic Australian genotypes with the aim of identifying key genes in subsequent research.

Material and methods

Plant material, growth conditions and salt treatments

Experiment 1

Five wild accessions chosen from two Australian endemic wild rice species, *O. meridionalis* and *O. australiensis*, were tested along with two cultivated varieties of *O. sativa*, Pokkali and IR29. The wild accessions were selected from a wide range of sites, including transiently saline waterways in the north and northwest of Australia. Approximately 30 genotypes were screened for symptoms and survival in preliminary experiments (unpublished data), exhibiting a wide spectrum of tolerance to 25–100 mM NaCl over a four-week treatment.

The initial testing led to a narrower selection of genotypes screened at Macquarie University, Sydney, Australia (lat. 33.7° S, long. 151.1° E) in spring 2016. Seeds were de-hulled and surface-sterilised by successive immersion in water (30 min), 4% commercial bleach (30 min) and at least five rinses with diH_2O. Seedlings were then germinated in petri dishes in the dark at 28 °C (*O. sativa*) and 36 °C (wild rice) and grown for a further 5 d at 28 °C. After 8 d, two to four seedlings per genotype were sown in a 1.5-L polyvinyl chloride (PVC) pot (with drainage holes) containing 1.3 L of locally sourced clay-loam, slow-release fertiliser (Nutricote Standard Blue, Yates, 0.04%) and placed in the greenhouse. Seedlings were thinned, leaving one uniformly sized and healthy seedling in each pot 15 d after transplanting (DAT).

Salt treatments were applied to the top of the pots gradually in three stages from 25 DAT (25, up to 40 and up to 80 mM daily increments). The final NaCl concentrations for the first screening were 0, 40 and 80 mM NaCl—a total electrolyte concentration resulting in an electrical conductivity (EC) of 0.0, 0.5, 4.5, and 8.7 dS m^{-1}, respectively. Plants were watered once a day with ~ 50 mL per pot of their respective salt concentration (including 0.4 g L^{-1} of Aquasol Soluble Fertiliser, Yates). A square aluminum tray was placed under each set of treatment pots and the drainage was collected every 3 d. Plants were exposed to salt treatments for 30 d in a controlled greenhouse with 30 °C/22 °C day/night temperature and relative humidity of 57% (± 9%, SD) during the day and 77% (± 2%, SD) at night.

A completely randomised design was used, with a minimum of five replicates (pots) for each plant genotype-treatment combination. The locations of the trays and of each pot within trays were changed randomly every 3 d to subject each one of the plants to the same conditions and to prevent neighbour effects. A few IR29 plants dehydrated two weeks after exposure to salt (80 mM NaCl treatment) and were removed from the statistical analysis.

Experiment 2

Seven lines of rice, including two cultivated *O. sativa* controls—Pokkali, a positive control (salt tolerant) and IR29, a negative control (salt sensitive)—were investigated at the four salt concentrations described above, with an additional salt treatment of 100 mM (EC = 10.5 dS m^{-1}). This experiment was performed in the South East Smarthouse at The Plant Accelerator (Australian Plant Phenomics Facility, University of Adelaide, Adelaide, Australia; lat. 34.9° S, long. 138.6° E) in the summer of 2017. The same greenhouse conditions and treatments were applied as in Experiment 1. The seedlings were sown and thinned following the same protocol as used in Experiment 1 in 2.5-L pots with 2.0–2.2 L of UC Davis-mix (2.5 g L^{-1} Mini Osmocote® 16–3-9 + te) and the surface was covered with white gravel (particle size ~ 2–5 mm) to minimise evaporation from the pot and to reduce algal growth. For the first 7 DAT, each pot was watered daily with ~ 100 mL from the top. The pots were placed on top of square containers (93 mm diameter, 50 mm height) to prevent water from spilling onto the conveyor system and to allow the drainage water to be collected.

Salt treatments were applied gradually in four steps from 22 DAT to the square container (25, up to 40, up to 80 and up to 100 mM daily increments). The holes in the pots allowed for the infiltration of salt solution into the soil through capillary action. The water level was maintained constant by weighing each plant and watering to a target volume of 600 mL. Daily imaging and watering were continued for 30 d after salt treatment until 30 d after salting (DAS). The same post-harvest parameters were measured as in Experiment 1.

Image-based high-throughput phenotyping was performed on rice genotypes selected from the wider group tested in initial screening experiment (spring 2016).

A split-unit design was performed concurrently, where 12 lanes × 14 positions (5–12, 15–20) with six replicates to assign the factorial set of treatments were occupied. Each replicate occupied two consecutive lanes and included all 28 rice line-treatment combinations. Each replicate comprised seven main units, each consisting of four carts arranged in a grid of two lanes × two positions. Thus, the 42 main units formed a grid of 6 reps × 7 main positions. The plant lines were assigned to main units using a 7 × 6 Youden square. The four salt treatments were assigned to the four carts within each main unit using a resolved incomplete block design for four treatments in blocks of size 2. The design was randomised using dae (Brien, 2018), a package for the R statistical computing environment (R Core Team, 2018).

Phenotyping of physiological traits

Gas exchange values

Plants were phenotyped throughout the experiment for growth parameters. Gas exchange parameters such as photosynthesis, stomatal conductance and transpiration were measured on DAS 29 and DAS 30 (for the first and second experiments, respectively) with an infrared, open gas exchange system (LI-6400, LICOR Inc., Lincoln, NE, USA). All gas measurements were completed on the same day between 10:00 am and 12:30 pm and were made on the youngest fully-expanded leaf (YFL) of each rice plant.

Growth and yield components

Plants were characterised for phenotypic responses to salinity stress on 30 d after salt application (DAS), the plants were harvested, and the following post-harvest parameters were determined. Shoot fresh weight (SFW) was measured for each plant immediately after harvest, as well as number of tillers. Plant shoots were dried at 65 °C in a ventilated oven for 48 h to constant weight and shoot dry weight (SDW) was measured.

Leaf chlorophyll determination

The YFL was collected from each plant on the day of harvest (DAS30); leaves were flash-frozen in liquid nitrogen after being washed with diH_2O. Chlorophyll was extracted using 95% ethanol and total chlorophyll was determined (Mackinney 1941). Chlorophyll concentrations at each salt level were normalised against control (non-salinised) levels.

Ion assay

The YFL of each plant was collected as described above. Samples were washed thoroughly and dried at 70 °C. Each sample was weighed and extracted with 10 ml 0.1 N acetic acid for every 10 mg of dried tissue. Samples were placed in a water bath at 90 °C for 3 h. Samples were diluted 10 times after the extracted tissues were cooled at room temperature. Sodium and potassium concentrations were measured using an Agilent 4200 Microwave Plasma Atomic Emission Spectrometer (Agilent Technologies, Melbourne, Australia).

Salinity tolerance estimation

Salinity tolerance (ST) was determined by the percentage ratio of mean shoot dry weight (80 mM NaCl) divided by mean shoot dry weight (no salt) [SDW (salt treatment))/ (SDW (control)) × 100]. Each plant was evaluated for seedling stage salinity tolerance based on visual symptoms using the International Rice Research Institute (IRRI) standard evaluation system (SES) scores (IRRI 2013).

RGB/fluorescence image capture and image analysis

Two types of non-destructive imaging systems were utilised to address our questions: RGB (red-green-blue)/visible spectrum and fluorescence (FLUO). Standard RGB images had a resolution of 8 M pixels, while fluorescence images had a resolution of 5 M pixels (Berger et al. 2012). However, in our experiment, some plants attained a physical height exceeding that of the field of view of the RGB camera (the RGB camera was closer to the plants than the fluorescence camera). Thus, we chose to use the projected shoot area (PSA) based on RGB images at the beginning of the experiment (DAS 4–19) and PSA based on fluorescence at the end (DAS 20 onwards). For the RGB images, PSA is the sum of the areas as measured (in kilopixels) from two side views at an angular separation of 90 degrees and a view from above; for the fluorescent images, PSA is the sum of the areas as measured (in kilopixels) from two side views at an angular separation of 90 degrees.

Consequently, a hybrid PSA trait was calculated using the RGB images for DAS 4–19 and the FLUO images for DAS 20 onwards. The PSA data from the FLUO images were transformed using the linear relationship between PSA from the RGB images and PSA from the FLUO images (for DAS 20). The conversion was made on the raw observations and then the new data were prepared for each plant as described below. Water levels were monitored and adjusted daily by the Scanalyzer 3D weighing and watering system (LemnaTec GmbH, Aachen, Germany), with pot weight before and after watering being recorded.

To screen for osmotic tolerance, plant growth rate after the addition of NaCl was determined using the hybrid PSA trait from DAS 2 to 30, where DAS 0 corresponded to the commencement of the salt treatments to generate the PSA of the plant. The results of the high-throughput screening focused on PSA and the absolute growth rate (AGR) and relative growth rate (RGR) derived for these plants. The traits were obtained as described (Al-Tamimi et al. 2016). The PSA AGR and PSA RGR were calculated from the PSA values by determining the difference between consecutive PSA and ln(PSA) values, respectively, and dividing by the time difference. Similarly, the daily water loss from each pot was obtained by subtracting the weight before watering in the current imaging day from the weight after watering on the previous imaging day. The PSA water use index (WUI) was calculated daily by dividing the PSA AGR by the water use. On the one occasion that water use values were negative due to leakage from a storm, values were replaced with blank values to avoid affecting the smoothed spline curve fitting.

Data preparation and statistical analysis

First experiment

Statistical significance of phenotypic traits was determined by Analysis of Variance (ANOVA) with Tukey HSD multiple comparison with significant values of $P \leq 0.05$ and $P \leq 0.01$. Pairwise comparisons were conducted using LSD-Test and Tukey adjustments to produce *p*-values for the significant differences of specific pairs using the R package ggplot2 (Wickham, 2009). A linear regression model was used to calculate the Salinity Tolerance (ST) against sodium and potassium concentrations and the corresponding r coefficients.

Second experiment

Data from the Smarthouse were first analysed using imageData (Brien, 2018) to determine subjectively the degree of smoothing required to produce growth curves using PSA values; this approach removed noise in the data while accurately capturing the underlying growth trajectories. PSA AGR and the PSA RGR were derived by fitting natural cubic smoothing splines to the data for each plant with different settings of the smoothing parameter degrees of freedom (df) (Al-Tamimi et al. 2016). A df value of five was chosen, as it gave the most satisfactory results over all three traits. The water use rate was also smoothed by fitting a spline using df = 5. After examination of the plots for the smoothed traits sPSA, sPSA AGR and sPSA RGR, we decided to investigate growth for six DAS endpoints (DAS 4, 9, 14, 19, 23 and 28) and thus the response of the rice plants to salt treatment was separated into five corresponding intervals.

Correlation analysis was performed on the biomass-related metrics (smoothed PSA 28 and 30 DAS) and manual measurements of SFW and SDW. Both SDW and SFW displayed a strong positive correlation with PSA, with the highest correlation between smoothed PSA and SDW ($r^2 = 0.966$, $P = 0.001$, $n = 168$) (Additional file 1: Figure S1) using the squared Pearson correlation coefficient. A similar strong positive correlation was found ($r^2 = 0.96$, $P = 0.001$, $n = 72$) in a previous study that measured the correlation between PSA and total plant area using a leaf area meter (LI-3100C; LI-COR) (Campbell et al. 2015). This validates our experimental set-up as suitable to monitor plant growth and physiological responses to salt treatments and indicates that PSA is an accurate and sensitive metric for assessing plant biomass accumulation in response to salinity.

To produce phenotypic means adjusted for the spatial variation in the Smarthouse, a mixed-model analysis was performed for each trait using the R package ASReml-R (Butler et al. 2009) and asremlPlus (Brien, 2018), both packages for the R statistical computing environment (R Core Team, 2018). The maximal mixed model used was described previously (Al-Tamimi et al. 2016).

Residual variances were tested using REML ratio tests with $\alpha = 0.05$ to test whether the differences were significant for both salinities and lines, for just one of them, or not at all. In order to reflect the results of these tests and to check that the assumptions underlying the analysis were met, the model was modified to residual-versus-fitted value plots and normal probability plots of the residuals inspected. Wald F-tests were conducted to check whether an interaction (between lines and salinity) was significant, for its main effects. The predicted means and standard errors were obtained for the selected model for salinity and lines effects. To compare a pair of predicted means the *p*-value for an approximate t-test was calculate from the predicted means and their standard errors. However, for cases in which the variances were unequal, these were computed for each prediction using the average variance of the pairwise differences over all pairwise differences in which the prediction was involved and are only approximate.

Results

First screening (experiment 1)

After 30 d of growth in non-salinised (control) conditions, *O. sativa*, *O. meridionalis* and *O. australiensis* shoot dry biomass ranged from 11.5 (IR29) to 22 g (Pokkali), with the exception of *Oa*-KR for which dry biomass reached 34 g by the end of the experiment. Average chlorophyll concentrations ranged from 1.67 to 3.94 mg g^{-1} (SDW), while mean net photosynthetic rates ranged from 14.9 to 19.9 μmol m^{-2} s^{-1} (Additional file 2: Table S1).

Relative to the non-salinised control plants, clear differences in phenotype became apparent after exposure to 40 and 80 mM NaCl. Visual symptoms across all six genotypes were assessed by SES, showing salt-induced injury when expressed relative to control plants (for which SES = 1.0; i.e. no loss of leaf function). In the oldest leaves of IR29, SES reached 5.4 at 40 mM and 8.3 at 80 mM NaCl, reflecting loss of function in all but the most recently expanded leaves (Fig. 1a). In the most salt-tolerant genotype (*Oa*-VR), SES was 1.8 at 40 mM and 2.4 at 80 mM NaCl. Chlorophyll concentrations followed an identical pattern (Fig. 1b), where in the salt-sensitive genotype (IR29) there was a 34% reduction at 40 mM and a 72% reduction at 80 mM NaCl, while in *Oa*-VR there was no change in chlorophyll concentration at 40 mM and a 19% reduction at 80 mM NaCl.

Seedling fresh and dry biomass were measured 30 DAS. Because of inherent variation in the growth rate of the wild species, biomass of plants treated with 40 and 80 mM NaCl are shown relative to control plants (Fig. 1c - dry weights; Additional file 2: Table S1). There was no growth penalty

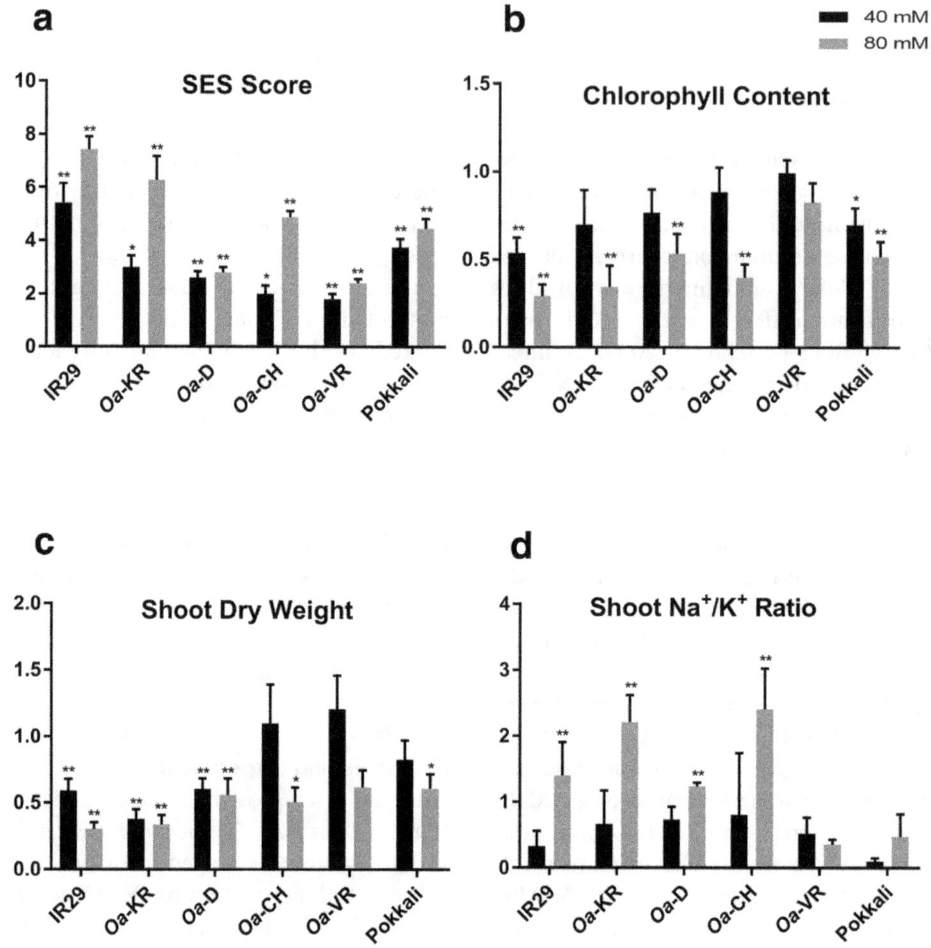

Fig. 1 **a** Standard Evaluation System (SES) scores [1-9]; **b** Normalized chlorophyll content (as a ratio of the control); **c** Normalized biomass growth by SDW (as a ratio of the control) and **d** Shoot Na^+/K^+ ratio of the four wild *Oryza* accessions and *O. sativa* controls: IR29 (salt sensitive) and Pokkali (salt tolerant). Trait means (± standard errors) are shown for each genotype under 40 and 80 mM NaCl (EC = 8.7 dS m^{-1}) at the seedling stage. For **a**, **b** and **c**, asterisks indicate significant differences from the non-salinised control for the same genotype, based on Student's t test (*$P < 0.05$, **$P < 0.01$). For **d**, asterisks indicate significant differences between 40 and 80 mM based on Student's t test (*$P < 0.05$, **$P < 0.01$) because the ratios (as used for **a** to **c**) were so low in non-salinised controls as to be negligible, whereas the increase in ratio from 40 to 80 mM was highly relevant salt tolerance differences between genotypes

in the two most tolerant wild rice genotypes (*Oa*-VR and *Oa*-CH) at 40 mM NaCl, with both being considerably more tolerant than the salt-tolerant *O. sativa* genotype, Pokkali. The most salt-sensitive wild rice line (*Oa*-D) was as susceptible to salt as IR29 at 40 mM NaCl. These data are consistent with visual symptoms, indicating that *Oa*-VR was the most salt-tolerant wild *Oryza* accession and *Oa*-D the least tolerant. Na:K ratio calculated at 40 and 80 mM NaCl (Fig. 1d) revealed the lowest Na:K ratios in *Oa*-VR and Pokkali while the other wild rice genotypes and IR29 had progressively higher ratios, reaching an average of 2.41 for *Oa*-CH.

Sodium and potassium ion concentrations were measured in the youngest fully expanded leaves, where tissues remained hydrated even in the salt-sensitive genotypes, as shown by the narrow range of variation in K^+ concentrations (Fig. 2). The relationships between ion concentrations and leaf biomass (as a percentage of controls) illustrate the strong negative relationship between Na^+ concentration and salinity tolerance, confirming that the exclusion of Na^+ conferred physiological tolerance (Fig. 2). The three most salt-sensitive genotypes had 300–500 µmol Na^+ g^{-1} (SDW) while the most salt-tolerant genotypes had up to three times less Na^+. A negative relationship between physiological tolerance (ST) and Na^+ concentrations in the youngest fully expanded leaves was clear when all genotypes were compared (Fig. 2). A weak positive relationship was recorded between K^+ concentrations in shoots and salinity tolerance. Notably, Na^+ concentrations in *Oa*-VR and Pokkali were lowest of all six genotypes (114 and 83 µmol g^{-1} (SDW), respectively) and when expressed on a tissue water basis (using the SFW/SDW ratio of 3.6 and 3.4,

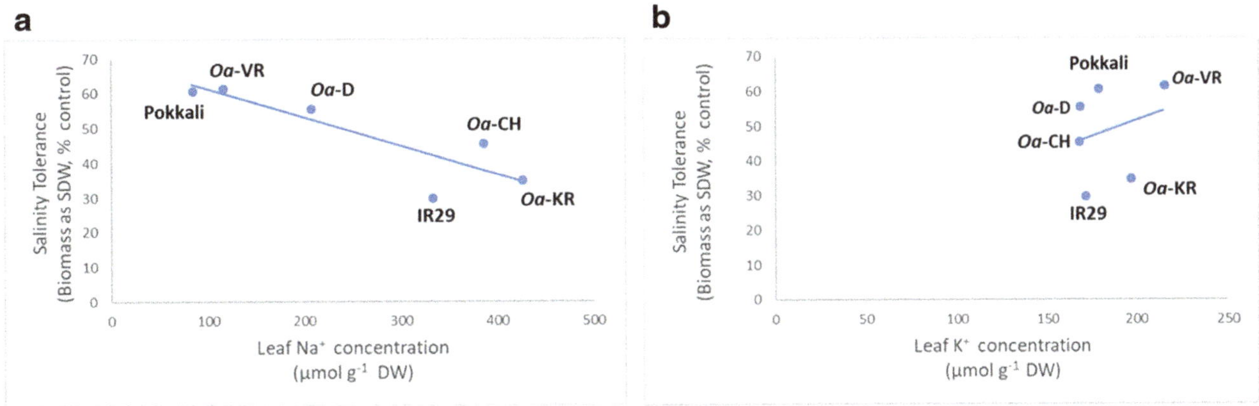

Fig. 2 Linear regression of Salinity Tolerance (ST) against: **a** leaf Na^+ concentrations [μmol Na^+ g^{-1} (SDW)] (R^2 = 0.75) and **b** leaf K^+ concentrations [μmol Na^+ g^{-1} (SDW)] (R^2 = 0.69). ST was calculated as the percentage ratio of mean shoot dry weight (salt treatment, 80 mM of NaCl) divided by mean shoot dry weight (control, no salt) [SDW (salt treatment))/(SDW (control)) x 100]

respectively) Na^+ concentrations were 34 and 44 μmol g^{-1} (FW), respectively; i.e. much lower than those in the soil solution in which they grew. *Oa*-VR accumulated 215 μmol K^+ g^{-1} (SDW), 20% more ($P < 0.05$) than the levels found in IR29 and *Oa*-D (171 and 168 μmol g^{-1} (SDW), respectively).

Depending upon the genotype, ion toxicity symptoms were first visible in leaves 7–15 DAS. Initially, salt-induced symptoms were always restricted to the older leaves but increased progressively in severity and extent until only the most recently emerged leaves were unaffected (data not shown).

Measurements at 80 mM NaCl established that the negative effects of salt were consistent across three vegetative traits—plant height, SDW and number of tillers (Additional file 3: Table S2). Furthermore, damage measured by SES scores correlated negatively with these traits, as well as photosynthetic rates ($P = 0.01$).

Plant accelerator (experiment 2)

There were no visual leaf symptoms or wilting in any genotype 4 d after salt was applied. Pokkali grew significantly faster (16.2 kpixels d^{-1}) than other lines over the first 9 d ($P < 0.05$) while IR29 grew slowest, in all treatments (Fig. 3; Additional file 4: Figure S2). The two wild rice species had the same relative growth rate at this earliest stage of salt treatment ($P > 0.05$), while Pokkali and IR29 grew significantly faster and slower, respectively (Additional file 5: Figure S3). Importantly, the average growth rates of the control plants during DAS 0 to 4 and 4 to 9 were significantly greater ($P < 0.05$) than any of the salt treatments (Fig. 3; Additional file 4: Figure S2). RGR in Pokkali declined steadily throughout the experiment, even in salt-treated plants (Additional file 4: Figure S2, Additional file 5: Figure S3), indicating that plants did not grow exponentially at any stage of the salt treatment. On the other hand, periods of exponential growth were observed in the other three genotypes, with exponential growth notably sustained in *Oa*-VR for the first 15 d of salt treatment (Additional file 5: Figure S3). After 23 DAS, RGR was lower (Pokkali, *Oa*-VR and *Oa*-D) or the same (IR29) in control plants when compared with salt-treated plants, which grew at 10% per day. These time-dependent shifts in the response of the genotypes to salinity were analysed using *p*-values for prediction mean differences within each interval identified in Fig. 3. While differential effects of salinity across genotypes were not seen in the absolute growth rate until plants had been exposed to salt for at least 19 d, salinity × genotype interactions were seen strongly in RGR from the beginning of the experiment. This is reflected in Additional file 5: Figure S3, where the changes in RGR in Pokkali plants reflected the vigorous canopy growth, early self-shading and distinctive, rapid canopy development rate compared with the other three genotypes tested.

There was a wide range of growth responses at each salt level in the seven genotypes imaged (Additional file 6: Figure S4), with IR29 notably the slowest growing genotype. Individual performances of the two *O. sativa* standard lines and two of the most contrasting *O. australiensis* accessions are represented at all four salt levels in Fig. 3. The reduction in shoot growth, as measured by PSA, was most pronounced at 80 and 100 mM NaCl, with smaller reductions at 40 mM NaCl (Fig. 3). By 12 DAS, non-salinised plants of all four genotypes were growing significantly faster than all salt-treated plants. Importantly, Pokkali, *Oa*-VR and *Oa*-D grew substantially faster than IR29: at 12 DAS, non-salinised control plants grew at 25.1, 13.8, 13.5 and 5.9 kilopixels d^{-1} (as measured by PSA) in the four genotypes, respectively. Pokkali, *Oa*-VR and *Oa*-D treated with 100 mM NaCl

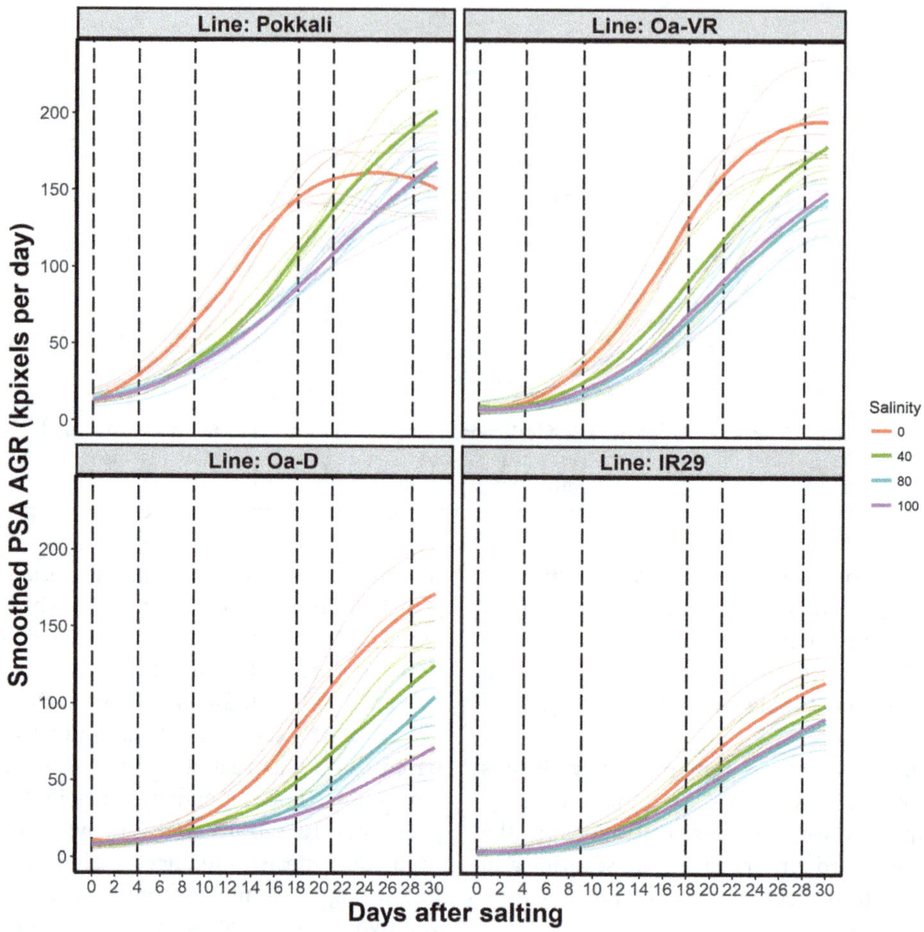

Fig. 3 Absolute growth rates of Pokkali, *Oa*-VR, *Oa*-D and IR29 from 0 to 30 DAS including non-salinised controls. Smoothed AGR values were derived from projected shoot area (PSA) values to which splines had been fitted. Thin lines represent individual plants. Bold lines represent the grand average of the six replicates plants for each treatment. The vertical broken lines represent the tested intervals

were reduced to 78–88% of the controls, while no effect of 100 mM NaCl could be detected in IR29 plants. Despite the reputation of IR29 as a salt-sensitive genotype, its inherently slow growth made responses to NaCl difficult to detect in the early stages of vegetative development (Additional file 5: Figure S3). The divergence in AGR between plants grown at 80 and 100 mM NaCl was notable, with Pokkali and *Oa*-VR plants growing at the same rate in these two highest salt treatments, while *Oa*-D plants grew significantly slower at 100 mM than at 80 mM NaCl (Fig. 3). Importantly, *Oa*-VR showed substantially less inhibition of growth in response to salinity when compared with *Oa*-D, supporting the observation from the first experiment that *Oa*-VR is the most salt tolerant of the wild rice accessions tested (Fig. 3). The most severe reduction in PSA across all genotypes tested in the Plant Accelerator was an *O. meridionalis* genotype (*Om*-T), where there was a 27% reduction after DAS9 and a further reduction of almost 20% by DAS18 in 100 mM NaCl.

Shoot images generated in the Plant Accelerator generated an estimate of *relative* leaf senescence using fluorescence optics, even though these values differ from visual analyses by SES, which showed that non-salinised leaves had not begun to senesce. However, the relative effects of NaCl on canopy development and the reported changes in senescence in salinised plants (Fig. 4) provide an accurate assessment of the impact of salt on *Oa*-VR and *Oa*-D (Hairmansis et al. 2014). Necrosis of older leaves was seen in the salt-sensitive genotype *Oa*-D after 30 d treatment with 40 mM NaCl, while the putatively salt-tolerant *Oa*-VR had minor leaf damage, even at 80 to 100 mM NaCl (Fig. 4). *Oa*-VR exhibited less than a 40% increase in relative senescence at 100 mM NaCl compared with the control, while an increase of more than 120% was recorded for *Oa*-D (Fig. 4). Furthermore, the impact of 100 mM NaCl on chlorophyll content was smaller in *Oa*-VR than in *Oa*-D (Fig. 4).

Compared with controls, WUI was impaired immediately after salt was applied (Fig. 5). While WUI

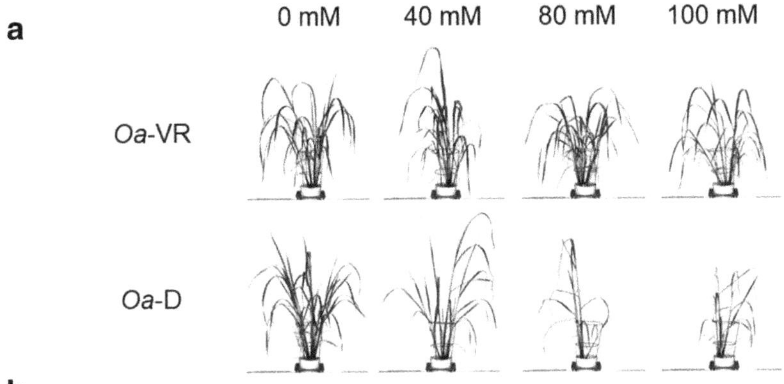

b

Line/Trait	Chlorophyll concentration [mg/g]			Average relative senescence		
	Non-salinised	100 mM	Reduction (%)	Non-salinised	100 mM	Increase (%)
***Oa*-VR**	1.90	0.96	49	0.026	0.036	37
***Oa*-D**	1.51	0.58	62	0.046	0.104	126
LSD (0.05)	0.50	0.13		0.01	0.02	

Fig. 4 a Phenotypic changes in response to the different salt treatments 30 days after salting for the salt-tolerant *Oa*-VR and the salt-sensitive (*Oa*-D). **b** Chlorophyll concentration and average relative senescence under non-salinised (0 mM) and salinised (100 mM NaCl) treatments for both tested genotypes

continued to increase in *Oa*-VR throughout the experiment at all salt levels (in *Oa*-D at 80 and 100 mM NaCl), it accelerated only after 14 d of salt treatment. Control plants used water more efficiently than salt-treated plants up until 18 DAS and 24 DAS in *Oa*-VR and *Oa*-D, respectively. At 100 mM NaCl, *Oa*-VR used water substantially more efficiently than *Oa*-D, with WUI 25% higher at 100 mM NaCl by the end of the experiment in *Oa*-VR.

Both Pokkali and *Oa*-VR had a 36% lower fresh biomass under the higher salt treatment (100 mM NaCl) compared with non-salinised controls, while higher reductions were recorded for IR29 and *Oa*-D (49 and 53%, respectively; Additional file 7: Table S3).

Discussion

Complementary approaches were taken to assess the salinity tolerance of lines/accessions of three rice species, *O. sativa*, *O. australiensis* and *O. meridionalis*. In a preliminary screening prior to these experiments, a survey of a wide range of wild *Oryza* accessions alongside Pokkali and IR29 produced a 'short-list' of five accessions chosen from *O. australiensis* and *O. meridionalis* that were selected for contrasting tolerance and sensitivity to salinity during early vegetative growth. The wild *Oryza* accessions chosen for this study evolved in geographically isolated populations, thereby broadening the range of genetic diversity and, with it, the opportunity to discover novel salt tolerance mechanisms (Menguer et al. 2017). However, the preliminary goal was to find contrasting salt tolerance within the same species in order to facilitate subsequent experiments involving mapping populations and comparative proteomics. In this paper, we report on one destructive experiment, with salt levels maintained at a steady state of 40 and 80 mM NaCl, and the second non-destructive experiment where soil was saturated initially with saline solution then followed by daily fresh water applications to replace evaporation and transpiration. The use of a series of images of plants in the Plant Accelerator gave a more dynamic picture of salinity tolerance than could be achieved by destructive measurements as in the first experiment. Ion concentrations in the YFL and phenotypic observations from the first experiment were seminal to developing a salt tolerance ranking.

Multiple strands of evidence from our data, including biomass, leaf visual symptoms, gas exchange and ion concentrations, confirm the wide range of tolerances to salt in the genotypes of wild and cultivated rice selected for these experiments. For example, chlorophyll levels were almost 50% lower in IR29 at 40 mM NaCl but were unaffected in *Oa*-VR, similar to contrasts in tolerance reported previously (Lutts et al. 1996), where 50 mM NaCl lowered chlorophyll levels by up to 70%. The criteria reported in Fig. 1 support the long-established view that Pokkali is highly tolerant to salt (Yeo et al. 1990) but make a case that the wild *O. australiensis* species (*Oa*-VR) has at least the same level of salt tolerance. In the first experiment, salt tolerance in *Oa*-VR was evident after 25 d of 80 mM NaCl, where shoot biomass was

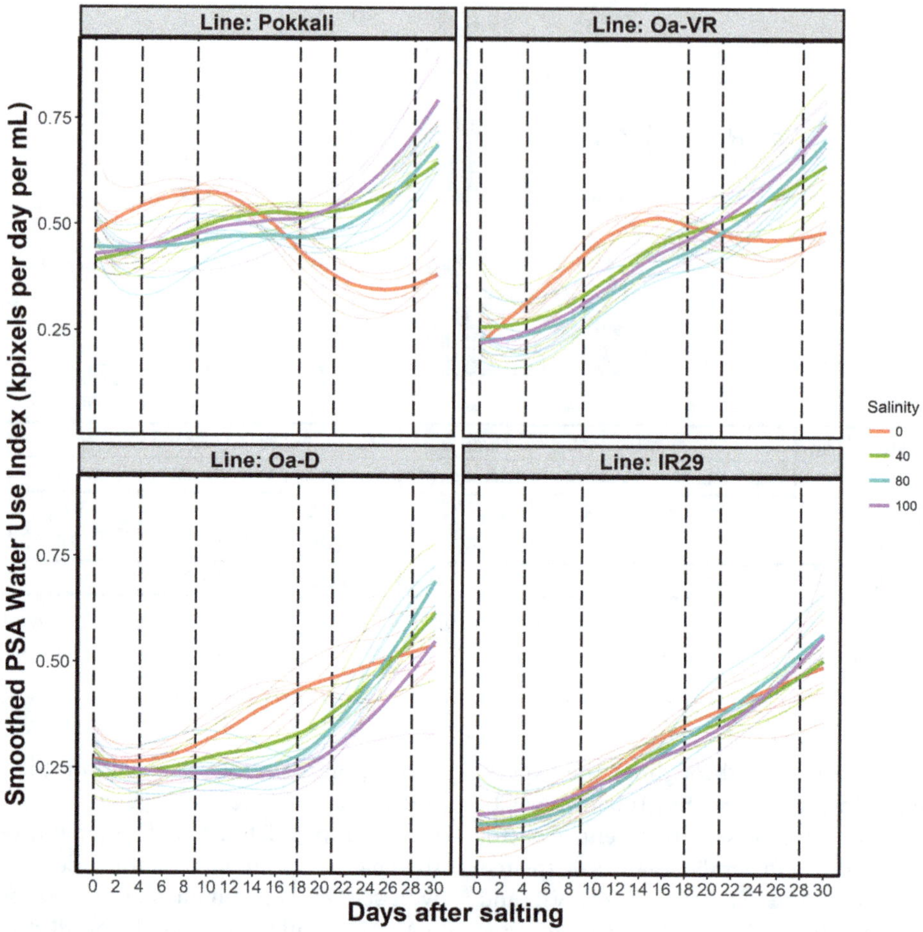

Fig. 5 Relationship between growth and water use during salt treatment. Smoothed PSA Water Use Index is shown for the selected genotypes under salt treatments and non-salinised control conditions. The values were obtained by dividing the total increase in sPSA for each interval by the total water loss in the same interval. Thin lines represent individual plants. Bold lines represent the grand average of the six replicates for each treatment. Vertical broken lines represent the tested intervals

reduced by 58% in Pokkali compared with controls, while the reduction in biomass in *Oa*-VR was marginally less (50%). Moreover, symptoms of leaf damage in *Oa*-VR due to NaCl were significantly less pronounced than those seen in Pokkali.

The additional level of salt tolerance found in *Oa*-VR offers a potential tool for crop improvement, especially in that *Oa*-VR is from a wild *Oryza* population with the unique EE genome (Jacquemin et al. 2013), and is thus phylogenetically remote from *O. sativa*; this enhances the possibility of identifying novel mechanisms of salt tolerance unique to *O. australiensis*. By contrast, IR29 is reputedly highly salt-sensitive (Martinez-Atienza et al. 2006; Islam et al. 2011). Surprisingly, for the most salt-sensitive of the wild rice genotypes (*Oa*-D and *Oa*-KR) in very moderate salinity (40 mM NaCl), biomass and ion concentrations were more strongly affected by salt than leaf symptoms, possibly indicating genotypic variation in tissue tolerance to NaCl, as reported earlier (Yeo et al. 1990). In reverse, the very slow absolute growth rates of IR29 appeared, paradoxically, to result in a small effect of salt on relative growth rates (Fig. 3) but much larger effects on senescence (Fig. 1a). This suggests that a range of performance criteria is essential to distinguish the intrinsic differences in salt tolerances in screening experiments. This underlines the polygenic nature of salt tolerance, where genes determining ion import, compartmentation and metabolic responses to salt are likely to play a collective role in physiological tolerance (Munns et al. 2008). Therefore, based on the overall indicators of salt tolerance and rates of shoot development, *Oa*-VR and *Oa*-D were chosen as complementary *O. australiensis* genotypes for image analysis (Fig. 4), representing contrasting tolerance to salt in otherwise indistinguishable *O. australiensis* accessions. While the salt-tolerant genotype (*Oa*-VR) is from the Northern Territory and the salt-sensitive accession is from the Kimberley region of Western Australia, there is no obvious basis for predicting their respective tolerances to salinity without a fine-scale investigation of the collection sites and the seasonal fluctuations in soil water content and soil chemistry.

The rate at which shoot growth responded to salt (Experiment 2), as well as the internal Na^+ and K^+ concentrations of young leaves (Experiment 1), provide insights into possible mechanisms of tolerance. In rice, only part of the Na^+ load reaching the leaves is taken up symplastically by the roots (Krishnamurthy et al. 2009), entering the transpiration stream and further regulated under the control of a suite of transporters. The low Na^+/K^+ ratios found in both *Oa*-VR and Pokkali (< 0.5) suggest that active mechanisms are in play to exclude Na^+, even when the external solution was fixed at 80 mM NaCl for 30 d. Early clues as to how this is achieved came from a QTL (Ren et al. 2005), now known to contain the OsHKT1;5 gene, which enhances Na^+ exclusion in rice (Hauser et al. 2010). Davenport et al. (2007) and others have established that the HKT1 transporters in *Arabidopsis* retrieve Na^+ from the xylem. In general, high-affinity K^+ uptake systems have now been shown to be pivotal for the management of salinity and deficiency symptoms in rice (Suzuki et al. 2016), as well as other species such as *Arabidopsis* and wheat (Byrt et al. 2007; Munns et al. 2008; Hauser et al. 2010). Further candidates such as the SOS1 transporter might also play a key part in the removal of Na^+ from the xylem stream (Shi et al. 2002). The complexity of the rice HKT transporters identified in *O. sativa* (Garciadeblás et al. 2003) has not yet been explored in a wider range of *Oryza* genetic backgrounds. The levels of tolerance reported for *O. australiensis* should stimulate an analysis of the expression of genes regulating Na^+ and K^+ transport and the functional properties of these transporters, which may have evolved in lineages of geographically isolated communities from the Australian savannah.

Sodium exclusion appeared to operate effectively in Pokkali and *Oa*-VR but failed in other wild rice accessions where Na^+/K^+ exceeded 2.0 in the most severe cases at 80 mM NaCl. An earlier study reported leaf Na^+/K^+ ratios of 4.4 in 21 *indica* rice lines after 48 d of about 35 mM NaCl (Asch et al. 2000), reinforcing the view that *Oa*-VR is tolerant to salt. Supporting this claim, Na^+ concentrations in Pokkali and *Oa*-VR calculated on a tissue-water basis were half those in the external solution when the roots were in an 80 mM solution. These contrasting degrees of Na^+ exclusion and the consequences for plant performance are illustrated by the strong relationship between ST and the accumulation of Na^+ (Fig. 2). Based on the observation that diminished apoplastic uptake of Na^+ in the roots of Pokkali (Krishnamurthy et al. 2011) enhances Na^+ exclusion, the degree of bypass flow in *Oa*-VR and the other genotypes in the current study is a priority for identifying the mechanism of salt tolerance. The consequences of Na^+ loads in leaves for shoot physiology (SES, chlorophyll content, photosynthesis and tiller development) was apparent for the wild *Oryza* species as well as the two *O. sativa* standard genotypes, with strong correlations between ion levels and leaf damage.

In the second experiment, relative growth rates could be observed continuously and non-destructively, revealing an impact of salt even in the first 4 DAS (Additional file 5 Figure S3). A binary impact of salt on plants is exerted through osmotic stress and ion toxicity (Greenway and Munns 1980). The long-term impact of salt in this 30-d salt treatment was primarily due to toxic effects of Na^+ rather than osmotic stress, which would have been most apparent in the earliest stages of the treatment period when tissue ion levels were lowest and osmotic adjustment was not yet established (Munns et al. 2016). The more salt-sensitive genotypes appeared to have less capacity to exclude salt, causing leaf Na^+ and K^+ concentrations to rise above parity and cause toxicity and metabolic impairment.

Water use efficiency was substantially greater in *Oa*-VR than *Oa*-D, particularly in the first two weeks after salt was applied, suggesting that the resilience of photosynthesis observed in salt-treated *Oa*-VR plants sustained growth (PSA) even as stomatal conductance fell by 60%. WUI values for *Oa*-D plants at 100 mM NaCl were notably lower than those at 40 and 80 mM NaCl, reflecting the progressively higher impact of NaCl on hydraulics in this sensitive genotype as concentrations increased from 40 to 100 mM NaCl. This trend of low WUI in salt-treated plants is consistent with previous studies of *indica* and *aus* rice (Al-Tamimi et al. 2016), as well as barley and wheat (Harris et al. 2010).

The effects of salt are dynamic, depending both upon relative growth rates and ion delivery and root:shoot ratios (Munns et al. 2016). Non-destructive measurements of growth showed that the relationship between control and salt-treated plants varied substantially over the time-course of treatment in all genotypes. This was partly due to the different developmental programs of each genotype, with Pokkali characterised by vigorous early growth and an early transition to flowering in non-saline conditions, when vegetative growth arrested; the transition to flowering was delayed in salt-treated plants. Such developmental effects are likely to be a factor in the impact of salinity on yield (Khatun et al. 1995). Among the wild rices, we have observed strong contrasts in photoperiod sensitivity between accessions, resulting in large differences in duration of vegetative growth. We speculate that this would affect the time-course of NaCl accumulation and its impact on biomass and grain yield.

Under paddy and rainfed conditions, salt levels in the root medium are unlikely to remain constant as they did in the treatment regime applied in the first experiment. This variation in salt load was better represented in the Plant Accelerator (Experiment 2), where soil was salinised and then transpired water replaced with fresh water to the soil surface daily. We contend that these contrasting regimes of salt application mimicked both steady-state and transient salinisation, including the salt loads imposed on rice paddies following spasmodic tidal surges. The ranking of salt-tolerance for both the *O. sativa* 'standard' genotypes and the four wild rice relatives was broadly maintained under the two experimental regimes we employed.

In this study, we explored the naturally occurring variation in salt tolerance among some of rice's wild relatives in comparisons to selected *O. sativa* cultivars. Despite the substantial genetic distance between *O. australiensis* (taxon E) and *Oryza sativa* (taxon A), several studies have managed to leap this species barrier, allowing these two species to be crossed (Morinaga et al. 1960; Nezu et al. 1960). Another study reported a rapid phenotype recovery of the recurrent parent after only two backcrosses (Multani et al. 1994). Using this backcrossing approach, *O. australiensis* accessions have been used in breeding programs as a source of tolerance to biotic stresses including bacterial blight resistance (Brar and Khush 1997), brown planthopper resistance (Jena et al. 2006) and blast resistance (Jeung et al. 2007; Suh et al. 2009). Our study highlights the potential use of the Australian wild-species alleles in breeding programs to exploit variations in abiotic stress generally and salinity tolerance in particular. However, harnessing alleles from wild relatives of rice that confer salt tolerance and applying them to modern cultivars remains a long-term objective until mechanisms of tolerance become clearer.

Additional files

Additional file 1: FigureS1. Relationships between Projected Shoot Area (kpixels) 28 and 30 days after salting with Fresh Weight and Dry Weight based on 168 individual plants using the fluorescence images. Squared Pearson correlation coefficients are given on the right.

Additional file 2: Table S1. Shoot dry weight, shoot fresh weight, chlorophyll concentration and photosynthetic rate for the four wild *Oryza* accessions and *O. sativa* controls.

Additional file 3: Table S2. Linear correlation (r values) between various physiological characteristics measured for the four wild *Oryza* accessions and *O. sativa* controls combined at seedling stage grown under 80 mM NaCl for 30 d. * = Significant at 5% level of probability and ** = Significant at 1% level of probability.

Additional file 4: Figure 2. Smoothed Projected Shoot Area (described by kpixels) of Absolute Growth Rates over six intervals within 0–28 days after salting. X-axis represents the salt levels and the error bars represent ±1/2 Confidence Interval.

Additional file 5: Figure S3. Smoothed Projected Shoot Area (described by kpixels) of Relative Growth Rates over the four salt treatments within 0–25 days after salting. Error bars represent ±1/2 Confidence Interval.

Additional file 6: Figure S4. Absolute growth rates of all tested genotypes from 0 to 30 DAS including non-salinised controls. Smoothed AGR values were derived from projected shoot area (PSA) values to which splines had been fitted. Thin lines represent individual plants. Bold lines represent the grand average of the six replicates plants for each treatment. The vertical broken lines represent the tested intervals.

Additional file 7: Table S3. Photosynthetic rate, stomatal conductance, number of tillers and shoot fresh weight of the four wild *Oryza* accessions and *O. sativa* controls. The first three traits were evaluated on 29 DAS while shoot fresh weight was measured on the termination of the experiment, on 30 DAS. Two measurements were excluded from the stomatal conductance analysis as they gave large negative values (− 30 and − 50). Reduction values were rounded to the nearest integer.

Abbreviations

AGT: Absolute Growth Rate; ANOVA: Analysis of Variance; DAS: Days After Salting; DAT: Days After Transplanting; DF: Degrees of Freedom; EC: Electrical Conductivity; FLUO: Fluorescence; IRRI: International Rice Research Institute; PSA: Projected Shoot Area; PVC: Polyvinyl Chloride; QTL: Quantitative Trait Locus; RGB: Red-Green-Blue; RGR: Relative Growth Rate; SDW: Shoot Dry Weight; SES: Standard Evaluation System; SFW: Shoot Fresh Weight; sPSA: Smoothed Projected Shoot Area; ST: Salinity Tolerance; WUI: Water Use Index; YFL: Youngest Fully Expanded Leaf

Acknowledgements

The authors acknowledge the financial support of the Australian Government National Collaborative Research Infrastructure Strategy (Australian Plant Phenomics Facility). The authors also acknowledge the use of the facilities, and scientific and technical assistance of the Australian Plant Phenomics Facility, which is supported by NCRIS. The authors would like to thank all staff from the Plant Accelerator at the University of Adelaide for support during the experiments. We also thank A/Prof. Stuart Roy for constructive comments on the manuscript.

Funding

The research reported in this publication was supported by funding from The Australian Plant Phenomics Facility. YY was supported by an International Postgraduate Research Scholarship.

Authors' contributions

YY designed and executed the first experiment. YY also phenotyped the plants (for both experiments), performed the data analyses for the first experiment and wrote the manuscript. CB designed the second experiment, performed the spatial correction, and conceived of and developed the statistical analyses for the phenotypic data of the second experiment. BB assisted with the phenotypic analyses and revised the manuscript. THR and BJA contributed to the original concept of the project and supervised the study. BJA conceived the project and its components and provided the genetic material. All authors read and contributed to the manuscript.

Competing interests

The authors declare that they have no competing interests.

Author details

[1]Sydney Institute of Agriculture, University of Sydney, Sydney, Australia. [2]School of Agriculture Food and Wine, University of Adelaide, Adelaide, Australia. [3]Australian Plant Phenomics Facility, The Plant Accelerator, Waite Research Institute, University of Adelaide, Adelaide, Australia. [4]Department of Biological Sciences, Macquarie University, Sydney, Australia.

References

Al-Tamimi N, Brien C, Oakey H (2016) Salinity tolerance loci revealed in rice using high-throughput non-invasive phenotyping. Nat Commun 7:13342

Asch F, Dingkuhn M, Dörffling K, Miezan K (2000) Leaf K / Na ratio predicts salinity induced yield loss in irrigated rice. Euphytica 113:109–118

Atieno J, Li Y, Langridge P (2017) Exploring genetic variation for salinity tolerance in chickpea using image-based phenotyping. Sci Rep 7:1–11

Atwell BJ, Wang H, Scafaro AP (2014) Could abiotic stress tolerance in wild relatives of rice be used to improve *Oryza sativa*? Plant Sci 215–216:48–58

Ballini E, Berruyer R, Morel JB (2007) Modern elite rice varieties of the "green revolution" have retained a large introgression from wild rice around the Pi33 rice blast resistance locus. New Phytol 175:340–350

Berger B, Bas De Regt MT (2012) High-throughput phenotyping in plants shoots. Methods Mol Biol 918:9–20

Brar DS, Khush GS (1997) Alien introgression in rice. Plant Mol Biol 35:35–47

Brien, C. J. (2018) dae: Functions useful in the design and ANOVA of experiments. Version 3.0-16

Brozynska M, Copetti D, Furtado A (2016) Sequencing of Australian wild rice genomes reveals ancestral relationships with domesticated rice. Plant Biotech J 15:1–10

Butler DG, Cullis BR, Gilmour AR, Gogel BJ (2009) Analysis of Mixed Models for S language environments: ASReml-R reference manual. Brisbane, DPI Publications

Byrt CS, Platten JD, Spielmeyer W (2007) HKT1;5-like cation transporters linked to Na^+ exclusion loci in wheat, *Nax2* and *Kna1*. Plant Physiol 143:1918–1928

Campbell MT, Du Q, Liu K (2017) A comprehensive image-based phenomic analysis reveals the complex genetic architecture of shoot growth dynamics in rice. Plant Genome 10:2

Campbell MT, Knecht AC, Berger B (2015) Integrating image-based phenomics and association analysis to dissect the genetic architecture of temporal salinity responses in rice. Plant Physiol 168:1476–1489

Davenport RJ, Muñoz-Mayor A, Jha D (2007) The Na^+ transporter AtHKT1;1 controls retrieval of Na^+ from the xylem in Arabidopsis. Plant Cell Environ 30:497–507

Flowers TJ (2004) Improving crop salt tolerance. J Exp Bot 55:307–319

Fukuda A, Nakamura A, Tagiri A (2004) Function, intracellular localization and the importance in salt tolerance of a vacuolar Na^+/H^+ antiporter from rice. Plant Cell Physiol 45:146–159

Garciadeblás B, Senn ME, Bañuelos MA, Rodríguez-Navarro A (2003) Sodium transport and HKT transporters: the rice model. Plant J 34:788–801

Grattan SR, Shannon MC, Roberts SR (2002) Rice is more sensitive to salinity than previously thought. Calif Agric 56:189–195

Greenway H, Munns R (1980) Mechanisms of salt tolerance in nonhalophytes. Annu Rev Plant Biol 31:149–190

Gregorio GB, Senadhira D (1993) Genetic analysis of salinity tolerance in rice (*Oryza sativa* L.). Theor Appl Genet 86:333–338

Hairmansis A, Berger B, Tester M, Roy SJ (2014) Image-based phenotyping for non-destructive screening of different salinity tolerance traits in rice. Rice 7:1–10

Harris BN, Sadras VO, Tester M (2010) A water-centred framework to assess the effects of salinity on the growth and yield of wheat and barley. Plant Soil 336:377–389

Hauser F, Horie T (2010) A conserved primary salt tolerance mechanism mediated by HKT transporters: a mechanism for sodium exclusion and maintenance of high K^+/Na^+ ratio in leaves during salinity stress. Plant Cell Environ 33:552–565

Henry RJ, Rice N, Waters DLE (2010) *Australian Oryza*: utility and conservation. Rice 3:235–241

IRRI (2013) Standard Evaluation System (SES) for Rice. International Rice Research Institute, Manila, p 38

Islam MR, Salam MA, Hassan L, Collard BCY, Singh RK, Gregorio GB (2011) QTL mapping for salinity tolerance in rice. Physiol Mol Biol Plants 23:137–146

Ismail AM, Horie T (2017) Molecular breeding approaches for improving salt tolerance. Annu Rev Plant Biol 68:1–30

Jacquemin J, Bhatia D, Singh K, Wing RA (2013) The international *Oryza* map alignment project: development of a genus-wide comparative genomics platform to help solve the 9 billion-people question. Curr Opin Plant Biol 16:147–156

Jena KK, Jeung JU, Lee JH (2006) High-resolution mapping of a new brown planthopper (BPH) resistance gene, *Bph18*(t), and marker-assisted selection for BPH resistance in rice (*Oryza sativa* L.). Theor Appl Genet 112:288–297

Jeung JU, Kim BR, Cho YC (2007) A novel gene, *Pi40*(t), linked to the DNA markers derived from NBS-LRR motifs confers broad spectrum of blast resistance in rice. Theor Appl Genet 115:1163–1177

Khatun S, Flowers TJ (1995) Effects of salinity on seed set in rice. Plant Cell Environ 18:61–67

Khush GS (1997) Origin, dispersal, cultivation and variation of rice. Plant Mol Biol 35:25–34

Khush GS (2005) What it will take to feed 5.0 billion rice consumers in 2030. Plant Mol Biol 59(1):–6

Krishnamurthy P, Ranathunge K, Franke R (2009) The role of root apoplastic transport barriers in salt tolerance of rice (*Oryza sativa* L.). Planta 230:119–134

Krishnamurthy P, Ranathunge K, Nayak S (2011) Root apoplastic barriers block Na^+ transport to shoots in rice (*Oryza sativa* L.). J Exp Bot 62:4215–4228

Lang N, Li Z, Buu B (2001) Microsatellite markers linked to salt tolerance in rice. Omonrice 9:9–21

Lutts S, Kinet JM, Bouharmont J (1995) Changes in plant response to NaCl during development of rice (*Oryza sativa* L.) varieties differing in salinity resistance. J Exp Bot 46:1843–1852

Lutts S, Kinet JM, Bouharmont J (1996) NaCl-induced senescence in leaves of rice (*Oryza sativa* L.) cultivars differing in salinity resistance. Ann Bot 78:389–398

Mackinney G (1941) Absorption of light by chlorophyll solutions. J Biol Chem 140:315–322

Martinez-Atienza J, Jiang X, Garciadeblas B (2006) Conservation of the salt overly sensitive pathway in rice. Plant Physiol 143:1001–1012

Menguer PK, Sperotto RA, Ricachenevsky FK (2017) A walk on the wild side : *Oryza* species as source for rice abiotic stress tolerance. Genet Mol Biol 40:238–252

Morinaga T, Kuriyama H (1960) Interspecific hybrids and genomic constitution of various species in the genus *Oryza*. Agric Hortic 35:1245–1247

Multani DS, Jena KK, Brar DS, de los Reyes BG, Angeles ER, Khush GS (1994) Development of monosomic alien addition lines and introgression of genes from *Oryza australiensis* Domin. to cultivated rice *O. sativa* L. Theor Appl Genet 88:102–109

Munns R, James RA, Gilliham M (2016) Tissue tolerance: an essential but elusive trait for salt-tolerant crops. Funct Plant Biol 43:1103–1113

Munns R, Tester M (2008) Mechanisms of salinity tolerance. Annu Rev Plant Biol 59:651–681

Nezu M, Katayama TC, Kihara H (1960) Genetic study of the genus *Oryza*. I. Crossability and chromosomal affinity among 17 species. Seiken Jiho 11:1–11

Ochiai K, Matoh T (2002) Characterization of the Na^+ delivery from roots to shoots in rice under saline stress: excessive salt enhances apoplastic transport in rice plants. Soil Sci Plant Nutr 48:371–378

Qadir M, Quillérou E, Nangia V (2014) Economics of salt-induced land degradation and restoration. Nat Resour Forum 38:282–295

R Core Team (2018) R: A language and environment for statistical computing. Vienna, Austria: R Foundation for Statistical Computing.

Rahman ML, Jiang W, Chu SH (2009) High-resolution mapping of two rice brown planthopper resistance genes, *Bph20*(t) and *Bph21*(t), originating from *Oryza minuta*. Theor Appl Genet 119:1237–1246

Ren Z-H, Gao J-P, Li L (2005) A rice quantitative trait locus for salt tolerance encodes a sodium transporter. Nat Genet 37:1141–1146

Sabouri H, Sabouri A (2008) New evidence of QTLs attributed to salinity tolerance in. African J Biotechnol 7:4376–4383

Scafaro AP, Gallé A, Van Rie J (2016) Heat tolerance in a wild *Oryza* species is attributed to maintenance of rubisco activation by a thermally stable rubisco activase ortholog. New Phytol 211:899–911

Scafaro AP, Haynes PA, Atwell BJ (2010) Physiological and molecular changes in *Oryza meridionalis* ng., a heat-tolerant species of wild rice. J Exp Bot 61:191–202

Shi H, Quintero FJ, Pardo JM, Zhu JK (2002) The putative plasma membrane Na^+/H^+ antiporter SOS1 controls long-distance Na^+ transport in plants. Plant Cell 14:465–477

Stein JC, Yu Y, Copetti D (2018) Genomes of 13 domesticated and wild rice relatives highlight genetic conservation, turnover and innovation across the genus *Oryza*. Nat Genet 50:285–296

Suh JP, Roh JH, Cho YC (2009) The pi40 gene for durable resistance to rice blast and molecular analysis of pi40-advanced backcross breeding lines. Phytopathology 99:243–250

Suzuki K, Costa A, Nakayama H (2016) OsHKT2;2/1-mediated Na^+ influx over K^+ uptake in roots potentially increases toxic Na^+ accumulation in a salt-tolerant landrace of rice Nona Bokra upon salinity stress. J Plant Res 129:67–77

Takagi H, Tamiru M, Abe A (2015) MutMap accelerates breeding of a salt-tolerant rice cultivar. Nat Biotechnol 33:445–449

Thomson MJ, de Ocampo M, Edgane J (2010) Characterizing the *Saltol* quantitative trait locus for salinity tolerance in rice. Rice 3:148–160

Wang W, Vinocur B, Altman A (2003) Plant responses to drought, salinity and extreme temperatures: towards genetic engineering for stress tolerance. Planta 218:1–14

Wickham H (2009) ggplot2: Create Elegant Data Visualisations Using the Grammar of Graphics. R package version 2.2.1.

Yadav R, Flowers TJ, Yeo A (1996) The involvement of the transpirational bypass flow in sodium uptake by high- and low-sodium-transporting lines of rice developed through intravarietal selection. Plant Cell Environ 19:329–336

Yao MZ, Wang JF, Chen HY, Zha HQ, Zhang HS (2005) Inheritance and QTL mapping of salt tolerance in rice. Rice Sci 12:25–32

Yeo AR, Yeo ME, Flowers SA, Flowers TJ (1990) Screening of rice (*Oryza sativa* L.) genotypes for physiological characters contributing to salinity resistance, and their relationship to overall performance. Theor Appl Genet 79:377–384

Yeo AR, Yeo ME, Flowers TJ (1987) The contribution of an apoplastic pathway to sodium uptake by rice roots in saline conditions. J Exp Bot 38:1141–1153

Zeng L, Shannon MC, Grieve CM (2002) Evaluation of salt tolerance in rice genotypes by multiple agronomic parameters. Euphytica:235–245

Zhu Q, Zheng X, Luo J (2007) Multilocus analysis of nucleotide variation of *Oryza sativa* and its wild relatives: severe bottleneck during domestication of rice. Mol Biol Evol 24:875–888

Permissions

All chapters in this book were first published in RICE, by Springer; hereby published with permission under the Creative Commons Attribution License or equivalent. Every chapter published in this book has been scrutinized by our experts. Their significance has been extensively debated. The topics covered herein carry significant findings which will fuel the growth of the discipline. They may even be implemented as practical applications or may be referred to as a beginning point for another development.

The contributors of this book come from diverse backgrounds, making this book a truly international effort. This book will bring forth new frontiers with its revolutionizing research information and detailed analysis of the nascent developments around the world.

We would like to thank all the contributing authors for lending their expertise to make the book truly unique. They have played a crucial role in the development of this book. Without their invaluable contributions this book wouldn't have been possible. They have made vital efforts to compile up to date information on the varied aspects of this subject to make this book a valuable addition to the collection of many professionals and students.

This book was conceptualized with the vision of imparting up-to-date information and advanced data in this field. To ensure the same, a matchless editorial board was set up. Every individual on the board went through rigorous rounds of assessment to prove their worth. After which they invested a large part of their time researching and compiling the most relevant data for our readers.

The editorial board has been involved in producing this book since its inception. They have spent rigorous hours researching and exploring the diverse topics which have resulted in the successful publishing of this book. They have passed on their knowledge of decades through this book. To expedite this challenging task, the publisher supported the team at every step. A small team of assistant editors was also appointed to further simplify the editing procedure and attain best results for the readers.

Apart from the editorial board, the designing team has also invested a significant amount of their time in understanding the subject and creating the most relevant covers. They scrutinized every image to scout for the most suitable representation of the subject and create an appropriate cover for the book.

The publishing team has been an ardent support to the editorial, designing and production team. Their endless efforts to recruit the best for this project, has resulted in the accomplishment of this book. They are a veteran in the field of academics and their pool of knowledge is as vast as their experience in printing. Their expertise and guidance has proved useful at every step. Their uncompromising quality standards have made this book an exceptional effort. Their encouragement from time to time has been an inspiration for everyone.

The publisher and the editorial board hope that this book will prove to be a valuable piece of knowledge for researchers, students, practitioners and scholars across the globe.

List of Contributors

Keita Tsukahara and Masanori Tamaoki
Center for Environmental Biology and Ecosystem Studies, National Institute for Environmental Studies, Onogawa 16-2, Tsukuba, Ibaraki 305-8506, Japan

Hiroko Sawada
Center for Environmental Biology and Ecosystem Studies, National Institute for Environmental Studies, Onogawa 16-2, Tsukuba, Ibaraki 305-8506, Japan
JSPS Research Fellow, Japan Society for the Promotion of Science, Chiyoda-ku, Tokyo 102-0083, Japan

Yoshihisa Kohno
Central Research Institute of Electric Power Industry, Abiko, Chiba 270-1194, Japan

Keitaro Suzuki
NARO Institute of Crop Science, Tsukuba, Ibaraki 305-8518, Japan

Nobuhiro Nagasawa
Department of Agribusiness, Akita Prefectural University, Minami akita-gun, Akita 010-0444, Japan

Kentaro Kaneko, Maiko Sasaki, Nanako Kuribayashi, Hiromu Suzuki, Takuya Inomata, Kimiko Itoh and Marouane Baslam
Graduate School of Science and Technology, Niigata University, Niigata 950-2181, Japan

Toshiaki Mitsui
Graduate School of Science and Technology, Niigata University, Niigata 950-2181, Japan
Department of Applied Biological Chemistry, Niigata University, Niigata 950-218, Japan

Yukiko Sasuga
Department of Applied Biological Chemistry, Niigata University, Niigata 950-218, Japan

Takeshi Shiraya
Department of Applied Biological Chemistry, Niigata University, Niigata 950-218, Japan
Niigata Crop Research Center, Niigata Agricultural Research Institute, Nagaoka 940-0826, Japan

B. P. Mallikarjuna Swamy, Ramil Mauleon, Wickneswari Ratnam, Ma. Teressa Sta. Cruz and Arvind Kumar
Plant Breeding Division, International Rice Research Institute (IRRI), Philippines

Noraziyah Abd Aziz Shamsudin
Plant Breeding Division, International Rice Research Institute (IRRI), Philippines
Faculty of Science and Technology, Universiti Kebangsaan Malaysia, 43600 Bangi, Selangor, Malaysia

Site Noorzuraini Abd Rahman
Plant Breeding Division, International Rice Research Institute (IRRI), Philippines MARDI, Seberang Perai, Pulau Pinang, Malaysia

Amitha Mithra SV, Chandra Prakash, Ramkumar MK and Nagendra Kumar Singh
ICAR-National Research Centre on Plant Biotechnology, New Delhi, India

Shanmugavadivel PS
ICAR-National Research Centre on Plant Biotechnology, New Delhi, India
Division of Plant Biotechnology, ICAR-Indian Institute of Pulses Research, Kanpur 208 024, India

Ratan Tiwari
ICAR-Indian Institute of Wheat and Barley Research, Karnal 132 001, India

Trilochan Mohapatra
Indian Council of Agricultural Research, Krishi Bhavan, New Delhi 110 001, India

Balwant Singh, Kabita Panda, Bikram Pratap Singh, Nisha Singh, Vandna Rai and Nagendra Kumar Singh
National Research Centre on Plant Biotechnology, Pusa Campus, New Delhi 110012, India

Shefali Mishra
National Research Centre on Plant Biotechnology, Pusa Campus, New Delhi 110012, India
Jacob School of Biotechnology and Bioengineering, Higginbottom Institute of Agriculture, Technology and Sciences, Allahabad 211007, India

Pragati Misra
Jacob School of Biotechnology and Bioengineering, Higginbottom Institute of Agriculture, Technology and Sciences, Allahabad 211007, India

Franca Locatelli, Monica Mattana and Annamaria Genga
Institute of Agricultural Biology and Biotechnology-National Research Council, via Bassini 15, 20133 Milan, Italy

Elena Baldoni
Institute of Agricultural Biology and Biotechnology-National Research Council, via Bassini 15, 20133 Milan, Italy
Dipartimento di Scienze Agrarie e Ambientali - Produzione, Territorio, Agroenergia, Università degli Studi di Milano, Via Celoria 2, 20133 Milan, Italy

Paolo Bagnaresi
Consiglio per la ricerca in agricoltura e l'analisi dell'economia agraria, Genomics Research Centre, Fiorenzuola d'Arda, Piacenza, Italy

Marydee Arceta, B. P. Mallikarjuna Swamy, Faty Diaw and R. K. Singh
International Rice Research Institute (IRRI), Philippines

M. Akhlasur Rahman
International Rice Research Institute (IRRI), Philippines
Bangladesh Rice Research Institute, Gazipur 1701, Bangladesh

M. Sazzadur Rahman
Bangladesh Rice Research Institute, Gazipur 1701, Bangladesh

Isaac Kofi Bimpong
Africa Rice Center, Sahel Regional Station, BP 96 St Louis, Senegal

J. B. Bizimana
IRRI-ESA Office, Bujumbura, Burundi

Evangeline D. Pascual
Institute of Biological Sciences, University of the Philippines at Los Baños, Laguna, Philippines

Alberto Casartelli
School of Agriculture, Food and Wine, Waite Campus, The University of Adelaide, Adelaide, SA, Australia

Sigrid Heuer
School of Agriculture, Food and Wine, Waite Campus, The University of Adelaide, Adelaide, SA, Australia
Rothamsted Research, Harpenden, UK

David Riewe and Thomas Altmann
Julius Kühn-Institute (JKI), Federal Research Centre for Cultivated Plants, Institute for Ecological Chemistry, Plant Analysis and Stored Product Protection, Berlin, Germany

Hans Michael Hubberten and Rainer Hoefgen
Max Planck Institute of Molecular Plant Physiology, Potsdam-Golm, Germany

Abdus Salam Khan
Department of Plant Breeding and Genetics, University of Agriculture, 38040, Faisalabad, Pakistan

Muhammad Abu Bakar Saddique
Department of Plant Breeding and Genetics, University of Agriculture, 38040, Faisalabad, Pakistan

Zulfiqar Ali
Department of Plant Breeding and Genetics, University of Agriculture, 38040, Faisalabad, Pakistan
Department of Plant Breeding and Genetics, Muhammad Nawaz Shareef University of Agriculture, 60000, Multan, Pakistan

Iqrar Ahmad Rana
Centre of Agricultural Biochemistry and Biotechnology, University of Agriculture, 38040, Faisalabad, Pakistan

Imran Haider Shamsi
Department of Agronomy, College of Agriculture and Biotechnology, Zhejiang University, 310058, Hangzhou, People's Republic of China

Jianping Liu, Xinjiao Sun, Feiyun Xu, Yingjiao Zhang, Qian Zhang, Rui Miao and Weifeng Xu
Center for Plant Water-use and Nutrition Regulation and College of Life Sciences, Joint International Research Laboratory of Water and Nutrient in Crop, Fujian Agriculture and Forestry University, Jinshan Fuzhou 350002, China

Jianhua Zhang
Department of Biology, Hong Kong Baptist University, Hong Kong, China

Jiansheng Liang
Department of Biology, Southern University of Science and Technology, Shenzhen 518055, China

Yuzhu Wang, Jinyu Shen, Jianfeng Yin and Yan Gao
Jiangsu Key Laboratory of Crop Genetics and Physiology/Co-Innovation Center for Modern Production Technology of Grain Crop, Yangzhou University, Yangzhou 225009, Jiangsu, China

Dongping Zhang
Jiangsu Key Laboratory of Crop Genetics and Physiology/Co-Innovation Center for Modern Production Technology of Grain Crop, Yangzhou University, Yangzhou 225009, Jiangsu, China
Department of Biology, Southern University of Science and Technology, Shenzhen 518055, China

Weifeng Xu
College of Life Sciences, Fujian Agriculture and Forestry University, Jinshan, Fuzhou 350002, China

Dahong Li
Department of Biological Engineering, Huanghuai University, Zhumadian 463000, Henan, China

Xu Yan and Jianwei Pan
MOE Key Laboratory of Cell Activities and Stress Adaptations, School of Life Sciences, Lanzhou University, Lanzhou 730000, China

Weihuai Pan
MOE Key Laboratory of Cell Activities and Stress Adaptations, School of Life Sciences, Lanzhou University, Lanzhou 730000, China
College of Life Sciences, Shaoxing University, Shaoxing 312000, Zhejiang, China

Jianxin Shou
College of Life Sciences, Shaoxing University, Shaoxing 312000, Zhejiang, China

Jinqiu Shen, Zhongzhong Zheng and Wenxiang Wang
College of Chemistry and Life Sciences, Zhejiang Normal University, Jinhua 321004, Zhejiang, China

Lixi Jiang
Department of Agronomy, College of Agriculture and Biotechnology, Zhejiang University, Hangzhou 310058, China

Hung-Chi Chen and Men-Chi Chang
Department of Agronomy, National Taiwan University, No. 1, Section 4, Roosevelt Road, Taipei, Taiwan, Republic of China

Wan-Hsing Cheng
Institute of Plant and Microbial Biology, Academia Sinica, Taipei, Taiwan, Republic of China

Chwan-Yang Hong
Department of Agricultural Chemistry, National Taiwan University, Taipei, Taiwan, Republic of China

Yu-Sen Chang
Department of Horticulture and Landscape Architecture, National Taiwan University, Taipei, Taiwan, Republic of China

Jiayan Liang, Shaoying Guo, Bo Sun, Qing Liu, Xionghui Chen, Haifeng Peng, Zemin Zhang and Qingjun Xie
State Key Laboratory for Conservation and Utilization of Subtropical Agro-Bioresources, Guangdong Provincial Key Laboratory of Plant Molecular Breeding, South China Agricultural University, Guangzhou 510642, China

Kathleen M. Brown
Department of Plant Science, The Pennsylvania State University, 102 Tyson Building, University Park, University Park, PA 16802-4200, USA

Mohamed Hazman
Department of Plant Science, The Pennsylvania State University, 102 Tyson Building, University Park, University Park, PA 16802-4200, USA
Agricultural Genetic Engineering Research Institute (AGERI), Agricultural Research Centre (ARC), 9 Gamma St., Giza 12619, Egypt

Yoav Yichie and Thomas H. Roberts
Sydney Institute of Agriculture, University of Sydney, Sydney, Australia

Chris Brien and Bettina Berger
School of Agriculture Food and Wine, University of Adelaide, Adelaide, Australia
Australian Plant Phenomics Facility, The Plant Accelerator, Waite Research Institute, University of Adelaide, Adelaide, Australia

Brian J. Atwell
Department of Biological Sciences, Macquarie University, Sydney, Australia

Antonio Costa de Oliveira, Luciano C da Maia, Daniel R Farias, Leomar G Woyann, Claudete C Mistura and Adriana P Soares-Bresolin
Plant Genomics and Breeding Center, Eliseu Maciel School of Agronomy, Federal University of Pelotas, 96010-610 Pelotas, RS, Brazil

Taciane Finatto
Plant Genomics and Breeding Center, Eliseu Maciel School of Agronomy, Federal University of Pelotas, 96010-610 Pelotas, RS, Brazil
Universidade Tecnológica Federal do Paraná, Campus Pato Branco, 85503-390 Pato Branco, PR, Brazil

Christel Llauro, Olivier Panaud and Nathalie Picault
Laboratoire Génome et Développement des Plantes, UMR 5096, Université de Perpignan Via Domitia, F-66860 Perpignan, France
CNRS, Laboratoire Génome et Développement des Plantes, UMR 5096, F-66860 Perpignan, France

Cristian Chaparro
Laboratoire Génome et Développement des Plantes, UMR 5096, Université de Perpignan Via Domitia, F-66860 Perpignan, France
CNRS, Laboratoire Génome et Développement des Plantes, UMR 5096, F-66860 Perpignan, France
Laboratoire Ecologie et Evolution des Interactions, UMR 5244, F-66860, Université de Perpignan Via Domitia, Perpignan, France

Silvas J Prince, R Beena and R Chandra Babu
Centre for Plant Molecular Biology and Biotechnology, Tamil Nadu Agricultural University (TNAU), Coimbatore 641 003, India

S Michael Gomez
International Center for Tropical Agriculture (CIAT), Colombia 6713, South America

S Senthivel
Agricultural Research Station, Tamil Nadu Agricultural University (TNAU), Paramakudi 623707, India

Index

A

Abiotic Stresses, 24, 48-49, 51, 69, 73-74, 81-82, 84-85, 102, 107, 111-112, 126, 129, 136-137, 140, 142, 153, 156-167, 175, 183, 185, 213-214

Allantoin, 101, 108, 111-116

Allele Mining, 49, 51-53, 62-63

Amino Acids And, 101, 114, 130

Amylopectin, 1, 3-7, 12, 19, 23-24

Amylose, 1, 3-4, 6-8, 10, 12, 19, 25-26

Anatomy, 197, 202-206, 208, 210-212

Association Mapping, 27-28, 32-39, 49, 61-63, 115

Aus-type Landraces, 101

Australian Native Rice, 213

C

Chaperones, 11, 15, 17, 25-26, 65

Circadian, 141-144, 152-155

Cloning, 25, 38, 114, 140, 156-157, 159, 164, 166-167, 184, 195

Crop Yield, 25, 40, 62, 85, 157, 186

D

Drought Tolerance, 27-28, 31-35, 37-39, 47, 50, 62-65, 73, 76, 81-83, 100-102, 107-108, 110-114, 116, 118, 124, 126-128, 154-155, 184-189, 193-195, 198, 207

Drought-sensitive Genotypes, 64-65

E

Endosperm, 1-3, 5-6, 8-13, 15, 17-19, 21, 23-26, 47, 172-173, 185

Environment Interactions, 38, 64, 84-85, 94, 97, 100

Epistatic Interaction Network, 42-43, 47-48

Ethylene Responsive Factor, 141, 148, 154

G

Genes, 2, 4, 7, 14-15, 17-21, 24-28, 30-31, 34-35, 37-69, 73-78, 80-83, 95-97, 99-100, 102, 111-114, 117-119, 127-128, 130, 136, 154, 156-157, 164-170, 172, 175, 177-178, 182, 225

Genetic Diversity, 27-29, 33, 36-39, 52, 63, 101, 109, 113, 214, 221

Genomics, 25, 37-39, 49-50, 63, 80-82, 99-100, 115, 154, 167, 184, 186, 195, 225

Grain Chalkiness, 6, 11-12, 19, 24-25

Grain Quality, 1-2, 7-11, 17, 21, 24-25, 31, 37, 40

Grain Yield, 1-3, 5, 7-8, 10, 12, 24, 27-28, 31, 34-39, 49, 62, 73, 81, 85, 101, 104, 110, 117-118, 127, 140, 155, 213, 223

H

Hasawi-aus Rice Landrace, 84

Heat Shock Protein, 11, 15

Heat Stress Treatment, 41, 133

Heat Tolerance, 13, 24, 35, 39-41, 43-46, 48-50, 129-133, 136-137, 140, 214, 225

Heavy Metal Toxicity, 156, 162, 164, 166

High-temperature Stress, 6-7, 11-15, 17-20, 25

Hormones, 64, 76, 112, 141, 152, 186, 194

Hydrogen Peroxide, 128-129, 139-140, 154

I

Inclusive Composite Interval Mapping, 42-43, 87-89, 98

J

Japonica Rice, 1, 6-7, 37, 50, 64-65, 78, 131, 137

L

Laser Ablation Tomography Images, 197, 208, 210

Leaf Morphology, 186, 191-194

Lignin, 64, 76, 78-79, 82, 197, 201-202, 204, 206, 208-211

M

Marker Assisted Selection, 40-41, 49, 52

Metabolites, 75, 101-115

Monodehydroascorbate Reductase, 129-130, 135, 139-140

O

Oryza Australiensis, 213, 225

Oryza Meridionalis, 213, 225

Osmotic Stress, 52, 64-69, 71, 73-76, 78-79, 81-83, 116-128, 155, 167-170, 185, 214, 223

Overexpression, 15, 20, 26, 38-39, 48-49, 74, 81, 112, 129-130, 132-137, 139-140, 142-143, 153-156, 159, 161-167, 185, 190, 195

Ozone Stress, 1-2, 7, 10

P

Panicle Exertion, 40

Phenotypic Variation, 40, 43-45, 82, 89, 211
Piriformospora Indica, 117, 127-128
Plant Height, 27, 36-37, 103-104, 117, 219
Plant Materials, 7, 21, 41, 85, 137, 142, 157, 183, 187, 198
Proline, 59, 73, 107-108, 111, 113-122, 124, 126-128, 141-142, 146, 154, 167
Putative Interactors, 141
Pyrroline-5-carboxylate Synthase, 117, 127

Q
Quantitative Trait Loci, 12, 24-25, 27, 38, 41, 49-50, 62-63, 84, 99-100, 102, 115

R
Recombinant Inbred Lines, 24, 40-41, 43, 50, 84-85, 99, 245-246
Reproductive Stage, 27-28, 32, 36, 38, 42, 49, 52, 62, 82, 85, 99, 127-128
Rna-seq Analysis, 64, 68, 190, 193

S
Salinity Stress, 40-41, 52, 62, 82, 84-86, 94-97, 99-100, 127, 137, 146, 148, 214, 216, 225
Salinity-sensitive Rice Varieties, 84
Salt Tolerance, 50-53, 56, 58-60, 62-63, 82, 86-87, 89, 94, 96-100, 130, 137, 139-141, 145-146, 148, 152, 154-155, 167, 170, 175, 177, 184-185, 192, 213-214, 218, 221-226
Seedling-stage Salinity Tolerance, 84-85, 92, 94-95, 97
Segregation Distortion Locus, 84
Segregation Distortion Markers, 84, 87, 99
Single Nucleotide Polymorphism, 34, 84, 100
Spikelet Sterility, 24, 40-41, 43-46, 50
Spikelets, 32, 35, 42, 101, 104, 115
Starch, 1-4, 6-13, 15, 18-21, 23-26, 47, 110, 116, 185
Starch Granule, 11, 13
Stomata, 129, 134-135, 137, 139-140
Stress Susceptibility Index, 42-43
Stress Tolerant Index, 42
Stress-induced Expression, 64-65
Stress-related Protein, 11

T
Tibetan Plateau Annual Wild Barley, 156-160, 164, 166-167
Total Antioxidant Capacity, 117-118, 120-122, 124, 127
Transcription Factors, 31, 34, 37-38, 46, 50, 64-65, 81-82, 118, 141, 153, 155, 169, 184-185

U
Uridine, 101, 108, 112, 114

www.ingramcontent.com/pod-product-compliance
Lightning Source LLC
LaVergne TN
LVHW080713210926
848485LV00003B/458